LINEAR ALGEBRA WITH APPLICATIONS

SECOND EDITION

LINEAR ALGEBRA WITH APPLICATIONS

Steven J. Leon

MACMILLAN PUBLISHING COMPANY

New York

COLLIER MACMILLAN PUBLISHERS

London

Macmillan Publishing Company
866 Third Avenue, New York, New York 10022

Collier Macmillan Canada, Inc.

Library of Congress Cataloging in Publication Data

Leon, Steven J.
  Linear algebra with applications.

  Bibliography: p.
  Includes index.
  1. Algebras, Linear.   I. Title.
QA184.L46 1986   512′.5   85-13673
ISBN-0-02-369810-1

Printing: 2 3 4 5 6 7 8   Year: 6 7 8 9 0 1 2 3

ISBN 0-02-369810-1

TO JUDITH RUSS LEON

PREFACE

This book is suitable for either a sophomore-level course or for a junior- or senior-level course. The student should have some familiarity with the basics of differential and integral calculus. This prerequisite can be met by either one semester or two quarters of elementary calculus. If the text is to be taught at the sophomore level, one should probably spend more time on the earlier chapters and omit many of the sections in the later chapters. On the other hand, if the course is to be taught on the junior or senior level, the instructor should probably spend less time on the early chapters and cover more of the later material. The explanations in the text are given in sufficient detail so that students at either level will have little trouble reading and understanding. To further aid the student, a large number of examples have been worked out completely. Applications have been scattered throughout the book rather than put together in a chapter at the end. In this way they can be used to motivate new material and to illustrate the relevance of the material that has just been presented. When applications are included at the end of a book, they are more likely to be omitted because of lack of time.

WHAT'S NEW IN THE SECOND EDITION? _____

During the past six years, the author has taught from the book nine times. He has had ample opportunity to carefully review and revise each section. Many improvements have been made in order to make the second edition more accessible to students.

The following are some of the major changes in the new edition.

1. All of the exercise sets have been expanded, some more than doubled. There are many new exercises at all levels. The exercises have generally been ordered so that the more routine problems are presented first and the more difficult theoretical exercises follow afterward.

2. There are more worked examples and more applications. Abstract concepts such as linear independence (Chapter 3) and similarity (Chapter 4) are introduced by worked examples. All of the applications in the first edition have been retained and additional applications have been included that show how to use linear algebra in cryptography, digital imaging, and numerical integration.

3. Chapter 4 on linear operators has been expanded from three sections to four. Much of the chapter has been rewritten. The subject is presented from a less abstract point of view. There are many more worked examples.

4. A discussion of the matrix exponential and its application to solving systems of linear differential equations has been added to Section 3 of Chapter 6. This is the student's first exposure to matrix functions and I believe it will be well received.

5. The discussion of positive definite matrices in Chapter 6 has been greatly expanded. Indeed, all of the material on positive definite matrices in the first edition has been moved up from Section 6 to Section 5. The new Section 6 deals with the properties of positive definite matrices in much more depth. Positive definite matrices arise in many modern applications of mathematics. This is an important subject; it deserves the expanded coverage.

GUIDE TO THE CHAPTERS

The book gives fairly complete coverage of a very broad subject. Consequently, there is probably more material included than can possibly be covered in a one-quarter or one-semester course. The instructor then has some freedom in the choice of topics and can design the course to meet the needs of the class. Some instructors may decide to emphasize the mathematical theory in Chapters 3 and 4 and others may decide to skip over these chapters and spend more time on the applied topics in Chapters 5 and 6. As a general rule, if you ask n mathematicians what should go into a course, you will get n different answers. Even if many of the topics in the book are omitted, the students should get a feeling for the overall scope of the subject matter. Furthermore, many of the students may use the book later as a reference and consequently may end up learning many of the omitted topics on their own. Sections designated by an asterisk in the table of contents are optional. These sections are not prerequisites to any of the following sections—they can be skipped over without any loss of continuity. The following is a guide to the various chapters in the book.

Chapter 1. The first two sections deal with systems of equations and the last three sections are concerned with matrices and matrix algebra. The block multiplication introduced in Section 5 is not used in the book until Chapter 6. Consequently, in sophomore level courses it may be possible to skip this section.

Chapter 2. This is a short chapter on determinants. The concept of the determinant is introduced by first studying the 2×2 case and then generalizing. In the first section the standard definition using permutations is presented. Those preferring the cofactor definition can skip over that material and go directly to the next subsection, where the cofactor expansion is presented. The properties of determinants presented in Section 2 can all be derived from the cofactor definition. Some instructors believe that Cramer's rule generally receives more attention than it merits. Therefore, Section 3 has been listed as optional.

Chapter 3. The basic theory of vector spaces is presented in this chapter. All five sections should be covered.

Chapter 4. Chapter 4 is devoted to linear transformations. Instructors wishing to present a more applied course may omit all or part of this chapter.

Chapter 5. This is a long chapter on orthogonality and its applications. Most of Sections 1, 2, 3, 5, 6, and 7 should be covered if at all possible. Section 4 on matrix norms could probably be omitted if one is pressed for time. This section did not receive an asterisk, since it is a prerequisite for Section 7 of Chapter 6 and the last five sections of Chapter 7. The last section is an optional section on orthogonal polynomials. Orthogonal polynomials play an important role in many areas of mathematics, but they never seem to get the attention they deserve in the standard courses.

Chapter 6. This chapter treats one of the most important subjects in linear algebra, eigenvalues. Sections 1 and 3 should definitely be covered. Section 2 presents one of the main applications of eigenvalues. If the instructor does not wish to cover the entire section, we recommend that the material through Example 1 be presented. Sections 4, 5, and 6 should be covered in junior or senior level courses but are optional for sophomore-level courses. Section 7 is optional.

Chapter 7. In a sense this whole chapter is optional, since in many cases there will not be enough time left in the semester to cover it. This is unfortunate because it may well be the most important chapter for students planning to work in industry. If there is only time to cover one topic, I would strongly recommend Section 7 on the singular value decomposition. This section is included in Chapter 7 because of its importance to numerical linear algebra, but it could just as well be covered in Chapter 6. Part of Section 9 is theoretical and could also be covered earlier, along with Section 7. Instructors who want to integrate numerics into the course may consider teaching the first three sections of this chapter immediately after completing Chapter 1. Section 4 could then be taught with the section on matrix norms in Chapter 5.

SUGGESTED COURSE OUTLINES

I. Two-Semester Sequence

In a two-semester sequence it is possible to cover all 41 sections of the text. I generally average about two days per section and only cover 35 sections. I omit the optional sections in Chapters 5, 6, and 7. I also skip the first three sections of Chapter 7, as they are usually covered in numerical analysis courses.

II. One-Semester Course

In a one-semester course most of the sections can be covered in one lecture. The longer sections require two lectures.

    A. Sophomore-Level Course

Chapter 1	Sections 1–4	6 lectures
Chapter 2	Sections 1–3	3 lectures
Chapter 3	Sections 1–5	8 lectures
Chapter 4	Sections 1–4	5 lectures

Chapter 5	Sections 1, 2, 3, 6, 7		8 lectures
Chapter 6	Sections 1, 2, 3		5 lectures
		Total	35 lectures

B. Junior- or Senior-Level Courses

The coverage in an upper division course is dependent on the background of the students. Below are two possible courses with 35 lectures each.

Course 1

Chapter 1	Sections 1–5	6 lectures
Chapter 2	Sections 1, 2	2 lectures
Chapter 3	Sections 1–5	6 lectures
Chapter 5	Sections 1–7	10 lectures
Chapter 6	Sections 1–6	9 lectures
	Section 7 if time allows	
Chapter 7	Section 7	2 lectures
	Part of Section 9 if time allows	

Course 2

Review of Topics in Chapters 1–3		4 lectures
Chapter 4	Sections 1–4	4 lectures
Chapter 5	Sections 1–7	10 lectures
Chapter 6	Sections 1–6	9 lectures
	Section 7 if time allows	
Chapter 7	Sections 4, 5, 7, 8, 9	8 lectures
	If time allows, Sections 1, 2, 3	

SUPPLEMENTARY MATERIALS

The Solutions Manual contains complete, worked-out solutions to all of the nonroutine exercises in the book. The manual also contains answers to all elementary exercises that were not already listed in the answer key in the textbook.

ACKNOWLEDGMENTS

The author would like to express his gratitude to the reviewers Wayne Barrett, Brigham Young University; George Davis, Georgia State University; and Donald Sherbert, University of Illinois; for their many helpful comments and suggestions. Nearly all of those suggestions were adopted in the second edition, and I believe they have resulted in a significant improvement in the book. A number of additional exercises for this edition were also provided by Wayne Barrett. Thanks to Warren Ferguson and others who taught from the first edition for their many helpful suggestions and to all those who helped uncover errata in the first edition. The author would again like to acknowledge the influence that Gene Golub and J. H. Wilkinson have had on this book. Thanks to Gerald Vieira and Judith Leon for their help with the proofreading. The revisions were all typed by Kate MacDougall; I would like to thank her for a very professional job. Finally, I would like to express my gratitude to the editorial, production, and sales staff at Macmillan for the work they have done on both the first and second editions.

S. L.

CONTENTS

xiii

LINEAR ALGEBRA WITH APPLICATIONS

1

MATRICES AND SYSTEMS OF EQUATIONS

INTRODUCTION

Probably the most important problem in mathematics is that of solving a system of linear equations. It would not be conservative to estimate that well over 75 percent of all mathematical problems encountered in scientific or industrial applications involve solving a linear system at some stage. Using the methods of modern mathematics, it is often possible to take a sophisticated problem and reduce it to a single system of linear equations. Linear systems arise in applications to such areas as business, economics, sociology, ecology, demography, genetics, electronics, engineering, and physics. It seems appropriate, then, that this book should begin with a section on linear systems.

1. SYSTEMS OF LINEAR EQUATIONS _____

A *linear equation in n unknowns* is an equation of the form

$$a_1x_1 + a_2x_2 + \cdots + a_nx_n = b$$

where $a_1, a_2, \ldots, a_n$ and b are real numbers and $x_1, x_2, \ldots, x_n$ are variables. A *linear system* of m equations in n unknowns is then a system of the form

(1)
$$
\begin{aligned}
a_{11}x_1 + a_{12}x_2 + \cdots + a_{1n}x_n &= b_1 \\
a_{21}x_1 + a_{22}x_2 + \cdots + a_{2n}x_n &= b_2 \\
&\vdots \\
a_{m1}x_1 + a_{m2}x_2 + \cdots + a_{mn}x_n &= b_m
\end{aligned}
$$

where the a_{ij}'s and the b_i's are all real numbers. We will refer to systems of the form (1) as $m \times n$ linear systems. The following are examples of linear systems:

(a) $x_1 + 2x_2 = 5$ (b) $x_1 - x_2 + x_3 = 2$ (c) $x_1 + x_2 = 2$

 $2x_1 + 3x_2 = 8$ $2x_1 + x_2 - x_3 = 4$ $x_1 - x_2 = 1$

 $x_1 = 4$

System (a) is a 2×2 system, (b) is a 2×3 system, and (c) is a 3×2 system.

By a solution to an $m \times n$ system, we mean an ordered n-tuple of numbers $(x_1, x_2, \ldots, x_n)$ that satisfies all the equations of the system. For example, the ordered pair $(1, 2)$ is a solution to system (a), since

$$1 \cdot (1) + 2 \cdot (2) = 5$$
$$2 \cdot (1) + 3 \cdot (2) = 8$$

The ordered triple $(2, 0, 0)$ is a solution to system (b), since

$$1 \cdot (2) - 1 \cdot (0) + 1 \cdot (0) = 2$$
$$2 \cdot (2) + 1 \cdot (0) - 1 \cdot (0) = 4$$

Actually, system (b) has many solutions. If α is any real number, it is easily seen that the ordered triple $(2, \alpha, \alpha)$ is a solution. However, system (c) has no solution. It follows from the third equation that the first coordinate of any solution would have to be 4. Using $x_1 = 4$ in the first two equations, we see that the second coordinate must satisfy

$$4 + x_2 = 2$$
$$4 - x_2 = 1$$

Since there is no real number that satisfies both of these equations, the system has no solution. If a linear system has no solution, we say that the system is *inconsistent*. Thus system (c) is inconsistent, while systems (a) and (b) are both consistent.

The set of all solutions to a linear system is called the *solution set* of the system. If a system is inconsistent, its solution set is empty. A consistent system will have a nonempty solution set. To solve a consistent system, one must find its solution set.

2 × 2 SYSTEMS

Let us examine geometrically a system of the form

$$a_{11}x_1 + a_{12}x_2 = b_1$$

$$a_{21}x_1 + a_{22}x_2 = b_2$$

Each equation can be represented graphically as a line in the plane. The ordered pair (x_1, x_2) will be a solution to the system if and only if it lies on both lines. For example, consider the three systems

(i) $x_1 + x_2 = 2$ (ii) $x_1 + x_2 = 2$ (iii) $x_1 + x_2 = 2$

$\quad\ x_1 - x_2 = 2$ $\quad\ \ x_1 + x_2 = 1$ $\quad\ \ -x_1 - x_2 = -2$

The two lines in system (i) intersect at the point $(2, 0)$. Thus $\{(2, 0)\}$ is the solution set to (i). In system (ii) the two lines are parallel. Therefore, system (ii) is inconsistent and hence the solution set is $\varnothing$. The two equations in system (iii) both represent the same line. Any point on that line will be a solution to the system (see Figure 1.1.1).

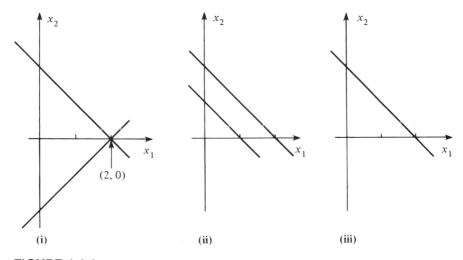

FIGURE 1.1.1

In general, there are three possibilities: the lines intersect at a point, they are parallel, or both equations represent the same line. The solution set then contains either one, zero, or infinitely many points.

The situation is similar for $m \times n$ systems. An $m \times n$ system may or may not be consistent. If it is consistent, then it must either have exactly one solution or infinitely many solutions. These are the only possibilities. We will see why this is so in Section 2 when we study the row echelon form. Of more immediate concern is the problem of finding all solutions to a given system. To tackle this problem, we introduce the notion of *equivalent systems*.

EQUIVALENT SYSTEMS

Consider the two systems

$$\text{(a)} \quad 3x_1 + 2x_2 - x_3 = -2 \qquad \text{(b)} \quad 3x_1 + 2x_2 - x_3 = -2$$
$$x_2 = 3 \qquad\qquad -3x_1 - x_2 + x_3 = 5$$
$$2x_3 = 4 \qquad\qquad 3x_1 + 2x_2 + x_3 = 2$$

System (a) is easy to solve because it is clear from the last two equations that $x_2 = 3$ and $x_3 = 2$. Using these values in the first equation, we get

$$3x_1 + 2 \cdot 3 - 2 = -2$$
$$x_1 = -2$$

Thus the solution to the system is $(-2, 3, 2)$. System (b) seems to be more difficult to solve. Actually, system (b) has the same solution as system (a). To see this, add the first two equations of the system

$$3x_1 + 2x_2 - x_3 = -2$$
$$\underline{-3x_1 - x_2 + x_3 = 5}$$
$$x_2 = 3$$

If (x_1, x_2, x_3) is any solution to (b), it must satisfy all the equations of the system. Thus it must satisfy any new equation formed by adding two of its equations. Therefore, x_2 must equal 3. Similarly, (x_1, x_2, x_3) must satisfy the new equation formed by subtracting the first equation from the third:

$$3x_1 + 2x_2 + x_3 = 2$$
$$\underline{3x_1 + 2x_2 - x_3 = -2}$$
$$2x_3 = 4$$

Therefore, any solution to system (b) must also be a solution to system (a). By a similar argument, it can be shown that any solution to (a) is also a solution to (b). This can be done by subtracting the first equation from the second:

$$x_2 \qquad = \quad 3$$

$$\underline{3x_1 + 2x_2 - x_3 = -2}$$

$$-3x_1 - \quad x_2 + x_3 = \quad 5$$

and by adding the first and third equations:

$$3x_1 + 2x_2 - \quad x_3 = -2$$

$$\underline{\qquad\qquad\quad 2x_3 = \quad 4}$$

$$3x_1 + 2x_2 + \quad x_3 = \quad 2$$

Thus (x_1, x_2, x_3) is a solution to system (b) if and only if it is a solution to system (a). Therefore, both systems have the same solution set, $\{(-2, 3, 2)\}$.

Definition.   Two systems of equations involving the same variables are said to be **equivalent** if they have the same solution set.

Clearly, if we interchange the order in which two equations of a system are written, this will have no effect on the solution set. The reordered system will be equivalent to the original system. For example, the systems

$$x_1 + 2x_2 = 4 \qquad\qquad 4x_1 + \quad x_2 = 6$$

$$3x_1 - \quad x_2 = 2 \qquad \text{and} \qquad 3x_1 - \quad x_2 = 2$$

$$4x_1 + \quad x_2 = 6 \qquad\qquad x_1 + 2x_2 = 4$$

clearly have the same solution set.

If one of the equations of a system is multiplied through by a nonzero real number, this will have no effect on the solution set and the new system will be equivalent to the original system. For example, the systems

$$x_1 + x_2 + \quad x_3 = 3 \qquad\qquad 2x_1 + 2x_2 + 2x_3 = 6$$

$$\text{and}$$

$$-2x_1 - x_2 + 4x_3 = 1 \qquad\qquad -2x_1 - \quad x_2 + 4x_3 = 1$$

are equivalent.

If a multiple of one equation is added to another equation, the new system will be equivalent to the original system. This follows since the n-tuple $(x_1, \ldots, x_n)$ will satisfy the two equations

$$a_{i1}x_1 + \cdots + a_{in}x_n = b_i$$

$$a_{j1}x_1 + \cdots + a_{jn}x_n = b_j$$

if and only if it satisfies the equations

$$a_{i1}x_1 + \cdots + a_{in}x_n = b_i$$

$$(a_{j1} + \alpha a_{i1})x_1 + \cdots + (a_{jn} + \alpha a_{in})x_n = b_j + \alpha b_i$$

To summarize, there are three operations that can be used on a system to obtain an equivalent system:

I. The order in which any two equations are written may be interchanged.
II. Both sides of an equation may be multiplied by the same nonzero real number.
III. A multiple of one equation may be added to another.

Given a system of equations, one can use these operations to obtain an equivalent system that is easier to solve.

$n \times n$ SYSTEMS

Let us restrict ourselves to $n \times n$ systems for the remainder of this section. We will show that if an $n \times n$ system has exactly one solution, then operations I and III can be used to obtain an equivalent "triangular system."

Definition. A system is said to be in **triangular form** if in the kth equation the coefficients of the first $k - 1$ variables are all zero and the coefficient of x_k is nonzero $(k = 1, \ldots, n)$.

EXAMPLE 1. The system

$$3x_1 + 2x_2 + x_3 = 1$$
$$x_2 - x_3 = 2$$
$$2x_3 = 4$$

is in triangular form, since in the second equation the coefficients are $0, 1, -1$, respectively, and in the third equation the coefficients are $0, 0, 2$, respectively. Because of the triangular form, this system is easy to solve. It follows from the third equation that $x_3 = 2$. Using this value in the second equation, we obtain

$$x_2 - 2 = 2 \quad \text{or} \quad x_2 = 4$$

Using $x_2 = 4$, $x_3 = 2$ in the first equation, we end up with

$$3x_1 + 2 \cdot 4 + 2 = 1$$
$$x_1 = -3$$

Thus the solution to the system is $(-3, 4, 2)$.   □

Any $n \times n$ triangular system can be solved in the same manner as the last example. First, the nth equation is solved for the value of x_n. This value is used in the $(n-1)$st equation to solve for x_{n-1}. The values x_n and x_{n-1} are used in the $(n-2)$nd equation to solve for x_{n-2}, and so on. We will refer to this method of solving a triangular system as *back substitution*.

EXAMPLE 2. Solve the system

$$2x_1 - x_2 + 3x_3 - 2x_4 = 1$$
$$x_2 - 2x_3 + 3x_4 = 2$$
$$4x_3 + 3x_4 = 3$$
$$4x_4 = 4$$

SOLUTION. Using back substitution, we obtain

$$4x_4 = 4 \qquad x_4 = 1$$
$$4x_3 + 3 \cdot 1 = 3 \qquad x_3 = 0$$
$$x_2 - 2 \cdot 0 + 3 \cdot 1 = 2 \qquad x_2 = -1$$
$$2x_1 - (-1) + 3 \cdot 0 - 2 \cdot 1 = 1 \qquad x_1 = 1$$

Thus the solution is $(1, -1, 0, 1)$. $\qquad\qquad\square$

If a system of equations is not triangular, we will use operations I and III to try to obtain an equivalent system that is in triangular form.

EXAMPLE 3. Solve the system

$$x_1 + 2x_2 + x_3 = 3$$
$$3x_1 - x_2 - 3x_3 = -1$$
$$2x_1 + 3x_2 + x_3 = 4$$

SOLUTION. Subtracting 3 times the first row from the second yields

$$-7x_2 - 6x_3 = -10$$

Subtracting 2 times the first row from the third row yields

$$x_2 \quad x_3 = -2$$

If the second and third equations of our system, respectively, are replaced by these new equations, we obtain the equivalent system

$$x_1 + 2x_2 + x_3 = 3$$
$$-7x_2 - 6x_3 = -10$$
$$-x_2 - x_3 = -2$$

If the third equation of this system is replaced by the sum of the third equation and $-\frac{1}{7}$ times the second equation, we end up with the triangular system

$$x_1 + 2x_2 + x_3 = 3$$
$$-7x_2 - 6x_3 = -10$$
$$-\tfrac{1}{7}x_3 = -\tfrac{4}{7}$$

Using back substitution, we get

$$x_3 = 4, \qquad x_2 = -2, \qquad x_1 = 3 \qquad \qquad \square$$

Let us look back at the system of equations in the last example. We can associate with that system a 3×3 array of numbers whose entries are the coefficients of the x_i's.

$$\begin{pmatrix} 1 & 2 & 1 \\ 3 & -1 & -3 \\ 2 & 3 & 1 \end{pmatrix}$$

We will refer to this array as the *coefficient matrix* of the system. The term *matrix* means simply a rectangular array of numbers. If we attach to the coefficient matrix an additional column whose entries are the numbers on the right-hand side of the system, we obtain the new matrix

$$\begin{pmatrix} 1 & 2 & 1 & 3 \\ 3 & -1 & -3 & -1 \\ 2 & 3 & 1 & 4 \end{pmatrix}$$

We will refer to this new matrix as the *augmented matrix*. In general, when an $m \times r$ matrix B is attached to an $m \times n$ matrix A in this way the augmented matrix is denoted by $(A|B)$. Thus, if

$$A = \begin{pmatrix} a_{11} & a_{12} & \cdots & a_{1n} \\ a_{21} & a_{22} & \cdots & a_{2n} \\ \vdots & & & \\ a_{m1} & a_{m2} & \cdots & a_{mn} \end{pmatrix} \qquad B = \begin{pmatrix} b_{11} & b_{12} & \cdots & b_{1r} \\ b_{21} & b_{22} & \cdots & b_{2r} \\ \vdots & & & \\ b_{m1} & b_{m2} & \cdots & b_{mr} \end{pmatrix}$$

then

$$(A|B) = \begin{pmatrix} a_{11} & \cdots & a_{1n} & b_{11} & \cdots & b_{1r} \\ \vdots & & & \vdots & & \\ a_{m1} & \cdots & a_{mn} & b_{m1} & \cdots & b_{mr} \end{pmatrix}$$

With each system of equations we may associate an augmented matrix of the form

$$\begin{pmatrix} a_{11} & \cdots & a_{1n} & b_1 \\ \vdots & & & \vdots \\ a_{m1} & \cdots & a_{mn} & b_m \end{pmatrix}$$

The system can be solved by performing operations on the augmented matrix. The x_i's are placeholders that can be omitted until the end of the computation. Corresponding to the three operations used to obtain equivalent systems, the following row operations may be applied to the augmented matrix.

Elementary Row Operations
I. Interchange two rows.
II. Multiply a row by a nonzero real number.
III. Add a multiple of one row to another row.

Returning to the example, we find that the first row is used to eliminate the elements in the first column of the remaining rows. We refer to the first row as the *pivotal row* and the entry 1 circled in the first row as the *pivot*.

$$
\begin{array}{l}
\text{pivot} \quad\; \rightarrow \\
\text{elements to} \; \rightarrow \\
\text{be eliminated} \; \rightarrow
\end{array}
\left(
\begin{array}{ccc|c}
\boxed{1} & 2 & 1 & 3 \\
\boxed{3} & -1 & -3 & -1 \\
\boxed{2} & 3 & 1 & 4
\end{array}
\right) \leftarrow \text{pivotal row}
$$

By using row operation III, 3 times the first row is subtracted from the second row and 2 times the first row is subtracted from the third. When this is done, we end up with the matrix

$$
\begin{pmatrix}
1 & 2 & 1 & 3 \\
0 & \boxed{-7} & -6 & -10 \\
0 & \boxed{-1} & -1 & -2
\end{pmatrix} \leftarrow \text{pivotal row}
$$

At this step we choose the second row as our new pivotal row and apply row operation III to eliminate the last element in the second column. We end up with the matrix

$$
\begin{pmatrix}
1 & 2 & 1 & 3 \\
0 & -7 & -6 & -10 \\
0 & 0 & -\frac{1}{7} & -\frac{4}{7}
\end{pmatrix}
$$

This is the augmented matrix for the triangular system, which is equivalent to the original system.

EXAMPLE 4.   Solve the system

$$
\begin{aligned}
- x_2 - x_3 + x_4 &= 0 \\
x_1 + x_2 + x_3 + x_4 &= 6 \\
2x_1 + 4x_2 + x_3 - 2x_4 &= -1 \\
3x_1 + x_2 - 2x_3 + 2x_4 &= 3
\end{aligned}
$$

SOLUTION. The augmented matrix for this system is

$$\begin{pmatrix} 0 & -1 & -1 & 1 & | & 0 \\ 1 & 1 & 1 & 1 & | & 6 \\ 2 & 4 & 1 & -2 & | & -1 \\ 3 & 1 & -2 & 2 & | & 3 \end{pmatrix}$$

Since it is not possible to eliminate any entries using 0 as a pivot element, we will use row operation I to interchange the first two rows of the augmented matrix. The new first row will be the pivotal row and the pivot element will be 1.

pivot element → $\begin{pmatrix} ① & 1 & 1 & 1 & | & 6 \\ 0 & -1 & -1 & 1 & | & 0 \\ \boxed{2} & 4 & 1 & -2 & | & -1 \\ \boxed{3} & 1 & -2 & 2 & | & 3 \end{pmatrix}$ ← pivotal row

Row operation III is then used twice to eliminate the two nonzero entries in the first column.

$$\begin{pmatrix} 1 & 1 & 1 & 1 & | & 6 \\ 0 & ⊖1 & -1 & 1 & | & 0 \\ 0 & \boxed{2} & -1 & -4 & | & -13 \\ 0 & \boxed{-2} & -5 & -1 & | & -15 \end{pmatrix}$$

Next, the second row is used as the pivotal row to eliminate the entries in the second column below the pivot element -1.

$$\begin{pmatrix} 1 & 1 & 1 & 1 & | & 6 \\ 0 & -1 & -1 & 1 & | & 0 \\ 0 & 0 & ⊖3 & -2 & | & -13 \\ 0 & 0 & \boxed{-3} & -3 & | & -15 \end{pmatrix}$$

Finally, the third row is used as the pivotal row to eliminate the last element in the third column.

$$\begin{pmatrix} 1 & 1 & 1 & 1 & | & 6 \\ 0 & -1 & -1 & 1 & | & 0 \\ 0 & 0 & -3 & -2 & | & -13 \\ 0 & 0 & 0 & -1 & | & -2 \end{pmatrix}$$

This augmented matrix represents a triangular system. The solution $(2, -1, 3, 2)$ is easily obtained using back substitution.   □

$n = 4$

Step 1

$$\left[\begin{array}{cccc|c} x & x & x & x & x \\ x & x & x & x & x \\ x & x & x & x & x \\ x & x & x & x & x \end{array}\right] \longrightarrow \left[\begin{array}{cccc|c} x & x & x & x & x \\ 0 & x & x & x & x \\ 0 & x & x & x & x \\ 0 & x & x & x & x \end{array}\right]$$

Step 2

$$\left[\begin{array}{cccc|c} x & x & x & x & x \\ 0 & x & x & x & x \\ 0 & x & x & x & x \\ 0 & x & x & x & x \end{array}\right] \longrightarrow \left[\begin{array}{cccc|c} x & x & x & x & x \\ 0 & x & x & x & x \\ 0 & 0 & x & x & x \\ 0 & 0 & x & x & x \end{array}\right]$$

Step 3

$$\left[\begin{array}{cccc|c} x & x & x & x & x \\ 0 & x & x & x & x \\ 0 & 0 & x & x & x \\ 0 & 0 & x & x & x \end{array}\right] \longrightarrow \left[\begin{array}{cccc|c} x & x & x & x & x \\ 0 & x & x & x & x \\ 0 & 0 & x & x & x \\ 0 & 0 & 0 & x & x \end{array}\right]$$

FIGURE 1.1.2

In general, if an $n \times n$ linear system can be reduced to triangular form, then it will have a unique solution that can be obtained by performing back substitution on the triangular system. We can think of the reduction process as an algorithm involving $n - 1$ steps. At the first step, a pivot element is chosen from among the nonzero entries in the first column of the matrix. The row containing the pivot element is called the *pivotal row*. We interchange rows (if necessary) so that the pivotal row is the new first row. Multiples of the pivotal row are then subtracted from each of the remaining $n - 1$ rows so as to obtain 0's in the $(2, 1), \ldots, (n, 1)$ positions. At the second step, a pivot element is chosen from the nonzero entries in column 2, rows 2 through n of the matrix. The row containing the pivot is then interchanged with the second row of the matrix and is used as the new pivotal row. Multiples of the pivotal row are then subtracted from the remaining $n - 2$ rows so as to eliminate all entries below the diagonal in the second column. The same procedure is repeated for columns 3 through $n - 1$. Note that at the second step row 1 and column 1 remain unchanged, at the third step the first two rows and first two columns remain unchanged, and so on. At each step the overall dimensions of the system are effectively reduced by one (see Figure 1.1.2).

If the elimination process can be carried out as described, we will arrive at an equivalent upper triangular system after $n - 1$ steps. However, the procedure will break down if, at any step, all possible choices for a pivot element are equal to 0. When this happens the alternative is to reduce the system to certain special echelon or staircase-shaped forms. These echelon forms will be studied in the next section. They will also be used for $m \times n$ systems, where $m \neq n$.

EXERCISES

1. Solve each of the following systems of equations.

(a) $x_1 - 3x_2 = 2$
$\qquad 2x_2 = 6$

(b) $x_1 + x_2 + x_3 = 8$
$\qquad 2x_2 + x_3 = 5$
$\qquad\qquad 3x_3 = 9$

(c) $x_1 + 2x_2 + 2x_3 + x_4 = 5$
$\qquad 3x_2 + x_3 - 2x_4 = 1$
$\qquad\qquad - x_3 + 2x_4 = -1$
$\qquad\qquad\qquad 4x_4 = 4$

(d) $x_1 + x_2 + x_3 + x_4 + x_5 = 5$
$\qquad 2x_2 + x_3 - 2x_4 + x_5 = 1$
$\qquad\qquad 4x_3 + x_4 - 2x_5 = 1$
$\qquad\qquad\qquad x_4 - 3x_5 = 0$
$\qquad\qquad\qquad\qquad 2x_5 = 2$

2. Write out the coefficient matrix for each of the systems in Exercise 1.

3. In each of the following systems, interpret each equation as a line in the plane. For each system, graph the lines and determine geometrically the number of solutions.

(a) $\quad x_1 + x_2 = 4$
$\qquad x_1 - x_2 = 2$

(b) $\quad x_1 + 2x_2 = 4$
$\qquad -2x_1 - 4x_2 = 4$

(c) $\quad 2x_1 - x_2 = 3$
$\qquad -4x_1 + 2x_2 = -6$

(d) $\quad x_1 + x_2 = 1$
$\qquad x_1 - x_2 = 1$
$\qquad -x_1 + 3x_2 = 3$

4. Write an augmented matrix for each of the systems in Exercise 3.

5. Write out the system of equations that corresponds to each of the following augmented matrices.

(a) $\begin{pmatrix} 3 & 2 & | & 8 \\ 1 & 5 & | & 7 \end{pmatrix}$

(b) $\begin{pmatrix} 5 & -2 & 1 & | & 3 \\ 2 & 3 & -4 & | & 0 \end{pmatrix}$

(c) $\begin{pmatrix} 2 & 1 & 4 & | & -1 \\ 4 & -2 & 3 & | & 4 \\ 5 & 2 & 6 & | & -1 \end{pmatrix}$

(d) $\begin{pmatrix} 4 & -3 & 1 & 2 & | & 4 \\ 3 & 1 & -5 & 6 & | & 5 \\ 1 & 1 & 2 & 4 & | & 8 \\ 5 & 1 & 3 & -2 & | & 7 \end{pmatrix}$

6. Solve each of the following systems.

(a) $\begin{aligned} x_1 - 2x_2 &= 5 \\ 3x_1 + x_2 &= 1 \end{aligned}$

(b) $\begin{aligned} 2x_1 + x_2 &= 8 \\ 4x_1 - 3x_2 &= 6 \end{aligned}$

(c) $\begin{aligned} 4x_1 + 3x_2 &= 4 \\ \tfrac{2}{3}x_1 + 4x_2 &= 3 \end{aligned}$

(d) $\begin{aligned} x_1 + 2x_2 - x_3 &= 1 \\ 2x_1 - x_2 + x_3 &= 3 \\ -x_1 + 2x_2 + 3x_3 &= 7 \end{aligned}$

(e) $\begin{aligned} 2x_1 + x_2 + 3x_3 &= 1 \\ 4x_1 + 3x_2 + 5x_3 &= 1 \\ 6x_1 + 5x_2 + 5x_3 &= -3 \end{aligned}$

(f) $\begin{aligned} 3x_1 + 2x_2 + x_3 &= 0 \\ -2x_1 + x_2 - x_3 &= 2 \\ 2x_1 - x_2 + 2x_3 &= -1 \end{aligned}$

(g) $\begin{aligned} \tfrac{1}{3}x_1 + \tfrac{2}{3}x_2 + 2x_3 &= -1 \\ x_1 + 2x_2 + \tfrac{3}{2}x_3 &= \tfrac{3}{2} \\ \tfrac{1}{2}x_1 + 2x_2 + \tfrac{12}{5}x_3 &= \tfrac{1}{10} \end{aligned}$

(h) $\begin{aligned} x_2 + x_3 + x_4 &= 0 \\ 3x_1 + 3x_3 - 4x_4 &= 7 \\ x_1 + x_2 + x_3 + 2x_4 &= 6 \\ 2x_1 + 3x_2 + x_3 + 3x_4 &= 6 \end{aligned}$

7. Given a system of the form

$$\begin{aligned} -m_1 x_1 + x_2 &= b_1 \\ -m_2 x_1 + x_2 &= b_2 \end{aligned}$$

where m_1, m_2, b_1, and b_2 are constants:

(a) Show that the system will have a unique solution if $m_1 \neq m_2$.

(b) If $m_1 = m_2$, show that the system will be consistent only if $b_1 = b_2$.

(c) Give a geometric interpretation to parts (a) and (b).

8. Consider a system of the form

$$\begin{aligned} a_{11}x_1 + a_{12}x_2 &= 0 \\ a_{21}x_1 + a_{22}x_2 &= 0 \end{aligned}$$

where a_{11}, a_{12}, a_{21}, and a_{22} are constants. Explain why a system of this form must be consistent.

9. Give a geometrical interpretation of a linear equation in three unknowns. Give a geometrical description of the possible solution sets for a 3×3 linear system.

2. ROW ECHELON FORM

In Section 1 we learned a method for reducing an $n \times n$ linear system to triangular form. This method, however, will fail if at any stage of the reduction process, all of the possible choices for a pivot element in a given column are 0.

EXAMPLE 1.   Consider the system represented by the augmented matrix

$$\begin{pmatrix} ① & 1 & 1 & 1 & 1 & 1 \\ -1 & -1 & 0 & 0 & 1 & -1 \\ -2 & -2 & 0 & 0 & 1 & 1 \\ 0 & 0 & 1 & 1 & 1 & -1 \\ 1 & 1 & 2 & 2 & 2 & 1 \end{pmatrix} \leftarrow \text{pivotal row}$$

If row operation III is used to eliminate the last four elements in the first column, the resulting matrix will be

$$\begin{pmatrix} 1 & 1 & 1 & 1 & 1 & 1 \\ 0 & 0 & ① & 1 & 2 & 0 \\ 0 & 0 & 2 & 2 & 3 & 3 \\ 0 & 0 & 1 & 1 & 1 & -1 \\ 0 & 0 & 1 & 1 & 1 & 0 \end{pmatrix} \leftarrow \text{pivotal row}$$

At this stage the reduction to triangular form breaks down. All four possible choices for the pivot element in the second column are 0. How do we proceed from here? Since our goal is to simplify the system as much as possible, it seems natural to move over to the third column and eliminate the last three entries.

$$\begin{pmatrix} 1 & 1 & 1 & 1 & 1 & 1 \\ 0 & 0 & 1 & 1 & 2 & 0 \\ 0 & 0 & 0 & 0 & -1 & 3 \\ 0 & 0 & 0 & 0 & -1 & -1 \\ 0 & 0 & 0 & 0 & -1 & 0 \end{pmatrix}$$

In the fourth column all the choices for a pivot element are 0; so again we move on to the next column. If we use the third row as the pivotal row, the last two entries in the fifth column are eliminated.

$$\begin{pmatrix} 1 & 1 & 1 & 1 & 1 & \bigm| & 1 \\ 0 & 0 & 1 & 1 & 2 & \bigm| & 0 \\ 0 & 0 & 0 & 0 & -1 & \bigm| & 3 \\ 0 & 0 & 0 & 0 & 0 & \bigm| & -4 \\ 0 & 0 & 0 & 0 & 0 & \bigm| & -3 \end{pmatrix}$$

The equations represented by the last two rows are

$$0x_1 + 0x_2 + 0x_3 + 0x_4 + 0x_5 = -4$$
$$0x_1 + 0x_2 + 0x_3 + 0x_4 + 0x_5 = -3$$

Since there are no 5-tuples that could possibly satisfy these equations, the system is inconsistent. Note that the coefficient matrix we end up with is not in triangular form; it is in staircase or echelon form. ☐

Suppose now that we change the right-hand side of the system in the last example so as to obtain a consistent system. For example, if we start with

$$\begin{pmatrix} 1 & 1 & 1 & 1 & 1 & \bigm| & 1 \\ -1 & -1 & 0 & 0 & 1 & \bigm| & -1 \\ -2 & -2 & 0 & 0 & 1 & \bigm| & 1 \\ 0 & 0 & 1 & 1 & 1 & \bigm| & 3 \\ 1 & 1 & 2 & 2 & 2 & \bigm| & 4 \end{pmatrix}$$

then the reduction process will yield the augmented matrix

$$\begin{pmatrix} 1 & 1 & 1 & 1 & 1 & \bigm| & 1 \\ 0 & 0 & 1 & 1 & 2 & \bigm| & 0 \\ 0 & 0 & 0 & 0 & -1 & \bigm| & 3 \\ 0 & 0 & 0 & 0 & 0 & \bigm| & 0 \\ 0 & 0 & 0 & 0 & 0 & \bigm| & 0 \end{pmatrix}$$

Clearly, the last two equations of the reduced system will be satisfied for any 5-tuple. Thus the solution set will be the set of all 5-tuples satisfying the first three equations.

(1)
$$x_1 + x_2 + x_3 + x_4 + x_5 = 1$$
$$x_3 + x_4 + 2x_5 = 0$$
$$-x_5 = 3$$

The variables corresponding to the first nonzero elements in each row of the augmented matrix will be referred to as *lead variables*. Thus x_1, x_3, and x_5 are the lead variables. The remaining variables corresponding to the columns skipped in the reduction process will be referred to as *free variables*. Thus x_2 and x_4 are the free variables. If we transfer the free variables over to the right-hand side in (1), we obtain the system

$$x_1 + x_3 + \ x_5 = 1 - x_2 - x_4$$
(2)
$$x_3 + 2x_5 = -x_4$$
$$-x_5 = 3$$

System (2) is triangular in the unknowns x_1, x_3, x_5. Thus for each pair of values assigned to x_2 and x_4, there will be a unique solution. For example, if $x_2 = x_4 = 0$, then $x_5 = -3$, $x_3 = 6$, $x_1 = -2$, and hence $(-2, 0, 6, 0, -3)$ is a solution to the system.

Definition. A matrix is said to be in **row echelon form** if
(1) The first nonzero entry in each row is 1.
(2) If row k does not consist entirely of zeros, the number of leading zero entries in row $k + 1$ is greater than the number of leading zero entries in row k.
(3) If there are rows whose entries are all zero, they are below the rows having nonzero entries.

EXAMPLE 2. The following matrices are in row echelon form.

$$\begin{pmatrix} 1 & 4 & 2 \\ 0 & 1 & 3 \\ 0 & 0 & 1 \end{pmatrix}, \quad \begin{pmatrix} 1 & 2 & 3 \\ 0 & 0 & 1 \\ 0 & 0 & 0 \end{pmatrix}, \quad \begin{pmatrix} 1 & 3 & 1 & 0 \\ 0 & 0 & 1 & 3 \\ 0 & 0 & 0 & 0 \end{pmatrix} \qquad \square$$

EXAMPLE 3. The following matrices are not in row echelon form.

$$\begin{pmatrix} 2 & 4 & 6 \\ 0 & 3 & 5 \\ 0 & 0 & 4 \end{pmatrix}, \quad \begin{pmatrix} 0 & 0 & 0 \\ 0 & 1 & 0 \end{pmatrix}, \quad \begin{pmatrix} 0 & 1 \\ 1 & 0 \end{pmatrix}$$

The first matrix does not satisfy condition 1. The second matrix fails to satisfy condition 3, and the third matrix fails to satisfy the second condition. $\qquad \square$

Definition. The process of using row operations I, II, and III to transform a linear system into one whose augmented matrix is in row echelon form is called **Gaussian elimination.**

Note that row operation II is necessary in order to scale the rows so that the lead coefficients are all 1. If the row echelon matrix contains a row of the form

$$(0 \quad 0 \quad \cdots \quad 0 | 1)$$

the system is inconsistent. Otherwise, the system will be consistent. If the system is consistent and the nonzero rows of the row echelon matrix form a triangular system, the system will have a unique solution.

OVERDETERMINED SYSTEMS

A linear system is said to be *overdetermined* if there are more equations than unknowns ($m > n$). Overdetermined systems are usually (but not always) inconsistent.

EXAMPLE 4

(a) $\begin{aligned} x_1 + x_2 &= 1 \\ x_1 - x_2 &= 3 \\ -x_1 + 2x_2 &= -2 \end{aligned}$

(b) $\begin{aligned} x_1 + 2x_2 + x_3 &= 1 \\ 2x_1 - x_2 + x_3 &= 2 \\ 4x_1 + 3x_2 + 3x_3 &= 4 \\ 2x_1 - x_2 + 3x_3 &= 5 \end{aligned}$

(c) $\begin{aligned} x_1 + 2x_2 + x_3 &= 1 \\ 2x_1 - x_2 + x_3 &= 2 \\ 4x_1 + 3x_2 + 3x_3 &= 4 \\ 3x_1 + x_2 + 2x_3 &= 3 \end{aligned}$

We will assume that, by now, the reader is familiar with the elimination process, and hence we will omit all the intermediate steps in reducing each of these systems.

System (a)

$$\left(\begin{array}{rr|r} 1 & 1 & 1 \\ 1 & -1 & 3 \\ -1 & 2 & -2 \end{array} \right) \rightarrow \left(\begin{array}{rr|r} 1 & 1 & 1 \\ 0 & 1 & -1 \\ 0 & 0 & 1 \end{array} \right)$$

It follows from the last row of the reduced matrix that the system is inconsistent. The three equations in system (a) represent lines in the plane. The first two lines intersect at the point $(2, -1)$. However, the third line does not pass through this point. Thus there are no points that lie on all three lines (see Figure 1.2.1).

System (b)

$$\left(\begin{array}{rrr|r} 1 & 2 & 1 & 1 \\ 2 & -1 & 1 & 2 \\ 4 & 3 & 3 & 4 \\ 2 & -1 & 3 & 5 \end{array} \right) \rightarrow \left(\begin{array}{rrr|r} 1 & 2 & 1 & 1 \\ 0 & 1 & \frac{1}{5} & 0 \\ 0 & 0 & 1 & \frac{3}{2} \\ 0 & 0 & 0 & 0 \end{array} \right)$$

Using back substitution, we see that system (b) has exactly one solution $(0.1, -0.3, 1.5)$. The solution is unique because the non-zero rows of the reduced matrix form a triangular system.

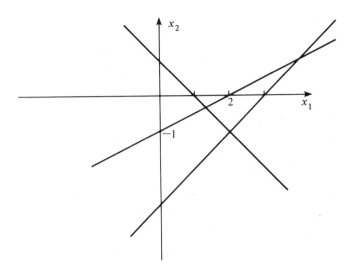

FIGURE 1.2.1

System (c)

$$\begin{pmatrix} 1 & 2 & 1 & | & 1 \\ 2 & -1 & 1 & | & 2 \\ 4 & 3 & 3 & | & 4 \\ 3 & 1 & 2 & | & 3 \end{pmatrix} \rightarrow \begin{pmatrix} 1 & 2 & 1 & | & 1 \\ 0 & 1 & \frac{1}{5} & | & 0 \\ 0 & 0 & 0 & | & 0 \\ 0 & 0 & 0 & | & 0 \end{pmatrix}$$

Solving for x_2 and x_1 in terms of x_3, we obtain

$$x_2 = -0.2x_3$$
$$x_1 = 1 - 2x_2 - x_3 = 1 - 0.6x_3$$

It follows that the solution set is the set of all ordered triples of the form $(1 - 0.6\alpha, -0.2\alpha, \alpha)$, where α is a real number. This system is consistent and has infinitely many solutions because of the free variable x_3.

UNDERDETERMINED SYSTEMS

A linear system is said to be *underdetermined* if there are fewer equations than unknowns ($m < n$). Although it is possible for underdetermined systems to be inconsistent, they are usually consistent with infinitely many solutions. It is not possible for an underdetermined system to have only one solution. The reason for this is that any row echelon form of the coefficient matrix will involve $r \leq m$ nonzero rows. Thus there will be r lead variables and $n - r$ free variables, where $n - r \geq n - m > 0$. If the system is consistent, we can assign the free variables arbitrary values and solve for the lead variables. Therefore, a consistent underdetermined system will have infinitely many solutions.

EXAMPLE 5

(a) $x_1 + 2x_2 + x_3 = -1$

$2x_1 + 4x_2 + 2x_3 = 3$

(b) $x_1 + x_2 + x_3 + x_4 + x_5 = 2$

$x_1 + x_2 + x_3 + 2x_4 + 2x_5 = 3$

$x_1 + x_2 + x_3 + 2x_4 + 3x_5 = 2$

System (a)

$$\left(\begin{array}{ccc|c} 1 & 2 & 1 & -1 \\ 2 & 4 & 2 & 3 \end{array}\right) \rightarrow \left(\begin{array}{ccc|c} 1 & 2 & 1 & -1 \\ 0 & 0 & 0 & 1 \end{array}\right)$$

Clearly, system (a) is inconsistent. We can think of the two equations in system (a) as representing planes in 3-space. Usually, two planes intersect in a line; however, in this case the planes are parallel.

System (b)

$$\left(\begin{array}{ccccc|c} 1 & 1 & 1 & 1 & 1 & 2 \\ 1 & 1 & 1 & 2 & 2 & 3 \\ 1 & 1 & 1 & 2 & 3 & 2 \end{array}\right) \rightarrow \left(\begin{array}{ccccc|c} 1 & 1 & 1 & 1 & 1 & 2 \\ 0 & 0 & 0 & 1 & 1 & 1 \\ 0 & 0 & 0 & 0 & 1 & -1 \end{array}\right)$$

System (b) is consistent, and since there are two free variables, the system will have infinitely many solutions. Often with systems like this it is convenient to continue with the elimination process until all the terms above each leading 1 are eliminated. Thus for system (b) we will continue and eliminate the first two entries in the fifth column and then the first element in the fourth column.

$$\left(\begin{array}{ccccc|c} 1 & 1 & 1 & 1 & 1 & 2 \\ 0 & 0 & 0 & 1 & 1 & 1 \\ 0 & 0 & 0 & 0 & 1 & -1 \end{array}\right)$$

$$\rightarrow \left(\begin{array}{ccccc|c} 1 & 1 & 1 & 1 & 0 & 3 \\ 0 & 0 & 0 & 1 & 0 & 2 \\ 0 & 0 & 0 & 0 & 1 & -1 \end{array}\right)$$

$$\rightarrow \left(\begin{array}{ccccc|c} 1 & 1 & 1 & 0 & 0 & 1 \\ 0 & 0 & 0 & 1 & 0 & 2 \\ 0 & 0 & 0 & 0 & 1 & -1 \end{array}\right)$$

If we put the free variables over on the right-hand side, it follows that

$$x_1 = 1 - x_2 - x_3$$

$$x_4 = 2$$

$$x_5 = -1$$

Thus, for any real numbers α and β, the 5-tuple $(1 - \alpha - \beta, \alpha, \beta, 2, -1)$ is a solution to the system. $\quad\square$

Definition.   A matrix is said to be in **reduced row echelon form** if:
(1) The matrix is in row echelon form.
(2) The first nonzero entry in each row is the only nonzero entry in its column.

The following matrices are in reduced row echelon form:

$$\begin{pmatrix} 1 & 0 \\ 0 & 1 \end{pmatrix}, \quad \begin{pmatrix} 1 & 0 & 0 & 3 \\ 0 & 1 & 0 & 2 \\ 0 & 0 & 1 & 1 \end{pmatrix}, \quad \begin{pmatrix} 0 & 1 & 2 & 0 \\ 0 & 0 & 0 & 1 \\ 0 & 0 & 0 & 0 \end{pmatrix}, \quad \begin{pmatrix} 1 & 2 & 0 & 1 \\ 0 & 0 & 1 & 3 \\ 0 & 0 & 0 & 0 \end{pmatrix}$$

The process of using elementary row operations to transform a matrix into reduced row echelon form is called *Gauss–Jordan reduction*.

EXAMPLE 6.   Use Gauss–Jordan reduction to solve the system

$$-2x_1 + x_2 - x_3 + 3x_4 = 0$$
$$2x_1 + 3x_2 + x_3 + x_4 = 0$$
$$4x_1 - x_2 - 3x_3 + 2x_4 = 0$$

SOLUTION

$$\begin{pmatrix} \boxed{-2} & 1 & -1 & 3 & | & 0 \\ \boxed{2} & 3 & 1 & 1 & | & 0 \\ \boxed{4} & -1 & -3 & 2 & | & 0 \end{pmatrix} \rightarrow \begin{pmatrix} 1 & -0.5 & \boxed{0.5} & -1.5 & | & 0 \\ 0 & 1 & 0 & 1 & | & 0 \\ 0 & 0 & \textcircled{1} & -1.4 & | & 0 \end{pmatrix} \begin{matrix} \text{row} \\ \text{echelon} \\ \text{form} \end{matrix}$$

$$\begin{pmatrix} -2 & 1 & -1 & 3 & | & 0 \\ 0 & \textcircled{4} & 0 & 4 & | & 0 \\ 0 & \boxed{1} & -5 & 8 & | & 0 \end{pmatrix} \begin{pmatrix} 1 & \boxed{-0.5} & 0 & -0.8 & | & 0 \\ 0 & \textcircled{1} & 0 & 1.0 & | & 0 \\ 0 & 0 & 1 & -1.4 & | & 0 \end{pmatrix}$$

$$\begin{pmatrix} -2 & 1 & -1 & 3 & | & 0 \\ 0 & 4 & 0 & 4 & | & 0 \\ 0 & 0 & -5 & 7 & | & 0 \end{pmatrix} \begin{pmatrix} 1 & 0 & 0 & -0.3 & | & 0 \\ 0 & 1 & 0 & 1.0 & | & 0 \\ 0 & 0 & 1 & -1.4 & | & 0 \end{pmatrix} \begin{matrix} \text{reduced} \\ \text{row echelon} \\ \text{form} \end{matrix}$$

If we set x_4 equal to any real number α, then $x_1 = 0.3\alpha$, $x_2 = -\alpha$, and $x_3 = 1.4\alpha$. Thus all ordered 4-tuples of the form $(0.3\alpha, -\alpha, 1.4\alpha, \alpha)$ are solutions to the system.   □

HOMOGENEOUS SYSTEMS

A system of linear equations is said to be *homogeneous* if the constants on the right-hand side are all zero. Homogeneous systems are always consistent. It is a trivial matter to find a solution; just set all the variables equal to zero. Thus, if an $m \times n$ homogeneous system has a unique solution, it must be the trivial solution $(0, 0, \ldots, 0)$. The homogeneous system in Example 6 consisted of $m = 3$ equations in $n = 4$

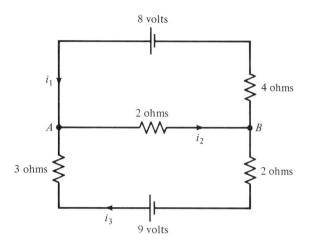

FIGURE 1.2.2

unknowns. In the case that $n > m$, there will always be free variables and consequently additional nontrivial solutions. This result has essentially been proved in our discussion of underdetermined systems, but because of its importance, we state it as a theorem.

Theorem 1.2.1.   *An $m \times n$ homogeneous system of linear equations has a nontrivial solution if $n > m$.*

Proof.   A homogeneous system is always consistent. The row echelon form of the matrix can have at most m nonzero rows. Thus there are at most m lead variables. Since there are n variables altogether and $n > m$, there must be some free variables. The free variables can be assigned arbitrary values. For each assignment of values to the free variables there is a solution to the system.   □

APPLICATION

In an electrical network it is possible to determine the amount of current in each branch in terms of the resistances and the voltages. In Figure 1.2.2 the ⊣⊢ represents a battery (measured in volts) that drives a charge and produces a current. The current will flow out of the terminal of the battery represented by the longer vertical line, that is, ⊏⊢. The symbol ⟋W⟍ represents a resistor. The resistances are measured in ohms. The capital letters represent nodes and the i's represent the currents between the nodes. The currents are measured in amperes. The arrows show the direction of the currents. If, however, one of the currents, say i_2, turns out to be negative, this would mean that the current along that branch is in the opposite direction of the arrow.

To determine the currents, *Kirchhoff's laws* are used:

1. At every node the sum of the incoming currents equals the sum of the outgoing currents.
2. Around every closed loop the algebraic sum of the voltage must equal the algebraic sum of the voltage drops.

The voltage drops E for each resistor are given by *Ohm's law*:

$$E = iR$$

where i represents the current in amperes and R the resistance in ohms.

Let us find the currents in the network pictured in Figure 1.2.2. From the first law, we have

$$i_1 - i_2 + i_3 = 0 \qquad \text{(node } A\text{)}$$

$$- i_1 + i_2 - i_3 = 0 \qquad \text{(node } B\text{)}$$

By the second law,

$$4i_1 + 2i_2 = 8 \qquad \text{(top loop)}$$

$$2i_2 + 5i_3 = 9 \qquad \text{(bottom loop)}$$

The network can be represented by the augmented matrix

$$\left(\begin{array}{ccc|c} 1 & -1 & 1 & 0 \\ -1 & 1 & -1 & 0 \\ 4 & 2 & 0 & 8 \\ 0 & 2 & 5 & 9 \end{array} \right)$$

This matrix is easily reduced to row echelon form

$$\left(\begin{array}{ccc|c} 1 & -1 & 1 & 0 \\ 0 & 1 & -\frac{2}{3} & \frac{4}{3} \\ 0 & 0 & 1 & 1 \\ 0 & 0 & 0 & 0 \end{array} \right)$$

Solving by back substitution, we see that $i_1 = 1$, $i_2 = 2$, and $i_3 = 1$.

EXERCISES

1. Which of the following matrices are in row echelon form? Which are in reduced row echelon form?

(a) $\begin{pmatrix} 1 & 2 & 3 & 4 \\ 0 & 0 & 1 & 2 \end{pmatrix}$

(b) $\begin{pmatrix} 1 & 0 & 0 \\ 0 & 0 & 0 \\ 0 & 0 & 1 \end{pmatrix}$

(c) $\begin{pmatrix} 1 & 3 & 0 \\ 0 & 0 & 1 \\ 0 & 0 & 0 \end{pmatrix}$

(d) $\begin{pmatrix} 0 & 1 \\ 0 & 0 \\ 0 & 0 \end{pmatrix}$

(e) $\begin{pmatrix} 1 & 1 & 1 \\ 0 & 1 & 2 \\ 0 & 0 & 3 \end{pmatrix}$ (f) $\begin{pmatrix} 1 & 4 & 6 \\ 0 & 0 & 1 \\ 0 & 1 & 3 \end{pmatrix}$

(g) $\begin{pmatrix} 1 & 0 & 0 & 1 & 2 \\ 0 & 1 & 0 & 2 & 4 \\ 0 & 0 & 1 & 3 & 6 \end{pmatrix}$ (h) $\begin{pmatrix} 0 & 1 & 3 & 4 \\ 0 & 0 & 1 & 3 \\ 0 & 0 & 0 & 0 \end{pmatrix}$

2. In each of the following, the augmented matrix is in row echelon form. For each case indicate whether or not the corresponding linear system is consistent. If the system has a unique solution, find it.

(a) $\left(\begin{array}{cc|c} 1 & 2 & 4 \\ 0 & 1 & 3 \\ 0 & 0 & 1 \end{array}\right)$ (b) $\left(\begin{array}{cc|c} 1 & 3 & 1 \\ 0 & 1 & -1 \\ 0 & 0 & 0 \end{array}\right)$

(c) $\left(\begin{array}{ccc|c} 1 & -2 & 4 & 1 \\ 0 & 0 & 1 & 3 \\ 0 & 0 & 0 & 0 \end{array}\right)$ (d) $\left(\begin{array}{ccc|c} 1 & -2 & 2 & -2 \\ 0 & 1 & -1 & 3 \\ 0 & 0 & 1 & 2 \end{array}\right)$

(e) $\left(\begin{array}{ccc|c} 1 & 3 & 2 & -2 \\ 0 & 0 & 1 & 4 \\ 0 & 0 & 0 & 1 \end{array}\right)$ (f) $\left(\begin{array}{ccc|c} 1 & -1 & 3 & 8 \\ 0 & 1 & 2 & 7 \\ 0 & 0 & 1 & 2 \\ 0 & 0 & 0 & 0 \end{array}\right)$

3. In each of the following, the augmented matrix is in reduced row echelon form. In each case find the solution set to the corresponding linear system.

(a) $\left(\begin{array}{ccc|c} 1 & 0 & 0 & -2 \\ 0 & 1 & 0 & 5 \\ 0 & 0 & 1 & 3 \end{array}\right)$ (b) $\left(\begin{array}{ccc|c} 1 & 4 & 0 & 2 \\ 0 & 0 & 1 & 3 \\ 0 & 0 & 0 & 1 \end{array}\right)$

(c) $\left(\begin{array}{ccc|c} 1 & -3 & 0 & 2 \\ 0 & 0 & 1 & -2 \\ 0 & 0 & 0 & 0 \end{array}\right)$ (d) $\left(\begin{array}{cccc|c} 1 & 2 & 0 & 1 & 5 \\ 0 & 0 & 1 & 3 & 4 \end{array}\right)$

(e) $\left(\begin{array}{cccc|c} 1 & 5 & -2 & 0 & 3 \\ 0 & 0 & 0 & 1 & 6 \\ 0 & 0 & 0 & 0 & 0 \\ 0 & 0 & 0 & 0 & 0 \end{array}\right)$ (f) $\left(\begin{array}{ccc|c} 0 & 1 & 0 & 2 \\ 0 & 0 & 1 & -1 \\ 0 & 0 & 0 & 0 \end{array}\right)$

4. For each of the following systems of equations, use Gaussian elimination to obtain an equivalent system whose coefficient matrix is in row echelon form.

(a) $\begin{aligned} x_1 - 2x_2 &= 3 \\ 2x_1 - x_2 &= 9 \end{aligned}$ (b) $\begin{aligned} 2x_1 - 3x_2 &= 5 \\ -4x_1 + 6x_2 &= 8 \end{aligned}$

(c) $\begin{aligned} x_1 + x_2 &= 0 \\ 2x_1 + 3x_2 &= 0 \\ 3x_1 - 2x_2 &= 0 \end{aligned}$ (d) $\begin{aligned} 3x_1 + 2x_2 - x_3 &= 4 \\ x_1 - 2x_2 + 2x_3 &= 1 \\ 11x_1 + 2x_2 + x_3 &= 14 \end{aligned}$

(e) $2x_1 + 3x_2 + x_3 = 1$

$\quad x_1 + x_2 + x_3 = 3$

$\quad 3x_1 + 4x_2 + 2x_3 = 4$

(f) $\quad x_1 - x_2 + 2x_3 = 4$

$\quad 2x_1 + 3x_2 - x_3 = 1$

$\quad 7x_1 + 3x_2 + 4x_3 = 7$

(g) $x_1 + x_2 + x_3 + x_4 = 0$

$\quad 2x_1 + 3x_2 - x_3 - x_4 = 2$

$\quad x_1 - x_2 + 2x_3 + 2x_4 = 3$

$\quad 2x_1 + 5x_2 - 2x_3 - 2x_4 = 4$

(h) $\quad x_1 - 2x_2 = 3$

$\quad 2x_1 + x_2 = 1$

$\quad -5x_1 + 8x_2 = 4$

(i) $\quad -x_1 + 2x_2 - x_3 = 2$

$\quad -2x_1 + 2x_2 + x_3 = 4$

$\quad 3x_1 + 2x_2 + 2x_3 = 5$

$\quad -3x_1 + 8x_2 + 5x_3 = 17$

(j) $\quad x_1 + 2x_2 - 3x_3 + x_4 = -2$

$\quad -x_1 - x_2 + 4x_3 - x_4 = 6$

$\quad -2x_1 - 4x_2 + 7x_3 - 2x_4 = 5$

(k) $x_1 + 3x_2 + x_3 + x_4 = 3$

$\quad 2x_1 - 2x_2 + x_3 + 2x_4 = 8$

$\quad x_1 + 11x_2 + 2x_3 + x_4 = 1$

(l) $x_1 - 3x_2 + x_3 = 1$

$\quad 2x_1 + x_2 - x_3 = 2$

$\quad x_1 + 4x_2 - 2x_3 = 1$

$\quad 5x_1 - 8x_2 + 2x_3 = 5$

5. For each of the systems in Exercise 4, indicate whether or not the system is consistent. If the system is consistent and involves no free variables, use back substitution to find the unique solution. If the system is consistent and there are free variables, transform it to reduced row echelon form and find all solutions.

6. Use Gauss–Jordan reduction to solve each of the following systems.

(a) $\quad x_1 + x_2 = -1$

$\quad 4x_1 - 3x_2 = 3$

(b) $x_1 + 3x_2 + x_3 + x_4 = 3$

$\quad 2x_1 - 2x_2 + x_3 + 2x_4 = 8$

$\quad 3x_1 + x_2 + 2x_3 - x_4 = -1$

(c) $x_1 + x_2 + x_3 = 0$

$\quad x_1 - x_2 - x_3 = 0$

(d) $\quad x_1 + x_2 + x_3 + x_4 = 0$

$\quad 2x_1 + x_2 - x_3 + 3x_4 = 0$

$\quad x_1 - 2x_2 + x_3 + x_4 = 0$

7. Let (c_1, c_2) be a solution to the 2×2 system
$$a_{11}x_1 + a_{12}x_2 = 0$$
$$a_{21}x_1 + a_{22}x_2 = 0$$
Show that for any real number α, the ordered pair $(\alpha c_1, \alpha c_2)$ is also a solution.

8. Determine the amount of each of the currents for the following networks.

(a)

(b)

(c)

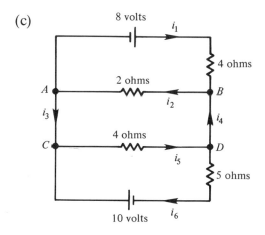

3. MATRIX ALGEBRA

In this section we define arithmetic operations with matrices and look at some of their algebraic properties. Matrices are one of the most powerful tools in mathematics. To use matrices effectively, we must be adept at matrix arithmetic.

The entries of a matrix are called *scalars*. They are usually either real or complex numbers. For the most part we will be working with matrices whose entries are real numbers. Unless otherwise stated, the reader may assume that the term *scalar* refers to a real number. There will, however, be some occasions (mainly in Chapter 6) when we will use the set of complex numbers as our scalar field.

If we wish to refer to matrices without specifically writing out all their entries, we will use capital letters A, B, C, and so on. In general, a_{ij} will denote the entry of the matrix A that is in the ith row and the jth column. Thus, if A is an $m \times n$ matrix, then

$$A = \begin{pmatrix} a_{11} & a_{12} \cdots a_{1n} \\ a_{21} & a_{22} \cdots a_{2n} \\ \vdots \\ a_{m1} & a_{m2} \cdots a_{mn} \end{pmatrix}$$

We will sometimes shorten this to $A = (a_{ij})$. Similarly, a matrix B may be referred to as (b_{ij}), a matrix C as (c_{ij}), and so on.

Definition.   Two $m \times n$ matrices A and B are said to be **equal** if $a_{ij} = b_{ij}$ for each i and j.

SCALAR MULTIPLICATION

If A is a matrix and α is a scalar, then αA is the matrix formed by multiplying each of the entries of A by α. For example, if

$$A = \begin{pmatrix} 4 & 8 & 2 \\ 6 & 8 & 10 \end{pmatrix}$$

then

$$\tfrac{1}{2}A = \begin{pmatrix} 2 & 4 & 1 \\ 3 & 4 & 5 \end{pmatrix} \quad \text{and} \quad 3A = \begin{pmatrix} 12 & 24 & 6 \\ 18 & 24 & 30 \end{pmatrix}$$

MATRIX ADDITION

If $A = (a_{ij})$ and $B = (b_{ij})$ are both $m \times n$ matrices, then the sum $A + B$ is the $m \times n$ matrix whose ijth entry is $a_{ij} + b_{ij}$ for each ordered pair (i, j). For example,

$$\begin{pmatrix} 3 & 2 & 1 \\ 4 & 5 & 6 \end{pmatrix} + \begin{pmatrix} 2 & 2 & 2 \\ 1 & 2 & 3 \end{pmatrix} = \begin{pmatrix} 5 & 4 & 3 \\ 5 & 7 & 9 \end{pmatrix}$$

$$\begin{pmatrix} 2 \\ 1 \\ 8 \end{pmatrix} + \begin{pmatrix} -8 \\ 3 \\ 2 \end{pmatrix} = \begin{pmatrix} 6 \\ 4 \\ 10 \end{pmatrix}$$

If we define $A - B$ to be $A + (-1)B$, then it turns out that $A - B$ is formed by subtracting the corresponding entry of B from each entry of A. Thus

$$\begin{pmatrix} 2 & 4 \\ 3 & 1 \end{pmatrix} - \begin{pmatrix} 4 & 5 \\ 2 & 3 \end{pmatrix} = \begin{pmatrix} 2 & 4 \\ 3 & 1 \end{pmatrix} + (-1)\begin{pmatrix} 4 & 5 \\ 2 & 3 \end{pmatrix}$$

$$= \begin{pmatrix} 2 & 4 \\ 3 & 1 \end{pmatrix} + \begin{pmatrix} -4 & -5 \\ -2 & -3 \end{pmatrix}$$

$$= \begin{pmatrix} 2-4 & 4-5 \\ 3-2 & 1-3 \end{pmatrix}$$

$$= \begin{pmatrix} -2 & -1 \\ 1 & -2 \end{pmatrix}$$

If O represents a matrix, with the same dimensions as A, whose entries are all 0, then

$$A + O = O + A = A$$

That is, the zero matrix acts as an additive identity on the set of all $m \times n$ matrices. Furthermore, each $m \times n$ matrix A has an additive inverse. Indeed,

$$A + (-1)A = O = (-1)A + A$$

It is customary to denote the additive inverse by $-A$. Thus

$$-A = (-1)A$$

MATRIX MULTIPLICATION

We have yet to define the most important operation, the multiplication of two matrices. Much of the motivation behind the definition comes from the applications to linear systems of equations. If we have a system of one linear equation in one unknown, it can be written in the form

(1) $$ax = b$$

We generally think of a, x and b as being scalars; however, they could also be treated as 1×1 matrices. More generally, given an $m \times n$

linear system

$$a_{11}x_1 + a_{12}x_2 + \cdots + a_{1n}x_n = b_1$$
$$a_{21}x_1 + a_{22}x_2 + \cdots + a_{2n}x_n = b_2$$
$$\vdots$$
$$a_{m1}x_1 + a_{m2}x_2 + \cdots + a_{mn}x_n = b_m$$

it is desirable to write the system in a form similar to (1), that is, as a matrix equation

$$AX = B$$

where $A = (a_{ij})$ is known, X is an $n \times 1$ matrix of unknowns, and B is an $m \times 1$ matrix representing the right-hand side of the system. Thus we set

$$A = \begin{pmatrix} a_{11} & a_{12} & \cdots & a_{1n} \\ a_{21} & a_{22} & \cdots & a_{2n} \\ \vdots & & & \\ a_{m1} & a_{m2} & \cdots & a_{mn} \end{pmatrix}, \qquad X = \begin{pmatrix} x_1 \\ x_2 \\ \vdots \\ x_n \end{pmatrix}, \qquad B = \begin{pmatrix} b_1 \\ b_2 \\ \vdots \\ b_m \end{pmatrix}$$

and

$$(2) \qquad AX = \begin{pmatrix} a_{11}x_1 + a_{12}x_2 + \cdots + a_{1n}x_n \\ a_{21}x_1 + a_{22}x_2 + \cdots + a_{2n}x_n \\ \vdots \\ a_{m1}x_1 + a_{m2}x_2 + \cdots + a_{mn}x_n \end{pmatrix}$$

Given an $m \times n$ matrix A and an $n \times 1$ matrix X it is possible to compute a product AX by (2). The product AX will be an $m \times 1$ matrix. The rule for determining the ith entry of AX is

$$a_{i1}x_1 + a_{i2}x_2 + \cdots + a_{in}x_n$$

Note that the ith entry is determined using only the ith row of A. The entries in that row are paired off with the corresponding entries of X and multiplied. The n products are then summed. Those readers familiar with dot products will recognize this as simply the dot product of the n-tuple corresponding to the ith row of A with the n-tuple corresponding to the matrix X.

ith entry of AX

$$\begin{pmatrix} x_1 \\ x_2 \\ \vdots \\ x_n \end{pmatrix} = a_{i1}x_1 + a_{i2}x_2 + \cdots + a_{in}x_n$$

$$(a_{i1} \quad a_{i2} \quad \cdots \quad a_{in})$$

In order to pair off the entries in this way, the number of columns of A must equal the number of rows of X. The entries of X can be either scalars or unknowns having scalar values.

EXAMPLE 1

$$A = \begin{pmatrix} 4 & 2 & 1 \\ 5 & 3 & 7 \end{pmatrix}, \qquad X = \begin{pmatrix} x_1 \\ x_2 \\ x_3 \end{pmatrix}$$

$$AX = \begin{pmatrix} 4x_1 + 2x_2 + x_3 \\ 5x_1 + 3x_2 + 7x_3 \end{pmatrix}$$ □

EXAMPLE 2

$$A = \begin{pmatrix} -3 & 1 \\ 2 & 5 \\ 4 & 2 \end{pmatrix}, \qquad X = \begin{pmatrix} 2 \\ 4 \end{pmatrix}$$

$$AX = \begin{pmatrix} -3 \cdot 2 + 1 \cdot 4 \\ 2 \cdot 2 + 5 \cdot 4 \\ 4 \cdot 2 + 2 \cdot 4 \end{pmatrix} = \begin{pmatrix} -2 \\ 24 \\ 16 \end{pmatrix}$$ □

EXAMPLE 3.   Write the following system of equations as a matrix equation $AX = B$.

$$3x_1 + 2x_2 + x_3 = 5$$
$$x_1 - 2x_2 + 5x_3 = -2$$
$$2x_1 + x_2 - 3x_3 = 1$$

SOLUTION

$$\begin{pmatrix} 3 & 2 & 1 \\ 1 & -2 & 5 \\ 2 & 1 & -3 \end{pmatrix} \begin{pmatrix} x_1 \\ x_2 \\ x_3 \end{pmatrix} = \begin{pmatrix} 5 \\ -2 \\ 1 \end{pmatrix}$$ □

More generally, it is possible to multiply a matrix A times a matrix B if the number of columns of A equals the number of rows of B. The first column of the product is determined by the first column of B, the second column by the second column of B, and so on. Thus, to determine the (i, j) entry of the product AB, we use the entries of the ith row of A and the jth column of B.

Definition.   If $A = (a_{ij})$ is an $m \times n$ matrix and $B = (b_{ij})$ is an $n \times r$ matrix, then the product $AB = C = (c_{ij})$ is the $m \times r$ matrix whose entries are defined by

$$c_{ij} = \sum_{k=1}^{n} a_{ik} b_{kj}$$

What this definition says is that to find the ijth element of the product, you take the ith row of A and the jth column of B, multiply the corresponding elements pairwise, and add the resulting numbers.

In order to pair off the elements in this manner, the number of columns of A must equal the number of rows of B. If this does not occur, the multiplication is impossible.

EXAMPLE 4.   If

$$A = \begin{pmatrix} -2 & 1 & 3 \\ 4 & 1 & 6 \end{pmatrix} \quad \text{and} \quad B = \begin{pmatrix} 3 & -2 \\ 2 & 4 \\ 1 & -3 \end{pmatrix}$$

then

$$AB = \begin{pmatrix} -2 \cdot 3 + 1 \cdot 2 + 3 \cdot 1 & -2 \cdot (-2) + 1 \cdot 4 + 3 \cdot (-3) \\ 4 \cdot 3 + 1 \cdot 2 + 6 \cdot 1 & 4 \cdot (-2) + 1 \cdot 4 + 6 \cdot (-3) \end{pmatrix}$$

$$= \begin{pmatrix} -1 & -1 \\ 20 & -22 \end{pmatrix}$$

and

$$BA = \begin{pmatrix} 3 \cdot (-2) + (-2) \cdot 4 & 3 \cdot 1 - 2 \cdot 1 & 3 \cdot 3 - 2 \cdot 6 \\ 2 \cdot (-2) + 4 \cdot 4 & 2 \cdot 1 + 4 \cdot 1 & 2 \cdot 3 + 4 \cdot 6 \\ 1 \cdot (-2) - 3 \cdot 4 & 1 \cdot 1 - 3 \cdot 1 & 1 \cdot 3 - 3 \cdot 6 \end{pmatrix}$$

$$= \begin{pmatrix} -14 & 1 & -3 \\ 12 & 6 & 30 \\ -14 & -2 & -15 \end{pmatrix} \qquad \qquad \Box$$

EXAMPLE 5.   If

$$A = \begin{pmatrix} 3 & 4 \\ 1 & 2 \end{pmatrix} \quad \text{and} \quad B = \begin{pmatrix} 1 & 2 \\ 4 & 5 \\ 3 & 6 \end{pmatrix}$$

then it is impossible to multiply A times B, since the number of columns of A does not equal the number of rows of B. However, it is possible to multiply B times A.

$$BA = \begin{pmatrix} 1 & 2 \\ 4 & 5 \\ 3 & 6 \end{pmatrix} \begin{pmatrix} 3 & 4 \\ 1 & 2 \end{pmatrix} = \begin{pmatrix} 5 & 8 \\ 17 & 26 \\ 15 & 24 \end{pmatrix} \qquad \qquad \Box$$

If A and B are both $n \times n$ matrices, then AB and BA will also be $n \times n$ matrices, but in general they will not be equal. *Multiplication of matrices is not commutative.*

EXAMPLE 6. If

$$A = \begin{pmatrix} 1 & 1 \\ 0 & 0 \end{pmatrix} \quad \text{and} \quad B = \begin{pmatrix} 1 & 1 \\ 2 & 2 \end{pmatrix}$$

then

$$AB = \begin{pmatrix} 1 & 1 \\ 0 & 0 \end{pmatrix}\begin{pmatrix} 1 & 1 \\ 2 & 2 \end{pmatrix} = \begin{pmatrix} 3 & 3 \\ 0 & 0 \end{pmatrix}$$

and

$$BA = \begin{pmatrix} 1 & 1 \\ 2 & 2 \end{pmatrix}\begin{pmatrix} 1 & 1 \\ 0 & 0 \end{pmatrix} = \begin{pmatrix} 1 & 1 \\ 2 & 2 \end{pmatrix}$$

and hence $AB \neq BA$. □

APPLICATION 1

Bob weighs 178 pounds. He wishes to lose weight through a program of dieting and exercise. After consulting Table 1 he sets up the exercise schedule in Table 2. How many calories will he burn up exercising each day if he follows this program?

TABLE 1. Calories Burned per Hour

	Exercise Activity			
Weight	Walking 2 mph	Running 5.5 mph	Bicycling 5.5 mph	Tennis (Moderate)
152	213	651	304	420
161	225	688	321	441
170	237	726	338	468
178	249	764	356	492

TABLE 2. Hours per Day for Each Activity

	Exercise Schedule			
	Walking	Running	Bicycling	Tennis
Monday	1.0	0.0	1.0	0.0
Tuesday	0.0	0.0	0.0	2.0
Wednesday	0.4	0.5	0.0	0.0
Thursday	0.0	0.0	0.5	2.0
Friday	0.4	0.5	0.0	0.0

SOLUTION.   The information pertaining to Bob is located in row four of Table 1. This information can be represented by a 4×1 matrix X. The information in Table 2 can be represented by a 5×4 matrix A. To answer the question, we simply calculate AX.

$$
\begin{pmatrix}
1.0 & 0.0 & 1.0 & 0.0 \\
0.0 & 0.0 & 0.0 & 2.0 \\
0.4 & 0.5 & 0.0 & 0.0 \\
0.0 & 0.0 & 0.5 & 2.0 \\
0.4 & 0.5 & 0.0 & 0.0
\end{pmatrix}
\begin{pmatrix}
249 \\
764 \\
356 \\
492
\end{pmatrix}
=
\begin{pmatrix}
605.0 \\
984.0 \\
481.6 \\
1162.0 \\
481.6
\end{pmatrix}
\quad
\begin{matrix}
Monday \\
Tuesday \\
Wednesday \\
Thursday \\
Friday
\end{matrix}
\qquad \square
$$

APPLICATION 2

A company manufactures three products. Its production expenses are divided into three categories. In each of these categories, an estimate is given for the cost of producing a single item of each of the products. An estimate is also made of the amount of each product to be produced per quarter. These estimates are given in the following tables.

**Production Costs per Item
(dollars)**

	Product		
Expenses	A	B	C
Raw materials	0.10	0.30	0.15
Labor	0.30	0.40	0.25
Overhead and miscellaneous	0.10	0.20	0.15

Amount Produced per Quarter

	Season			
Product	Summer	Fall	Winter	Spring
A	4000	4500	4500	4000
B	2000	2600	2400	2200
C	5800	6200	6000	6000

The company would like to present at their stockholders meeting a single table showing the total costs for each quarter in each of the three categories: raw materials, labor, and overhead.

SOLUTION. Let us consider the problem in terms of matrices. Each of the two tables can be represented by a matrix.

$$M = \begin{pmatrix} 0.10 & 0.30 & 0.15 \\ 0.30 & 0.40 & 0.25 \\ 0.10 & 0.20 & 0.15 \end{pmatrix}$$

and

$$P = \begin{pmatrix} 4000 & 4500 & 4500 & 4000 \\ 2000 & 2600 & 2400 & 2200 \\ 5800 & 6200 & 6000 & 6000 \end{pmatrix}$$

If we form the product MP, the first column of MP will represent the costs for the summer quarter.

Raw materials: $(0.10)(4000) + (0.30)(2000) + (0.15)(5800) = 1870$

Labor: $(0.30)(4000) + (0.40)(2000) + (0.25)(5800) = 3450$

Overhead and miscellaneous: $(0.10)(4000) + (0.20)(2000) + (0.15)(5800) = 1670$

The costs for the fall quarter are given in the second column of MP.

Raw materials: $(0.10)(4500) + (0.30)(2600) + (0.15)(6200) = 2160$

Labor: $(0.30)(4500) + (0.40)(2600) + (0.25)(6200) = 3940$

Overhead and miscellaneous: $(0.10)(4500) + (0.20)(2600) + (0.15)(6200) = 1900$

Columns 3 and 4 of MP represent the costs for the winter and spring quarters.

$$MP = \begin{pmatrix} 1870 & 2160 & 2070 & 1960 \\ 3450 & 3940 & 3810 & 3580 \\ 1670 & 1900 & 1830 & 1740 \end{pmatrix}$$

The entries in row 1 of MP represent the total cost of raw materials for each of the four quarters. The entries in rows 2 and 3 represent total cost for labor and overhead, respectively, for each of the four quarters. The yearly expenses in each category may be obtained by adding the entries in each row. The numbers in each of the columns may be added to obtain the total production costs for each quarter. The following table summarizes the total production costs.

	Season				
	Summer	Fall	Winter	Spring	Year
Raw materials	1,870	2,160	2,070	1,960	8,060
Labor	3,450	3,940	3,810	3,580	14,780
Overhead and miscellaneous	1,670	1,900	1,830	1,740	7,140
Total production costs	6,990	8,000	7,710	7,280	29,980

NOTATIONAL RULES

Just as in ordinary algebra, if an expression involves both multiplication and addition and there are no parentheses to indicate the order of the operations, multiplications are carried out before additions. This is true for both scalar and matrix multiplications. For example, if

$$A = \begin{pmatrix} 3 & 4 \\ 1 & 2 \end{pmatrix}, \qquad B = \begin{pmatrix} 1 & 3 \\ 2 & 1 \end{pmatrix}, \qquad \text{and} \qquad C = \begin{pmatrix} -2 & 1 \\ 3 & 2 \end{pmatrix}$$

then

$$A + BC = \begin{pmatrix} 3 & 4 \\ 1 & 2 \end{pmatrix} + \begin{pmatrix} 7 & 7 \\ -1 & 4 \end{pmatrix} = \begin{pmatrix} 10 & 11 \\ 0 & 6 \end{pmatrix}$$

and

$$3A + B = \begin{pmatrix} 9 & 12 \\ 3 & 6 \end{pmatrix} + \begin{pmatrix} 1 & 3 \\ 2 & 1 \end{pmatrix} = \begin{pmatrix} 10 & 15 \\ 5 & 7 \end{pmatrix}$$

ALGEBRAIC RULES

The following theorem provides some useful rules for doing matrix arithmetic.

Theorem 1.3.1. *Each of the following statements is valid for any scalars α and β and for any matrices A, B, and C for which the indicated operations are defined.*

(1) $A + B = B + A$
(2) $(A + B) + C = A + (B + C)$
(3) $(AB)C = A(BC)$
(4) $A(B + C) = AB + AC$
(5) $(A + B)C = AC + BC$
(6) $(\alpha\beta)A = \alpha(\beta A)$
(7) $\alpha(AB) = (\alpha A)B = A(\alpha B)$
(8) $(\alpha + \beta)A = \alpha A + \beta A$
(9) $\alpha(A + B) = \alpha A + \alpha B$

We will prove two of the rules and leave the rest for the reader to verify.

Proof of (4).   Assume that $A = (a_{ij})$ is an $m \times n$ matrix and $B = (b_{ij})$ and $C = (c_{ij})$ are both $n \times r$ matrices. Let $D = A(B + C)$ and $E = AB + AC$. It follows that

$$d_{ij} = \sum_{k=1}^{n} a_{ik}(b_{kj} + c_{kj})$$

and

$$e_{ij} = \sum_{k=1}^{n} a_{ik}b_{kj} + \sum_{k=1}^{n} a_{ik}c_{kj}$$

But

$$\sum_{k=1}^{n} a_{ik}(b_{kj} + c_{kj}) = \sum_{k=1}^{n} a_{ik}b_{kj} + \sum_{k=1}^{n} a_{ik}c_{kj}$$

so that $d_{ij} = e_{ij}$ and hence $A(B + C) = AB + AC$.   □

Proof of (3).   Let A be an $m \times n$ matrix, B an $n \times r$ matrix, and C an $r \times s$ matrix. Let $D = AB$ and $E = BC$. We must show that $DC = AE$. By the definition of matrix multiplication,

$$d_{il} = \sum_{k=1}^{n} a_{ik}b_{kl} \qquad \text{and} \qquad e_{kj} = \sum_{l=1}^{r} b_{kl}c_{lj}$$

The ijth term of DC is

$$\sum_{l=1}^{r} d_{il}c_{lj} = \sum_{l=1}^{r} \left(\sum_{k=1}^{n} a_{ik}b_{kl} \right) c_{lj}$$

and the ijth entry of AE is

$$\sum_{k=1}^{n} a_{ik}e_{kj} = \sum_{k=1}^{n} a_{ik} \left(\sum_{l=1}^{r} b_{kl}c_{lj} \right)$$

Since

$$\sum_{l=1}^{r} \left(\sum_{k=1}^{n} a_{ik}b_{kl} \right) c_{lj} = \sum_{l=1}^{r} \sum_{k=1}^{n} a_{ik}b_{kl}c_{lj} = \sum_{k=1}^{n} a_{ik} \left(\sum_{l=1}^{r} b_{kl}c_{lj} \right)$$

it follows that

$$(AB)C = DC = AE = A(BC)$$ □

EXAMPLE 7. If

$$A = \begin{pmatrix} 1 & 2 \\ 3 & 4 \end{pmatrix}, \qquad B = \begin{pmatrix} 2 & 1 \\ -3 & 2 \end{pmatrix}, \qquad \text{and} \qquad C = \begin{pmatrix} 1 & 0 \\ 2 & 1 \end{pmatrix}$$

verify that $A(BC) = (AB)C$ and $A(B + C) = AB + AC$.

SOLUTION

$$A(BC) = \begin{pmatrix} 1 & 2 \\ 3 & 4 \end{pmatrix}\begin{pmatrix} 4 & 1 \\ 1 & 2 \end{pmatrix} = \begin{pmatrix} 6 & 5 \\ 16 & 11 \end{pmatrix}$$

$$(AB)C = \begin{pmatrix} -4 & 5 \\ -6 & 11 \end{pmatrix}\begin{pmatrix} 1 & 0 \\ 2 & 1 \end{pmatrix} = \begin{pmatrix} 6 & 5 \\ 16 & 11 \end{pmatrix}$$

Thus

$$A(BC) = \begin{pmatrix} 6 & 5 \\ 16 & 11 \end{pmatrix} = (AB)C$$

$$A(B + C) = \begin{pmatrix} 1 & 2 \\ 3 & 4 \end{pmatrix}\begin{pmatrix} 3 & 1 \\ -1 & 3 \end{pmatrix} = \begin{pmatrix} 1 & 7 \\ 5 & 15 \end{pmatrix}$$

$$AB + AC = \begin{pmatrix} -4 & 5 \\ -6 & 11 \end{pmatrix} + \begin{pmatrix} 5 & 2 \\ 11 & 4 \end{pmatrix} = \begin{pmatrix} 1 & 7 \\ 5 & 15 \end{pmatrix}$$

Therefore,

$$A(B + C) = AB + AC \qquad\qquad \square$$

Notation. Since $(AB)C = A(BC)$, one may simply omit the parentheses and write ABC. The same is true for a product of four or more matrices. In the case where an $n \times n$ matrix is multiplied by itself a number of times, it is convenient to use exponential notation. Thus, if k is a positive integer, then

$$A^k = \underbrace{AA \cdots A}_{k \text{ times}}$$

EXAMPLE 8.   If

$$A = \begin{pmatrix} 1 & 1 \\ 1 & 1 \end{pmatrix}$$

then

$$A^2 = \begin{pmatrix} 1 & 1 \\ 1 & 1 \end{pmatrix}\begin{pmatrix} 1 & 1 \\ 1 & 1 \end{pmatrix} = \begin{pmatrix} 2 & 2 \\ 2 & 2 \end{pmatrix}$$

$$A^3 = AAA = AA^2 = \begin{pmatrix} 1 & 1 \\ 1 & 1 \end{pmatrix}\begin{pmatrix} 2 & 2 \\ 2 & 2 \end{pmatrix} = \begin{pmatrix} 4 & 4 \\ 4 & 4 \end{pmatrix}$$

and in general

$$A^n = \begin{pmatrix} 2^{n-1} & 2^{n-1} \\ 2^{n-1} & 2^{n-1} \end{pmatrix} \qquad\qquad \square$$

APPLICATION 3

In a certain town 30 percent of the married women get divorced each year and 20 percent of the single women get married each year. There are 8000 married women and 2000 single women. Assuming that the total population of women remains constant, how many married women and how many single women will there be after 1 year? After 2 years?

SOLUTION.   Form a matrix A as follows. The entries in the first row of A will be the percent of married and single women, respectively, that are married after 1 year. The entries in the second row will be the percent of women who are single after 1 year. Thus

$$A = \begin{pmatrix} 0.70 & 0.20 \\ 0.30 & 0.80 \end{pmatrix}$$

If we let $X = \begin{pmatrix} 8000 \\ 2000 \end{pmatrix}$, then the number of married and single women after 1 year can be computed by multiplying A times X.

$$AX = \begin{pmatrix} 0.70 & 0.20 \\ 0.30 & 0.80 \end{pmatrix} \begin{pmatrix} 8000 \\ 2000 \end{pmatrix} = \begin{pmatrix} 6000 \\ 4000 \end{pmatrix}$$

After 1 year there will be 6000 married women and 4000 single women. To find the number of married and single women after 2 years, we compute

$$A^2 X = A(AX) = \begin{pmatrix} 0.70 & 0.20 \\ 0.30 & 0.80 \end{pmatrix} \begin{pmatrix} 6000 \\ 4000 \end{pmatrix} = \begin{pmatrix} 5000 \\ 5000 \end{pmatrix}$$

After 2 years half of the women will be married and half will be single.   □

Given an $m \times n$ matrix A, it is often useful to form a new $n \times m$ matrix whose columns are the rows of A.

Definition.   The **transpose** of an $m \times n$ matrix A is the $n \times m$ matrix B defined by

$$(3) \qquad\qquad\qquad b_{ji} = a_{ij}$$

for $j = 1, \ldots, n$ and $i = 1, \ldots, m$. The transpose of A is denoted by A^T.

It follows from (3) that the jth row of A^T has the same entries, respectively, as the jth column of A, and the ith column of A^T has the same entries, respectively, as the ith row of A.

EXAMPLE 9

(a) If $A = \begin{pmatrix} 1 & 2 & 3 \\ 4 & 5 & 6 \end{pmatrix}$, then $A^T = \begin{pmatrix} 1 & 4 \\ 2 & 5 \\ 3 & 6 \end{pmatrix}$.

(b) If $B = \begin{pmatrix} -3 & 2 & 1 \\ 4 & 3 & 2 \\ 1 & 2 & 5 \end{pmatrix}$, then $B^T = \begin{pmatrix} -3 & 4 & 1 \\ 2 & 3 & 2 \\ 1 & 2 & 5 \end{pmatrix}$.

(c) If $C = \begin{pmatrix} 1 & 2 \\ 2 & 3 \end{pmatrix}$, then $C^T = \begin{pmatrix} 1 & 2 \\ 2 & 3 \end{pmatrix} = C$.

The matrix C is its own transpose.   □

Definition.   An $n \times n$ matrix A is said to be **symmetric** if $A^T = A$.

The following are examples of symmetric matrices:

$$\begin{pmatrix} 1 & 0 \\ 0 & -4 \end{pmatrix} \qquad \begin{pmatrix} 2 & 3 & 4 \\ 3 & 1 & 5 \\ 4 & 5 & 3 \end{pmatrix} \qquad \begin{pmatrix} 0 & 1 & 2 \\ 1 & 1 & -2 \\ 2 & -2 & -3 \end{pmatrix}$$

There are four basic algebraic rules involving transposes.

1. $(A^T)^T = A$
2. $(\alpha A)^T = \alpha A^T$
3. If A and B are both $m \times n$ matrices, then $(A + B)^T = A^T + B^T$.
4. If A is an $m \times n$ matrix and B is an $n \times r$ matrix, then $(AB)^T = B^T A^T$.

We will prove the fourth rule and leave the first three for the reader to verify. Let $C = AB$ and denote the ijth entries of A^T, B^T, and C^T by a_{ij}^T, b_{ij}^T, and c_{ij}^T, respectively. Thus

$$c_{ij}^T = c_{ji}, \qquad a_{ij}^T = a_{ji}, \qquad b_{ij}^T = b_{ji}$$

The ijth term of $B^T A^T$ is given by

$$\sum_{k=1}^{n} b_{ik}^T a_{kj}^T = \sum_{k=1}^{n} a_{jk} b_{ki}$$

The ijth term of $C^T = (AB)^T$ is given by

$$c_{ij}^T = c_{ji} = \sum_{k=1}^{n} a_{jk} b_{ki}$$

Therefore, $(AB)^T = B^T A^T$.

EXERCISES

1. If

$$A = \begin{pmatrix} 3 & 1 & 4 \\ -2 & 0 & 1 \\ 1 & 2 & 2 \end{pmatrix} \qquad \text{and} \qquad B = \begin{pmatrix} 1 & 0 & 2 \\ -3 & 1 & 1 \\ 2 & -4 & 1 \end{pmatrix}$$

compute:

(a) $2A$ (b) $A + B$ (c) $2A - 3B$
(d) $(2A)^T - (3B)^T$ (e) AB (f) BA
(g) $A^T B^T$ (h) $(BA)^T$

2. For each of the following pairs of matrices, determine whether or not it is possible to multiply the first matrix times the second. If it is possible, perform the multiplication.

(a) $\begin{pmatrix} 3 & 5 & 1 \\ -2 & 0 & 2 \end{pmatrix}\begin{pmatrix} 2 & 1 \\ 1 & 3 \\ 4 & 1 \end{pmatrix}$ (b) $\begin{pmatrix} 4 & -2 \\ 6 & -4 \\ 8 & -6 \end{pmatrix}(1 \quad 2 \quad 3)$

(c) $\begin{pmatrix} 1 & 4 & 3 \\ 0 & 1 & 4 \\ 0 & 0 & 2 \end{pmatrix}\begin{pmatrix} 3 & 2 \\ 1 & 1 \\ 4 & 5 \end{pmatrix}$ (d) $\begin{pmatrix} 4 & 6 \\ 2 & 1 \end{pmatrix}\begin{pmatrix} 3 & 1 & 5 \\ 4 & 1 & 6 \end{pmatrix}$

(e) $\begin{pmatrix} 4 & 6 & 1 \\ 2 & 1 & 1 \end{pmatrix}\begin{pmatrix} 3 & 1 & 5 \\ 4 & 1 & 6 \end{pmatrix}$

3. For which of the pairs in Exercise 2 is it possible to multiply the second matrix times the first, and what would the dimension of the product be?

4. Write each of the following systems of equations as a matrix equation.

(a) $3x_1 + 2x_2 = 1$ (b) $x_1 + x_2 \qquad\quad = 5$

$\quad 2x_1 - 3x_2 = 5$ $\qquad 2x_1 + x_2 - x_3 = 6$

$\qquad\qquad\qquad\qquad\qquad 3x_1 - 2x_2 + 2x_3 = 7$

(c) $2x_1 + x_2 + x_3 = 4$

$\quad x_1 - x_2 + 2x_3 = 2$

$\quad 3x_1 - 2x_2 - x_3 = 0$

5. If $A = \begin{pmatrix} 3 & 4 \\ 1 & 1 \\ 2 & 7 \end{pmatrix}$, verify that:

(a) $5A = 3A + 2A$ (b) $6A = 3(2A)$ (c) $(A^T)^T = A$

6. If

$$A = \begin{pmatrix} 4 & 1 & 6 \\ 2 & 3 & 5 \end{pmatrix} \quad \text{and} \quad B = \begin{pmatrix} 1 & 3 & 0 \\ -2 & 2 & -4 \end{pmatrix}$$

verify that:
(a) $A + B = B + A$
(b) $3(A + B) = 3A + 3B$
(c) $(A + B)^T = A^T + B^T$

7. If

$$A = \begin{pmatrix} 2 & 1 \\ 6 & 3 \\ -2 & 4 \end{pmatrix} \quad \text{and} \quad B = \begin{pmatrix} 2 & 4 \\ 1 & 6 \end{pmatrix}$$

verify that:
(a) $3(AB) = (3A)B = A(3B)$
(b) $(AB)^T = B^T A^T$

8. If

$$A = \begin{pmatrix} 2 & 4 \\ 1 & 3 \end{pmatrix}, \qquad B = \begin{pmatrix} -2 & 1 \\ 0 & 4 \end{pmatrix}, \qquad \text{and} \qquad C = \begin{pmatrix} 3 & 1 \\ 2 & 1 \end{pmatrix}$$

verify that:
(a) $(A + B) + C = A + (B + C)$
(b) $(AB)C = A(BC)$
(c) $A(B + C) = AB + AC$
(d) $(A + B)C = AC + BC$

9. Prove the associative law of multiplication for 2×2 matrices; that is, let

$$A = \begin{pmatrix} a_{11} & a_{12} \\ a_{21} & a_{22} \end{pmatrix}, \qquad B = \begin{pmatrix} b_{11} & b_{12} \\ b_{21} & b_{22} \end{pmatrix}, \qquad \text{and} \qquad C = \begin{pmatrix} c_{11} & c_{12} \\ c_{21} & c_{22} \end{pmatrix}$$

and show that

$$(AB)C = A(BC)$$

10. Find 2×2 matrices A and B that are both not the zero matrix for which $AB = O$.

11. Let

$$A = \begin{pmatrix} \frac{1}{2} & -\frac{1}{2} \\ -\frac{1}{2} & \frac{1}{2} \end{pmatrix}$$

Compute A^2 and A^3. What will A^n turn out to be?

12. Let

$$A = \begin{pmatrix} \frac{1}{2} & -\frac{1}{2} & -\frac{1}{2} & -\frac{1}{2} \\ -\frac{1}{2} & \frac{1}{2} & -\frac{1}{2} & -\frac{1}{2} \\ -\frac{1}{2} & -\frac{1}{2} & \frac{1}{2} & -\frac{1}{2} \\ -\frac{1}{2} & -\frac{1}{2} & -\frac{1}{2} & \frac{1}{2} \end{pmatrix}$$

Compute A^2 and A^3. What will A^{2n} and A^{2n+1} turn out to be?

13. Let

$$A = \begin{pmatrix} 0 & 1 & 0 & 0 \\ 0 & 0 & 1 & 0 \\ 0 & 0 & 0 & 1 \\ 0 & 0 & 0 & 0 \end{pmatrix}$$

Show that $A^n = O$ for $n \geq 4$.

14. Let A be an $m \times n$ matrix.
(a) Explain why the matrix multiplications $A^T A$ and $A A^T$ are possible.
(b) Show that $A^T A$ and $A A^T$ are both symmetric.

15. Let A and B be symmetric $n \times n$ matrices. Prove that $AB = BA$ if and only if AB is also symmetric.

16. Suppose that in Application 2, Bob loses 8 pounds. If he continues the same exercise program, how many calories will he burn up each day?

17. In Application 3, how many married women and how many single women will there be after 3 years?

18. Let A be a 2×2 matrix with $a_{11} \neq 0$ and let $\alpha = a_{21}/a_{11}$. Show that A can be factored into a product of the form

$$\begin{pmatrix} 1 & 0 \\ \alpha & 1 \end{pmatrix} \begin{pmatrix} a_{11} & a_{12} \\ 0 & b \end{pmatrix}$$

What is the value of b?

4. SPECIAL TYPES OF MATRICES

In this section we look at special types of $n \times n$ matrices, such as triangular matrices, diagonal matrices, and elementary matrices. These special types of matrices play an important role in the solution of matrix equations. We begin by considering a special matrix I that acts like a multiplicative identity, that is,

$$IA = AI = A$$

for any $n \times n$ matrix A. We also discuss the existence and computation of multiplicative inverses.

THE IDENTITY MATRIX

One very important matrix is the $n \times n$ matrix I with 1's on the diagonal and 0's off the diagonal. Thus $I = (\delta_{ij})$, where

$$\delta_{ij} = \begin{cases} 1 & \text{if} \quad i = j \\ 0 & \text{if} \quad i \neq j \end{cases}$$

If A is any $n \times n$ matrix, $AI = IA = A$. The matrix I acts as an identity for the multiplication of $n \times n$ matrices and consequently is referred to as the *identity matrix*. For example,

$$\begin{pmatrix} 1 & 0 & 0 \\ 0 & 1 & 0 \\ 0 & 0 & 1 \end{pmatrix} \begin{pmatrix} 3 & 4 & 1 \\ 2 & 6 & 3 \\ 0 & 1 & 8 \end{pmatrix} = \begin{pmatrix} 3 & 4 & 1 \\ 2 & 6 & 3 \\ 0 & 1 & 8 \end{pmatrix}$$

and

$$\begin{pmatrix} 3 & 4 & 1 \\ 2 & 6 & 3 \\ 0 & 1 & 8 \end{pmatrix} \begin{pmatrix} 1 & 0 & 0 \\ 0 & 1 & 0 \\ 0 & 0 & 1 \end{pmatrix} = \begin{pmatrix} 3 & 4 & 1 \\ 2 & 6 & 3 \\ 0 & 1 & 8 \end{pmatrix}$$

In general, if B is any $m \times n$ matrix and C is any $n \times r$ matrix, then

$$BI = B \quad \text{and} \quad IC = C$$

Notation.   The set of all n-tuples of real numbers is called *Euclidean n-space* and is usually denoted by R^n. The elements of R^n are called *vectors*. Note, however, that the solution to the matrix equation $AX = B$ will be an $n \times 1$ matrix rather than an n-tuple. In general, when working with matrix equations it is more convenient to think of R^n as consisting of column vectors ($n \times 1$ matrices) rather than row vectors ($1 \times n$ matrices). The standard notation for a column vector is a boldface lowercase letter.

$$\mathbf{x} = \begin{pmatrix} x_1 \\ x_2 \\ \vdots \\ x_n \end{pmatrix} \qquad \mathbf{x}^T = (x_1, \ldots, x_n)$$

$$\text{column vector} \qquad \text{row vector}$$

Following this convention, we will use the notation $A\mathbf{x} = \mathbf{b}$, rather than $AX = B$, to represent a linear system of equations.

Given an $m \times n$ matrix A it is often necessary to refer to a particular column. The jth column vector of A will be denoted by $\mathbf{a}_j$. Thus

$$\mathbf{a}_j = \begin{pmatrix} a_{1j} \\ a_{2j} \\ \vdots \\ a_{mj} \end{pmatrix} \qquad j = 1, \ldots, n$$

We can express A in terms of its column vectors as follows:

$$A = (\mathbf{a}_1, \mathbf{a}_2, \ldots, \mathbf{a}_n)$$

Similarly, if B is an $n \times r$ matrix, then $B = (\mathbf{b}_1, \mathbf{b}_2, \ldots, \mathbf{b}_r)$, where

$$\mathbf{b}_j = (b_{1j}, \ldots, b_{nj})^T \qquad j = 1, \ldots, r$$

The only exception to this notation is in the case of the identity matrix I. The standard notation for the jth column vector of I is $\mathbf{e}_j$ rather than $\mathbf{i}_j$. Thus the $n \times n$ identity matrix can be written

$$I = (\mathbf{e}_1, \ldots, \mathbf{e}_n)$$

DIAGONAL AND TRIANGULAR MATRICES

In order to solve an $m \times n$ system $A\mathbf{x} = \mathbf{b}$, we will apply matrix operations to both sides to obtain an equivalent system $U\mathbf{x} = \mathbf{c}$, where the matrix U is in a simpler form. In particular, if A is $n \times n$, the new system will be easy to solve if U is "triangular" or "diagonal."

An $n \times n$ matrix A is said to be *upper triangular* if $a_{ij} = 0$ for $i > j$ and *lower triangular* if $a_{ij} = 0$ for $i < j$. Also, A is said to be *triangular* if it is either upper triangular or lower triangular. For example, the 3×3 matrices

$$\begin{pmatrix} 3 & 2 & 1 \\ 0 & 2 & 1 \\ 0 & 0 & 5 \end{pmatrix} \quad \text{and} \quad \begin{pmatrix} 1 & 0 & 0 \\ 6 & 2 & 0 \\ 1 & 4 & 3 \end{pmatrix}$$

are both triangular. The first is upper triangular and the second is lower triangular.

A triangular matrix may have 0's on the diagonal; however, for a linear system $A\mathbf{x} = \mathbf{b}$ to be in triangular form, the coefficient matrix A must be triangular with nonzero diagonal entries.

An $n \times n$ matrix A is *diagonal* if $a_{ij} = 0$ whenever $i \neq j$. The matrices

$$\begin{pmatrix} 1 & 0 \\ 0 & 2 \end{pmatrix} \quad \begin{pmatrix} 1 & 0 & 0 \\ 0 & 3 & 0 \\ 0 & 0 & 1 \end{pmatrix} \quad \begin{pmatrix} 0 & 0 & 0 \\ 0 & 2 & 0 \\ 0 & 0 & 0 \end{pmatrix}$$

are all diagonal. A diagonal matrix is both upper triangular and lower triangular.

MATRIX INVERSION

Definition. A matrix A is said to be **nonsingular** or **invertible** if there exists a matrix B such that $AB = BA = I$. The matrix B is said to be a multiplicative inverse of A.

If B and C are both multiplicative inverses of A, then

$$B = BI = B(AC) = (BA)C = IC = C$$

Thus a matrix can have at most one multiplicative inverse. We will refer to the multiplicative inverse of a nonsingular matrix A as simply the **inverse of A** and denote it by A^{-1}.

EXAMPLE 1. The matrices

$$\begin{pmatrix} 2 & 4 \\ 3 & 1 \end{pmatrix} \quad \text{and} \quad \begin{pmatrix} -\frac{1}{10} & \frac{2}{5} \\ \frac{3}{10} & -\frac{1}{5} \end{pmatrix}$$

are inverses of each other, since

$$\begin{pmatrix} 2 & 4 \\ 3 & 1 \end{pmatrix} \begin{pmatrix} -\frac{1}{10} & \frac{2}{5} \\ \frac{3}{10} & -\frac{1}{5} \end{pmatrix} = \begin{pmatrix} 1 & 0 \\ 0 & 1 \end{pmatrix}$$

and

$$\begin{pmatrix} -\frac{1}{10} & \frac{2}{5} \\ \frac{3}{10} & -\frac{1}{5} \end{pmatrix} \begin{pmatrix} 2 & 4 \\ 3 & 1 \end{pmatrix} = \begin{pmatrix} 1 & 0 \\ 0 & 1 \end{pmatrix}$$

□

EXAMPLE 2. The triangular matrices

$$\begin{pmatrix} 1 & 2 & 3 \\ 0 & 1 & 4 \\ 0 & 0 & 1 \end{pmatrix} \qquad \text{and} \qquad \begin{pmatrix} 1 & -2 & 5 \\ 0 & 1 & -4 \\ 0 & 0 & 1 \end{pmatrix}$$

are inverses, since

$$\begin{pmatrix} 1 & 2 & 3 \\ 0 & 1 & 4 \\ 0 & 0 & 1 \end{pmatrix} \begin{pmatrix} 1 & -2 & 5 \\ 0 & 1 & -4 \\ 0 & 0 & 1 \end{pmatrix} = \begin{pmatrix} 1 & 0 & 0 \\ 0 & 1 & 0 \\ 0 & 0 & 1 \end{pmatrix}$$

and

$$\begin{pmatrix} 1 & -2 & 5 \\ 0 & 1 & -4 \\ 0 & 0 & 1 \end{pmatrix} \begin{pmatrix} 1 & 2 & 3 \\ 0 & 1 & 4 \\ 0 & 0 & 1 \end{pmatrix} = \begin{pmatrix} 1 & 0 & 0 \\ 0 & 1 & 0 \\ 0 & 0 & 1 \end{pmatrix}$$

□

EXAMPLE 3. The matrix

$$A = \begin{pmatrix} 1 & 0 \\ 0 & 0 \end{pmatrix}$$

has no inverse. Indeed, if B is any 2×2 matrix, then

$$BA = \begin{pmatrix} b_{11} & b_{12} \\ b_{21} & b_{22} \end{pmatrix} \begin{pmatrix} 1 & 0 \\ 0 & 0 \end{pmatrix} = \begin{pmatrix} b_{11} & 0 \\ b_{21} & 0 \end{pmatrix}$$

Thus BA cannot equal I.

□

Definition. An $n \times n$ matrix is said to be **singular** if it does not have a multiplicative inverse.

EQUIVALENT SYSTEMS

Given an $m \times n$ linear system $A\mathbf{x} = \mathbf{b}$, we can obtain an equivalent system by multiplying both sides of the equation by a nonsingular $m \times m$ matrix M.

(1) $$A\mathbf{x} = \mathbf{b}$$

(2) $$MA\mathbf{x} = M\mathbf{b}$$

Clearly, any solution to (1) will also be a solution to (2). On the other hand, if $\hat{\mathbf{x}}$ is a solution to (2), then

$$M^{-1}(MA\hat{\mathbf{x}}) = M^{-1}(M\mathbf{b})$$

$$A\hat{\mathbf{x}} = \mathbf{b}$$

so the two systems are equivalent.

To obtain an equivalent system that is easier to solve, we can apply a sequence of nonsingular matrices $E_1, \ldots, E_k$ to both sides of the equation $A\mathbf{x} = \mathbf{b}$ to obtain a simpler system

$$U\mathbf{x} = \mathbf{c}$$

where $U = E_k \cdots E_1 A$ and $\mathbf{c} = E_k \cdots E_2 E_1 \mathbf{b}$. The new system will be equivalent to the original provided that $M = E_k \cdots E_1$ is nonsingular. However, M is the product of nonsingular matrices. The following theorem shows that any product of nonsingular matrices is nonsingular.

Theorem 1.4.1.   *If A and B are nonsingular $n \times n$ matrices, then AB is also nonsingular and $(AB)^{-1} = B^{-1}A^{-1}$.*

Proof

$$(B^{-1}A^{-1})AB = B^{-1}(A^{-1}A)B = B^{-1}B = I$$

$$(AB)(B^{-1}A^{-1}) = A(BB^{-1})A^{-1} = AA^{-1} = I \qquad \square$$

It follows by induction that if $E_1, \ldots, E_k$ are all nonsingular, then the product $E_1 E_2 \cdots E_k$ is nonsingular and

$$(E_1 E_2 \cdots E_k)^{-1} = E_k^{-1} \cdots E_2^{-1} E_1^{-1}$$

We will show next that any of the three elementary row operations can be accomplished by multiplying A on the left by a nonsingular matrix.

ELEMENTARY MATRICES

A matrix obtained from the identity matrix I by the performance of one elementary row operation is called an *elementary* matrix.

There are three types of elementary matrices corresponding to the three types of elementary row operations.

Type I.   An elementary matrix of type I is a matrix obtained by interchanging two rows of I.

EXAMPLE 4.   Let

$$E_1 = \begin{pmatrix} 0 & 1 & 0 \\ 1 & 0 & 0 \\ 0 & 0 & 1 \end{pmatrix}$$

E_1 is an elementary matrix of type I, since it was obtained by interchanging the first two rows of I. Let A be a 3×3 matrix.

$$E_1 A = \begin{pmatrix} 0 & 1 & 0 \\ 1 & 0 & 0 \\ 0 & 0 & 1 \end{pmatrix} \begin{pmatrix} a_{11} & a_{12} & a_{13} \\ a_{21} & a_{22} & a_{23} \\ a_{31} & a_{32} & a_{33} \end{pmatrix} = \begin{pmatrix} a_{21} & a_{22} & a_{23} \\ a_{11} & a_{12} & a_{13} \\ a_{31} & a_{32} & a_{33} \end{pmatrix}$$

$$AE_1 = \begin{pmatrix} a_{11} & a_{12} & a_{13} \\ a_{21} & a_{22} & a_{23} \\ a_{31} & a_{32} & a_{33} \end{pmatrix} \begin{pmatrix} 0 & 1 & 0 \\ 1 & 0 & 0 \\ 0 & 0 & 1 \end{pmatrix} = \begin{pmatrix} a_{12} & a_{11} & a_{13} \\ a_{22} & a_{21} & a_{23} \\ a_{32} & a_{31} & a_{33} \end{pmatrix}$$

Multiplying A on the left by E_1 interchanges the first and second rows of A. Right multiplication of A by E_1 is equivalent to the elementary column operation of interchanging the first and second columns. $\square$

Type II. An elementary matrix of type II is a matrix obtained by multiplying a row of I by a nonzero constant.

EXAMPLE 5

$$E_2 = \begin{pmatrix} 1 & 0 & 0 \\ 0 & 1 & 0 \\ 0 & 0 & 3 \end{pmatrix}$$

is an elementary matrix of type II.

$$\begin{pmatrix} 1 & 0 & 0 \\ 0 & 1 & 0 \\ 0 & 0 & 3 \end{pmatrix} \begin{pmatrix} a_{11} & a_{12} & a_{13} \\ a_{21} & a_{22} & a_{23} \\ a_{31} & a_{32} & a_{33} \end{pmatrix} = \begin{pmatrix} a_{11} & a_{12} & a_{13} \\ a_{21} & a_{22} & a_{23} \\ 3a_{31} & 3a_{32} & 3a_{33} \end{pmatrix}$$

$$\begin{pmatrix} a_{11} & a_{12} & a_{13} \\ a_{21} & a_{22} & a_{23} \\ a_{31} & a_{32} & a_{33} \end{pmatrix} \begin{pmatrix} 1 & 0 & 0 \\ 0 & 1 & 0 \\ 0 & 0 & 3 \end{pmatrix} = \begin{pmatrix} a_{11} & a_{12} & 3a_{13} \\ a_{21} & a_{22} & 3a_{23} \\ a_{31} & a_{32} & 3a_{33} \end{pmatrix}$$

Multiplication on the left by E_2 performs the elementary row operation of multiplying the third row by 3, while multiplication on the right by E_2 performs the elementary column operation of multiplying the third column by 3. $\square$

Type III. An elementary matrix of type III is a matrix obtained from I by adding a multiple of one row to another row.

EXAMPLE 6

$$E_3 = \begin{pmatrix} 1 & 0 & 3 \\ 0 & 1 & 0 \\ 0 & 0 & 1 \end{pmatrix}$$

is an elementary matrix of type III. If A is a 3×3 matrix, then

$$E_3 A = \begin{pmatrix} a_{11} + 3a_{31} & a_{12} + 3a_{32} & a_{13} + 3a_{33} \\ a_{21} & a_{22} & a_{23} \\ a_{31} & a_{32} & a_{33} \end{pmatrix}$$

$$AE_3 = \begin{pmatrix} a_{11} & a_{12} & 3a_{11} + a_{13} \\ a_{21} & a_{22} & 3a_{21} + a_{23} \\ a_{31} & a_{32} & 3a_{31} + a_{33} \end{pmatrix}$$

Multiplication on the left by E_3 adds 3 times the third row to the first row. Multiplication on the right adds 3 times the first column to the third column. $\square$

In general, suppose that E is an $n \times n$ elementary matrix. We can think of E as being obtained from I by either a row operation or a column operation. If A is any $n \times r$ matrix, then premultiplying A by E has the effect of performing that same row operation on A. If B is an $m \times n$ matrix, postmultiplying B by E is equivalent to performing that same column operation on B.

Theorem 1.4.2.   *If E is an elementary matrix, E is invertible and E^{-1} is an elementary matrix of the same type.*

Proof.   If E is the elementary matrix of type I formed from I by interchanging the ith and jth rows, then E can be transformed back into I by interchanging these same rows again. Thus $EE = I$ and hence E is its own inverse. If E is the elementary matrix of type II formed by multiplying the ith row of I by a nonzero scalar α, then E can be transformed into the identity by multiplying either its ith row or its ith column by $1/\alpha$. Thus

$$E^{-1} = \begin{pmatrix} 1 & & & & & & & \\ & \ddots & & & & & O & \\ & & 1 & & & & & \\ & & & 1/\alpha & & & & \\ & & & & 1 & & & \\ & O & & & & \ddots & & \\ & & & & & & 1 \end{pmatrix} \quad i\text{th row}$$

Finally, suppose that E is the elementary matrix of type III formed from I by adding m times the ith row to the jth row.

$$E = \begin{pmatrix} 1 & & & & & & \\ \vdots & \ddots & & & & O & \\ 0 & \cdots & 1 & & & & \\ \vdots & & & \ddots & & & \\ 0 & \cdots & m & \cdots & 1 & & \\ \vdots & & & & & \ddots & \\ 0 & \cdots & 0 & \cdots & 0 & \cdots & 1 \end{pmatrix} \begin{matrix} \\ \\ i\text{th row} \\ \\ j\text{th row} \\ \\ \end{matrix}$$

E can be transformed back into I by either subtracting m times the ith row from the jth row or by subtracting m times the jth column from the ith column. Thus

$$E^{-1} = \begin{pmatrix} 1 & & & & & & \\ \vdots & \ddots & & & & O & \\ 0 & \cdots & 1 & & & & \\ \vdots & & & \ddots & & & \\ 0 & \cdots & -m & \cdots & 1 & & \\ \vdots & & & & & \ddots & \\ 0 & \cdots & 0 & \cdots & 0 & \cdots & 1 \end{pmatrix} \qquad \square$$

Definition.   A matrix B is **row equivalent** to A if there exists a finite sequence $E_1, E_2, \ldots, E_k$ of elementary matrices such that

$$B = E_k E_{k-1} \cdots E_1 A$$

In other words, B is row equivalent to A if B can be obtained from A by a finite number of row operations. In view of Theorem 1.4.2, it follows that:
  (i) If A is row equivalent to B, then B is row equivalent to A.
  (ii) If A is row equivalent to B, and B is row equivalent to C, then A is row equivalent to C.

The details of the proofs of (i) and (ii) are left as an exercise for the reader.

Theorem 1.4.3.   *The following are equivalent*:
  (a) *A is nonsingular.*
  (b) *$A\mathbf{x} = \mathbf{0}$ has only the trivial solution $\mathbf{0}$.*
  (c) *A is row equivalent to I.*

Proof.   We prove first that statement (a) implies statement (b). If A is nonsingular and $\hat{\mathbf{x}}$ is a solution to $A\mathbf{x} = \mathbf{0}$, then

$$\hat{\mathbf{x}} = I\hat{\mathbf{x}} = (A^{-1}A)\hat{\mathbf{x}} = A^{-1}(A\hat{\mathbf{x}}) = A^{-1}\mathbf{0} = \mathbf{0}$$

Thus $A\mathbf{x} = \mathbf{0}$ has only the trivial solution. Next we show that statement (b) implies statement (c). If we use elementary row operations, the system can be transformed into the form $U\mathbf{x} = \mathbf{0}$, where U is in row echelon form. If one of the diagonal elements of U were 0, the last row of U would consist entirely of 0's. But then $A\mathbf{x} = \mathbf{0}$ would be equivalent to a system with more unknowns than equations and hence by Theorem 1.2.1 would have a nontrivial solution. Thus U must be a triangular matrix with diagonal elements all equal to 1. It follows then that I is the reduced row echelon form of A and hence A is row equivalent to I.

Finally, we will show that statement (c) implies statement (a). If A is row equivalent to I, there exist elementary matrices $E_1, E_2, \ldots, E_k$ such that

$$A = E_k E_{k-1} \cdots E_1 I = E_k E_{k-1} \cdots E_1$$

But since E_i is invertible, $i = 1, \ldots, k$, the product $E_k E_{k-1} \cdots E_1$ is also invertible. Hence A is nonsingular and

$$A^{-1} = \left(E_k E_{k-1} \cdots E_1 \right)^{-1} = E_1^{-1} E_2^{-1} \cdots E_k^{-1} \qquad \square$$

Corollary 1.4.4. *The system of equations $A\mathbf{x} = \mathbf{b}$ has a unique solution if and only if A is nonsingular.*

Proof. If A is nonsingular, then $A^{-1}\mathbf{b}$ is the only solution to $A\mathbf{x} = \mathbf{b}$. Conversely, suppose that $A\mathbf{x} = \mathbf{b}$ has a unique solution $\hat{\mathbf{x}}$. If A is singular, $A\mathbf{x} = \mathbf{0}$ has a solution $\mathbf{z} \neq \mathbf{0}$. Let $\mathbf{y} = \hat{\mathbf{x}} + \mathbf{z}$. Clearly, $\mathbf{y} \neq \hat{\mathbf{x}}$ and

$$A\mathbf{y} = A(\hat{\mathbf{x}} + \mathbf{z}) = A\hat{\mathbf{x}} + A\mathbf{z} = \mathbf{b} + \mathbf{0} = \mathbf{b}$$

Thus $\mathbf{y}$ is also a solution to $A\mathbf{x} = \mathbf{b}$, which is a contradiction. Therefore, if $A\mathbf{x} = \mathbf{b}$ has a unique solution, then A must be nonsingular. $\square$

If A is nonsingular, A is row equivalent to I, so there exist elementary matrices $E_1, \ldots, E_k$ such that

$$E_k E_{k-1} \cdots E_1 A = I$$

Multiplying both sides of this equation on the right by A^{-1}, one obtains

$$E_k E_{k-1} \cdots E_1 I = A^{-1}$$

Thus the same series of elementary row operations that transform a nonsingular matrix A into I will transform I into A^{-1}. This gives us a method for computing A^{-1}. If we augment A by I and perform the elementary row operations that transform A into I on the augmented matrix, then I will be transformed into A^{-1}. That is, the reduced row echelon form of the augmented matrix $(A|I)$ will be $(I|A^{-1})$.

EXAMPLE 7. Compute A^{-1} if

$$A = \begin{pmatrix} 1 & 4 & 3 \\ -1 & -2 & 0 \\ 2 & 2 & 3 \end{pmatrix}$$

SOLUTION

$$\begin{pmatrix} 1 & 4 & 3 & 1 & 0 & 0 \\ -1 & -2 & 0 & 0 & 1 & 0 \\ 2 & 2 & 3 & 0 & 0 & 1 \end{pmatrix} \rightarrow \begin{pmatrix} 1 & 4 & 0 & \frac{1}{2} & -\frac{3}{2} & -\frac{1}{2} \\ 0 & 2 & 0 & \frac{1}{2} & -\frac{1}{2} & -\frac{1}{2} \\ 0 & 0 & 6 & 1 & 3 & 1 \end{pmatrix}$$

$$\begin{pmatrix} 1 & 4 & 3 & 1 & 0 & 0 \\ 0 & 2 & 3 & 1 & 1 & 0 \\ 0 & -6 & -3 & -2 & 0 & 1 \end{pmatrix} \begin{pmatrix} 1 & 0 & 0 & -\frac{1}{2} & -\frac{1}{2} & \frac{1}{2} \\ 0 & 2 & 0 & \frac{1}{2} & -\frac{1}{2} & -\frac{1}{2} \\ 0 & 0 & 6 & 1 & 3 & 1 \end{pmatrix}$$

$$\begin{pmatrix} 1 & 4 & 3 & 1 & 0 & 0 \\ 0 & 2 & 3 & 1 & 1 & 0 \\ 0 & 0 & 6 & 1 & 3 & 1 \end{pmatrix} \begin{pmatrix} 1 & 0 & 0 & -\frac{1}{2} & -\frac{1}{2} & \frac{1}{2} \\ 0 & 1 & 0 & \frac{1}{4} & -\frac{1}{4} & -\frac{1}{4} \\ 0 & 0 & 1 & \frac{1}{6} & \frac{1}{2} & \frac{1}{6} \end{pmatrix}$$

Thus

$$A^{-1} = \begin{pmatrix} -\frac{1}{2} & -\frac{1}{2} & \frac{1}{2} \\ \frac{1}{4} & -\frac{1}{4} & -\frac{1}{4} \\ \frac{1}{6} & \frac{1}{2} & \frac{1}{6} \end{pmatrix}$$

$\square$

EXAMPLE 8. Solve the system

$$x_1 + 4x_2 + 3x_3 = 12$$
$$-x_1 - 2x_2 \qquad\quad = -12$$
$$2x_1 + 2x_2 + 3x_3 = 8$$

The coefficient matrix of this system is the matrix A of the last example. The solution to the system then is

$$\mathbf{x} = A^{-1}\mathbf{b} = \begin{pmatrix} -\frac{1}{2} & -\frac{1}{2} & \frac{1}{2} \\ \frac{1}{4} & -\frac{1}{4} & -\frac{1}{4} \\ \frac{1}{6} & \frac{1}{2} & \frac{1}{6} \end{pmatrix} \begin{pmatrix} 12 \\ -12 \\ 8 \end{pmatrix} = \begin{pmatrix} 4 \\ 4 \\ -\frac{8}{3} \end{pmatrix}$$

$\square$

EXERCISES

1. Which of the following are elementary matrices? Classify each elementary matrix by type.

(a) $\begin{pmatrix} 0 & 1 \\ 1 & 0 \end{pmatrix}$ (b) $\begin{pmatrix} 2 & 0 \\ 0 & 3 \end{pmatrix}$

(c) $\begin{pmatrix} 1 & 0 & 0 \\ 0 & 1 & 0 \\ 5 & 0 & 1 \end{pmatrix}$ (d) $\begin{pmatrix} 1 & 0 & 0 \\ 0 & 5 & 0 \\ 0 & 0 & 1 \end{pmatrix}$

2. Find the inverse of each of the matrices in Exercise 1. For each elementary matrix, verify that its inverse is an elementary matrix of the same type.

3. For each of the following pairs of matrices, find an elementary matrix E such that $EA = B$.

(a) $A = \begin{pmatrix} 2 & -1 \\ 5 & 3 \end{pmatrix}$ $B = \begin{pmatrix} -4 & 2 \\ 5 & 3 \end{pmatrix}$

(b) $A = \begin{pmatrix} 2 & 1 & 3 \\ -2 & 4 & 5 \\ 3 & 1 & 4 \end{pmatrix}$ $B = \begin{pmatrix} 2 & 1 & 3 \\ 3 & 1 & 4 \\ -2 & 4 & 5 \end{pmatrix}$

(c) $A = \begin{pmatrix} 4 & -2 & 3 \\ 1 & 0 & 2 \\ -2 & 3 & 1 \end{pmatrix}$ $B = \begin{pmatrix} 4 & -2 & 3 \\ 1 & 0 & 2 \\ 0 & 3 & 5 \end{pmatrix}$

4. For each of the following pairs of matrices, find an elementary matrix E such that $AE = B$.

(a) $A = \begin{pmatrix} 4 & 1 & 3 \\ 2 & 1 & 4 \\ 1 & 3 & 2 \end{pmatrix}$ $B = \begin{pmatrix} 3 & 1 & 4 \\ 4 & 1 & 2 \\ 2 & 3 & 1 \end{pmatrix}$

(b) $A = \begin{pmatrix} 2 & 4 \\ 1 & 6 \end{pmatrix}$ $B = \begin{pmatrix} 2 & -2 \\ 1 & 3 \end{pmatrix}$

(c) $A = \begin{pmatrix} 4 & -2 & 3 \\ -2 & 4 & 2 \\ 6 & 1 & -2 \end{pmatrix}$ $B = \begin{pmatrix} 2 & -2 & 3 \\ -1 & 4 & 2 \\ 3 & 1 & -2 \end{pmatrix}$

5. Let

$$A = \begin{pmatrix} 1 & 0 & 1 \\ 3 & 3 & 4 \\ 2 & 2 & 3 \end{pmatrix}$$

(a) Verify that

$$A^{-1} = \begin{pmatrix} 1 & 2 & -3 \\ -1 & 1 & -1 \\ 0 & -2 & 3 \end{pmatrix}$$

(b) Use A^{-1} to solve $A\mathbf{x} = \mathbf{b}$ for the following choices of $\mathbf{b}$:

(i) $\mathbf{b} = (1, 1, 1)^T$

(ii) $\mathbf{b} = (1, 2, 3)^T$

(iii) $\mathbf{b} = (-2, 1, 0)^T$

6. Find the inverse of each of the following matrices.

(a) $\begin{pmatrix} -1 & 1 \\ 1 & 0 \end{pmatrix}$ (b) $\begin{pmatrix} 2 & 5 \\ 1 & 3 \end{pmatrix}$

(c) $\begin{pmatrix} 2 & 6 \\ 3 & 8 \end{pmatrix}$ (d) $\begin{pmatrix} 3 & 0 \\ 9 & 3 \end{pmatrix}$

(e) $\begin{pmatrix} 1 & 1 & 1 \\ 0 & 1 & 1 \\ 0 & 0 & 1 \end{pmatrix}$ (f) $\begin{pmatrix} 2 & 0 & 5 \\ 0 & 3 & 0 \\ 1 & 0 & 3 \end{pmatrix}$

(g) $\begin{pmatrix} -1 & -3 & -3 \\ 2 & 6 & 1 \\ 3 & 8 & 3 \end{pmatrix}$ (h) $\begin{pmatrix} 1 & 0 & 1 \\ -1 & 1 & 1 \\ -1 & -2 & -3 \end{pmatrix}$

7. Let

$$A = \begin{pmatrix} a_{11} & a_{12} \\ a_{21} & a_{22} \end{pmatrix}$$

Show that if $d = a_{11}a_{22} - a_{21}a_{12} \neq 0$, then

$$A^{-1} = \frac{1}{d} \begin{pmatrix} a_{22} & -a_{12} \\ -a_{21} & a_{11} \end{pmatrix}$$

8. Let A be a nonsingular matrix. Show that A^{-1} is also nonsingular and $(A^{-1})^{-1} = A$.

9. Prove that if A is invertible, then A^T is invertible and

$$(A^T)^{-1} = (A^{-1})^T.$$

[**Hint:** $(AB)^T = B^T A^T$]

10. Show that the product of two $n \times n$ diagonal matrices is a diagonal matrix.

11. Is the transpose of an elementary matrix an elementary matrix of the same type? Is the product of two elementary matrices an elementary matrix?

12. Let A and B be $n \times n$ matrices and let $C = AB$. Prove that if B is singular, then C must be singular.

[**Hint:** Use Theorem 1.4.3.]

13. Let U be an $n \times n$ upper triangular matrix with nonzero diagonal entries.
(a) Explain why U must be nonsingular.
(b) Explain why U^{-1} must be upper triangular.

14. Prove that if A is row equivalent to B, then B is row equivalent to A.

15. (a) Prove that if A is row equivalent to B and B is row equivalent to C, then A is row equivalent to C.
(b) Prove that any two nonsingular $n \times n$ matrices are row equivalent.

16. Prove that B is row equivalent to A if and only if there exists a nonsingular matrix M such that $B = MA$.

5. PARTITIONED MATRICES

Often it is useful to think of a matrix as being composed of a number of submatrices. A matrix A can be partitioned into smaller matrices by drawing horizontal lines between the rows and vertical lines between the columns. For example, let

$$A = \begin{pmatrix} 1 & -2 & 4 & 1 & 3 \\ 2 & 1 & 1 & 1 & 1 \\ 3 & 3 & 2 & -1 & 2 \\ 4 & 6 & 2 & 2 & 4 \end{pmatrix}$$

If lines are drawn between the second and third rows and between the third and fourth columns, then A will be divided into four submatrices, A_{11}, A_{12}, A_{21}, and A_{22}.

$$\begin{pmatrix} A_{11} & A_{12} \\ A_{21} & A_{22} \end{pmatrix} = \left(\begin{array}{ccc|cc} 1 & -2 & 4 & 1 & 3 \\ 2 & 1 & 1 & 1 & 1 \\ \hline 3 & 3 & 2 & -1 & 2 \\ 4 & 6 & 2 & 2 & 4 \end{array} \right)$$

One useful way of partitioning a matrix is into columns. For example, if

$$B = \begin{pmatrix} -1 & 2 & 1 \\ 2 & 3 & 1 \\ 1 & 4 & 1 \end{pmatrix}$$

we can partition B into three column submatrices:

$$B = (\mathbf{b}_1, \mathbf{b}_2, \mathbf{b}_3) = \left(\begin{array}{c|c|c} -1 & 2 & 1 \\ 2 & 3 & 1 \\ 1 & 4 & 1 \end{array} \right)$$

Suppose that we are given a matrix

$$A = \begin{pmatrix} 1 & 3 & 1 \\ 2 & 1 & -2 \end{pmatrix}$$

and we wish to compute the second column of AB without computing the entire product. Since the second column of AB is determined by $\mathbf{b}_2$, one need only calculate

$$A\mathbf{b}_2 = \begin{pmatrix} 1 & 3 & 1 \\ 2 & 1 & -2 \end{pmatrix} \begin{pmatrix} 2 \\ 3 \\ 4 \end{pmatrix} = \begin{pmatrix} 15 \\ -1 \end{pmatrix}$$

In general, if A is an $m \times n$ matrix and $B = (\mathbf{b}_1, \ldots, \mathbf{b}_r)$ is an $n \times r$ matrix, then the jth column of AB will be $A\mathbf{b}_j$. This is a direct consequence of the definition of matrix multiplication. If $C = AB$, then

$$c_{ij} = \sum_{k=1}^{n} a_{ik} b_{kj}$$

and hence

$$\mathbf{c}_j = \begin{pmatrix} \Sigma a_{1k} b_{kj} \\ \Sigma a_{2k} b_{kj} \\ \vdots \\ \Sigma a_{mk} b_{kj} \end{pmatrix} = A\mathbf{b}_j$$

Thus

$$\boxed{AB = (A\mathbf{b}_1, A\mathbf{b}_2, \ldots, A\mathbf{b}_r)}$$

In particular

$$(\mathbf{a}_1, \ldots, \mathbf{a}_n) = A = AI = (A\mathbf{e}_1, \ldots, A\mathbf{e}_n)$$

Let A be an $m \times n$ matrix. Corresponding to the partition $(\mathbf{a}_1, \ldots, \mathbf{a}_n)$ of A into column submatrices, one can partition A^T into row submatrices

$$A^T = \begin{pmatrix} \mathbf{a}_1^T \\ \mathbf{a}_2^T \\ \vdots \\ \mathbf{a}_n^T \end{pmatrix}$$

If C is an $m \times r$ matrix, we can form the product $A^T C$. The ith row of the product is determined by the ith row of A^T. Thus the ith row of $A^T C$ will be $\mathbf{a}_i^T C$. In general, the product can be partitioned into rows as follows:

$$(1) \qquad A^{T}C = \begin{pmatrix} \mathbf{a}_1^{T}C \\ \mathbf{a}_2^{T}C \\ \vdots \\ \mathbf{a}_n^{T}C \end{pmatrix}$$

The rows of A^{T} can be represented by the $\mathbf{a}_i^{T}$'s. Unfortunately, there is no standard notation available to represent the rows of A. Note that

$$(2) \qquad A^{T} = \left(A^{T}\mathbf{e}_1, \ldots, A^{T}\mathbf{e}_m \right)$$

Transposing both sides of (2), we get

$$A = \begin{pmatrix} \mathbf{e}_1^{T}A \\ \mathbf{e}_2^{T}A \\ \vdots \\ \mathbf{e}_m^{T}A \end{pmatrix}$$

Thus the jth row vector of A is $\mathbf{e}_j^{T}A$.

If B is an $n \times r$ matrix, then, corresponding to (1), we can partition AB into rows:

$$AB = \begin{pmatrix} \mathbf{e}_1^{T}A \\ \mathbf{e}_2^{T}A \\ \vdots \\ \mathbf{e}_m^{T}A \end{pmatrix} B = \begin{pmatrix} \left(\mathbf{e}_1^{T}A\right)B \\ \left(\mathbf{e}_2^{T}A\right)B \\ \vdots \\ \left(\mathbf{e}_m^{T}A\right)B \end{pmatrix}$$

BLOCK MULTIPLICATION

Let A be an $m \times n$ matrix and B an $n \times r$ matrix. It is often useful to partition A and B and express the product in terms of the submatrices of A and B. Consider the following four cases:

1. $B = \begin{pmatrix} B_1 & B_2 \end{pmatrix}$, where B_1 is an $n \times t$ matrix and B_2 is an $n \times (r - t)$ matrix.

$$AB = A\left(\mathbf{b}_1, \ldots, \mathbf{b}_t, \mathbf{b}_{t+1}, \ldots, \mathbf{b}_r\right)$$

$$= \left(A\mathbf{b}_1, \ldots, A\mathbf{b}_t, A\mathbf{b}_{t+1}, \ldots, A\mathbf{b}_r \right)$$

$$= \begin{pmatrix} AB_1 & AB_2 \end{pmatrix}$$

Thus

$$A\begin{pmatrix} B_1 & B_2 \end{pmatrix} = \begin{pmatrix} AB_1 & AB_2 \end{pmatrix}$$

2. $A = \begin{pmatrix} A_1 \\ A_2 \end{pmatrix}$, where A_1 is a $k \times n$ matrix and A_2 is an $(m - k) \times n$ matrix.

$$AB = \begin{pmatrix} \mathbf{e}_1^T A \\ \vdots \\ \mathbf{e}_k^T A \\ \mathbf{e}_{k+1}^T A \\ \vdots \\ \mathbf{e}_m^T A \end{pmatrix} B = \begin{pmatrix} \mathbf{e}_1^T AB \\ \vdots \\ \mathbf{e}_k^T AB \\ \mathbf{e}_{k+1}^T AB \\ \vdots \\ \mathbf{e}_m^T AB \end{pmatrix} = \begin{pmatrix} A_1 B \\ A_2 B \end{pmatrix}$$

Thus

$$\begin{pmatrix} A_1 \\ A_2 \end{pmatrix} B = \begin{pmatrix} A_1 B \\ A_2 B \end{pmatrix}$$

3. $A = \begin{pmatrix} A_1 & A_2 \end{pmatrix}$ and $B = \begin{pmatrix} B_1 \\ B_2 \end{pmatrix}$, where A_1 is an $m \times s$ matrix, A_2 is an $m \times (n - s)$ matrix, B_1 is an $s \times r$ matrix, and B_2 is an $(n - s) \times r$ matrix. If $C = AB$, then

$$c_{ij} = \sum_{l=1}^{n} a_{il} b_{lj}$$

$$= \sum_{l=1}^{s} a_{il} b_{lj} + \sum_{l=s+1}^{n} a_{il} b_{lj}$$

Thus c_{ij} is the sum of the ijth entry of $A_1 B_1$ and the ijth entry of $A_2 B_2$.

$$AB = C = A_1 B_1 + A_2 B_2$$

$$\begin{pmatrix} A_1 & A_2 \end{pmatrix} \begin{pmatrix} B_1 \\ B_2 \end{pmatrix} = A_1 B_1 + A_2 B_2$$

4. Let A and B both be partitioned as follows:

$$A = \left(\begin{array}{c|c} A_{11} & A_{12} \\ \hline A_{21} & A_{22} \end{array} \right) \begin{array}{l} k \\ m - k \end{array} \qquad B = \left(\begin{array}{c|c} B_{11} & B_{12} \\ \hline B_{21} & B_{22} \end{array} \right) \begin{array}{l} s \\ n - s \end{array}$$
$$\qquad\quad\; s \quad\; n - s \qquad\qquad\qquad\quad t \quad\; r - t$$

Let

$$A_1 = \begin{pmatrix} A_{11} \\ A_{21} \end{pmatrix} \qquad A_2 = \begin{pmatrix} A_{12} \\ A_{22} \end{pmatrix}$$

$$B_1 = (B_{11} \quad B_{12}) \qquad B_2 = (B_{21} \quad B_{22})$$

It follows from case 3 that

$$AB = \begin{pmatrix} A_1 & A_2 \end{pmatrix} \begin{pmatrix} B_1 \\ B_2 \end{pmatrix} = A_1 B_1 + A_2 B_2$$

It follows from cases 1 and 2 that

$$A_1 B_1 = \begin{pmatrix} A_{11} \\ A_{21} \end{pmatrix} B_1 = \begin{pmatrix} A_{11} B_1 \\ A_{21} B_1 \end{pmatrix} = \begin{pmatrix} A_{11} B_{11} & A_{11} B_{12} \\ A_{21} B_{11} & A_{21} B_{12} \end{pmatrix}$$

$$A_2 B_2 = \begin{pmatrix} A_{12} \\ A_{22} \end{pmatrix} B_2 = \begin{pmatrix} A_{12} B_2 \\ A_{22} B_2 \end{pmatrix} = \begin{pmatrix} A_{12} B_{21} & A_{12} B_{22} \\ A_{22} B_{21} & A_{22} B_{22} \end{pmatrix}$$

Therefore,

$$\begin{pmatrix} A_{11} & A_{12} \\ A_{21} & A_{22} \end{pmatrix} \begin{pmatrix} B_{11} & B_{12} \\ B_{21} & B_{22} \end{pmatrix} = \begin{pmatrix} A_{11} B_{11} + A_{12} B_{21} & A_{11} B_{12} + A_{12} B_{22} \\ A_{21} B_{11} + A_{22} B_{21} & A_{21} B_{12} + A_{22} B_{22} \end{pmatrix}$$

In general, if the blocks have the proper dimensions, the block multiplication can be carried out in the same manner as ordinary matrix multiplication. If

$$A = \begin{pmatrix} A_{11} & \cdots & A_{1t} \\ \vdots & & \\ A_{s1} & \cdots & A_{st} \end{pmatrix} \qquad \text{and} \qquad B = \begin{pmatrix} B_{11} & \cdots & B_{1r} \\ \vdots & & \\ B_{t1} & \cdots & B_{tr} \end{pmatrix}$$

then

$$AB = \begin{pmatrix} C_{11} & \cdots & C_{1r} \\ \vdots & & \\ C_{s1} & \cdots & C_{sr} \end{pmatrix}$$

where

$$C_{ij} = \sum_{k=1}^{t} A_{ik} B_{kj}$$

The multiplication can be carried out in this manner only if the number of columns of A_{ik} equals the number of rows of B_{kj} for each k.

EXAMPLE 1. Let

$$A = \begin{pmatrix} 1 & 1 & 1 & 1 \\ 2 & 2 & 1 & 1 \\ 3 & 3 & 2 & 2 \end{pmatrix}$$

and

$$B = \begin{pmatrix} B_{11} & B_{12} \\ B_{21} & B_{22} \end{pmatrix} = \begin{pmatrix} 1 & 1 & | & 1 & 1 \\ 1 & 2 & | & 1 & 1 \\ \overline{3} & 1 & | & 1 & 1 \\ 3 & 2 & | & 1 & 2 \end{pmatrix}$$

Partition A into four blocks and perform the block multiplication.

SOLUTION.   Since each B_{kj} has two rows, the A_{ik}'s must each have two columns. Thus there are two possibilities:

(i)
$$\begin{pmatrix} A_{11} & A_{12} \\ A_{21} & A_{22} \end{pmatrix} = \begin{pmatrix} 1 & 1 & | & 1 & 1 \\ 2 & 2 & | & 1 & 1 \\ 3 & 3 & | & 2 & 2 \end{pmatrix}$$

in which case

$$\begin{pmatrix} 1 & 1 & | & 1 & 1 \\ 2 & 2 & | & 1 & 1 \\ 3 & 3 & | & 2 & 2 \end{pmatrix}\begin{pmatrix} 1 & 1 & | & 1 & 1 \\ 1 & 2 & | & 1 & 1 \\ \overline{3} & 1 & | & 1 & 1 \\ 3 & 2 & | & 1 & 2 \end{pmatrix} = \begin{pmatrix} 8 & 6 & | & 4 & 5 \\ 10 & 9 & | & 6 & 7 \\ 18 & 15 & | & 10 & 12 \end{pmatrix}$$

or

(ii)
$$\begin{pmatrix} A_{11} & A_{12} \\ A_{21} & A_{22} \end{pmatrix} = \begin{pmatrix} 1 & 1 & | & 1 & 1 \\ 2 & 2 & | & 1 & 1 \\ \overline{3} & 3 & | & 2 & 2 \end{pmatrix}$$

in which case

$$\begin{pmatrix} 1 & 1 & | & 1 & 1 \\ 2 & 2 & | & 1 & 1 \\ \overline{3} & 3 & | & 2 & 2 \end{pmatrix}\begin{pmatrix} 1 & 1 & | & 1 & 1 \\ 1 & 2 & | & 1 & 1 \\ \overline{3} & 1 & | & 1 & 1 \\ 3 & 2 & | & 1 & 2 \end{pmatrix} = \begin{pmatrix} 8 & 6 & | & 4 & 5 \\ 10 & 9 & | & 6 & 7 \\ \overline{18} & 15 & | & 10 & 12 \end{pmatrix} \quad \square$$

EXAMPLE 2. Let A be an $n \times n$ matrix of the form

$$\begin{pmatrix} A_{11} & O \\ O & A_{22} \end{pmatrix}$$

where A_{11} is a $k \times k$ matrix $(k < n)$. Show that A is nonsingular if and only if A_{11} and A_{22} are nonsingular.

SOLUTION.   If A_{11} and A_{22} are nonsingular, then

$$\begin{pmatrix} A_{11}^{-1} & O \\ O & A_{22}^{-1} \end{pmatrix}\begin{pmatrix} A_{11} & O \\ O & A_{22} \end{pmatrix} = \begin{pmatrix} I_k & O \\ O & I_{n-k} \end{pmatrix} = I$$

and

$$\begin{pmatrix} A_{11} & O \\ O & A_{22} \end{pmatrix} \begin{pmatrix} A_{11}^{-1} & O \\ O & A_{22}^{-1} \end{pmatrix} = \begin{pmatrix} I_k & O \\ O & I_{n-k} \end{pmatrix} = I$$

so A is nonsingular and

$$A^{-1} = \begin{pmatrix} A_{11}^{-1} & O \\ O & A_{22}^{-1} \end{pmatrix}$$

Conversely, if A is nonsingular, then let $B = A^{-1}$ and partition B in the same manner as A. Since

$$BA = I = AB$$

it follows that

$$\begin{pmatrix} B_{11} & B_{12} \\ B_{21} & B_{22} \end{pmatrix} \begin{pmatrix} A_{11} & O \\ O & A_{22} \end{pmatrix} = \begin{pmatrix} I_k & O \\ O & I_{n-k} \end{pmatrix} = \begin{pmatrix} A_{11} & O \\ O & A_{22} \end{pmatrix} \begin{pmatrix} B_{11} & B_{12} \\ B_{21} & B_{22} \end{pmatrix}$$

$$\begin{pmatrix} B_{11}A_{11} & B_{12}A_{22} \\ B_{21}A_{11} & B_{22}A_{22} \end{pmatrix} = \begin{pmatrix} I_k & O \\ O & I_{n-k} \end{pmatrix} = \begin{pmatrix} A_{11}B_{11} & A_{11}B_{12} \\ A_{22}B_{21} & A_{22}B_{22} \end{pmatrix}$$

Thus

$$B_{11}A_{11} = I_k = A_{11}B_{11}$$
$$B_{22}A_{22} = I_{n-k} = A_{22}B_{22}$$

and hence A_{11} and A_{22} are both nonsingular with inverses B_{11} and B_{22}, respectively.   □

EXERCISES

1. Let A be a nonsingular $n \times n$ matrix. Perform the following multiplications.

(a) $A^{-1}(A \quad I)$ (b) $\begin{pmatrix} A \\ I \end{pmatrix} A^{-1}$ (c) $(A \quad I)^T (A \quad I)$

(d) $(A \quad I)(A \quad I)^T$ (e) $\begin{pmatrix} A^{-1} \\ I \end{pmatrix} (A \quad I)$

2. Let $B = A^T A$. Show that $b_{ij} = \mathbf{a}_i^T \mathbf{a}_j$.

3. Let

$$A = \begin{pmatrix} 1 & 1 \\ 2 & -1 \end{pmatrix} \quad \text{and} \quad B = \begin{pmatrix} 2 & 1 \\ 1 & 3 \end{pmatrix}$$

(a) Calculate $A\mathbf{b}_1$ and $A\mathbf{b}_2$.
(b) Calculate $(1 \quad 1)B$ and $(2 \quad -1)B$.
(c) Multiply AB and verify that its column vectors are the vectors in part (a) and its row vectors are the vectors in part (b).

4. Let

$$I = \begin{pmatrix} 1 & 0 \\ 0 & 1 \end{pmatrix} \qquad E = \begin{pmatrix} 0 & 1 \\ 1 & 0 \end{pmatrix} \qquad O = \begin{pmatrix} 0 & 0 \\ 0 & 0 \end{pmatrix}$$

$$C = \begin{pmatrix} 1 & 0 \\ -1 & 1 \end{pmatrix} \qquad D = \begin{pmatrix} 2 & 0 \\ 0 & 2 \end{pmatrix}$$

and

$$B = \begin{pmatrix} B_{11} & B_{12} \\ B_{21} & B_{22} \end{pmatrix} = \left(\begin{array}{cc|cc} 1 & 1 & 1 & 1 \\ 1 & 2 & 1 & 1 \\ \hline 3 & 1 & 1 & 1 \\ 3 & 2 & 1 & 2 \end{array} \right)$$

Perform each of the following block multiplications.

(a) $\begin{pmatrix} O & I \\ I & O \end{pmatrix} \begin{pmatrix} B_{11} & B_{12} \\ B_{21} & B_{22} \end{pmatrix}$ (b) $\begin{pmatrix} C & O \\ O & C \end{pmatrix} \begin{pmatrix} B_{11} & B_{12} \\ B_{21} & B_{22} \end{pmatrix}$

(c) $\begin{pmatrix} D & O \\ O & I \end{pmatrix} \begin{pmatrix} B_{11} & B_{12} \\ B_{21} & B_{22} \end{pmatrix}$ (d) $\begin{pmatrix} E & O \\ O & E \end{pmatrix} \begin{pmatrix} B_{11} & B_{12} \\ B_{21} & B_{22} \end{pmatrix}$

5. Perform each of the following block multiplications.

(a) $\left(\begin{array}{ccc|c} 1 & 1 & 1 & -1 \\ 2 & 1 & 2 & -1 \end{array} \right) \left(\begin{array}{ccc} 4 & -2 & 1 \\ 2 & 3 & 1 \\ 1 & 1 & 2 \\ 1 & 2 & 3 \end{array} \right)$

(b) $\left(\begin{array}{cc} 4 & -2 \\ 2 & 3 \\ 1 & 1 \\ \hline 1 & 2 \end{array} \right) \left(\begin{array}{ccc|c} 1 & 1 & 1 & -1 \\ 2 & 1 & 2 & -1 \end{array} \right)$

(c) $\left(\begin{array}{cc|cc} \frac{3}{5} & -\frac{4}{5} & 0 & 0 \\ \frac{4}{5} & \frac{3}{5} & 0 & 0 \\ \hline 0 & 0 & 1 & 0 \end{array} \right) \left(\begin{array}{cc|c} \frac{3}{5} & \frac{4}{5} & 0 \\ -\frac{4}{5} & \frac{3}{5} & 0 \\ \hline 0 & 0 & 1 \\ 0 & 0 & 0 \end{array} \right)$

(d) $\left(\begin{array}{ccc|cc} 0 & 0 & 1 & 0 & 0 \\ 0 & 1 & 0 & 0 & 0 \\ 1 & 0 & 0 & 0 & 0 \\ \hline 0 & 0 & 0 & 0 & 1 \\ 0 & 0 & 0 & 1 & 0 \end{array} \right) \left(\begin{array}{cc} 1 & -1 \\ 2 & -2 \\ 3 & -3 \\ \hline 4 & -4 \\ 5 & -5 \end{array} \right)$

6. Let

$$A = \begin{pmatrix} A_{11} & A_{12} \\ A_{21} & A_{22} \end{pmatrix} \qquad \text{and} \qquad A^T = \begin{pmatrix} A_{11}^T & A_{21}^T \\ A_{12}^T & A_{22}^T \end{pmatrix}$$

Is it possible to perform the block multiplications of AA^T and $A^T A$? Explain.

7. Let

$$A = \begin{pmatrix} A_{11} & A_{12} \\ O & A_{22} \end{pmatrix}$$

where all four blocks are $n \times n$ matrices.

(a) If A_{11} and A_{22} are nonsingular, show that A must also be nonsingular and that A^{-1} must be of the form

$$\left(\begin{array}{c|c} A_{11}^{-1} & C \\ \hline O & A_{22}^{-1} \end{array} \right)$$

(b) Determine C.

8. Let

$$A = \begin{pmatrix} O & I \\ B & O \end{pmatrix}$$

where all four submatrices are $k \times k$. Determine A^2 and A^4.

9. Let I denote the $n \times n$ identity matrix. Find a block form for the inverse of each of the following $2n \times 2n$ matrices.

(a) $\begin{pmatrix} O & I \\ I & O \end{pmatrix}$ (b) $\begin{pmatrix} I & O \\ B & I \end{pmatrix}$

10. Let

$$A = \begin{pmatrix} A_{11} & A_{12} \\ A_{21} & A_{22} \end{pmatrix}$$

where A_{11} is a $k \times k$ nonsingular matrix. Show that A can be factored into a product

$$\begin{pmatrix} I & O \\ B & I \end{pmatrix} \begin{pmatrix} A_{11} & A_{12} \\ O & C \end{pmatrix}$$

where

$$B = A_{21}A_{11}^{-1} \quad \text{and} \quad C = A_{22} - A_{21}A_{11}^{-1}A_{12}$$

11. Let A be an $n \times n$ matrix and $\mathbf{x} \in R^n$. Partition A into columns and $\mathbf{x}$ into rows and perform the block multiplication of A times $\mathbf{x}$.

12. Show that if A is an $n \times n$ matrix with the property $A\mathbf{x} = \mathbf{0}$ for all $\mathbf{x} \in R^n$, then $A = O$.

[***Hint:*** Let $\mathbf{x} = \mathbf{e}_j$ for $j = 1, \ldots, n$.]

13. Let B and C be $n \times n$ matrices with the property $B\mathbf{x} = C\mathbf{x}$ for all $\mathbf{x} \in R^n$. Show that $B = C$.

14. Consider a system of the form

$$\begin{pmatrix} A & \mathbf{a} \\ \mathbf{c}^T & \beta \end{pmatrix} \begin{pmatrix} \mathbf{x} \\ x_{n+1} \end{pmatrix} = \begin{pmatrix} \mathbf{b} \\ b_{n+1} \end{pmatrix}$$

where A is a nonsingular $n \times n$ matrix and $\mathbf{a}$, $\mathbf{b}$, and $\mathbf{c}$ are vectors in R^n.

(a) Multiply both sides of the system by

$$\begin{pmatrix} A^{-1} & \mathbf{0} \\ -\mathbf{c}^T A^{-1} & 1 \end{pmatrix}$$

to obtain an equivalent triangular system.

(b) Set $\mathbf{y} = A^{-1}\mathbf{a}$ and $\mathbf{z} = A^{-1}\mathbf{b}$. Show that if $\beta - \mathbf{c}^T\mathbf{y} \neq 0$, then the solution of the system is given by

$$x_{n+1} = \frac{b_{n+1} - \mathbf{c}^T\mathbf{z}}{\beta - \mathbf{c}^T\mathbf{y}}$$

$$\mathbf{x} = \mathbf{z} - x_{n+1}\mathbf{y}$$

2

DETERMINANTS

With each square matrix it is possible to associate a real number called the determinant of the matrix. The value of this number will tell us whether or not the matrix is singular.

In Section 1 the definition of the determinant of a matrix is given. Although the value of the determinant can be calculated using the definition, it is generally more convenient to use some other method that involves fewer arithmetic operations. One such method is the method of cofactors. This method is probably the simplest method for evaluating determinants of 3×3 matrices. It also works well for larger matrices that have many zero entries. In Section 2 we study properties of determinants and derive an elimination method for evaluating determinants. The elimination method is generally the simplest method to use for evaluating the determinant of an $n \times n$ matrix when $n > 3$. In Section 3 we see how determinants can be applied to solving $n \times n$ linear systems and how they can be used to calculate the inverse of a matrix. Further applications of determinants are presented in Chapters 3 and 6.

1. THE DETERMINANT OF A MATRIX

With each $n \times n$ matrix A it is possible to associate a scalar, $\det(A)$, whose value will tell us whether or not the matrix is nonsingular. Before proceeding to the general definition let us consider the two simplest cases.

Case 1. $A = (a)$ is a 1×1 matrix. In this case A will have a multiplicative inverse if and only if $a \neq 0$. Thus we define

$$\det(A) = a$$

Case 2. 2×2 matrices. Let

$$A = \begin{pmatrix} a_{11} & a_{12} \\ a_{21} & a_{22} \end{pmatrix}$$

By Theorem 1.4.3, A will be nonsingular if and only if it is row equivalent to I. If $a_{21} = 0$, then A will be row equivalent to the identity if and only if a_{11} and a_{22} are both nonzero or, equivalently,

$$a_{11}a_{22} \neq 0$$

If $a_{21} \neq 0$, we can test whether or not A is row equivalent to I by performing the following operations:

(i) Multiply the first row of A by $-a_{21}$:

$$\begin{pmatrix} -a_{11}a_{21} & -a_{12}a_{21} \\ a_{21} & a_{22} \end{pmatrix}$$

(ii) Add a_{11} times the second row to the new first row:

$$\begin{pmatrix} 0 & a_{11}a_{22} - a_{12}a_{21} \\ a_{21} & a_{22} \end{pmatrix}$$

Since $a_{21} \neq 0$, the resulting matrix will be row equivalent to I if and only if

$$a_{11}a_{22} - a_{12}a_{21} \neq 0$$

Thus, if A is any 2×2 matrix and we define

$$\det(A) = a_{11}a_{22} - a_{12}a_{21}$$

then A is nonsingular if and only if $\det(A) \neq 0$.

Notation.   One can refer to the determinant of a specific matrix by enclosing the array between vertical lines. For example, if

$$A = \begin{pmatrix} 3 & 4 \\ 2 & 1 \end{pmatrix}$$

then

$$\begin{vmatrix} 3 & 4 \\ 2 & 1 \end{vmatrix}$$

represents the determinant of A.

Let us now define the determinant of a general $n \times n$ matrix. There are two possible ways to define determinants. We will give the standard definition and present the alternative cofactor definition as a theorem. Those preferring the other alternative may opt to skip the following discussion and use Theorem 2.1.1 as a definition. In order to understand the standard definition, some preliminary definitions are necessary.

Definition. Let S be a finite set. An ordering of the elements of S is said to be a **permutation** of S.

If a set S has n elements, there are $n!$ permutations of S. Each permutation can be represented by an ordered n-tuple. For example, if $S = \{1, 2, 3\}$, there are six possible permutations.

$$(1, 2, 3), \quad (1, 3, 2), \quad (2, 1, 3), \quad (2, 3, 1), \quad (3, 1, 2), \quad (3, 2, 1)$$

Let π_n denote the set of all permutations of $\{1, 2, \ldots, n\}$. An element $\mathbf{k} = (k_1, k_2, \ldots, k_n)$ of π_n is said to have an *inversion* if $k_i > k_j$ for $i < j$ (i.e., a larger number precedes a smaller number). A permutation is said to be *even* if it has an even number of inversions and it is said to be *odd* if it has an odd number of inversions. In π_3 the permutations $(1, 2, 3)$, $(2, 3, 1)$, and $(3, 1, 2)$ are even and the permutations $(1, 3, 2)$, $(2, 1, 3)$, and $(3, 2, 1)$ are odd. For each permutation $\mathbf{k}$ in π_n, we define

$$\delta(\mathbf{k}) = \begin{cases} 0 & \text{if} \quad \mathbf{k} \text{ is even} \\ 1 & \text{if} \quad \mathbf{k} \text{ is odd} \end{cases}$$

EXAMPLE 1. The permutation $\mathbf{k} = (3, 2, 5, 4, 1)$ has six inversions and consequently is even. The permutation $\mathbf{j} = (2, 1, 5, 3, 4)$ has three inversions and hence is odd. Thus $\delta(\mathbf{k}) = 0$ and $\delta(\mathbf{j}) = 1$. $\square$

The determinant of an $n \times n$ matrix A is defined as a sum of products involving the entries of A. Each term summed will be a product of n entries of A. Each product will have exactly one entry from each row of A and exactly one entry from each column of A. We can order the terms in each product so that the element from the first row of A is first, the entry from the second row is second, and so on. Thus each of the products will have the form

$$a_{1k_1} a_{2k_2} \cdots a_{nk_n}$$

where $(k_1, k_2, \ldots, k_n)$ is an ordering of the columns of A. We assign to each such product either a $+$ or $-$ sign, depending on whether the permutation $(k_1, k_2, \ldots, k_n)$ is even or odd. For every permutation $\mathbf{k}$ in π_n we can form a signed product in this manner. The determinant of A is defined to be the sum of all such signed products.

Definition.   Let $A = (a_{ij})$ be an $n \times n$ matrix. The **determinant of** A, denoted $\det(A)$, is defined by

$$\det(A) = \sum (-1)^{\delta(\mathbf{k})} a_{1k_1} a_{2k_2} \cdots a_{nk_n}$$

where the sum is taken over all $\mathbf{k} = (k_1, \ldots, k_n)$ in π_n.

The determinant of a 1×1 matrix (a_{11}) is just a_{11}. If A is a 2×2 matrix, then

$$\det(A) = (-1)^0 a_{11}a_{22} + (-1)^1 a_{12}a_{21} = a_{11}a_{22} - a_{12}a_{21}$$

This formula is easy to remember because it involves only subtracting the product of the off-diagonal elements from the product of the diagonal elements. If A is a 3×3 matrix, then

(1)
$$\begin{aligned}
\det(A) = {} & a_{11}a_{22}a_{33} - a_{11}a_{23}a_{32} - a_{12}a_{21}a_{33} \\
& + a_{12}a_{23}a_{31} + a_{13}a_{21}a_{32} - a_{13}a_{22}a_{31}
\end{aligned}$$

The formula for the determinant of a 4×4 matrix would involve the sum of 24 products of four elements.

COFACTOR EXPANSION OF DET(A)

To simplify matters, it is convenient to introduce the notation of minors and cofactors.

Definition.   Let $A = (a_{ij})$ be an $n \times n$ matrix. Let M_{ij} be the $(n-1) \times (n-1)$ matrix obtained from A by deleting the row and column containing a_{ij}. The determinant of M_{ij} is called the **minor** of a_{ij}. We define the **cofactor** A_{ij} of a_{ij} by

$$A_{ij} = (-1)^{i+j} \det(M_{ij})$$

Returning to the 3×3 case, if we rewrite (1) in the form

$$\begin{aligned}
\det(A) = {} & a_{11}(a_{22}a_{33} - a_{23}a_{32}) - a_{12}(a_{21}a_{33} - a_{23}a_{31}) \\
& + a_{13}(a_{21}a_{32} - a_{22}a_{31})
\end{aligned}$$

we have

$$\det(A) = a_{11}\det(M_{11}) - a_{12}\det(M_{12}) + a_{13}\det(M_{13})$$

or

(2) $$\det(A) = a_{11}A_{11} + a_{12}A_{12} + a_{13}A_{13}$$

Equation (2) is called the *cofactor expansion* of $\det(A)$ using the first row of A. Actually, any row of A can be used for the cofactor expansion. We could just as well have rewritten (1) in the form

$$\det(A) = -a_{21}(a_{12}a_{33} - a_{13}a_{32}) + a_{22}(a_{11}a_{33} - a_{13}a_{31})$$
$$- a_{23}(a_{11}a_{32} - a_{12}a_{31})$$
$$= a_{21}A_{21} + a_{22}A_{22} + a_{23}A_{23}$$

In a similar manner, it can be shown that

$$\det(A) = a_{31}A_{31} + a_{32}A_{32} + a_{33}A_{33}$$

The determinant of A can also be expanded using the columns of A. Rewriting (1) in the form

$$\det(A) = a_{11}(a_{22}a_{33} - a_{23}a_{32}) - a_{21}(a_{12}a_{33} - a_{13}a_{32})$$
$$+ a_{31}(a_{12}a_{23} - a_{13}a_{22})$$

we see that

$$\det(A) = a_{11}A_{11} + a_{21}A_{21} + a_{31}A_{31}$$

The reader may verify that $\det(A)$ can also be expressed as a cofactor expansion using the second and third columns.

We state the following theorem without proof.

Theorem 2.1.1. *If A is an $n \times n$ matrix with $n \geq 2$, then $\det(A)$ can be expressed as a cofactor expansion using any row or column of A.*

$$\det(A) = a_{i1}A_{i1} + a_{i2}A_{i2} + \cdots + a_{in}A_{in}$$
$$= a_{1j}A_{1j} + a_{2j}A_{2j} + \cdots + a_{nj}A_{nj}$$

for $i = 1, \ldots, n$ and $j = 1, \ldots, n$

EXAMPLE 2. Use cofactors to evaluate

$$\begin{vmatrix} 3 & 2 & 4 \\ 1 & -2 & 3 \\ 2 & 3 & 2 \end{vmatrix}$$

SOLUTION. Expanding along the first row, we get

$$\begin{vmatrix} 3 & 2 & 4 \\ 1 & -2 & 3 \\ 2 & 3 & 2 \end{vmatrix} = 3\begin{vmatrix} -2 & 3 \\ 3 & 2 \end{vmatrix} - 2\begin{vmatrix} 1 & 3 \\ 2 & 2 \end{vmatrix} + 4\begin{vmatrix} 1 & -2 \\ 2 & 3 \end{vmatrix}$$
$$= 3 \cdot (-13) - 2 \cdot (-4) + 4 \cdot 7$$
$$= -3 \qquad \qquad \square$$

The cofactor expansion of a 4×4 determinant will involve four 3×3 determinants. One can often save work by expanding along the row or column that contains the most zeros. For example, to evaluate

$$\begin{vmatrix} 0 & 2 & 3 & 0 \\ 0 & 4 & 5 & 0 \\ 0 & 1 & 0 & 3 \\ 2 & 0 & 1 & 3 \end{vmatrix}$$

one would expand down the first column. The first three terms will drop out, leaving

$$-2 \begin{vmatrix} 2 & 3 & 0 \\ 4 & 5 & 0 \\ 1 & 0 & 3 \end{vmatrix} = -2 \cdot 3 \cdot \begin{vmatrix} 2 & 3 \\ 4 & 5 \end{vmatrix} = 12$$

Using cofactors to evaluate determinants is generally more efficient than using the definition, since the cofactor method involves fewer arithmetic operations. The cofactor method is particularly well suited for sparse matrices (i.e., matrices with many zero entries).

Theorem 2.1.2.   *If A is an $n \times n$ matrix, then $\det(A^T) = \det(A)$.*

Proof.   The proof is by induction on n. Clearly, the result holds if $n = 1$, since a 1×1 matrix is necessarily symmetric. Assume that the result holds for all $k \times k$ matrices and that A is a $(k + 1) \times (k + 1)$ matrix. Expanding $\det(A)$ along the first row of A, we get

$$\det(A) = a_{11} \det(M_{11}) - a_{12} \det(M_{12}) + - \cdots \pm a_{1,k+1} \det(M_{1,k+1})$$

Since the M_{ij}'s are all $k \times k$ matrices, it follows from the induction hypothesis that

(3)

$$\det(A) = a_{11} \det(M_{11}^T) - a_{12} \det(M_{12}^T) + - \cdots \pm a_{1,k+1} \det(M_{1,k+1}^T)$$

The right-hand side of (3) is just the expansion by minors of $\det(A^T)$ using the first column of A^T. Therefore,

$$\det(A^T) = \det(A) \qquad \qquad \Box$$

Theorem 2.1.3.   *If A is an $n \times n$ triangular matrix, the determinant of A equals the product of the diagonal elements of A.*

Proof.   In view of Theorem 2.1.2, it suffices to prove the theorem for lower triangular matrices. The result follows easily using the cofactor expansion and induction on n. The details of this are left for the reader as an exercise. $\qquad \Box$

In the next section we look at the effect of row operations on the value of the determinant. This will allow us to make use of Theorem

2.1.3 to derive a third and more efficient method for computing the value of a determinant.

EXERCISES

1. Determine whether the following permutations in π_5 are odd or even.

(a) $(5, 1, 2, 3, 4)$ (b) $(5, 2, 3, 4, 1)$ (c) $(5, 4, 3, 2, 1)$
(d) $(2, 1, 3, 5, 4)$ (e) $(3, 2, 1, 4, 5)$

2. Use the definition of $\det(A)$ to evaluate

$$\begin{vmatrix} 3 & 0 & 0 \\ 1 & 2 & 0 \\ 5 & 6 & 4 \end{vmatrix}$$

3. Evaluate the following determinants.

(a) $\begin{vmatrix} 3 & 5 \\ -2 & -3 \end{vmatrix}$ (b) $\begin{vmatrix} 5 & -2 \\ -8 & 4 \end{vmatrix}$

(c) $\begin{vmatrix} 3 & 1 & 2 \\ 2 & 4 & 5 \\ 2 & 4 & 5 \end{vmatrix}$ (d) $\begin{vmatrix} 4 & 3 & 0 \\ 3 & 1 & 2 \\ 5 & -1 & -4 \end{vmatrix}$

(e) $\begin{vmatrix} 1 & 3 & 2 \\ 4 & 1 & -2 \\ 2 & 1 & 3 \end{vmatrix}$ (f) $\begin{vmatrix} 2 & -1 & 2 \\ 1 & 3 & 2 \\ 5 & 1 & 6 \end{vmatrix}$

(g) $\begin{vmatrix} 2 & 0 & 0 & 1 \\ 0 & 1 & 0 & 0 \\ 1 & 6 & 2 & 0 \\ 1 & 1 & -2 & 3 \end{vmatrix}$ (h) $\begin{vmatrix} 2 & 1 & 2 & 1 \\ 3 & 0 & 1 & 1 \\ -1 & 2 & -2 & 1 \\ -3 & 2 & 3 & 1 \end{vmatrix}$

4. Evaluate the following determinants by inspection.

(a) $\begin{vmatrix} 3 & 5 \\ 2 & 4 \end{vmatrix}$ (b) $\begin{vmatrix} 2 & 0 & 0 \\ 4 & 1 & 0 \\ 7 & 3 & -2 \end{vmatrix}$

(c) $\begin{vmatrix} 3 & 0 & 0 \\ 2 & 1 & 1 \\ 1 & 2 & 2 \end{vmatrix}$ (d) $\begin{vmatrix} 4 & 0 & 2 & 1 \\ 5 & 0 & 4 & 2 \\ 2 & 0 & 3 & 4 \\ 1 & 0 & 2 & 3 \end{vmatrix}$

5. Evaluate the following determinant. Write your answer as a polynomial in x.

$$\begin{vmatrix} a - x & b & c \\ 1 & -x & 0 \\ 0 & 1 & -x \end{vmatrix}$$

6. Find all values of λ for which the following determinant will equal 0.

$$\begin{vmatrix} 2 - \lambda & 4 \\ 3 & 3 - \lambda \end{vmatrix}$$

7. Write out the details of the proof of Theorem 2.1.3.

8. Does $\det(A + B) = \det(A) + \det(B)$? Justify your answer.

9. Let A be a symmetric tridiagonal matrix (i.e., A is symmetric and $a_{ij} = 0$ whenever $|i - j| > 1$). Let B be the matrix formed from A by deleting the first two rows and columns. Show that

$$\det(A) = a_{11} \det(M_{11}) - a_{12}^2 \det(B)$$

10. Prove that if a row or a column of an $n \times n$ matrix consists entirely of zeros, then $\det(A) = 0$.

11. Use mathematical induction to prove that if A is an $(n + 1) \times (n + 1)$ matrix with two identical rows, then $\det(A) = 0$.

12. Let A, B, and C be three $n \times n$ matrices such that for some k,

$$a_{kj} = b_{kj} + c_{kj} \qquad j = 1, \ldots, n$$

and

$$a_{ij} = b_{ij} = c_{ij} \qquad \text{whenever} \quad i \neq k$$

Show that

$$\det(A) = \det(B) + \det(C)$$

2. PROPERTIES OF DETERMINANTS

In this section we consider the effect of row operations on the determinant of a matrix. From this we will see that the value of the determinant actually tells us whether or not the matrix is singular. We will also learn a method for evaluating determinants using row operations. We begin by examining the effects of row operations II and III on the value of the determinant.

ROW OPERATION II

A row of A is multiplied by a nonzero constant. Let E denote the elementary matrix of type II formed from I by multiplying the ith row by the nonzero constant α. If $\det(EA)$ is expanded by cofactors along the ith row, then

$$\det(EA) = \alpha a_{i1}A_{i1} + \alpha a_{i2}A_{i2} + \cdots + \alpha a_{in}A_{in}$$
$$= \alpha(a_{i1}A_{i1} + a_{i2}A_{i2} + \cdots + a_{in}A_{in})$$
$$= \alpha\det(A)$$

In particular,

$$\det(E) = \det(EI) = \alpha\det(I) = \alpha$$

and hence

$$\det(EA) = \alpha\det(A) = \det(E)\det(A)$$

ROW OPERATION III

A multiple of one row is added to another row. Let E be the elementary matrix of type III formed from I by adding c times the ith row to the jth row. Since E is triangular and its diagonal elements are all 1, it follows that $\det(E) = 1$. We will show that

$$\det(EA) = \det(A) = \det(E)\det(A)$$

By definition,

$$\det(EA) = \sum_{\mathbf{k}\in\pi_n}(-1)^{\delta(\mathbf{k})}\left(a_{1k_1}a_{2k_2}\cdots(a_{jk_j} + ca_{ik_j})\cdots a_{nk_n}\right)$$

$$= \sum_{\mathbf{k}\in\pi_n}(-1)^{\delta(\mathbf{k})}\left(a_{1k_1}\cdots a_{nk_n}\right)$$

$$+ c\sum_{\mathbf{k}\in\pi_n}(-1)^{\delta(\mathbf{k})}\left(a_{1k_1}\cdots a_{ik_i}\cdots a_{ik_j}\cdots a_{nk_n}\right)$$

$$= \det(A) + c\det(A^*)$$

where

$$A^* = \begin{pmatrix} a_{11} & a_{12} & \cdots & a_{1n} \\ \vdots & & & \\ a_{i1} & a_{i2} & \cdots & a_{in} \\ \vdots & & & \\ a_{i1} & a_{i2} & \cdots & a_{in} \\ \vdots & & & \\ a_{n1} & a_{n2} & \cdots & a_{nn} \end{pmatrix}$$

Since the ith and jth rows of A^* are the same, it follows that $\det(A^*) = 0$ (see Exercise 11 of Section 1). Thus

$$\det(EA) = \det(A) = \det(E)\det(A)$$

This property can also be derived using cofactors. After c times the ith row is added to the jth row, one could expand along the new jth row. The result will follow using Lemma 2.3.1 from the next section.

ROW OPERATION I

To see the effects of row operation I, we note that this operation can be accomplished using row operations II and III. We illustrate how this is done for 3×3 matrices.

$$A = \begin{pmatrix} a_{11} & a_{12} & a_{13} \\ a_{21} & a_{22} & a_{23} \\ a_{31} & a_{32} & a_{33} \end{pmatrix}$$

Subtracting row 3 from row 2 yields

$$A^{(1)} = \begin{pmatrix} a_{11} & a_{12} & a_{13} \\ a_{21} - a_{31} & a_{22} - a_{32} & a_{23} - a_{33} \\ a_{31} & a_{32} & a_{33} \end{pmatrix}$$

Next, the second row of $A^{(1)}$ is added to the third row:

$$A^{(2)} = \begin{pmatrix} a_{11} & a_{12} & a_{13} \\ a_{21} - a_{31} & a_{22} - a_{32} & a_{23} - a_{33} \\ a_{21} & a_{22} & a_{23} \end{pmatrix}$$

Subtracting row 3 from row 2, we get

$$A^{(3)} = \begin{pmatrix} a_{11} & a_{12} & a_{13} \\ -a_{31} & -a_{32} & -a_{33} \\ a_{21} & a_{22} & a_{23} \end{pmatrix}$$

Since all of these matrices have been formed using row operation III, it follows that

$$\det(A) = \det(A^{(1)}) = \det(A^{(2)}) = \det(A^{(3)})$$

Finally, if the second row of A is multiplied through by -1, one obtains

$$A^{(4)} = \begin{pmatrix} a_{11} & a_{12} & a_{13} \\ a_{31} & a_{32} & a_{33} \\ a_{21} & a_{22} & a_{23} \end{pmatrix}$$

Since row operation II was used, it follows that

$$\det(A^{(4)}) = -1 \det(A^{(3)}) = -\det(A)$$

$A^{(4)}$ is just the matrix obtained by interchanging the second and third rows of A.

This same argument can be applied to $n \times n$ matrices to show that whenever two rows are switched the sign of the determinant is changed. Thus if A is $n \times n$ and E_{ij} is the $n \times n$ elementary matrix formed by interchanging the ith and jth rows of I, then

$$\det(E_{ij}A) = -\det(A)$$

In particular,

$$\det(E_{ij}) = \det(E_{ij}I) = -\det(I) = -1$$

Thus for any elementary matrix E of type I,

$$\det(EA) = -\det(A) = \det(E)\det(A)$$

In summation, if E is an elementary matrix, then

$$\det(EA) = \det(E)\det(A)$$

where

$$\det(E) = \begin{cases} -1 & \text{if} \quad E \text{ is of type I} \\ \alpha \neq 0 & \text{if} \quad E \text{ is of type II} \\ 1 & \text{if} \quad E \text{ is of type III} \end{cases}$$

Similar results hold for column operations. Indeed, if E is an elementary matrix, then

$$\det(AE) = \det((AE)^T) = \det(E^T A^T)$$
$$= \det(E^T)\det(A^T) = \det(E)\det(A)$$

Note that all elementary matrices have nonzero determinants. This observation can be used to prove the following theorem.

Theorem 2.2.1. *An $n \times n$ matrix A is singular if and only if*

$$\det(A) = 0$$

Proof. The matrix A can be reduced to row echelon form with a finite number of row operations. Thus

$$U = E_k E_{k-1} \cdots E_1 A$$

where U is in row echelon form and the E_i's are all elementary matrices.

$$\det(U) = \det(E_k E_{k-1} \cdots E_1 A)$$
$$= \det(E_k)\det(E_{k-1}) \cdots \det(E_1)\det(A)$$

Since the determinants of the E_i's are all nonzero, it follows that $\det(A) = 0$ if and only if $\det(U) = 0$. If A is singular, then U has a row consisting entirely of zeros and hence $\det(U) = 0$. If A is nonsingular, U is triangular with 1's along the diagonal and hence $\det(U) = 1$. $\quad\square$

From the foregoing proof we can obtain a method for computing $\det(A)$. Reduce A to row echelon form.

$$U = E_k E_{k-1} \cdots E_1 A$$

If the last row of U consists entirely of zeros, A is singular and $\det(A) = 0$. Otherwise, A is nonsingular and

$$\det(A) = \left[\det(E_k)\det(E_{k-1}) \cdots \det(E_1)\right]^{-1}$$

Actually, if A is nonsingular, it is simpler to reduce A to triangular form. This can be done using only row operations I and III. Thus

$$T = E_m E_{m-1} \cdots E_1 A$$

and hence

$$\det(A) = \pm \det(T) = \pm t_{11} t_{22} \cdots t_{nn}$$

The sign will be positive if row operation I has been used an even number of times and negative otherwise.

EXAMPLE 1.   Evaluate

$$\begin{vmatrix} 2 & 1 & 3 \\ 4 & 2 & 1 \\ 6 & -3 & 4 \end{vmatrix}$$

SOLUTION

$$\begin{vmatrix} 2 & 1 & 3 \\ 4 & 2 & 1 \\ 6 & -3 & 4 \end{vmatrix} = \begin{vmatrix} 2 & 1 & 3 \\ 0 & 0 & -5 \\ 0 & -6 & -5 \end{vmatrix}$$

$$= (-1) \begin{vmatrix} 2 & 1 & 3 \\ 0 & -6 & -5 \\ 0 & 0 & -5 \end{vmatrix}$$

$$= (-1)(2)(-6)(-5)$$

$$= -60 \qquad \qquad \square$$

We now have three methods for evaluating the determinant of an $n \times n$ matrix A. If $n > 3$ and A has nonzero entries, elimination is the most efficient method in the sense that it involves less arithmetic operations than the other two methods. In Table 2.2.1 the number of arithmetic operations involved in each of the methods is given for $n = 2, 3, 4, 5$. It is not difficult to derive general formulas for the

TABLE 2.2.1

n	Additions	Multipli-cations	Additions	Multipli-cations	Additions	Multipli-cations and Divisions
2	1	2	1	2	1	3
3	5	12	5	9	5	10
4	23	72	23	40	14	23
5	119	480	119	205	30	44
	Definition		Cofactors		Elimination	

number of operations in each of the methods (see Exercises 10, 11, and 12).

We have seen that for any elementary matrix E,

$$\det(EA) = \det(E)\det(A) = \det(AE)$$

This is a special case of the following theorem.

Theorem 2.2.2.   *If A and B are $n \times n$ matrices, then $\det(AB) = \det(A)\det(B)$.*

Proof.   If B is singular, it follows from Theorem 1.4.3 that AB is also singular (see Exercise 12 of Chapter 1, Section 4), and hence

$$\det(AB) = 0 = \det(A)\det(B)$$

If B is nonsingular, B can be written as a product of elementary matrices. We have already seen that the result holds for elementary matrices. Thus

$$\begin{aligned}
\det(AB) &= \det(AE_k E_{k-1} \cdots E_1) \\
&= \det(A)\det(E_k)\det(E_{k-1}) \cdots \det(E_1) \\
&= \det(A)\det(E_k E_{k-1} \cdots E_1) \\
&= \det(A)\det(B) \qquad\qquad\qquad\qquad \square
\end{aligned}$$

If A is singular, the computed value of $\det(A)$ using exact arithmetic must be 0. This result, however, is unlikely if the computations are done by computer. Because of the finite number system of the computer, roundoff errors are usually unavoidable. Consequently, it is more likely that the computed value of $\det(A)$ will only be near 0. Because of roundoff errors, it is virtually impossible to determine that a matrix is singular from computer computations. In computer applications it is often more meaningful to ask whether a matrix is "close" to being singular. In general, the value of $\det(A)$ is not a good indicator of nearness to singularity. In Chapter 7 we will discuss how to determine whether or not a matrix is close to being singular.

EXERCISES

1. Evaluate each of the following determinants by inspection.

(a) $\begin{vmatrix} 0 & 0 & 3 \\ 0 & 4 & 1 \\ 2 & 3 & 1 \end{vmatrix}$
(b) $\begin{vmatrix} 1 & 1 & 1 & 3 \\ 0 & 3 & 1 & 1 \\ 0 & 0 & 2 & 2 \\ -1 & -1 & -1 & 2 \end{vmatrix}$

(c) $\begin{vmatrix} 0 & 0 & 0 & 1 \\ 1 & 0 & 0 & 0 \\ 0 & 1 & 0 & 0 \\ 0 & 0 & 1 & 0 \end{vmatrix}$

2. Let

$$A = \begin{pmatrix} 0 & 1 & 2 & 3 \\ 1 & 1 & 1 & 1 \\ -2 & -2 & 3 & 3 \\ 1 & 2 & -2 & -3 \end{pmatrix}$$

(a) Use the elimination method to evaluate $\det(A)$.

(b) Use the value of $\det(A)$ to evaluate

$$\begin{vmatrix} 0 & 1 & 2 & 3 \\ -2 & -2 & 3 & 3 \\ 1 & 2 & -2 & -3 \\ 1 & 1 & 1 & 1 \end{vmatrix} + \begin{vmatrix} 0 & 1 & 2 & 3 \\ 1 & 1 & 1 & 1 \\ -1 & -1 & 4 & 4 \\ 2 & 3 & -1 & -2 \end{vmatrix}$$

3. For each of the following, compute the determinant and state whether the matrix is singular or nonsingular.

(a) $\begin{pmatrix} 3 & 1 \\ 6 & 2 \end{pmatrix}$ (b) $\begin{pmatrix} 3 & 1 \\ 4 & 2 \end{pmatrix}$ (c) $\begin{pmatrix} 3 & 3 & 1 \\ 0 & 1 & 2 \\ 0 & 2 & 3 \end{pmatrix}$

(d) $\begin{pmatrix} 2 & 1 & 1 \\ 4 & 3 & 5 \\ 2 & 1 & 2 \end{pmatrix}$ (e) $\begin{pmatrix} 2 & -1 & 3 \\ -1 & 2 & -2 \\ 1 & 4 & 0 \end{pmatrix}$

(f) $\begin{pmatrix} 1 & 1 & 1 & 1 \\ 2 & -1 & 3 & 2 \\ 0 & 1 & 2 & 1 \\ 0 & 0 & 7 & 3 \end{pmatrix}$

4. Let A be an $n \times n$ matrix and α a scalar. Show that

$$\det(\alpha A) = \alpha^n \det(A)$$

5. Let A be a nonsingular matrix. Show that

$$\det(A^{-1}) = \frac{1}{\det(A)}$$

6. Let A and B be 3×3 matrices with $\det(A) = 4$ and $\det(B) = 5$. Find the value of:

(a) $\det(AB)$
(b) $\det(3A)$
(c) $\det(2AB)$
(d) $\det(A^{-1}B)$

7. Suppose that a 3×3 matrix A factors into a product

$$\begin{pmatrix} 1 & 0 & 0 \\ l_{21} & 1 & 0 \\ l_{31} & l_{32} & 1 \end{pmatrix} \begin{pmatrix} u_{11} & u_{12} & u_{13} \\ 0 & u_{22} & u_{23} \\ 0 & 0 & u_{33} \end{pmatrix}$$

Determine the value of $\det(A)$.

8. Let A and B be $n \times n$ matrices. Prove that the product AB is nonsingular if and only if A and B are both nonsingular.

9. Let A and B be $n \times n$ matrices. Prove that if $AB = I$, then $BA = I$.

10. How many additions and how many multiplications are necessary to evaluate the determinant of an $n \times n$ matrix directly using the definition of the determinant?

11. Show that evaluating the determinant of an $n \times n$ matrix by cofactors involves $(n! - 1)$ additions and $\sum_{k=1}^{n-1} n!/k!$ multiplications.

12. Show that the elimination method of computing the value of the determinant of an $n \times n$ matrix involves $[n(n-1)(2n-1)]/6$ additions and $[(n-1)(n^2 + n + 3)]/3$ multiplications and divisions.

*3. CRAMER'S RULE

In this section we learn a method for computing the inverse of a nonsingular matrix A using determinants. We also learn a method for solving $A\mathbf{x} = \mathbf{b}$ using determinants. Both methods depend upon the following lemma.

Lemma 2.3.1.   *Let A be an $n \times n$ matrix. If A_{jk} denotes the cofactor of a_{jk} for $k = 1, \ldots, n$, then*

$$(1) \quad a_{i1}A_{j1} + a_{i2}A_{j2} + \cdots + a_{in}A_{jn} = \begin{cases} \det(A) & \text{if} \quad i = j \\ 0 & \text{if} \quad i \neq j \end{cases}$$

Proof.   If $i = j$, (1) is just the cofactor expansion of $\det(A)$ along the ith row of A. To prove (1) in the case $i \neq j$, let A^* be the matrix obtained by replacing the jth row of A by the ith row of A.

$$A^* = \begin{pmatrix} a_{11} & a_{12} & \cdots & a_{1n} \\ \vdots & & & \\ a_{i1} & a_{i2} & \cdots & a_{in} \\ \vdots & & & \\ a_{i1} & a_{i2} & \cdots & a_{in} \\ \vdots & & & \\ a_{n1} & a_{n2} & \cdots & a_{nn} \end{pmatrix} \begin{matrix} \\ \\ j\text{th row} \\ \\ \\ \\ \\ \end{matrix}$$

Since two rows of A^* are the same, its determinant must be zero. It follows from the cofactor expansion of $\det(A^*)$ along the jth row that

$$0 = \det(A^*) = a_{i1}A_{j1}^* + a_{i2}A_{j2}^* + \cdots + a_{in}A_{jn}^*$$
$$= a_{i1}A_{j1} + a_{i2}A_{j2} + \cdots + a_{in}A_{jn} \qquad \square$$

THE ADJOINT OF A MATRIX

Let A be an $n \times n$ matrix. We can define a new matrix called the *adjoint* of A by

$$\text{adj } A = \begin{pmatrix} A_{11} & A_{21} & \cdots & A_{n1} \\ A_{12} & A_{22} & \cdots & A_{n2} \\ \vdots & & & \\ A_{1n} & A_{2n} & \cdots & A_{nn} \end{pmatrix}$$

Thus to form the adjoint, one must replace each term by its cofactor and then transpose the resulting matrix. It follows from (1) that

$$A(\text{adj } A) = \det(A)I$$

If A is nonsingular, $\det(A)$ is a nonzero scalar and we may write

$$A\left(\frac{1}{\det(A)}\text{adj } A\right) = I$$

Thus

$$\boxed{A^{-1} = \frac{1}{\det(A)}\text{adj } A}$$

EXAMPLE 1.   For a 2×2 matrix

$$\text{adj } A = \begin{pmatrix} a_{22} & -a_{12} \\ -a_{21} & a_{11} \end{pmatrix}$$

If A is nonsingular, then

$$A^{-1} = \frac{1}{a_{11}a_{22} - a_{12}a_{21}}\begin{pmatrix} a_{22} & -a_{12} \\ -a_{21} & a_{11} \end{pmatrix} \qquad \square$$

EXAMPLE 2.   Let

$$A = \begin{pmatrix} 2 & 1 & 2 \\ 3 & 2 & 2 \\ 1 & 2 & 3 \end{pmatrix}$$

Compute adj A and A^{-1}.

SOLUTION

$$\text{adj } A = \begin{pmatrix} \begin{vmatrix} 2 & 2 \\ 2 & 3 \end{vmatrix} & -\begin{vmatrix} 3 & 2 \\ 1 & 3 \end{vmatrix} & \begin{vmatrix} 3 & 2 \\ 1 & 2 \end{vmatrix} \\ -\begin{vmatrix} 1 & 2 \\ 2 & 3 \end{vmatrix} & \begin{vmatrix} 2 & 2 \\ 1 & 3 \end{vmatrix} & -\begin{vmatrix} 2 & 1 \\ 1 & 2 \end{vmatrix} \\ \begin{vmatrix} 1 & 2 \\ 2 & 2 \end{vmatrix} & -\begin{vmatrix} 2 & 2 \\ 3 & 2 \end{vmatrix} & \begin{vmatrix} 2 & 1 \\ 3 & 2 \end{vmatrix} \end{pmatrix}^T$$

$$= \begin{pmatrix} 2 & 1 & -2 \\ -7 & 4 & 2 \\ 4 & -3 & 1 \end{pmatrix}$$

$$A^{-1} = \frac{1}{\det(A)} \text{adj } A$$

$$= \tfrac{1}{5} \begin{pmatrix} 2 & 1 & -2 \\ -7 & 4 & 2 \\ 4 & -3 & 1 \end{pmatrix} \qquad \square$$

Using the formula

$$A^{-1} = \frac{1}{\det(A)} \text{adj } A$$

we can derive a rule for representing the solution to the system $A\mathbf{x} = \mathbf{b}$ in terms of determinants.

Theorem 2.3.2 (Cramer's Rule).   *Let A be an $n \times n$ nonsingular matrix and let $\mathbf{b} \in R^n$. Let A_i be the matrix obtained by replacing the ith column of A by $\mathbf{b}$. If $\mathbf{x}$ is the unique solution to $A\mathbf{x} = \mathbf{b}$, then*

$$x_i = \frac{\det(A_i)}{\det(A)} \qquad for \quad i = 1, 2, \ldots, n$$

Proof.   Since

$$\mathbf{x} = A^{-1}\mathbf{b} = \frac{1}{\det(A)}(\text{adj } A)\mathbf{b}$$

it follows that

$$x_i = \frac{b_1 A_{1i} + b_2 A_{2i} + \cdots + b_n A_{ni}}{\det(A)}$$

$$= \frac{\det(A_i)}{\det(A)} \qquad \square$$

EXAMPLE 3. Use Cramer's rule to solve

$$x_1 + 2x_2 + x_3 = 5$$
$$2x_1 + 2x_2 + x_3 = 6$$
$$x_1 + 2x_2 + 3x_3 = 9$$

SOLUTION

$$\det(A) = \begin{vmatrix} 1 & 2 & 1 \\ 2 & 2 & 1 \\ 1 & 2 & 3 \end{vmatrix} = -4 \qquad \det(A_1) = \begin{vmatrix} 5 & 2 & 1 \\ 6 & 2 & 1 \\ 9 & 2 & 3 \end{vmatrix} = -4$$

$$\det(A_2) = \begin{vmatrix} 1 & 5 & 1 \\ 2 & 6 & 1 \\ 1 & 9 & 3 \end{vmatrix} = -4 \qquad \det(A_3) = \begin{vmatrix} 1 & 2 & 5 \\ 2 & 2 & 6 \\ 1 & 2 & 9 \end{vmatrix} = -8$$

Therefore,

$$x_1 = \frac{-4}{-4} = 1, \qquad x_2 = \frac{-4}{-4} = 1, \qquad x_3 = \frac{-8}{-4} = 2 \qquad \square$$

Cramer's rule gives us a convenient method for writing down the solution to an $n \times n$ system of linear equations in terms of determinants. To compute the solution, however, one must evaluate $n + 1$ determinants of order n. Evaluating even two of these determinants generally involves more computation than solving the system using Gaussian elimination.

APPLICATION

A common way of sending a coded message is to assign an integer value to each letter of the alphabet and to send the message as a string of integers. For example, the message

SEND MONEY

might be coded as

5, 8, 10, 21, 7, 2, 10, 8, 3

Here the S is represented by a 5, the E by an 8, and so on. Unfortunately, this type of code is generally quite easy to break. We can, however, disguise the message even further using matrix multiplications. If A is a matrix whose entries are all integers and whose determinant is ± 1, then since $A^{-1} = \pm \operatorname{adj} A$, the entries of A^{-1} will be integers. We can use such a matrix to transform the message. The transformed message will be more difficult to decipher. To illustrate the technique, let

$$A = \begin{pmatrix} 1 & 2 & 1 \\ 2 & 5 & 3 \\ 2 & 3 & 2 \end{pmatrix}$$

The coded message is put into the columns of a matrix B having three rows.

$$B = \begin{pmatrix} 5 & 21 & 10 \\ 8 & 7 & 8 \\ 10 & 2 & 3 \end{pmatrix}$$

The product

$$AB = \begin{pmatrix} 1 & 2 & 1 \\ 2 & 5 & 3 \\ 2 & 3 & 2 \end{pmatrix} \begin{pmatrix} 5 & 21 & 10 \\ 8 & 7 & 8 \\ 10 & 2 & 3 \end{pmatrix} = \begin{pmatrix} 31 & 37 & 29 \\ 80 & 83 & 69 \\ 54 & 67 & 50 \end{pmatrix}$$

gives the coded message to be sent out:

$$31, \quad 80, \quad 54, \quad 37, \quad 83, \quad 67, \quad 29, \quad 69, \quad 50$$

The person receiving the message can decode it by multiplying by A^{-1}:

$$\begin{pmatrix} 1 & -1 & 1 \\ 2 & 0 & -1 \\ -4 & 1 & 1 \end{pmatrix} \begin{pmatrix} 31 & 37 & 29 \\ 80 & 83 & 69 \\ 54 & 67 & 50 \end{pmatrix} = \begin{pmatrix} 5 & 21 & 10 \\ 8 & 7 & 8 \\ 10 & 2 & 3 \end{pmatrix}$$

To construct a coding matrix A, we can begin with the identity I and successively apply row operation III, being careful to add integer multiples of one row to another. Row operation I can also be used. The resulting matrix A will have integer entries and since

$$\det(A) = \pm\det(I) = \pm1$$

A^{-1} will also have integer entries.

This coding technique lends itself well to generalizations involving block multiplication. For example, if we represent the message by three vectors

$$\mathbf{b}_1 = \begin{pmatrix} 5 \\ 8 \\ 10 \end{pmatrix}, \qquad \mathbf{b}_2 = \begin{pmatrix} 21 \\ 7 \\ 2 \end{pmatrix}, \qquad \mathbf{b}_3 = \begin{pmatrix} 10 \\ 8 \\ 3 \end{pmatrix}$$

and construct 3×3 matrices A_1, A_2, A_3 whose entries are integers and whose determinants are ±1, then we can transform the message by multiplying

$$\begin{pmatrix} A_1 & O & O \\ O & A_2 & O \\ O & O & A_3 \end{pmatrix} \begin{pmatrix} \mathbf{b}_1 \\ \mathbf{b}_2 \\ \mathbf{b}_3 \end{pmatrix} = \begin{pmatrix} \mathbf{c}_1 \\ \mathbf{c}_2 \\ \mathbf{c}_3 \end{pmatrix}$$

The message is sent out as

$$\left(\mathbf{c}_1^T, \mathbf{c}_2^T, \mathbf{c}_3^T \right)$$

It can be decoded by multiplying

$$\begin{pmatrix} A_1^{-1} & O & O \\ O & A_2^{-1} & O \\ O & O & A_3^{-1} \end{pmatrix} \begin{pmatrix} \mathbf{c}_1 \\ \mathbf{c}_2 \\ \mathbf{c}_3 \end{pmatrix}$$

REFERENCE

1. Robert Hansen, *Two-Year College Mathematics Journal*, *13*(1), 1982.

EXERCISES

1. For each of the following compute (i) $\det(A)$, (ii) adj A, and (iii) A^{-1}.

(a) $A = \begin{pmatrix} 1 & 2 \\ 3 & -1 \end{pmatrix}$ (b) $A = \begin{pmatrix} 3 & 1 \\ 2 & 4 \end{pmatrix}$

(c) $A = \begin{pmatrix} 1 & 3 & 1 \\ 2 & 1 & 1 \\ -2 & 2 & -1 \end{pmatrix}$ (d) $A = \begin{pmatrix} 1 & 1 & 1 \\ 0 & 1 & 1 \\ 0 & 0 & 1 \end{pmatrix}$

2. Use Cramer's rule to solve each of the following systems.

(a) $\begin{aligned} x_1 + 2x_2 &= 3 \\ 3x_1 - x_2 &= 1 \end{aligned}$ (b) $\begin{aligned} 2x_1 + 3x_2 &= 2 \\ 3x_1 + 2x_2 &= 5 \end{aligned}$

(c) $\begin{aligned} 2x_1 + x_2 - 3x_3 &= 0 \\ 4x_1 + 5x_2 + x_3 &= 8 \\ -2x_1 - x_2 + 4x_3 &= 2 \end{aligned}$ (d) $\begin{aligned} x_1 + 3x_2 + x_3 &= 1 \\ 2x_1 + x_2 + x_3 &= 5 \\ -2x_1 + 2x_2 - x_3 &= -8 \end{aligned}$

(e) $\begin{aligned} x_1 + x_2 \qquad\qquad &= 0 \\ x_2 + x_3 - 2x_4 &= 1 \\ x_1 \qquad + 2x_3 + x_4 &= 0 \\ x_1 + x_2 \qquad + x_4 &= 0 \end{aligned}$

3. Given

$$A = \begin{pmatrix} 1 & 2 & 3 \\ 2 & 3 & 4 \\ 3 & 4 & 5 \end{pmatrix}$$

(a) Compute the determinant of A. Is A nonsingular?

(b) Compute adj A and the product A adj A.

4. If A is singular, what can you say about the product A adj A?

5. Let **x** and **y** be elements of R^3 and let **z** be the vector in R^3 whose coordinates are defined by

$$z_1 = \begin{vmatrix} x_2 & x_3 \\ y_2 & y_3 \end{vmatrix}, \qquad z_2 = -\begin{vmatrix} x_1 & x_3 \\ y_1 & y_3 \end{vmatrix}, \qquad z_3 = \begin{vmatrix} x_1 & x_2 \\ y_1 & y_2 \end{vmatrix}$$

Show that

$$\mathbf{z}^T\mathbf{x} = \mathbf{z}^T\mathbf{y} = 0$$

6. Let B_j denote the matrix obtained by replacing the jth column of the identity matrix with a vector $\mathbf{b} = (b_1, \ldots, b_n)^T$. Show that

$$\det(B_j) = b_j \qquad \text{for} \quad j = 1, \ldots, n$$

7. Let A be an $n \times n$ matrix with $n > 1$. Show that

$$\det(\text{adj}\, A) = (\det(A))^{n-1}$$

8. Show that if A is nonsingular, then $\text{adj}\, A$ is nonsingular and

$$(\text{adj}\, A)^{-1} = \det(A^{-1})A = \text{adj}\, A^{-1}$$

9. Show that if A is singular, then $\text{adj}\, A$ is also singular.

10. Show that if $\det(A) = 1$, then

$$\text{adj}(\text{adj}\, A) = A$$

11. Suppose that Q is a matrix with the property $Q^{-1} = Q^T$. Show that

$$q_{ij} = \frac{Q_{ij}}{\det(Q)}$$

12. In coding a message, a blank space was represented by 0, an A by 1, a B by 2, a C by 3, and so on. The message was transformed using the matrix

$$A = \begin{pmatrix} -1 & -1 & 2 & 0 \\ 1 & 1 & -1 & 0 \\ 0 & 0 & -1 & 1 \\ 1 & 0 & 0 & -1 \end{pmatrix}$$

and sent out as

$$-19, \quad 19, \quad 25, \quad -21, \quad 0, \quad 18, \quad -18, \quad 15,$$
$$3, \quad 10, \quad -8, \quad 3, \quad -2, \quad 20, \quad -7, \quad 12$$

What was the message?

3

VECTOR SPACES

The operations of addition and scalar multiplication are used in many diverse contexts in mathematics. Regardless of the context, however, these operations usually obey the same set of arithmetic rules. Thus a general theory of mathematical systems involving addition and scalar multiplication will have application to many areas in mathematics. Mathematical systems of this form are called vector spaces or linear spaces. In this chapter the definition of a vector space is given and some of the general theory of vector spaces is developed.

1. VECTOR SPACES: DEFINITION AND EXAMPLES

In this section we present the formal definition of a vector space. Before doing this, however, it is instructive to look at a number of examples. We begin with the Euclidean vector spaces R^n.

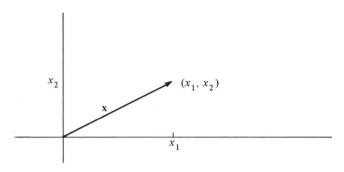

FIGURE 3.1.1

EUCLIDEAN VECTOR SPACES

Perhaps the most elementary vector spaces are the Euclidean vector spaces R^n, $n = 1, 2, \ldots$. For simplicity, let us consider first R^2. Non-zero vectors in R^2 can be represented geometrically by directed line segments. This geometric representation will help us to visualize how the operations of scalar multiplication and addition work in R^2. Given a nonzero vector $\mathbf{x} = \begin{pmatrix} x_1 \\ x_2 \end{pmatrix}$, we can associate it with the line segment in the plane from $(0, 0)$ to (x_1, x_2) (see Figure 3.1.1). If we equate line segments that have the same length and direction (Figure 3.1.2), $\mathbf{x}$ can be represented by any line segment from (a, b) to $(a + x_1, b + x_2)$. For example, the vector $\mathbf{x} = \begin{pmatrix} 2 \\ 1 \end{pmatrix}$ in R^2 could just as well be represented by the directed line segment from $(2, 2)$ to $(4, 3)$, or from $(-1, -1)$ to $(1, 0)$, as shown in Figure 3.1.3.

FIGURE 3.1.2

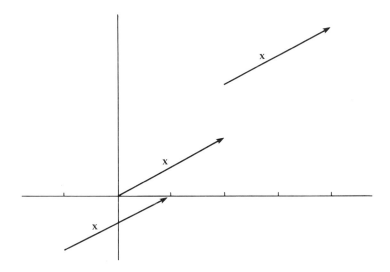

FIGURE 3.1.3

We can think of the Euclidean length of a vector $\mathbf{x} = \begin{pmatrix} x_1 \\ x_2 \end{pmatrix}$ as the length of any directed line segment representing $\mathbf{x}$. The length of the line segment from $(0,0)$ to (x_1, x_2) is $\sqrt{x_1^2 + x_2^2}$ (see Figure 3.1.4). For each vector $\mathbf{x} = \begin{pmatrix} x_1 \\ x_2 \end{pmatrix}$ and each scalar α, the product $\alpha\mathbf{x}$ is defined by

$$\alpha \begin{pmatrix} x_1 \\ x_2 \end{pmatrix} = \begin{pmatrix} \alpha x_1 \\ \alpha x_2 \end{pmatrix}$$

For example, as shown in Figure 3.1.5, if $\mathbf{x} = \begin{pmatrix} 2 \\ 1 \end{pmatrix}$, then

$$3\mathbf{x} = \begin{pmatrix} 6 \\ 3 \end{pmatrix}, \qquad -\mathbf{x} = \begin{pmatrix} -2 \\ -1 \end{pmatrix}, \qquad -2\mathbf{x} = \begin{pmatrix} -4 \\ -2 \end{pmatrix}$$

FIGURE 3.1.4

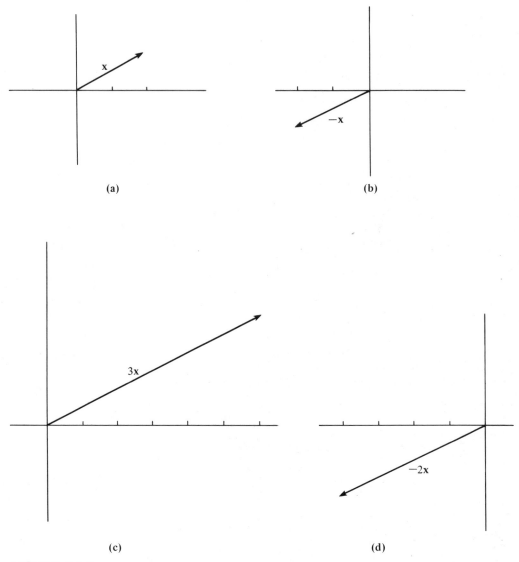

(a)   (b)

(c)   (d)

FIGURE 3.1.5

The vector $3\mathbf{x}$ is in the same direction as $\mathbf{x}$, but its length is three times that of $\mathbf{x}$. The vector $-\mathbf{x}$ has the same length as $\mathbf{x}$, but it points in the opposite direction. The vector $-2\mathbf{x}$ is twice as long as $\mathbf{x}$ and it points in the same direction as $-\mathbf{x}$. The sum of two vectors $\mathbf{u} = \begin{pmatrix} u_1 \\ u_2 \end{pmatrix}$ and $\mathbf{v} = \begin{pmatrix} v_1 \\ v_2 \end{pmatrix}$ is defined by

$$\mathbf{u} + \mathbf{v} = \begin{pmatrix} u_1 + v_1 \\ u_2 + v_2 \end{pmatrix}$$

Note that if $\mathbf{v}$ is placed at the terminal point of $\mathbf{u}$, then $\mathbf{u} + \mathbf{v}$ is

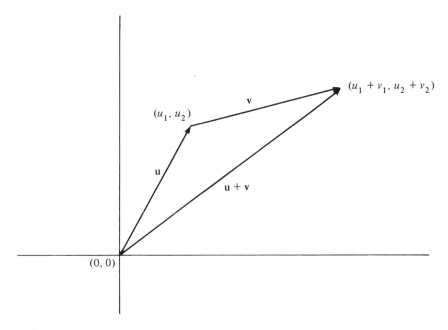

FIGURE 3.1.6

represented by the directed line segment from the initial point of $\mathbf{u}$ to the terminal point of $\mathbf{v}$ (Figure 3.1.6). If both $\mathbf{u}$ and $\mathbf{v}$ are placed at the origin and a parallelogram is formed as in Figure 3.1.7, the diagonals of the parallelogram will represent the sum $\mathbf{u} + \mathbf{v}$ and the difference $\mathbf{v} - \mathbf{u}$. In a similar manner, vectors in R^3 can be represented by directed line segments in 3-space (see Figure 3.1.8).

FIGURE 3.1.7

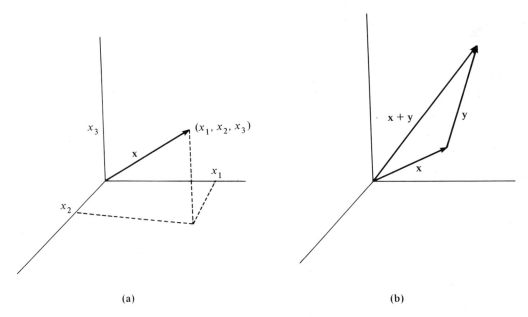

(a)   (b)

FIGURE 3.1.8

In general, scalar multiplication and addition in R^n are defined by

$$\alpha \mathbf{x} = \begin{pmatrix} \alpha x_1 \\ \alpha x_2 \\ \vdots \\ \alpha x_n \end{pmatrix} \quad \text{and} \quad \mathbf{x} + \mathbf{y} = \begin{pmatrix} x_1 + y_1 \\ x_2 + y_2 \\ \vdots \\ x_n + y_n \end{pmatrix}$$

for any $\mathbf{x}, \mathbf{y} \in R^n$ and any scalar α.

THE VECTOR SPACE $R^{m \times n}$

We can also view R^n as the set of all $n \times 1$ matrices with real entries. The addition and scalar multiplication of vectors in R^n is just the usual addition and scalar multiplication of matrices. More generally, let $R^{m \times n}$ denote the set of all $m \times n$ matrices with real entries. If $A = (a_{ij})$ and $B = (b_{ij})$, the sum $A + B$ is defined to be the $m \times n$ matrix $C = (c_{ij})$, where $c_{ij} = a_{ij} + b_{ij}$. Given a scalar α, one can define αA to be the $m \times n$ matrix whose ijth entry is αa_{ij}. Thus, by defining operations on the set $R^{m \times n}$, we have created a mathematical system. The operations of addition and scalar multiplication on $R^{m \times n}$ obey certain arithmetic rules. These rules form the axioms that are used to define the concept of a vector space.

Definition.   Let V be a set on which the operations of addition and scalar multiplication are defined. By this we mean that with each pair of elements **x** and **y** in V, one can associate a unique element $\mathbf{x} + \mathbf{y}$ that is also in V, and with each element **x** in V and each scalar α, one can associate a unique element $\alpha\mathbf{x}$ in V. The set V together with the operations of addition and scalar multiplication is said to form a **vector space** if the following axioms are satisfied.

A1. $\mathbf{x} + \mathbf{y} = \mathbf{y} + \mathbf{x}$ for any **x** and **y** in V.

A2. $(\mathbf{x} + \mathbf{y}) + \mathbf{z} = \mathbf{x} + (\mathbf{y} + \mathbf{z})$ for any $\mathbf{x}, \mathbf{y}, \mathbf{z}$ in V.

A3. There exists an element **0** in V such that $\mathbf{x} + \mathbf{0} = \mathbf{x}$ for each $\mathbf{x} \in V$.

A4. For each $\mathbf{x} \in V$, there exists an element $-\mathbf{x}$ in V such that $\mathbf{x} + (-\mathbf{x}) = \mathbf{0}$.

A5. $\alpha(\mathbf{x} + \mathbf{y}) = \alpha\mathbf{x} + \alpha\mathbf{y}$ for each real number α and any **x** and **y** in V.

A6. $(\alpha + \beta)\mathbf{x} = \alpha\mathbf{x} + \beta\mathbf{x}$ for any real numbers α and β and any $\mathbf{x} \in V$.

A7. $(\alpha\beta)\mathbf{x} = \alpha(\beta\mathbf{x})$ for any real numbers α and β and any $\mathbf{x} \in V$.

A8. $1 \cdot \mathbf{x} = \mathbf{x}$ for all $\mathbf{x} \in V$.

The elements of V are called *vectors* and are usually denoted by letters from the end of the alphabet: **u**, **v**, **w**, **x**, **y**, and **z**. The real numbers are called *scalars*. The symbol **0** was used in order to distinguish the zero vector from the scalar 0. In some contexts, complex numbers are used for scalars. However, in this text, scalars will usually be real numbers. Often the term *real vector space* is used to indicate that the set of scalars is the set of real numbers.

The reader may verify that R^n and $R^{m \times n}$ with the usual addition and scalar multiplication of matrices are both vector spaces. There are a number of other important examples of vector spaces.

THE VECTOR SPACE $C[a, b]$

Let $C[a, b]$ denote the set of all real-valued functions that are defined and continuous on the closed interval $[a, b]$. In this case our universal set is a set of functions. Thus our vectors are the functions in $C[a, b]$. The sum $f + g$ of two functions in $C[a, b]$ is defined by

$$(f + g)(x) = f(x) + g(x)$$

for all x in $[a, b]$. The new function $f + g$ is an element of $C[a, b]$, since the sum of two continuous functions is continuous. If f is a function in $C[a, b]$ and α is a real number, define αf by

$$(\alpha f)(x) = \alpha f(x)$$

for all x in $[a, b]$. Clearly, αf is in $C[a, b]$ since a constant times a continuous function is always continuous. Thus on $C[a, b]$ we have defined the operations of addition and scalar multiplication. The reader may verify that these operations satisfy the eight axioms of a vector space.

THE VECTOR SPACE P_n

Let P_n denote the set of all polynomials of degree less than n. Define $p + q$ and αp by

$$(p + q)(x) = p(x) + q(x)$$

and

$$(\alpha p)(x) = \alpha p(x)$$

for all real numbers x. It is easily verified that axioms A1 through A8 hold. Thus P_n with the standard addition and scalar multiplication of functions is a vector space.

We close this section with a theorem that states three more fundamental properties of vector spaces.

Theorem 3.1.1. *If V is a vector space and $\mathbf{x}$ is any element of V, then*
  (i) $0\mathbf{x} = \mathbf{0}$
  (ii) $\mathbf{x} + \mathbf{y} = \mathbf{0}$ *implies that* $\mathbf{y} = -\mathbf{x}$ (*i.e., the additive inverse of* $\mathbf{x}$ *is unique*).
  (iii) $(-1)\mathbf{x} = -\mathbf{x}$

Proof. It follows from axioms A6 and A8 that

$$\mathbf{x} = 1\mathbf{x} = (1 + 0)\mathbf{x} = 1\mathbf{x} + 0\mathbf{x} = \mathbf{x} + 0\mathbf{x}$$

Thus

(A2) $-\mathbf{x} + \mathbf{x} = -\mathbf{x} + (\mathbf{x} + 0\mathbf{x}) = (-\mathbf{x} + \mathbf{x}) + 0\mathbf{x}$

(A1, A3, and A4) $\mathbf{0} = \mathbf{0} + 0\mathbf{x} = 0\mathbf{x}$

To prove (ii), suppose that $\mathbf{x} + \mathbf{y} = \mathbf{0}$. Then

$$-\mathbf{x} = -\mathbf{x} + \mathbf{0} = -\mathbf{x} + (\mathbf{x} + \mathbf{y})$$

Therefore,

(A1, A2, A3, and A4)

$$-\mathbf{x} = (-\mathbf{x} + \mathbf{x}) + \mathbf{y} = \mathbf{0} + \mathbf{y} = \mathbf{y}$$

Finally, to prove (iii), note that

[(i) and A6] $\mathbf{0} = 0\mathbf{x} = (1 + (-1))\mathbf{x} = 1\mathbf{x} + (-1)\mathbf{x}$

Thus

(A8) $\mathbf{x} + (-1)\mathbf{x} = \mathbf{0}$

and it follows from part (ii) that

$$(-1)\mathbf{x} = -\mathbf{x} \qquad \square$$

EXERCISES

1. Let C be the set of complex numbers. Define addition on C by
$$(a + bi) + (c + di) = (a + c) + (b + d)i$$
and define scalar multiplication by
$$\alpha(a + bi) = \alpha a + \alpha bi$$
for all real numbers α. Show that C is a vector space with these operations.

2. Let V be the set of all ordered pairs of real numbers with addition defined by
$$(x_1, x_2) + (y_1, y_2) = (x_1 + y_1, x_2 + y_2)$$
and scalar multiplication defined by
$$\alpha(x_1, x_2) = (\alpha x_1, x_2)$$
Is V a vector space with these operations? Justify your answer.

3. Show that $R^{m \times n}$ with the usual addition and scalar multiplication of matrices satisfies the eight axioms of a vector space.

4. Show that $C[a, b]$ with the usual scalar multiplication and addition of functions satisfies the eight axioms of a vector space.

5. Let P be the set of all polynomials. Show that P with the usual addition and scalar multiplication of functions forms a vector space.

6. Show that the element $\mathbf{0}$ in a vector space is unique.

7. Let $\mathbf{x}$, $\mathbf{y}$, and $\mathbf{z}$ be vectors in a vector space V. Prove that if
$$\mathbf{x} + \mathbf{y} = \mathbf{x} + \mathbf{z}$$
then $\mathbf{y} = \mathbf{z}$.

8. Let V be a vector space and let $\mathbf{x} \in V$.
(a) Show that $\beta \mathbf{0} = \mathbf{0}$ for each scalar β.
(b) Show that if $\alpha \mathbf{x} = \mathbf{0}$, then either $\alpha = 0$ or $\mathbf{x} = \mathbf{0}$.

9. Let S be the set of all ordered pairs of real numbers. Define scalar multiplication and addition on S by
$$\alpha(x_1, x_2) = (\alpha x_1, \alpha x_2)$$
$$(x_1, x_2) + (y_1, y_2) = (x_1 + y_1, 0)$$
Show that S is not a vector space. Which of the eight axioms fail to hold?

10. Let S denote the set of all infinite sequences of real numbers with scalar multiplication and addition defined by

$$\alpha\{a_n\} = \{\alpha a_n\}$$

$$\{a_n\} + \{b_n\} = \{a_n + b_n\}$$

Show that S is a vector space.

11. One can define a one-to-one correspondence between the elements of P_n and R^n by

$$p(x) = a_1 + a_2 x + \cdots + a_n x^{n-1} \leftrightarrow (a_1, \ldots, a_n)^T = \mathbf{a}$$

Show that if $p \leftrightarrow \mathbf{a}$ and $q \leftrightarrow \mathbf{b}$, then
(a) $\alpha p \leftrightarrow \alpha \mathbf{a}$ for any scalar α.
(b) $p + q \leftrightarrow \mathbf{a} + \mathbf{b}$.
[In general, two vector spaces are said to be *isomorphic* if their elements can be put into a one-to-one correspondence that is preserved under scalar multiplication and addition as in (a) and (b)].

2. SUBSPACES

Given a vector space V, it is often possible to form another vector space by taking a subset S of V and using the operations of V. Since V is a vector space, the operations of addition and scalar multiplication always produce another vector in V. In order for a new system using a subset S of V as its universal set to be a vector space, the set S must be closed under the operations of addition and scalar multiplication; that is, the sum of two elements of S must always be an element of S and the product of a scalar and an element of S must always be an element of S.

EXAMPLE 1. Let $S = \left\{ \begin{pmatrix} x_1 \\ x_2 \end{pmatrix} \middle| x_2 = 2x_1 \right\}$. S is a subset of R^2. If $\begin{pmatrix} c \\ 2c \end{pmatrix}$ is any element of S and α is any scalar, then

$$\alpha \begin{pmatrix} c \\ 2c \end{pmatrix} = \begin{pmatrix} \alpha c \\ 2\alpha c \end{pmatrix}$$

which is an element of S. If $\begin{pmatrix} a \\ 2a \end{pmatrix}$ and $\begin{pmatrix} b \\ 2b \end{pmatrix}$ are any two elements of S, their sum

$$\begin{pmatrix} a + b \\ 2a + 2b \end{pmatrix} = \begin{pmatrix} a + b \\ 2(a + b) \end{pmatrix}$$

is also an element of S. It is easily seen that the mathematical system consisting of the set S (instead of R^2), together with the operations from R^2, is itself a vector space.　　　□

Definition. If S is a nonempty subset of a vector space V, and S satisfies the following conditions:

　　(i) $\alpha \mathbf{x} \in S$ whenever $\mathbf{x} \in S$ for any scalar α

　　(ii) $\mathbf{x} + \mathbf{y} \in S$ whenever $\mathbf{x} \in S$ and $\mathbf{y} \in S$

then S is said to be a **subspace** of V.

Condition (i) says that S is closed under scalar multiplication; that is, whenever an element of S is multiplied by a scalar, the result is an element of S. Condition (ii) says that S is closed under addition; that is, the sum of two elements of S is always an element of S. Thus, if we do arithmetic using the operations from V and the elements of S, we will always end up with elements of S. A subspace of V, then, is a subset S that is closed under the operations of V.

Let S be a subspace of a vector space V. Using the operations of addition and scalar multiplication as defined on V, we can form a new mathematical system with S as the universal set. It is easily seen that all eight axioms will remain valid for this new system. Axioms A3 and A4 follow from Theorem 3.1.1 and condition (i) of the definition of a subspace. The remaining six axioms are valid for any elements of V, so in particular they are valid for the elements of S. Thus every subspace is a vector space in its own right.

EXAMPLE 2. Let $S = \{(x_1, x_2, x_3)^T | x_1 = x_2\}$. It follows that S is a subspace of R^3, since

　　(i) If $\mathbf{x} = (a, a, b)^T \in S$, then

$$\alpha \mathbf{x} = (\alpha a, \alpha a, \alpha b)^T \in S$$

　　(ii) If $(a, a, b)^T$ and $(c, c, d)^T$ are arbitrary elements of S, then

$$(a, a, b)^T + (c, c, d)^T = (a + c, a + c, b + d)^T \in S \qquad □$$

EXAMPLE 3. Let $S = \left\{ \begin{pmatrix} x \\ 1 \end{pmatrix} \middle| x \text{ is a real number} \right\}$. S is not a subspace of R^2. In this case both conditions fail. S is not closed under scalar multiplication, since $\alpha \begin{pmatrix} x \\ 1 \end{pmatrix} \notin S$ unless $\alpha = 1$. S is not closed under addition, since

$$\begin{pmatrix} x \\ 1 \end{pmatrix} + \begin{pmatrix} y \\ 1 \end{pmatrix} = \begin{pmatrix} x + y \\ 2 \end{pmatrix} \notin S \qquad □$$

EXAMPLE 4. Let V be any vector space. Let $S = \{\mathbf{0}\}$. We leave it to the reader to verify that S is a subspace of V. The reader may also verify that V is a subspace of itself.　　　□

EXAMPLE 5.   Let S be the set of all polynomials of degree less than n with the property $p(0) = 0$. We claim S is a subspace of P_n. This follows, since:

(i) If $p(x) \in S$ and α is a scalar, then

$$\alpha p(0) = \alpha \cdot 0 = 0$$

and hence $\alpha p \in S$.

(ii) If $p(x)$ and $q(x)$ are elements of S, then

$$(p + q)(0) = p(0) + q(0) = 0 + 0 = 0$$

and hence $p + q \in S$.   □

EXAMPLE 6.   Let $C^n[a, b]$ be the set of all functions f that have a continuous nth derivative on $[a, b]$. We leave it to the reader to verify that $C^n[a, b]$ is a subspace of $C[a, b]$.   □

EXAMPLE 7.   Let A be the set of all f in $C^2[a, b]$ such that

$$f''(x) + f(x) = 0$$

for all x in $[a, b]$. If $f \in A$ and α is any scalar, then for any $x \in [a, b]$,

$$(\alpha f)''(x) + (\alpha f)(x) = \alpha f''(x) + \alpha f(x)$$
$$= \alpha(f''(x) + f(x)) = \alpha \cdot 0 = 0$$

Thus $\alpha f \in A$. If f and g are both in A, then

$$(f + g)''(x) + (f + g)(x) = f''(x) + g''(x) + f(x) + g(x)$$
$$= [f''(x) + f(x)] + [g''(x) + g(x)]$$
$$= 0 + 0 = 0$$

Thus the set of all solutions on $[a, b]$ to the differential equation $y'' + y = 0$ forms a subspace of $C^2[a, b]$. If we note that $f(x) = \sin x$ and $g(x) = \cos x$ are both in A, it follows that any function of the form $c_1 \sin x + c_2 \cos x$ must also be in A. One can easily verify that functions of this form are solutions to $y'' + y = 0$.   □

THE NULLSPACE OF A MATRIX

Let A be an $m \times n$ matrix. Let $N(A)$ denote the set of all solutions to the homogeneous system $A\mathbf{x} = \mathbf{0}$. Thus

$$N(A) = \{\mathbf{x} \in R^n | A\mathbf{x} = \mathbf{0}\}$$

We claim that $N(A)$ is a subspace of R^n. If $\mathbf{x} \in N(A)$ and α is a scalar, then

$$A(\alpha\mathbf{x}) = \alpha A\mathbf{x} = \alpha\mathbf{0} = \mathbf{0}$$

and hence $\alpha \mathbf{x} \in N(A)$. If $\mathbf{x}$ and $\mathbf{y}$ are elements of $N(A)$, then

$$A(\mathbf{x} + \mathbf{y}) = A\mathbf{x} + A\mathbf{y} = \mathbf{0} + \mathbf{0} = \mathbf{0}$$

Therefore, $\mathbf{x} + \mathbf{y} \in N(A)$. It follows then that $N(A)$ is a subspace of R^n. The set of all solutions to the homogeneous system $A\mathbf{x} = \mathbf{0}$ forms a subspace of R^n. The subspace $N(A)$ is called the *nullspace* of A.

EXAMPLE 8. Determine $N(A)$ if

$$A = \begin{pmatrix} 1 & 1 & 1 & 0 \\ 2 & 1 & 0 & 1 \end{pmatrix}$$

SOLUTION. Using Gauss–Jordan reduction to solve $A\mathbf{x} = \mathbf{0}$, we obtain

$$\begin{pmatrix} 1 & 1 & 1 & 0 & | & 0 \\ 2 & 1 & 0 & 1 & | & 0 \end{pmatrix} \rightarrow \begin{pmatrix} 1 & 0 & -1 & 1 & | & 0 \\ 0 & -1 & -2 & 1 & | & 0 \end{pmatrix}$$
$$\begin{pmatrix} 1 & 1 & 1 & 0 & | & 0 \\ 0 & -1 & -2 & 1 & | & 0 \end{pmatrix} \quad \begin{pmatrix} 1 & 0 & -1 & 1 & | & 0 \\ 0 & 1 & 2 & -1 & | & 0 \end{pmatrix}$$

The reduced row echelon form involves two free variables, x_3 and x_4.

$$x_1 = x_3 - x_4$$
$$x_2 = -2x_3 + x_4$$

Thus, if we set $x_3 = \alpha$ and $x_4 = \beta$, then

$$\mathbf{x} = \begin{pmatrix} \alpha - \beta \\ -2\alpha + \beta \\ \alpha \\ \beta \end{pmatrix} = \alpha \begin{pmatrix} 1 \\ -2 \\ 1 \\ 0 \end{pmatrix} + \beta \begin{pmatrix} -1 \\ 1 \\ 0 \\ 1 \end{pmatrix}$$

is a solution to $A\mathbf{x} = \mathbf{0}$. The vector space $N(A)$ consists of all vectors of the form

$$\alpha \begin{pmatrix} 1 \\ -2 \\ 1 \\ 0 \end{pmatrix} + \beta \begin{pmatrix} -1 \\ 1 \\ 0 \\ 1 \end{pmatrix}$$

where α and β are scalars. □

Definition. Let $\mathbf{v}_1, \mathbf{v}_2, \ldots, \mathbf{v}_n$ be vectors in a vector space V. A sum of the form $\alpha_1 \mathbf{v}_1 + \alpha_2 \mathbf{v}_2 + \cdots + \alpha_n \mathbf{v}_n$, where $\alpha_1, \ldots, \alpha_n$ are scalars, is called a **linear combination** of $\mathbf{v}_1, \mathbf{v}_2, \ldots, \mathbf{v}_n$. The set of all linear combinations of $\mathbf{v}_1, \mathbf{v}_2, \ldots, \mathbf{v}_n$ is called the **span** of $\mathbf{v}_1, \ldots, \mathbf{v}_n$. The span of $\mathbf{v}_1, \ldots, \mathbf{v}_n$ will be denoted by $\mathbb{S}(\mathbf{v}_1, \ldots, \mathbf{v}_n)$.

In Example 8, we saw that the nullspace of A was the span of the vectors $(1, -2, 1, 0)^T$ and $(-1, 1, 0, 1)^T$.

EXAMPLE 9.   In R^3 the span of $\mathbf{e}_1$ and $\mathbf{e}_2$ is the set of all vectors of the form

$$\alpha\mathbf{e}_1 + \beta\mathbf{e}_2 = \begin{pmatrix} \alpha \\ \beta \\ 0 \end{pmatrix}$$

The reader may verify that $\mathbb{S}(\mathbf{e}_1, \mathbf{e}_2)$ is a subspace of R^3. The span of $\mathbf{e}_1, \mathbf{e}_2, \mathbf{e}_3$ is the set of all vectors of the form

$$\alpha_1\mathbf{e}_1 + \alpha_2\mathbf{e}_2 + \alpha_3\mathbf{e}_3 = \begin{pmatrix} \alpha_1 \\ \alpha_2 \\ \alpha_3 \end{pmatrix}$$

Thus $\mathbb{S}(\mathbf{e}_1, \mathbf{e}_2, \mathbf{e}_3) = R^3$.   □

Theorem 3.2.1.   *If* $\mathbf{v}_1, \mathbf{v}_2, \ldots, \mathbf{v}_n$ *are elements of a vector space* V, *then* $\mathbb{S}(\mathbf{v}_1, \mathbf{v}_2, \ldots, \mathbf{v}_n)$ *is a subspace of* V.

Proof.   Let β be a scalar and let $\mathbf{v} = \alpha_1\mathbf{v}_1 + \alpha_2\mathbf{v}_2 + \cdots + \alpha_n\mathbf{v}_n$ be an arbitrary element of $\mathbb{S}(\mathbf{v}_1, \mathbf{v}_2, \ldots, \mathbf{v}_n)$. Since

$$\beta\mathbf{v} = (\beta\alpha_1)\mathbf{v}_1 + (\beta\alpha_2)\mathbf{v}_2 + \cdots + (\beta\alpha_n)\mathbf{v}_n$$

it follows that $\beta\mathbf{v} \in \mathbb{S}(\mathbf{v}_1, \ldots, \mathbf{v}_n)$. Next we must show that any sum of elements of $\mathbb{S}(\mathbf{v}_1, \ldots, \mathbf{v}_n)$ is in $\mathbb{S}(\mathbf{v}_1, \ldots, \mathbf{v}_n)$. Let $\mathbf{v} = \alpha_1\mathbf{v}_1 + \cdots + \alpha_n\mathbf{v}_n$ and $\mathbf{w} = \beta_1\mathbf{v}_1 + \cdots + \beta_n\mathbf{v}_n$,

$$\mathbf{v} + \mathbf{w} = (\alpha_1 + \beta_1)\mathbf{v}_1 + \cdots + (\alpha_n + \beta_n)\mathbf{v}_n \in \mathbb{S}(\mathbf{v}_1, \ldots, \mathbf{v}_n)$$

Therefore, $\mathbb{S}(\mathbf{v}_1, \ldots, \mathbf{v}_n)$ is a subspace of V.   □

We will refer to $\mathbb{S}(\mathbf{v}_1, \ldots, \mathbf{v}_n)$ as the subspace spanned by $\mathbf{v}_1, \mathbf{v}_2, \ldots, \mathbf{v}_n$. It may happen that $\mathbb{S}(\mathbf{v}_1, \ldots, \mathbf{v}_n) = V$, in which case we say that $\mathbf{v}_1, \ldots, \mathbf{v}_n$ span V or that $\{\mathbf{v}_1, \ldots, \mathbf{v}_n\}$ is a spanning set for V. Thus we have the following definition.

Definition.   The set $\{\mathbf{v}_1, \ldots, \mathbf{v}_n\}$ is a **spanning set** for V if and only if every vector in V can be written as a linear combination of $\mathbf{v}_1, \mathbf{v}_2, \ldots, \mathbf{v}_n$.

EXAMPLE 10.   Which of the following are spanning sets for R^3?
(a) $\{\mathbf{e}_1, \mathbf{e}_2, \mathbf{e}_3, (1, 2, 3)^T\}$
(b) $\{(1, 1, 1)^T, (1, 1, 0)^T, (1, 0, 0)^T\}$
(c) $\{(1, 0, 1)^T, (0, 1, 0)^T\}$
(d) $\{(1, 2, 4)^T, (2, 1, 3)^T, (4, -1, 1)^T\}$

SOLUTION.   To determine whether a set spans R^3, one must determine whether an arbitrary vector $(a, b, c)^T$ in R^3 can be written as a linear combination of the vectors in the set. In part (a) it is easily seen that $(a, b, c)^T$ can be written as

$$(a, b, c)^T = a\mathbf{e}_1 + b\mathbf{e}_2 + c\mathbf{e}_3 + 0(1, 2, 3)^T$$

For part (b) we must determine whether or not it is possible to find constants $\alpha_1, \alpha_2, \alpha_3$ such that

$$\begin{pmatrix} a \\ b \\ c \end{pmatrix} = \alpha_1 \begin{pmatrix} 1 \\ 1 \\ 1 \end{pmatrix} + \alpha_2 \begin{pmatrix} 1 \\ 1 \\ 0 \end{pmatrix} + \alpha_3 \begin{pmatrix} 1 \\ 0 \\ 0 \end{pmatrix}$$

This leads to the system of equations

$$\alpha_1 + \alpha_2 + \alpha_3 = a$$
$$\alpha_1 + \alpha_2 \qquad = b$$
$$\alpha_1 \qquad\qquad = c$$

Since the coefficient matrix of the system is nonsingular, the system has a unique solution. In fact, we find that

$$\begin{pmatrix} \alpha_1 \\ \alpha_2 \\ \alpha_3 \end{pmatrix} = \begin{pmatrix} c \\ b - c \\ a - b \end{pmatrix}$$

Thus

$$\begin{pmatrix} a \\ b \\ c \end{pmatrix} = c \begin{pmatrix} 1 \\ 1 \\ 1 \end{pmatrix} + (b - c) \begin{pmatrix} 1 \\ 1 \\ 0 \end{pmatrix} + (a - b) \begin{pmatrix} 1 \\ 0 \\ 0 \end{pmatrix}$$

so the three vectors span R^3.

For part (c), one should note that linear combinations of $(1, 0, 1)^T$ and $(0, 1, 0)^T$ produce vectors of the form $(\alpha, \beta, \alpha)^T$. Thus any vector $(a, b, c)^T$ in R^3 where $a \neq c$ would not be in the span of these two vectors.

Part (d) can be done in the same manner as part (b). If

$$\begin{pmatrix} a \\ b \\ c \end{pmatrix} = \alpha_1 \begin{pmatrix} 1 \\ 2 \\ 4 \end{pmatrix} + \alpha_2 \begin{pmatrix} 2 \\ 1 \\ 3 \end{pmatrix} + \alpha_3 \begin{pmatrix} 4 \\ -1 \\ 1 \end{pmatrix}$$

then

$$\alpha_1 + 2\alpha_2 + 4\alpha_3 = a$$
$$2\alpha_1 + \alpha_2 - \alpha_3 = b$$
$$4\alpha_1 + 3\alpha_2 + \alpha_3 = c$$

In this case, however, the coefficient matrix is singular. Gaussian elimination will yield a system of the form

$$\alpha_1 + \alpha_2 + 4\alpha_3 = a$$
$$\alpha_2 + 3\alpha_3 = \frac{2a - b}{3}$$
$$0 = 2a - 3c + 5b$$

If

$$2a - 3c + 5b \neq 0$$

then the system is inconsistent. Hence, for most choices of a, b, c, it is impossible to express $(a, b, c)^T$ as a linear combination of $(1, 2, 4)^T, (2, 1, 3)^T, (4, -1, 1)^T$. The vectors do not span R^3. □

EXAMPLE 11. The vectors $1 - x^2$, $x + 2$, and x^2 span P_3. If $ax^2 + bx + c$ is any polynomial in P_3, then it is possible to find scalars α_1, α_2, and α_3 such that

$$ax^2 + bx + c = \alpha_1(1 - x^2) + \alpha_2(x + 2) + \alpha_3 x^2$$

Indeed,

$$\alpha_1(1 - x^2) + \alpha_2(x + 2) + \alpha_3 x^2$$
$$= (\alpha_3 - \alpha_1)x^2 + \alpha_2 x + (\alpha_1 + 2\alpha_2)$$

Setting

$$\alpha_3 - \alpha_1 = a$$
$$\alpha_2 = b$$
$$\alpha_1 + 2\alpha_2 = c$$

and solving yields $\alpha_1 = c - 2b$, $\alpha_2 = b$, and $\alpha_3 = a + c - 2b$. □

In Example 10(a) we saw that the vectors $e_1, e_2, e_3, (1, 2, 3)^T$ span R^3. Clearly, R^3 could be spanned using only the vectors e_1, e_2, e_3. The vector $(1, 2, 3)^T$ is really not necessary. In the next section we will consider the problem of finding minimal spanning sets for a vector space V (i.e., spanning sets that contain the smallest possible number of vectors).

EXERCISES

1. Determine whether or not the following are subspaces of R^2.
   (a) $\{(x_1, x_2)^T | x_1 + x_2 = 0\}$
   (b) $\{(x_1, x_2)^T | x_1 x_2 = 0\}$
   (c) $\{(x_1, x_2)^T | x_1 = 3x_2\}$
   (d) $\{(x_1, x_2)^T | x_1 = 3x_2 + 1\}$

2. Determine whether or not the following sets are subspaces of R^3.
   (a) $\{(x_1, x_2, x_3)^T | x_1 + x_3 = 1\}$
   (b) $\{(x_1, x_2, x_3)^T | x_1 = x_2 = x_3\}$
   (c) $\{(x_1, x_2, x_3)^T | x_3 = x_1 + x_2\}$
   (d) $\{(x_1, x_2, x_3)^T | x_3 = x_1^2 + x_2^2\}$

3. Determine whether or not the following are subspaces of $R^{2 \times 2}$
   (a) The set of all 2×2 diagonal matrices
   (b) The set of all 2×2 lower triangular matrices
   (c) The set of all 2×2 matrices A such that $a_{12} = 1$
   (d) The set of all 2×2 matrices B such that $b_{11} = 0$
   (e) The set of all symmetric 2×2 matrices
   (f) The set of all nonsingular 2×2 matrices

4. Determine the nullspace of each of the following matrices.

   (a) $\begin{pmatrix} 2 & 1 \\ 3 & 2 \end{pmatrix}$
   (b) $\begin{pmatrix} 1 & 2 & -3 & -1 \\ -2 & -4 & 6 & 3 \end{pmatrix}$

   (c) $\begin{pmatrix} 1 & 3 & -4 \\ 2 & -1 & -1 \\ -1 & -3 & 4 \end{pmatrix}$
   (d) $\begin{pmatrix} 1 & 1 & -1 & 2 \\ 2 & 2 & -3 & 1 \\ -1 & -1 & 0 & -5 \end{pmatrix}$

5. Determine whether or not the following are subspaces of P_4. (Be careful!)
   (a) The set of polynomials in P_4 of even degree
   (b) The set of all polynomials of degree 3
   (c) The set of all polynomials $p(x)$ in P_4 such that $p(0) = 0$
   (d) The set of all polynomials in P_4 having at least one real root

6. Determine whether or not the following are subspaces of $C[-1, 1]$.
   (a) The set of functions f in $C[-1, 1]$ such that $f(-1) = f(1)$
   (b) The set of odd functions in $C[-1, 1]$
   (c) The set of continuous nondecreasing functions on $[-1, 1]$
   (d) The set of functions f in $C[-1, 1]$ such that $f(-1) = 0$ *and* $f(1) = 0$
   (e) The set of functions f in $C[-1, 1]$ such that $f(-1) = 0$ *or* $f(1) = 0$

7. Determine whether or not the following are spanning sets for R^2.

   (a) $\left\{ \begin{pmatrix} 2 \\ 1 \end{pmatrix}, \begin{pmatrix} 3 \\ 2 \end{pmatrix} \right\}$
   (b) $\left\{ \begin{pmatrix} 2 \\ 3 \end{pmatrix}, \begin{pmatrix} 4 \\ 6 \end{pmatrix} \right\}$

   (c) $\left\{ \begin{pmatrix} -2 \\ 1 \end{pmatrix}, \begin{pmatrix} 1 \\ 3 \end{pmatrix}, \begin{pmatrix} 2 \\ 4 \end{pmatrix} \right\}$
   (d) $\left\{ \begin{pmatrix} -1 \\ 2 \end{pmatrix}, \begin{pmatrix} 1 \\ -2 \end{pmatrix}, \begin{pmatrix} 2 \\ -4 \end{pmatrix} \right\}$

   (e) $\left\{ \begin{pmatrix} 1 \\ 2 \end{pmatrix}, \begin{pmatrix} -1 \\ 1 \end{pmatrix} \right\}$

8. Which of the following are spanning sets for R^3? Justify your answers.
   (a) $\{(1, 0, 0)^T, (0, 1, 1)^T, (1, 0, 1)^T\}$
   (b) $\{(1, 0, 0)^T, (0, 1, 1)^T, (1, 0, 1)^T, (1, 2, 3)^T\}$

    (c) $\{(2, 1, -2)^T, (3, 2, -2)^T, (2, 2, 0)^T\}$
    (d) $\{(2, 1, -2)^T, (-2, -1, 2)^T, (4, 2, -4)^T\}$
    (e) $\{(1, 1, 3)^T, (0, 2, 1)^T\}$

9. Which of the following are spanning sets for P_3? Justify your answers.
    (a) $\{1, x^2, x^2 - 2\}$ (b) $\{2, x^2, x, 2x + 3\}$
    (c) $\{x + 2, x + 1, x^2 - 1\}$ (d) $\{x + 2, x^2 - 1\}$

10. In $R^{2 \times 2}$ let

$$E_{11} = \begin{pmatrix} 1 & 0 \\ 0 & 0 \end{pmatrix}, \qquad E_{12} = \begin{pmatrix} 0 & 1 \\ 0 & 0 \end{pmatrix}$$

$$E_{21} = \begin{pmatrix} 0 & 0 \\ 1 & 0 \end{pmatrix}, \qquad E_{22} = \begin{pmatrix} 0 & 0 \\ 0 & 1 \end{pmatrix}$$

Show that $E_{11}, E_{12}, E_{21}, E_{22}$ span $R^{2 \times 2}$.

11. Let S be the vector space of infinite sequences defined in Exercise 10 of Section 1. Let S_0 be the set of $\{a_n\}$ with the property $a_n \to 0$ as $n \to \infty$. Show that S_0 is a subspace of S.

12. Prove that if S is a subspace of R^1, either $S = \{0\}$ or $S = R^1$.

13. Let A be an $n \times n$ matrix. Prove that the following statements are equivalent.
    (a) $N(A) = \{0\}$
    (b) A is nonsingular.
    (c) For each $\mathbf{b} \in R^n$, the system $A\mathbf{x} = \mathbf{b}$ has a unique solution.

14. Let U and V be subspaces of a vector space W. Prove that $U \cap V$ is also a subspace of W.

15. Let S be the subspace of R^2 spanned by $\mathbf{e}_1$ and let T be the subspace of R^2 spanned by $\mathbf{e}_2$. Is $S \cup T$ a subspace of R^2? Explain.

16. Let U and V be subspaces of a vector space W. Define
    $U + V = \{\mathbf{z} | \mathbf{z} = \mathbf{u} + \mathbf{v} \text{ where } \mathbf{u} \in U \text{ and } \mathbf{v} \in V\}$
Show that $U + V$ is a subspace of W.

3. LINEAR INDEPENDENCE

In this section we look more closely at the structure of vector spaces. To begin with, we restrict ourselves to vector spaces that can be generated from a finite set of elements. Each vector in the vector space can be built up from the elements in this generating set using only the operations of addition and scalar multiplication. The generating set is usually referred to as a spanning set. In particular, it is desirable to find a "minimal" spanning set. To do this, it is necessary to consider how the vectors in the collection "depend" on each other. Consequently, we introduce the concepts of linear dependence and linear independence. These simple concepts provide the keys to understanding the structure of vector spaces.

Consider the following vectors in R^3:

$$\mathbf{x}_1 = \begin{pmatrix} 1 \\ -1 \\ 2 \end{pmatrix}, \qquad \mathbf{x}_2 = \begin{pmatrix} -2 \\ 3 \\ 1 \end{pmatrix}, \qquad \mathbf{x}_3 = \begin{pmatrix} -1 \\ 3 \\ 8 \end{pmatrix}$$

Let S be the subspace of R^3 spanned by $\mathbf{x}_1, \mathbf{x}_2, \mathbf{x}_3$. Actually, S can be represented in terms of the two vectors $\mathbf{x}_1$ and $\mathbf{x}_2$, since the vector $\mathbf{x}_3$ is already in the span of $\mathbf{x}_1$ and $\mathbf{x}_2$.

(1) $$\mathbf{x}_3 = 3\mathbf{x}_1 + 2\mathbf{x}_2$$

Any linear combination of $\mathbf{x}_1, \mathbf{x}_2, \mathbf{x}_3$ can be reduced to a linear combination of $\mathbf{x}_1$ and $\mathbf{x}_2$:

$$\alpha_1\mathbf{x}_1 + \alpha_2\mathbf{x}_2 + \alpha_3\mathbf{x}_3 = \alpha_1\mathbf{x}_1 + \alpha_2\mathbf{x}_2 + \alpha_3(3\mathbf{x}_1 + 2\mathbf{x}_2)$$
$$= (\alpha_1 + 3\alpha_3)\mathbf{x}_1 + (\alpha_2 + 2\alpha_3)\mathbf{x}_2$$

Thus

$$S = \mathbb{S}\{\mathbf{x}_1, \mathbf{x}_2, \mathbf{x}_3\} = \mathbb{S}\{\mathbf{x}_1, \mathbf{x}_2\}$$

Equation (1) can be rewritten in the form

(2) $$3\mathbf{x}_1 + 2\mathbf{x}_2 - 1\mathbf{x}_3 = \mathbf{0}$$

Since the three coefficients in (2) are nonzero, one could solve for any vector in terms of the other two.

$$\mathbf{x}_1 = -\tfrac{2}{3}\mathbf{x}_2 + \tfrac{1}{3}\mathbf{x}_3, \qquad \mathbf{x}_2 = -\tfrac{3}{2}\mathbf{x}_1 + \tfrac{1}{2}\mathbf{x}_3, \qquad \mathbf{x}_3 = 3\mathbf{x}_1 + 2\mathbf{x}_2$$

It follows that

$$\mathbb{S}\{\mathbf{x}_1, \mathbf{x}_2, \mathbf{x}_3\} = \mathbb{S}\{\mathbf{x}_2, \mathbf{x}_3\} = \mathbb{S}\{\mathbf{x}_1, \mathbf{x}_3\} = \mathbb{S}\{\mathbf{x}_1, \mathbf{x}_2\}$$

Because of the dependency relation (2), the subspace S can be represented as the span of any two of the given vectors.

On the other hand, no such dependency relationship exists between $\mathbf{x}_1$ and $\mathbf{x}_2$. Indeed, if there were scalars c_1 and c_2, not both 0, such that

(3) $$c_1\mathbf{x}_1 + c_2\mathbf{x}_2 = \mathbf{0}$$

then we could solve for one of the vectors in terms of the other.

$$\mathbf{x}_1 = -\frac{c_2}{c_1}\mathbf{x}_2 \quad (c_1 \neq 0) \qquad \text{or} \qquad \mathbf{x}_2 = -\frac{c_1}{c_2}\mathbf{x}_1 \quad (c_2 \neq 0)$$

Clearly, neither of the two vectors in question is a multiple of the other.

Thus $\mathbb{S}\{\mathbf{x}_1\}$ and $\mathbb{S}\{\mathbf{x}_2\}$ are both proper subspaces of $\mathbb{S}\{\mathbf{x}_1, \mathbf{x}_2\}$ and the only way that (3) can hold is if $c_1 = c_2 = 0$.

One can generalize this example by making the following observations.

(i) If $\mathbf{v}_1, \mathbf{v}_2, \ldots, \mathbf{v}_n$ span a vector space V and one of these vectors can be written as a linear combination of the other $n - 1$ vectors, then those $n - 1$ vectors span V.

(ii) Given n vectors $\mathbf{v}_1, \ldots, \mathbf{v}_n$, it is possible to write one of the vectors as a linear combination of the other $n - 1$ vectors if and only if there exists scalars $c_1, \ldots, c_n$ not all zero such that

$$c_1\mathbf{v}_1 + c_2\mathbf{v}_2 + \cdots + c_n\mathbf{v}_n = \mathbf{0}$$

Proof of (i). Suppose that $\mathbf{v}_n$ can be written as a linear combination of $\mathbf{v}_1, \mathbf{v}_2, \ldots, \mathbf{v}_{n-1}$.

$$\mathbf{v}_n = \beta_1\mathbf{v}_1 + \beta_2\mathbf{v}_2 + \cdots + \beta_{n-1}\mathbf{v}_{n-1}$$

Let $\mathbf{v}$ be any element of V. Since $\mathbf{v}_1, \ldots, \mathbf{v}_n$ span V, we can write

$$\mathbf{v} = \alpha_1\mathbf{v}_1 + \alpha_2\mathbf{v}_2 + \cdots + \alpha_{n-1}\mathbf{v}_{n-1} + \alpha_n\mathbf{v}_n$$
$$= \alpha_1\mathbf{v}_1 + \alpha_2\mathbf{v}_2 + \cdots + \alpha_{n-1}\mathbf{v}_{n-1} + \alpha_n(\beta_1\mathbf{v}_1 + \cdots + \beta_{n-1}\mathbf{v}_{n-1})$$
$$= (\alpha_1 + \alpha_n\beta_1)\mathbf{v}_1 + (\alpha_2 + \alpha_n\beta_2)\mathbf{v}_2 + \cdots + (\alpha_{n-1} + \alpha_n\beta_{n-1})\mathbf{v}_{n-1}$$

Thus any vector $\mathbf{v}$ in V can be written as a linear combination of $\mathbf{v}_1, \mathbf{v}_2, \ldots, \mathbf{v}_{n-1}$, and hence these vectors span V. $\qquad\square$

Proof of (ii). Suppose that one of the vectors $\mathbf{v}_1, \mathbf{v}_2, \ldots, \mathbf{v}_n$, say $\mathbf{v}_n$, can be written as a linear combination of the others.

$$\mathbf{v}_n = \alpha_1\mathbf{v}_1 + \alpha_2\mathbf{v}_2 + \cdots + \alpha_{n-1}\mathbf{v}_{n-1}$$

If we set $c_i = \alpha_i$ for $i = 1, \ldots, n - 1$ and $c_n = -1$, then it follows that

$$\sum_{i=1}^{n} c_i\mathbf{v}_i = \sum_{i=1}^{n-1} \alpha_i\mathbf{v}_i - \sum_{i=1}^{n-1} \alpha_i\mathbf{v}_i = \mathbf{0}$$

Conversely, if

$$c_1\mathbf{v}_1 + c_2\mathbf{v}_2 + \cdots + c_n\mathbf{v}_n = \mathbf{0}$$

and at least one of the c_i's, say c_n, is nonzero, then

$$\mathbf{v}_n = \frac{-c_1}{c_n}\mathbf{v}_1 + \frac{-c_2}{c_n}\mathbf{v}_2 + \cdots + \frac{-c_{n-1}}{c_n}\mathbf{v}_{n-1} \qquad\square$$

Definition. The vectors $\mathbf{v}_1, \mathbf{v}_2, \ldots, \mathbf{v}_n$ in a vector space V are said to be **linearly independent** if

$$c_1\mathbf{v}_1 + c_2\mathbf{v}_2 + \cdots + c_n\mathbf{v}_n = \mathbf{0}$$

implies that all of the scalars $c_1, \ldots, c_n$ must equal 0.

It follows from (i) and (ii) that if $\{\mathbf{v}_1, \mathbf{v}_2, \ldots, \mathbf{v}_n\}$ is a minimal spanning set, then $\mathbf{v}_1, \mathbf{v}_2, \ldots, \mathbf{v}_n$ are linearly independent. Conversely, if $\mathbf{v}_1, \ldots, \mathbf{v}_n$ are linearly independent and span V, then $\{\mathbf{v}_1, \ldots, \mathbf{v}_n\}$ is a minimal spanning set for V (see Exercise 14). A minimal spanning set is called a *basis*. The concept of a basis will be studied in more detail in the next section.

EXAMPLE 1. The vectors $\begin{pmatrix} 1 \\ 1 \end{pmatrix}$ and $\begin{pmatrix} 1 \\ 2 \end{pmatrix}$ are linearly independent, since if

$$c_1 \begin{pmatrix} 1 \\ 1 \end{pmatrix} + c_2 \begin{pmatrix} 1 \\ 2 \end{pmatrix} = \begin{pmatrix} 0 \\ 0 \end{pmatrix}$$

then

$$c_1 + c_2 = 0$$
$$c_1 + 2c_2 = 0$$

and the only solution to this system is $c_1 = 0, c_2 = 0$. $\square$

Definition. The vectors $\mathbf{v}_1, \mathbf{v}_2, \ldots, \mathbf{v}_n$ in a vector space V are said to be **linearly dependent** if there exist scalars $c_1, c_2, \ldots, c_n$ not all zero such that

$$c_1 \mathbf{v}_1 + c_2 \mathbf{v}_2 + \cdots + c_n \mathbf{v}_n = \mathbf{0}$$

EXAMPLE 2. Let $\mathbf{x} = (1, 2, 3)^T$. The vectors $\mathbf{e}_1, \mathbf{e}_2, \mathbf{e}_3, \mathbf{x}$ are linearly dependent, since

$$\mathbf{e}_1 + 2\mathbf{e}_2 + 3\mathbf{e}_3 - \mathbf{x} = \mathbf{0}$$

(In this case $c_1 = 1$, $c_2 = 2$, $c_3 = 3$, $c_4 = -1$.) $\square$

If $\mathbf{x}$ and $\mathbf{y}$ are linearly dependent in R^2, then

$$c_1 \mathbf{x} + c_2 \mathbf{y} = \mathbf{0}$$

where c_1 and c_2 are not both 0. If, say, $c_1 \neq 0$, we can write

$$\mathbf{x} = -\frac{c_2}{c_1} \mathbf{y}$$

If two vectors in R^2 are linearly dependent, one of the vectors can be written as a scalar multiple of the other. Thus, if both vectors are placed at the origin, they will lie along the same line (see Figure 3.3.1). If

$$\mathbf{x} = \begin{pmatrix} x_1 \\ x_2 \\ x_3 \end{pmatrix} \qquad \text{and} \qquad \mathbf{y} = \begin{pmatrix} y_1 \\ y_2 \\ y_3 \end{pmatrix}$$

are linearly independent in R^3, then the points (x_1, x_2, x_3) and

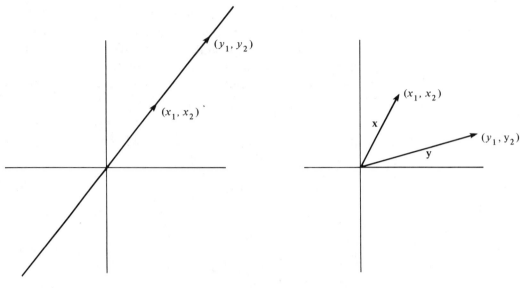

(a) **x** and **y** linearly dependent.

(b) **x** and **y** linearly independent.

FIGURE 3.3.1

(y_1, y_2, y_3) do not lie on the same line through the origin in 3-space. Since $(0, 0, 0)$, (x_1, x_2, x_3), and (y_1, y_2, y_3) are not collinear, they determine a plane. If (z_1, z_2, z_3) lies on this plane, the vector $\mathbf{z} = (z_1, z_2, z_3)^T$ can be written as a linear combination of **x** and **y**, and hence **x**, **y**, and **z** are linearly dependent. If (z_1, z_2, z_3) does not lie on the plane, the three vectors will be linearly independent (see Figure 3.3.2).

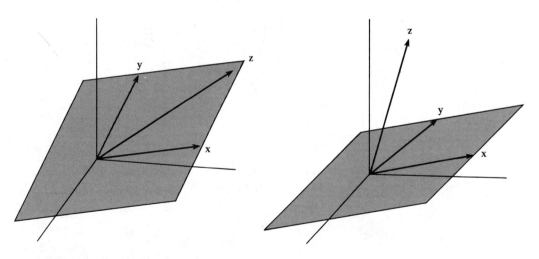

(a) **x**, **y**, **z** linearly dependent.

(b) **x**, **y**, **z** linearly independent.

FIGURE 3.3.2

EXAMPLE 3. Which of the following collections of vectors are linearly independent in R^3?

(a) $(1,1,1)^T, (1,1,0)^T, (1,0,0)^T$.

(b) $(1,0,1)^T, (0,1,0)^T$.

(c) $(1,2,4)^T, (2,1,3)^T, (4,-1,1)^T$.

SOLUTION

(a) These vectors are linearly independent. If

$$c_1(1,1,1)^T + c_2(1,1,0)^T + c_3(1,0,0)^T = (0,0,0)^T$$

then

$$c_1 + c_2 + c_3 = 0$$
$$c_1 + c_2 \quad\quad = 0$$
$$c_1 \quad\quad\quad\quad = 0$$

and the only solution to this system is $c_1 = 0$, $c_2 = 0$, $c_3 = 0$.

(b) If

$$c_1(1,0,1)^T + c_2(0,1,0)^T = (0,0,0)^T$$

then

$$(c_1, c_2, c_1)^T = (0,0,0)^T$$

so $c_1 = c_2 = 0$. Therefore, the two vectors are linearly independent.

(c) If

$$c_1(1,2,4)^T + c_2(2,1,3)^T + c_3(4,-1,1)^T = (0,0,0)^T$$

then

$$c_1 + 2c_2 + 4c_3 = 0$$
$$2c_1 + c_2 - c_3 = 0$$
$$4c_1 + 3c_2 + c_3 = 0$$

The coefficient matrix of this system is singular. Hence the system has nontrivial solutions, and hence the vectors are linearly dependent. □

Notice in Example 3, parts (a) and (c), that it was necessary to solve a 3×3 system to determine whether or not the three vectors were linearly independent. In part (a), where the coefficient matrix was nonsingular, the vectors were linearly independent, while in part (c), where the coefficient matrix was singular, the vectors were linearly dependent. This illustrates a special case of the following theorem.

Theorem 3.3.1. *Let* $\mathbf{x}_1, \mathbf{x}_2, \ldots, \mathbf{x}_n$ *be n vectors in R^n with* $\mathbf{x}_i = (x_{1i}, x_{2i}, \ldots, x_{ni})^T$ *for $i = 1, \ldots, n$. Let $X = (x_{ij})$; that is, the $\mathbf{x}_i$'s form* *the columns of X. The vectors* $x_1, x_2, \ldots, x_n$ *will be linearly dependent if and only if X is singular.*

Proof. The equation

$$c_1x_1 + c_2x_2 + \cdots + c_nx_n = 0$$

is equivalent to the system of equations

$$c_1x_{11} + c_2x_{12} + \cdots + c_nx_{1n} = 0$$
$$c_1x_{21} + c_2x_{22} + \cdots + c_nx_{2n} = 0$$
$$\vdots$$
$$c_1x_{n1} + c_2x_{n2} + \cdots + c_nx_{nn} = 0$$

If we let $c = (c_1, c_2, \ldots, c_n)^T$, the system can be written as a matrix equation

$$Xc = 0$$

This equation will have a nontrivial solution if and only if X is singular. Thus $x_1, \ldots, x_n$ will be linearly dependent if and only if X is singular. $\square$

One can use Theorem 3.3.1 to test whether n vectors are linearly independent in R^n. One simply forms a matrix X whose columns are the vectors being tested. To determine whether or not X is singular, one can calculate $\det(X)$. If $\det(X) = 0$, the vectors are linearly dependent. If $\det(X) \neq 0$, the vectors are linearly independent.

EXAMPLE 4. Determine whether or not the vectors $(4, 2, 3)^T$, $(2, 3, 1)^T$, and $(2, -5, 3)^T$ are linearly dependent.

SOLUTION.   Since

$$\begin{vmatrix} 4 & 2 & 2 \\ 2 & 3 & -5 \\ 3 & 1 & 3 \end{vmatrix} = 0$$

the vectors are linearly dependent. $\square$

Determinants can also be used to help determine if a set of vectors is linearly independent in $C^{(n-1)}[a, b]$. Indeed, let $f_1, f_2, \ldots, f_n$ be elements of $C^{(n-1)}[a, b]$. If these vectors are linearly dependent, then there exist scalars $c_1, c_2, \ldots, c_n$ not all zero such that

(4) $$c_1f_1(x) + c_2f_2(x) + \cdots + c_nf_n(x) = 0$$

for each x in $[a, b]$. Taking the derivative with respect to x of both sides of (4) yields

$$c_1f_1'(x) + c_2f_2'(x) + \cdots + c_nf_n'(x) = 0$$

If we continue taking derivatives of both sides, we end up with the system

$$
\begin{aligned}
c_1 f_1(x) &+ c_2 f_2(x) + \cdots + c_n f_n(x) = 0 \\
c_1 f_1'(x) &+ c_2 f_2'(x) + \cdots + c_n f_n'(x) = 0 \\
&\;\;\vdots \\
c_1 f_1^{(n-1)}(x) &+ c_2 f_2^{(n-1)}(x) + \cdots + c_n f_n^{(n-1)}(x) = 0
\end{aligned}
$$

Thus, for each fixed x in $[a, b]$, the matrix equation

$$
(5) \quad
\begin{pmatrix}
f_1(x) & f_2(x) & \cdots & f_n(x) \\
f_1'(x) & f_2'(x) & \cdots & f_n'(x) \\
\vdots & & & \\
f_1^{(n-1)}(x) & f_2^{(n-1)}(x) & \cdots & f_n^{(n-1)}(x)
\end{pmatrix}
\begin{pmatrix}
\alpha_1 \\ \alpha_2 \\ \vdots \\ \alpha_n
\end{pmatrix}
=
\begin{pmatrix}
0 \\ 0 \\ \vdots \\ 0
\end{pmatrix}
$$

has a nontrivial solution $(c_1, c_2, \ldots, c_n)^T$. Thus, if $f_1, \ldots, f_n$ are linearly dependent in $C^{(n-1)}[a, b]$, then for each fixed x in $[a, b]$, the coefficient matrix of the system (5) is singular. If the matrix is singular, its determinant is zero.

Definition. Let $f_1, f_2, \ldots, f_n$ be functions in $C^{(n-1)}[a, b]$ and define the function $W[f_1, f_2, \ldots, f_n](x)$ on $[a, b]$ by

$$
W[f_1, f_2, \ldots, f_n](x) =
\begin{vmatrix}
f_1(x) & f_2(x) & \cdots & f_n(x) \\
f_1'(x) & f_2'(x) & \cdots & f_n'(x) \\
\vdots & & & \\
f_1^{(n-1)}(x) & f_2^{(n-1)}(x) & \cdots & f_n^{(n-1)}(x)
\end{vmatrix}
$$

The function $W[f_1, f_2, \ldots, f_n]$ is called the **Wronskian** of $f_1, f_2, \ldots, f_n$.

Theorem 3.3.2. *Let $f_1, f_2, \ldots, f_n$ be elements of $C^{(n-1)}[a, b]$. If there exists a point x_0 in $[a, b]$ such that $W[f_1, f_2, \ldots, f_n](x_0) \neq 0$, then $f_1, f_2, \ldots, f_n$ are linearly independent.*

Proof. If $f_1, \ldots, f_n$ were linearly dependent, then by the preceding discussion the coefficient matrix in (5) would be singular for each x in $[a, b]$ and hence $W[f_1, f_2, \ldots, f_n](x)$ would be identically zero on $[a, b]$. $\square$

If $f_1, f_2, \ldots, f_n$ are linearly independent in $C^{(n-1)}[a, b]$, they will also be linearly independent in $C[a, b]$.

EXAMPLE 5. Show that e^x and e^{-x} are linearly independent in $C(-\infty, \infty)$.

SOLUTION

$$W[e^x, e^{-x}] = \begin{vmatrix} e^x & e^{-x} \\ e^x & -e^{-x} \end{vmatrix} = -2$$

Since $W[e^x, e^{-x}]$ is not identically zero, e^x and e^{-x} are linearly independent.   □

EXAMPLE 6.   It is easily seen that the functions x^2 and $x|x|$ are linearly independent in $C[-1, 1]$. If

$$c_1 x^2 + c_2 x|x| = 0$$

for all x in $[-1, 1]$, then in particular for $x = 1$ and $x = -1$, we have

$$c_1 + c_2 = 0$$
$$c_1 - c_2 = 0$$

and hence $c_1 = c_2 = 0$. The Wronskian, however, is identically zero.

$$W[x^2, x|x|](x) = \begin{vmatrix} x^2 & x|x| \\ 2x & 2|x| \end{vmatrix} = 0$$

This example shows that the converse to Theorem 3.3.2 is not valid.   □

EXAMPLE 7.   Show that the vectors $1, x, x^2, x^3$ are linearly independent in P_4.

SOLUTION

$$W[1, x, x^2, x^3] = \begin{vmatrix} 1 & x & x^2 & x^3 \\ 0 & 1 & 2x & 3x^2 \\ 0 & 0 & 2 & 6x \\ 0 & 0 & 0 & 6 \end{vmatrix} = 12$$

Since $W[1, x, x^2, x^3] \neq 0$, the vectors are linearly independent.   □

EXERCISES

1. Determine whether or not the following vectors are linearly independent in R^2.

(a) $\begin{pmatrix} 2 \\ 1 \end{pmatrix}, \begin{pmatrix} 3 \\ 2 \end{pmatrix}$ (b) $\begin{pmatrix} 2 \\ 3 \end{pmatrix}, \begin{pmatrix} 4 \\ 6 \end{pmatrix}$

(c) $\begin{pmatrix} -2 \\ 1 \end{pmatrix}, \begin{pmatrix} 1 \\ 3 \end{pmatrix}, \begin{pmatrix} 2 \\ 4 \end{pmatrix}$ (d) $\begin{pmatrix} -1 \\ 2 \end{pmatrix}, \begin{pmatrix} 1 \\ -2 \end{pmatrix}, \begin{pmatrix} 2 \\ -4 \end{pmatrix}$

(e) $\begin{pmatrix} 1 \\ 2 \end{pmatrix}, \begin{pmatrix} -1 \\ 1 \end{pmatrix}$

2. Determine whether or not the following vectors are linearly independent in R^3.

(a) $\begin{pmatrix} 1 \\ 0 \\ 0 \end{pmatrix}, \begin{pmatrix} 0 \\ 1 \\ 1 \end{pmatrix}, \begin{pmatrix} 1 \\ 0 \\ 1 \end{pmatrix}$ (b) $\begin{pmatrix} 1 \\ 0 \\ 0 \end{pmatrix}, \begin{pmatrix} 0 \\ 1 \\ 1 \end{pmatrix}, \begin{pmatrix} 1 \\ 0 \\ 1 \end{pmatrix}, \begin{pmatrix} 1 \\ 2 \\ 3 \end{pmatrix}$

(c) $\begin{pmatrix} 2 \\ 1 \\ -2 \end{pmatrix}, \begin{pmatrix} 3 \\ 2 \\ -2 \end{pmatrix}, \begin{pmatrix} 2 \\ 2 \\ 0 \end{pmatrix}$ (d) $\begin{pmatrix} 2 \\ 1 \\ -2 \end{pmatrix}, \begin{pmatrix} -2 \\ -1 \\ 2 \end{pmatrix}, \begin{pmatrix} 4 \\ 2 \\ -4 \end{pmatrix}$

(e) $\begin{pmatrix} 1 \\ 1 \\ 3 \end{pmatrix}, \begin{pmatrix} 0 \\ 2 \\ 1 \end{pmatrix}$

3. For each of the sets of vectors in Exercise 2, describe geometrically the span of the given vectors.

4. Determine whether or not the following are linearly independent in $R^{2 \times 2}$.

(a) $\begin{pmatrix} 1 & 0 \\ 1 & 1 \end{pmatrix}, \begin{pmatrix} 0 & 1 \\ 0 & 0 \end{pmatrix}$ (b) $\begin{pmatrix} 1 & 0 \\ 0 & 1 \end{pmatrix}, \begin{pmatrix} 0 & 1 \\ 0 & 0 \end{pmatrix}, \begin{pmatrix} 0 & 0 \\ 1 & 0 \end{pmatrix}$

(c) $\begin{pmatrix} 1 & 0 \\ 0 & 1 \end{pmatrix}, \begin{pmatrix} 0 & 1 \\ 0 & 0 \end{pmatrix}, \begin{pmatrix} 2 & 3 \\ 0 & 2 \end{pmatrix}$

5. Determine whether or not the following vectors are linearly independent in P_3.
(a) $1, x^2, x^2 - 2$ (b) $2, x^2, x, 2x + 3$
(c) $x + 2, x + 1, x^2 - 1$ (d) $x + 2, x^2 - 1$

6. For each of the following, show that the given vectors are linearly independent in $C[-1, 1]$.
(a) $\cos \pi x, \sin \pi x$ (b) $x^{3/2}, x^{5/2}$
(c) $1, e^x + e^{-x}, e^x - e^{-x}$ (d) e^x, e^{-x}, e^{2x}

7. Determine whether or not the vectors $\cos x, 1, \sin^2(x/2)$ are linearly independent in $C[-\pi, \pi]$.

8. Consider the vectors $\cos(x + \alpha)$ and $\sin x$ in $C[-\pi, \pi]$. For what values of α will the two vectors be linearly dependent? Give a graphical interpretation of your answer.

9. Given the functions $2x$ and $|x|$:
(a) Show that these vectors are linearly independent in $C[-1, 1]$.
(b) Show that the vectors are linearly dependent in $C[0, 1]$.

10. Prove that any finite set of vectors that contains the zero vector must be linearly dependent.

11. Let $\mathbf{v}_1, \ldots, \mathbf{v}_n$ be linearly independent in V and let
$$\mathbf{v} = \alpha_1 \mathbf{v}_1 + \cdots + \alpha_n \mathbf{v}_n$$
Suppose that $\mathbf{v}$ can also be written as
$$\mathbf{v} = \beta_1 \mathbf{v}_1 + \cdots + \beta_n \mathbf{v}_n$$
Show that
$$\beta_j = \alpha_j \quad \text{for} \quad j = 1, \ldots, n.$$

12. Let $\mathbf{x}_1, \ldots, \mathbf{x}_k$ be linearly independent vectors in R^n and let A be a nonsingular $n \times n$ matrix. Define $\mathbf{y}_i = A\mathbf{x}_i$ for $i = 1, \ldots, k$. Show that $\mathbf{y}_1, \ldots, \mathbf{y}_k$ are linearly independent.

13. Let $\{\mathbf{v}_1, \ldots, \mathbf{v}_n\}$ be a spanning set for a vector space V and let $\mathbf{v}$ be any element of V. Show that $\mathbf{v}, \mathbf{v}_1, \ldots, \mathbf{v}_n$ are linearly dependent.

14. Let $\mathbf{v}_1, \mathbf{v}_2, \ldots, \mathbf{v}_n$ be linearly independent vectors in a vector space V. Show that $\mathbf{v}_2, \ldots, \mathbf{v}_n$ cannot span V.

4. BASIS AND DIMENSION

In Section 3 we showed that a spanning set for a vector space is minimal if its elements are linearly independent. The elements of a minimal spanning set form the basic building blocks for the whole vector space, and consequently we say they form a "basis" for the vector space.

Definition. The vectors $\mathbf{v}_1, \mathbf{v}_2, \ldots, \mathbf{v}_n$ form a **basis** for a vector space V if and only if
(i) $\mathbf{v}_1, \ldots, \mathbf{v}_n$ are linearly independent.
(ii) $\mathbf{v}_1, \ldots, \mathbf{v}_n$ span V.

EXAMPLE 1. The *standard basis* for R^3 is $\{\mathbf{e}_1, \mathbf{e}_2, \mathbf{e}_3\}$ but there are many other bases that could be used such as $\{(1, 1, 1)^T, (1, 1, 0)^T, (1, 0, 1)^T\}$ or $\{(1, 2, 3)^T, (0, 1, 1)^T, (2, 0, 1)^T\}$. We will see shortly that any basis for R^3 must have exactly three elements. □

EXAMPLE 2. In $R^{2 \times 2}$ consider the set $\{E_{11}, E_{12}, E_{21}, E_{22}\}$, where

$$E_{11} = \begin{pmatrix} 1 & 0 \\ 0 & 0 \end{pmatrix}, \qquad E_{12} = \begin{pmatrix} 0 & 1 \\ 0 & 0 \end{pmatrix}$$

$$E_{21} = \begin{pmatrix} 0 & 0 \\ 1 & 0 \end{pmatrix}, \qquad E_{22} = \begin{pmatrix} 0 & 0 \\ 0 & 1 \end{pmatrix}$$

If

$$c_1 E_{11} + c_2 E_{12} + c_3 E_{21} + c_4 E_{22} = O$$

then

$$\begin{pmatrix} c_1 & c_2 \\ c_3 & c_4 \end{pmatrix} = \begin{pmatrix} 0 & 0 \\ 0 & 0 \end{pmatrix}$$

so $c_1 = c_2 = c_3 = c_4 = 0$. Therefore, $E_{11}, E_{12}, E_{21}, E_{22}$ are linearly independent. If A is in $R^{2\times 2}$ then

$$A = a_{11} E_{11} + a_{12} E_{12} + a_{21} E_{21} + a_{22} E_{22}$$

Thus $E_{11}, E_{12}, E_{21}, E_{22}$ span $R^{2\times 2}$ and hence form a basis for $R^{2\times 2}$.   □

In many applications it is necessary to find a particular subspace of a vector space V. This can be done by finding the basis elements of the subspace. For example, to find all solutions to the system

$$x_1 + x_2 + x_3 \qquad\; = 0$$
$$2x_1 + x_2 \qquad\; + x_4 = 0$$

one must find the nullspace of the matrix

$$A = \begin{pmatrix} 1 & 1 & 1 & 0 \\ 2 & 1 & 0 & 1 \end{pmatrix}$$

In Example 8 of Section 2, we saw that $N(A)$ is the subspace of R^4 spanned by the vectors

$$\begin{pmatrix} 1 \\ -2 \\ 1 \\ 0 \end{pmatrix} \quad \text{and} \quad \begin{pmatrix} -1 \\ 1 \\ 0 \\ 1 \end{pmatrix}$$

Since these two vectors are linearly independent, they form a basis for $N(A)$.

Theorem 3.4.1.   *If* $v_1, v_2, \ldots, v_n$ *form a basis for a vector space* V, *then any collection of* m *vectors in* V, *where* $m > n$, *is linearly dependent.*

Proof.   Let $u_1, u_2, \ldots, u_m$ be m vectors in V where $m > n$. Then since $v_1, v_2, \ldots, v_n$ span V, we have

$$u_i = a_{i1} v_1 + a_{i2} v_2 + \cdots + a_{in} v_n \qquad \text{for} \qquad i = 1, 2, \ldots, m$$

A linear combination $c_1 u_1 + c_2 u_2 + \cdots + c_m u_m$ can be written in the form

$$c_1 \sum_{j=1}^{n} a_{1j} v_j + c_2 \sum_{j=1}^{n} a_{2j} v_j + \cdots + c_m \sum_{j=1}^{n} a_{mj} v_j$$

Rearranging terms, we see that

$$c_1\mathbf{u}_1 + c_2\mathbf{u}_2 + \cdots + c_m\mathbf{u}_m = \sum_{j=1}^{n}\left(\sum_{i=1}^{m} a_{ij}c_i\right)\mathbf{v}_j$$

Now consider the system of equations

$$\sum_{i=1}^{m} a_{ij}c_i = 0 \qquad j = 1, 2, \ldots, n$$

This is a homogeneous system with more unknowns than equations. Therefore, by Theorem 1.2.1, the system has a nontrivial solution $(\hat{c}_1, \hat{c}_2, \ldots, \hat{c}_m)$. But then

$$\hat{c}_1\mathbf{u}_1 + \hat{c}_2\mathbf{u}_2 + \cdots + \hat{c}_m\mathbf{u}_m = \sum_{j=1}^{n} 0\,\mathbf{v}_j = \mathbf{0}$$

and hence $\mathbf{u}_1, \mathbf{u}_2, \ldots, \mathbf{u}_m$ are linearly dependent. □

Corollary 3.4.2. *If $\{\mathbf{v}_1, \ldots, \mathbf{v}_n\}$ and $\{\mathbf{u}_1, \ldots, \mathbf{u}_m\}$ are both bases for a vector space V, then $n = m$.*

Proof. Let $\mathbf{v}_1, \mathbf{v}_2, \ldots, \mathbf{v}_n$ and $\mathbf{u}_1, \mathbf{u}_2, \ldots, \mathbf{u}_m$ both be bases for V. Since $\mathbf{u}_1, \mathbf{u}_2, \ldots, \mathbf{u}_m$ are linearly independent, it follows from Theorem 3.4.1 that $m \leq n$. By the same reasoning $\mathbf{u}_1, \mathbf{u}_2, \ldots, \mathbf{u}_m$ form a basis for V and $\mathbf{v}_1, \mathbf{v}_2, \ldots, \mathbf{v}_n$ are linearly independent, so $n \leq m$. □

Definition. Let V be a vector space. If V has a basis consisting of n vectors, we say that V has **dimension** n. The subspace $\{\mathbf{0}\}$ of V is said to have dimension 0. V is said to be **finite-dimensional** if there is a finite set of vectors that spans V; otherwise, we say that V is **infinite-dimensional**.

If $\mathbf{x}$ is a nonzero vector in R^3, then $\mathbf{x}$ spans a one-dimensional subspace $\mathcal{S}(\mathbf{x}) = \{\alpha\mathbf{x}|\alpha$ is a scalar$\}$. A vector $(a, b, c)^T$ will be in $\mathcal{S}(\mathbf{x})$ if and only if the point (a, b, c) is on the line determined by $(0, 0, 0)$ and (x_1, x_2, x_3). Thus a one-dimensional subspace of R^3 can be represented geometrically by a line through the origin (Figure 3.4.1). If $\mathbf{x}$ and $\mathbf{y}$ are linearly independent in R^3, then $\mathcal{S}(\mathbf{x}, \mathbf{y}) = \{\alpha\mathbf{x} + \beta\mathbf{y}|\alpha$ and β are scalars$\}$ is a two-dimensional subspace of R^3. A vector $(a, b, c)^T$ will be in $\mathcal{S}(\mathbf{x}, \mathbf{y})$ if and only if (a, b, c) lies on the plane determined by $(0, 0, 0)$, (x_1, x_2, x_3) and (y_1, y_2, y_3). Thus we can think of a two-dimensional subspace of R^3 as a plane through the origin. If $\mathbf{x}$, $\mathbf{y}$, and $\mathbf{z}$ are linearly independent in R^3, they form a basis for R^3 and $\mathcal{S}(\mathbf{x}, \mathbf{y}, \mathbf{z}) = R^3$. Thus any fourth point $(a, b, c)^T$ must lie in $\mathcal{S}(\mathbf{x}, \mathbf{y}, \mathbf{z})$.

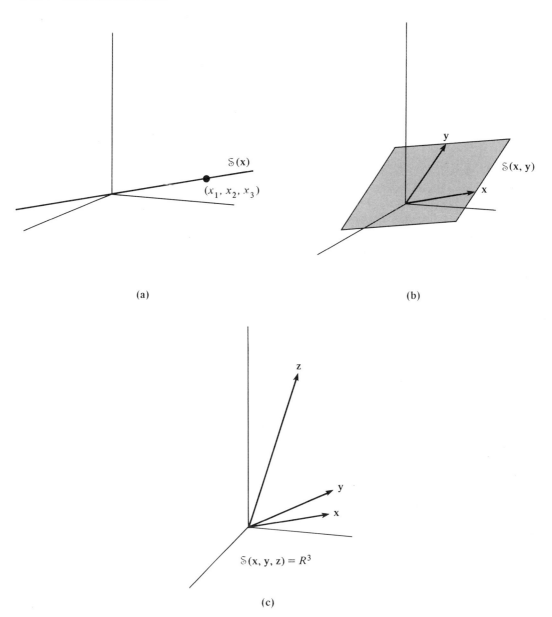

FIGURE 3.4.1

EXAMPLE 3.   Let P be the vector space of all polynomials. We claim that P is infinite-dimensional. If P were finite-dimensional, say of dimension n, any set of $n + 1$ vectors would be linearly dependent. However, $1, x, x^2, \ldots, x^n$ are linearly independent, since $W[1, x, x^2, \ldots, x^n] > 0$. Therefore, P cannot be of dimension n. Since n was arbitrary, P must be infinite-dimensional. The same argument shows that $C[a, b]$ is infinite-dimensional.   □

Theorem 3.4.3. *If V is a vector space of dimension n > 0:*
(I) *Any set of n linearly independent vectors spans V*
(II) *Any n vectors that span V are linearly independent*

Proof. To prove (I), suppose that $v_1, \ldots, v_n$ are linearly independent and v is any other vector in V. It follows from Theorem 3.4.1 that $v_1, v_2, \ldots, v_n, v$ are linearly dependent. Thus there exist scalars $c_1, c_2, \ldots, c_n, c_{n+1}$ not all zero such that

$$(1) \qquad c_1 v_1 + c_2 v_2 + \cdots + c_n v_n + c_{n+1} v = 0$$

The scalar c_{n+1} cannot be zero, for then (1) would imply that $v_1, \ldots, v_n$ are linearly dependent. Thus (1) can be solved for v.

$$v = \alpha_1 v_1 + \alpha_2 v_2 + \cdots + \alpha_n v_n$$

where $\alpha_i = -c_i/c_{n+1}$ for $i = 1, 2, \ldots, n$. Since v was an arbitrary vector in V, it follows that $v_1, v_2, \ldots, v_n$ span V.

To prove (II), suppose that $v_1, \ldots, v_n$ span V. If $v_1, \ldots, v_n$ are linearly dependent, then one of the v_i's, say v_n, can be written as a linear combination of the others. It follows that $v_1, \ldots, v_{n-1}$ will still span V. If $v_1, \ldots, v_{n-1}$ are linearly dependent, we can eliminate another vector and still have a spanning set. We can continue eliminating vectors in this way until we arrive at a linearly independent spanning set with $k < n$ elements. But this contradicts dim $V = n$. Therefore, $v_1, \ldots, v_n$ must be linearly independent.   □

EXAMPLE 4. Show that $\left\{ \begin{pmatrix} 1 \\ 2 \\ 3 \end{pmatrix}, \begin{pmatrix} -2 \\ 1 \\ 0 \end{pmatrix}, \begin{pmatrix} 1 \\ 0 \\ 1 \end{pmatrix} \right\}$ is a basis for R^3.

SOLUTION. Since dim $R^3 = 3$, we need only show that these three vectors are linearly independent. That follows since

$$\begin{vmatrix} 1 & -2 & 1 \\ 2 & 1 & 0 \\ 3 & 0 & 1 \end{vmatrix} = 2$$
□

We close this section with two final observations. These are given in the following theorem.

Theorem 3.4.4. *If V is a vector space of dimension n, then*:
(i) *No set of less than n vectors can span V.*
(ii) *Any subset of less than n linearly independent vectors can be extended to form a basis for V.*

Observation (i) follows by the same reasoning that was used to prove (II) of Theorem 3.4.3. To prove (ii), suppose that $v_1, \ldots, v_k$ are linearly independent and $k < n$. It follows from (i) that $\mathcal{S}(\mathbf{v}_1, \ldots, \mathbf{v}_k)$ is a proper subspace of V and hence there exists a vector $\mathbf{v}_{k+1}$ that is in V but not in $\mathcal{S}(\mathbf{v}_1, \ldots, \mathbf{v}_k)$. It follows that the vectors $\mathbf{v}_1, \mathbf{v}_2, \ldots, \mathbf{v}_k, \mathbf{v}_{k+1}$ are linearly independent. If $k + 1 < n$, then in the same manner $\{\mathbf{v}_1, \ldots, \mathbf{v}_k, \mathbf{v}_{k+1}\}$ can be extended to a set of $k + 2$ linearly independent vectors. This process may be continued until a set $\{\mathbf{v}_1, \mathbf{v}_2, \ldots, \mathbf{v}_k, \mathbf{v}_{k+1}, \ldots, \mathbf{v}_n\}$ of n linearly independent vectors is obtained.

EXERCISES

1. In Exercise 1 of Section 3, indicate whether or not the given vectors form a basis for R^2.

2. In Exercise 2 of Section 3, indicate whether or not the given vectors form a basis for R^3.

3. Given the vectors

$$\mathbf{x}_1 = \begin{pmatrix} 2 \\ 1 \end{pmatrix}, \qquad \mathbf{x}_2 = \begin{pmatrix} 4 \\ 3 \end{pmatrix}, \qquad \mathbf{x}_3 = \begin{pmatrix} 7 \\ -3 \end{pmatrix}$$

(a) Show that $\mathbf{x}_1$ and $\mathbf{x}_2$ form a basis for R^2.
(b) Why must $\mathbf{x}_1, \mathbf{x}_2, \mathbf{x}_3$ be linearly dependent?
(c) What is the dimension of $\mathcal{S}(\mathbf{x}_1, \mathbf{x}_2, \mathbf{x}_3)$?

4. Given the vectors

$$\mathbf{x}_1 = \begin{pmatrix} 3 \\ -2 \\ 4 \end{pmatrix}, \qquad \mathbf{x}_2 = \begin{pmatrix} -3 \\ 2 \\ -4 \end{pmatrix}, \qquad \mathbf{x}_3 = \begin{pmatrix} -6 \\ 4 \\ -8 \end{pmatrix}$$

What is the dimension of $\mathcal{S}(\mathbf{x}_1, \mathbf{x}_2, \mathbf{x}_3)$?

5. Given

$$\mathbf{x}_1 = \begin{pmatrix} 2 \\ 1 \\ 3 \end{pmatrix}, \qquad \mathbf{x}_2 = \begin{pmatrix} 3 \\ -1 \\ 4 \end{pmatrix}, \qquad \mathbf{x}_3 = \begin{pmatrix} 2 \\ 6 \\ 4 \end{pmatrix}$$

(a) Show that $\mathbf{x}_1, \mathbf{x}_2, \mathbf{x}_3$ are linearly dependent.
(b) Show that $\mathbf{x}_1$ and $\mathbf{x}_2$ are linearly independent.
(c) What is the dimension of $\mathcal{S}(\mathbf{x}_1, \mathbf{x}_2, \mathbf{x}_3)$?
(d) Give a geometric description of $\mathcal{S}(\mathbf{x}_1, \mathbf{x}_2, \mathbf{x}_3)$.

6. In Exercise 2 of Section 2, some of the sets formed subspaces of R^3. In each of these cases find a basis for the subspace and determine its dimension.

7. In Exercise 3 of Section 2, some of the sets formed subspaces of $R^{2\times 2}$. In each of these cases find a basis for the subspace and determine its dimension.

8. In $C[-\pi, \pi]$ find the dimension of the subspace spanned by $1, \cos 2x, \cos^2 x$.

9. In each of the following, find the dimension of the subspace of P_3 spanned by the given vectors.
   (a) $x, x - 1, x^2 + 1$ (b) $x, x - 1, x^2 + 1, x^2 - 1$
   (c) $x^2, x^2 - x - 1, x + 1$ (d) $2x, x - 2$

10. Let S be the subspace of P_3 consisting of all polynomials $p(x)$ such that $p(0) = 0$ and let T be the subspace of all polynomials $q(x)$ such that $q(1) = 0$. Find a basis for S, a basis for T, and a basis for $S \cap T$.

11. In R^4 let U be the subspace of all vectors of the form $(u_1, u_2, 0, 0)^T$ and let V be the subspace of all vectors of the form $(0, v_2, v_3, 0)^T$. What are the dimensions of $U, V, U \cap V, U + V$? Find a basis for each of these four subspaces.

12. Is it possible to find a pair of two-dimensional subspaces U and V of R^3 such that $U \cap V = \{\mathbf{0}\}$? Prove your answer.

   [**Hint:** Let $\{\mathbf{u}_1, \mathbf{u}_2\}$ and $\{\mathbf{v}_1, \mathbf{v}_2\}$ be bases for U and V, respectively. Show that $\mathbf{u}_1, \mathbf{u}_2, \mathbf{v}_1, \mathbf{v}_2$ are linearly dependent.] Give a geometrical interpretation of your conclusion.

5. THE ROW SPACE AND COLUMN SPACE

If A is an $m \times n$ matrix, each row of A is an n-tuple of real numbers and hence can be considered as a vector in $R^{1\times n}$. The m vectors corresponding to the rows of A will be referred to as the *row vectors* of A. Similarly, each column of A can be considered as a vector in R^m and one can associate n *column vectors* with the matrix A.

Definition. If A is an $m \times n$ matrix, the subspace of $R^{1\times n}$ spanned by the row vectors of A is called the **row space** of A. The subspace of R^m spanned by the column vectors of A is called the **column space** of A.

EXAMPLE 1.   Let

$$A = \begin{pmatrix} 1 & 0 & 0 \\ 0 & 1 & 0 \end{pmatrix}$$

The row space of A is the set of all 3-tuples of the form

$$\alpha(1,0,0) + \beta(0,1,0) = (\alpha, \beta, 0)$$

The column space of A is the set of all vectors of the form

$$\alpha\begin{pmatrix} 1 \\ 0 \end{pmatrix} + \beta\begin{pmatrix} 0 \\ 1 \end{pmatrix} + \gamma\begin{pmatrix} 0 \\ 0 \end{pmatrix} = \begin{pmatrix} \alpha \\ \beta \end{pmatrix}$$

Thus the row space of A is a two-dimensional subspace of $R^{1 \times 3}$ and the column space of A is R^2.   □

Theorem 3.5.1.   *Two row equivalent matrices have the same row space.*

Proof.   If B is row equivalent to A, then B can be formed from A by a finite sequence of row operations. Thus the row vectors of B must be linear combinations of the row vectors of A. Consequently, the row space of B must be a subspace of the row space of A. Since A is row equivalent to B, by the same reasoning, the row space of A is a subspace of the row space of B.   □

Definition.   The **rank** of a matrix A is the dimension of the row space of A.

To determine the rank of a matrix, one can reduce the matrix to row echelon form. The nonzero rows of the row echelon matrix will form a basis for the row space.

EXAMPLE 2.   Let

$$A = \begin{pmatrix} 1 & -2 & 3 \\ 2 & -5 & 1 \\ 1 & -4 & -7 \end{pmatrix}$$

Reducing A to row echelon form, one obtains the matrix

$$U = \begin{pmatrix} 1 & -2 & 3 \\ 0 & 1 & 5 \\ 0 & 0 & 0 \end{pmatrix}$$

Clearly, $(1, -2, 3)$ and $(0, 1, 5)$ form a basis for the row space of U. Since U and A are row equivalent, they have the same row space and hence the rank of A is 2.   □

LINEAR SYSTEMS

The concepts of row space and column space are useful in the study of linear systems. The system $A\mathbf{x} = \mathbf{b}$ can be written in the form

$$
(1) \qquad x_1 \begin{pmatrix} a_{11} \\ a_{21} \\ \vdots \\ a_{m1} \end{pmatrix} + x_2 \begin{pmatrix} a_{12} \\ a_{22} \\ \vdots \\ a_{m2} \end{pmatrix} + \cdots + x_n \begin{pmatrix} a_{1n} \\ a_{2n} \\ \vdots \\ a_{mn} \end{pmatrix} = \begin{pmatrix} b_1 \\ b_2 \\ \vdots \\ b_m \end{pmatrix}
$$

It follows from (1) that the system $A\mathbf{x} = \mathbf{b}$ will be consistent if and only if $\mathbf{b}$ can be written as a linear combination of the column vectors of A. Thus $A\mathbf{x} = \mathbf{b}$ is consistent if and only if $\mathbf{b}$ is in the column space of A. If $\mathbf{b}$ is replaced by the zero vector, then (1) becomes

$$
(2) \qquad x_1 \mathbf{a}_1 + x_2 \mathbf{a}_2 + \cdots + x_n \mathbf{a}_n = \mathbf{0}
$$

It follows from (2) that the system $A\mathbf{x} = \mathbf{0}$ will have only the trivial solution $\mathbf{x} = \mathbf{0}$ if and only if the column vectors of A are linearly independent.

Theorem 3.5.2.   *Let A be an $m \times n$ matrix. The linear system $A\mathbf{x} = \mathbf{b}$ is consistent for every $\mathbf{b} \in R^m$ if and only if the column vectors of A span R^m. The system $A\mathbf{x} = \mathbf{b}$ has at most one solution for every $\mathbf{b} \in R^m$ if and only if the column vectors of A are linearly independent.*

Proof.   We have seen that the system $A\mathbf{x} = \mathbf{b}$ is consistent if and only if $\mathbf{b}$ is in the column space of A. It follows that $A\mathbf{x} = \mathbf{b}$ will be consistent for every $\mathbf{b} \in R^m$ if and only if the column vectors of A span R^m. To prove the second statement, note that if $A\mathbf{x} = \mathbf{b}$ has at most one solution for every $\mathbf{b}$, then in particular the system $A\mathbf{x} = \mathbf{0}$ can have only the trivial solution, and hence the column vectors of A must be linearly independent. Conversely, if the column vectors of A are linearly independent, $A\mathbf{x} = \mathbf{0}$ has only the trivial solution. Now if $\mathbf{x}_1$ and $\mathbf{x}_2$ were both solutions to $A\mathbf{x} = \mathbf{b}$, then $\mathbf{x}_1 - \mathbf{x}_2$ would be a solution to $A\mathbf{x} = \mathbf{0}$,

$$
A(\mathbf{x}_1 - \mathbf{x}_2) = A\mathbf{x}_1 - A\mathbf{x}_2 = \mathbf{b} - \mathbf{b} = \mathbf{0}
$$

It follows that $\mathbf{x}_1 - \mathbf{x}_2 = \mathbf{0}$, and hence $\mathbf{x}_1$ must equal $\mathbf{x}_2$.　　□

Let A be an $m \times n$ matrix. If the column vectors of A span R^m, then n must be greater than or equal to m, since no set of less than m vectors could span R^m. If the columns of A are linearly independent, then n must be less than or equal to m, since every set of more than m vectors in R^m is linearly dependent. Thus, if the column vectors of A form a basis for R^m, then n must equal m.

Corollary 3.5.3. *An $n \times n$ matrix A is nonsingular if and only if the column vectors of A form a basis for R^n.*

In general, the rank and the dimension of the nullspace always add up to the number of columns of the matrix. The dimension of the nullspace of a matrix is called the *nullity* of the matrix.

Theorem 3.5.4. *If A is an $m \times n$ matrix, then the rank of A plus the nullity of A equals n.*

Proof. Let U be the reduced row echelon form of A. The system $A\mathbf{x} = \mathbf{0}$ is equivalent to the system $U\mathbf{x} = \mathbf{0}$. If A has rank r, then U will have r nonzero rows, and consequently the system $U\mathbf{x} = \mathbf{0}$ will involve r lead variables and $n - r$ free variables. The dimension of $N(A)$ will equal the number of free variables. $\qquad\square$

EXAMPLE 3. Let

$$A = \begin{pmatrix} 1 & 2 & -1 & 1 \\ 2 & 4 & -3 & 0 \\ 1 & 2 & 1 & 5 \end{pmatrix}$$

Find a basis for the row space of A and a basis for $N(A)$. Verify that $\dim N(A) = n - r$.

SOLUTION. The reduced row echelon form of A is given by

$$U = \begin{pmatrix} 1 & 2 & 0 & 3 \\ 0 & 0 & 1 & 2 \\ 0 & 0 & 0 & 0 \end{pmatrix}$$

Thus $\{(1, 2, 0, 3), (0, 0, 1, 2)\}$ is a basis for the row space of A and A has rank 2. Since the systems $A\mathbf{x} = \mathbf{0}$ and $U\mathbf{x} = \mathbf{0}$ are equivalent, it follows that $\mathbf{x}$ is in $N(A)$ if and only if

$$x_1 + 2x_2 + \qquad 3x_4 = 0$$
$$x_3 + 2x_4 = 0$$

The lead variables x_1 and x_3 can be solved for in terms of the free variables x_2 and x_4.

$$x_1 = -2x_2 - 3x_4$$
$$x_3 = -2x_4$$

Let $x_2 = \alpha$ and $x_4 = \beta$. It follows that $N(A)$ consists of all vectors of the form

$$\begin{pmatrix} x_1 \\ x_2 \\ x_3 \\ x_4 \end{pmatrix} = \begin{pmatrix} -2\alpha - 3\beta \\ \alpha \\ -2\beta \\ \beta \end{pmatrix} = \alpha \begin{pmatrix} -2 \\ 1 \\ 0 \\ 0 \end{pmatrix} + \beta \begin{pmatrix} -3 \\ 0 \\ -2 \\ 1 \end{pmatrix}$$

The vectors $(-2, 1, 0, 0)^T$ and $(-3, 0, -2, 1)^T$ form a basis for $N(A)$. Note that

$$n - r = 4 - 2 = 2 = \dim N(A) \qquad \square$$

THE COLUMN SPACE

The matrices A and U in Example 3 have different column spaces; however, their column vectors satisfy the same dependency relations. For the matrix U, the column vectors $\mathbf{u}_1$ and $\mathbf{u}_3$ are linearly independent, while

$$\mathbf{u}_2 = 2\mathbf{u}_1$$
$$\mathbf{u}_4 = 3\mathbf{u}_1 + 2\mathbf{u}_3$$

The same relations hold for the columns of A. The vectors $\mathbf{a}_1$ and $\mathbf{a}_3$ are linearly independent, while

$$\mathbf{a}_2 = 2\mathbf{a}_1$$
$$\mathbf{a}_4 = 3\mathbf{a}_1 + 2\mathbf{a}_3$$

In general, if A is an $m \times n$ matrix and U is the row echelon form of A, then since $A\mathbf{x} = \mathbf{0}$ if and only if $U\mathbf{x} = \mathbf{0}$, their column vectors satisfy the same dependency relations. We will use this to prove that the dimension of the column space of A is equal to the dimension of the row space of A.

Theorem 3.5.5.   *If A is an $m \times n$ matrix, the dimension of the row space of A equals the dimension of the column space of A.*

Proof.   If A is an $m \times n$ matrix of rank r, the row echelon form U of A will have r lead 1's. The columns of U corresponding to the lead 1's will be linearly independent. They do not, however, form a basis for the column space of A, since in general A and U will have different column spaces. Let U_L denote the matrix obtained from U by deleting all of the columns corresponding to the free variables. Delete the same columns from A and denote the new matrix by A_L. The matrices A_L and U_L are row equivalent. Thus, if $\mathbf{x}$ is a solution to $A_L\mathbf{x} = \mathbf{0}$, then $\mathbf{x}$ must also be a solution to $U_L\mathbf{x} = \mathbf{0}$. Since the columns of U_L are linearly independent $\mathbf{x}$ must equal $\mathbf{0}$. It follows from the remarks preceding Theorem 3.5.2 that the columns of A_L are linearly independent. Since A_L has r columns, the dimension of the column space of A is at least r.

We have proved that for any matrix the dimension of the column space is greater than or equal to the dimension of the row space. Applying this result to the matrix A^T, we see that

$$\dim(\text{row space of } A) = \dim(\text{column space of } A^T)$$
$$\geq \dim(\text{row space of } A^T)$$
$$= \dim(\text{column space of } A)$$

Thus, for any matrix A, the dimension of the row space must equal the dimension of the column space.   □

One can use the row echelon form U of A to find a basis for the column space of A. One need only determine the columns of U which correspond to the lead 1's. Those same columns of A will be linearly independent and form a basis for the column space of A.

EXAMPLE 4. Let

$$A = \begin{pmatrix} 1 & -2 & 1 & 1 & 2 \\ -1 & 3 & 0 & 2 & -2 \\ 0 & 1 & 1 & 3 & 4 \\ 1 & 2 & 5 & 13 & 5 \end{pmatrix}$$

The row echelon form of A is given by

$$U = \begin{pmatrix} 1 & -2 & 1 & 1 & 2 \\ 0 & 1 & 1 & 3 & 0 \\ 0 & 0 & 0 & 0 & 1 \\ 0 & 0 & 0 & 0 & 0 \end{pmatrix}$$

The lead 1's occur in the first, second, and fifth columns. Thus

$$\mathbf{a}_1 = \begin{pmatrix} 1 \\ -1 \\ 0 \\ 1 \end{pmatrix}, \qquad \mathbf{a}_2 = \begin{pmatrix} -2 \\ 3 \\ 1 \\ 2 \end{pmatrix}, \qquad \mathbf{a}_5 = \begin{pmatrix} 2 \\ -2 \\ 4 \\ 5 \end{pmatrix}$$

form a basis for the column space of A.   □

EXAMPLE 5. Find the dimension of the subspace of R^4 spanned by

$$\mathbf{x}_1 = \begin{pmatrix} 1 \\ 2 \\ -1 \\ 0 \end{pmatrix}, \qquad \mathbf{x}_2 = \begin{pmatrix} 2 \\ 5 \\ -3 \\ 2 \end{pmatrix}, \qquad \mathbf{x}_3 = \begin{pmatrix} 2 \\ 4 \\ -2 \\ 0 \end{pmatrix}, \qquad \mathbf{x}_4 = \begin{pmatrix} 3 \\ 8 \\ -5 \\ 4 \end{pmatrix}$$

SOLUTION. The subspace $S(\mathbf{x}_1, \mathbf{x}_2, \mathbf{x}_3, \mathbf{x}_4)$ is the same as the column space of the matrix

$$X = \begin{pmatrix} 1 & 2 & 2 & 3 \\ 2 & 5 & 4 & 8 \\ -1 & -3 & -2 & -5 \\ 0 & 2 & 0 & 4 \end{pmatrix}$$

The row echelon form of X is

$$\begin{pmatrix} 1 & 2 & 2 & 3 \\ 0 & 1 & 0 & 2 \\ 0 & 0 & 0 & 0 \\ 0 & 0 & 0 & 0 \end{pmatrix}$$

The first two columns x_1, x_2 of X will form a basis for the column space of X. Thus $\dim S(x_1, x_2, x_3, x_4) = 2$. $\square$

EXERCISES

1. For each of the following matrices, find a basis for the row space, a basis for the column space, and a basis for the nullspace.

(a) $\begin{pmatrix} 1 & 3 & 2 \\ 2 & 1 & 4 \\ 4 & 7 & 8 \end{pmatrix}$

(b) $\begin{pmatrix} -3 & 1 & 3 & 4 \\ 1 & 2 & -1 & -2 \\ -3 & 8 & 4 & 2 \end{pmatrix}$

(c) $\begin{pmatrix} 1 & 3 & -2 & 1 \\ 2 & 1 & 3 & 2 \\ 3 & 4 & 5 & 6 \end{pmatrix}$

2. In each of the following, determine the dimension of the subspace of R^3 spanned by the given vectors.

(a) $\begin{pmatrix} 1 \\ -2 \\ 2 \end{pmatrix}, \begin{pmatrix} 2 \\ -2 \\ 4 \end{pmatrix}, \begin{pmatrix} -3 \\ 3 \\ 6 \end{pmatrix}$

(b) $\begin{pmatrix} 1 \\ 1 \\ 1 \end{pmatrix}, \begin{pmatrix} 1 \\ 2 \\ 3 \end{pmatrix}, \begin{pmatrix} 2 \\ 3 \\ 1 \end{pmatrix}$

(c) $\begin{pmatrix} 1 \\ -1 \\ 2 \end{pmatrix}, \begin{pmatrix} -2 \\ 2 \\ -4 \end{pmatrix}, \begin{pmatrix} 3 \\ -2 \\ 5 \end{pmatrix}, \begin{pmatrix} 2 \\ -1 \\ 3 \end{pmatrix}$

3. Construct 3×3 matrices of rank 0, rank 1, rank 2, and rank 3.

4. How many solutions will the linear system $Ax = b$ have if b is in the column space of A and the column vectors of A are linearly independent? Explain.

5. Let A and B be row equivalent matrices.
   (a) Show that the dimension of the column space of A equals the dimension of the column space of B.
   (b) Are the column spaces of the two matrices necessarily the same? Justify your answer.

6. Prove Corollary 3.5.3.

7. Suppose that A and B are $n \times n$ matrices with the property that $Ax = Bx$ for all $x \in R^n$.
   (a) Show that $N(A - B) = R^n$.
   (b) Show that $A - B$ must have rank 0 and consequently that $A = B$.

8. Let A and B be $n \times n$ matrices. Show that $AB = O$ if and only if the column space of B is a subspace of the nullspace of A.

9. Let $A \in R^{m \times n}$, $\mathbf{b} \in R^m$ and let $\hat{\mathbf{x}}$ be a particular solution to the system $A\mathbf{x} = \mathbf{b}$. Prove the following.
   (a) A vector $\mathbf{y}$ in R^n will be a solution to $A\mathbf{x} = \mathbf{b}$ if and only if $\mathbf{y} = \hat{\mathbf{x}} + \mathbf{z}$, where $\mathbf{z} \in N(A)$.
   (b) If $N(A) = \{\mathbf{0}\}$, then the solution $\hat{\mathbf{x}}$ is unique.

10. Let $\mathbf{x}$ and $\mathbf{y}$ be nonzero vectors in R^m and R^n, respectively, and let $A = \mathbf{x}\mathbf{y}^T$.
    (a) Show that $\{\mathbf{x}\}$ is a basis for the column space of A and that $\{\mathbf{y}^T\}$ is a basis for the row space of A.
    (b) What is the dimension of $N(A)$?

11. Let $A \in R^{m \times n}$, $B \in R^{n \times r}$, and $C = AB$.
    (a) Show that the column space of C is a subspace of the column space of A.
    (b) Show that the row space of C is a subspace of the row space of B.

12. Let $A \in R^{m \times n}$, $B \in R^{n \times r}$, and $C = AB$.
    (a) Show that if A and B both have linearly independent column vectors, then the column vectors of C will also be linearly independent.
    (b) Show that if A and B both have linearly independent row vectors, then the row vectors of C will also be linearly independent.

    [**Hint:** Apply part (a) to C^T.]

13. Let $A \in R^{m \times n}$, $B \in R^{n \times r}$, and $C = AB$.
    (a) Show that if the column vectors of B are linearly dependent, then the column vectors of C must be linearly dependent.
    (b) Show that if the row vectors of A are linearly dependent, then the row vectors of C are linearly dependent.

    [**Hint:** Apply part (a) to C^T.]

14. An $m \times n$ matrix A is said to have a *right inverse* if there exists an $n \times m$ matrix C such that $AC = I_m$. A is said to have a *left inverse* if there exists an $n \times m$ matrix D such that $DA = I_n$.
    (a) If A has a right inverse, show that the column vectors of A span R^m.
    (b) Is it possible for an $m \times n$ matrix to have a right inverse if $n < m$? $n \geq m$? Explain.

15. Prove: If A is an $m \times n$ matrix and the column vectors of A span R^m, then A has a right inverse.

[***Hint:*** Let $\mathbf{e}_j$ denote the jth column of I_m and solve $A\mathbf{x} = \mathbf{e}_j$ for $j = 1, \ldots, m$.]

16. Show that a matrix B has a left inverse if and only if B^T has a right inverse.

17. Let B be an $n \times m$ matrix whose columns are linearly independent. Show that B has a left inverse.

18. Prove that if a matrix B has a left inverse, the columns of B are linearly independent.

19. If a matrix U is in row echelon form, show that the nonzero row vectors of U form a basis for the row space of U.

4

LINEAR TRANSFORMATIONS

Linear mappings from one vector space to another play an important role in mathematics. This chapter provides an introduction to the theory of such mappings. In Section 1 the definition of a linear transformation is given and a number of examples are presented. In Section 2 it is shown that each linear transformation L mapping an n-dimensional vector space V into an m-dimensional vector space W can be represented by an $m \times n$ matrix A. Thus one can work with the matrix A in place of the operator L. In Section 3 we consider the problem of changing coordinate systems, that is, the problem of switching from one ordered basis to another. If L maps V into itself, the matrix representing L will depend on the ordered basis chosen for V. Thus L may be represented by a matrix A with respect to one ordered basis and by another matrix B with respect to another ordered basis. In Section 4 we consider the relationship between different matrices which represent the same linear operator. In many applications it is desirable to choose the basis for V so that the matrix representing the linear transformation is either diagonal or in some other simple form.

1. DEFINITION AND EXAMPLES

In the study of vector spaces the most important types of mappings are linear transformations.

Definition. A mapping L from a vector space V into a vector space W is said to be a **linear transformation** or a **linear operator** if

$$(1) \qquad L(\alpha \mathbf{v}_1 + \beta \mathbf{v}_2) = \alpha L(\mathbf{v}_1) + \beta L(\mathbf{v}_2)$$

for all $\mathbf{v}_1, \mathbf{v}_2 \in V$ and for all scalars α and β.

If L is a linear transformation mapping a vector space V into W, it follows from (1) that

$$(2) \qquad L(\mathbf{v}_1 + \mathbf{v}_2) = L(\mathbf{v}_1) + L(\mathbf{v}_2) \qquad (\alpha = \beta = 1)$$

and

$$(3) \qquad L(\alpha \mathbf{v}) = \alpha L(\mathbf{v}) \qquad (\mathbf{v} = \mathbf{v}_1, \ \beta = 0)$$

Conversely, if L satisfies (2) and (3), then

$$L(\alpha \mathbf{v}_1 + \beta \mathbf{v}_2) = L(\alpha \mathbf{v}_1) + L(\beta \mathbf{v}_2)$$
$$= \alpha L(\mathbf{v}_1) + \beta L(\mathbf{v}_2)$$

Thus L is a linear operator on V if and only if L satisfies (2) and (3).

Let us now consider some examples of linear transformations. We begin with linear operators that map R^2 into itself. In this case it is easier to see the geometric effect of the operator.

LINEAR OPERATORS ON R^2

EXAMPLE 1. Let L be the operator defined by

$$L(\mathbf{x}) = 3\mathbf{x}$$

for each $\mathbf{x} \in R^2$. Since

$$L(\alpha \mathbf{x}) = 3(\alpha \mathbf{x}) = \alpha(3\mathbf{x}) = \alpha L(\mathbf{x})$$

and

$$L(\mathbf{x} + \mathbf{y}) = 3(\mathbf{x} + \mathbf{y}) = (3\mathbf{x}) + (3\mathbf{y}) = L(\mathbf{x}) + L(\mathbf{y})$$

it follows that L is a linear transformation. We can think of L as a stretching by a factor of 3 (see Figure 4.1.1). In general, if α is a positive scalar, the linear transformation $F(\mathbf{x}) = \alpha \mathbf{x}$ can be thought of as a stretching or shrinking by a factor of α. □

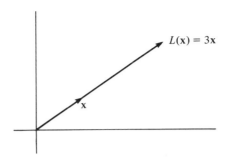

FIGURE 4.1.1

EXAMPLE 2. Consider the mapping L defined by

$$L(\mathbf{x}) = x_1 \mathbf{e}_1$$

for each $\mathbf{x} \in R^2$. Thus, if $\mathbf{x} = (x_1, x_2)^T$, then $L(\mathbf{x}) = (x_1, 0)^T$. If $\mathbf{y} = (y_1, y_2)^T$, then

$$\alpha \mathbf{x} + \beta \mathbf{y} = \begin{pmatrix} \alpha x_1 + \beta y_1 \\ \alpha x_2 + \beta y_2 \end{pmatrix}$$

and it follows that

$$\begin{aligned}
L(\alpha \mathbf{x} + \beta \mathbf{y}) &= (\alpha x_1 + \beta y_1) \mathbf{e}_1 \\
&= \alpha(x_1 \mathbf{e}_1) + \beta(y_1 \mathbf{e}_1) \\
&= \alpha L(\mathbf{x}) + \beta L(\mathbf{y})
\end{aligned}$$

Thus L is a linear transformation. We can think of L as a projection onto the x_1 axis (see Figure 4.1.2). $\qquad \square$

EXAMPLE 3. Let L be the operator defined by

$$L(\mathbf{x}) = (x_1, -x_2)^T$$

FIGURE 4.1.2

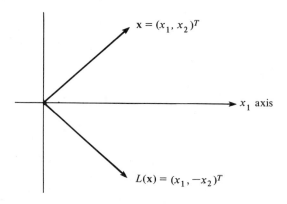

FIGURE 4.1.3

for each $\mathbf{x} = (x_1, x_2)^T$ in R^2. Since

$$L(\alpha\mathbf{x} + \beta\mathbf{y}) = \begin{pmatrix} \alpha x_1 + \beta y_1 \\ -(\alpha x_2 + \beta y_2) \end{pmatrix}$$

$$= \alpha\begin{pmatrix} x_1 \\ -x_2 \end{pmatrix} + \beta\begin{pmatrix} y_1 \\ -y_2 \end{pmatrix}$$

$$= \alpha L(\mathbf{x}) + \beta L(\mathbf{y})$$

it follows that L is a linear transformation. The operator L has the effect of reflecting a vector $\mathbf{x}$ about the x_1 axis (see Figure 4.1.3). □

EXAMPLE 4. The operator L defined by

$$L(\mathbf{x}) = (-x_2, x_1)^T$$

is linear, since

$$L(\alpha\mathbf{x} + \beta\mathbf{y}) = \begin{pmatrix} -(\alpha x_2 + \beta y_2) \\ \alpha x_1 + \beta y_1 \end{pmatrix}$$

$$= \alpha\begin{pmatrix} -x_2 \\ x_1 \end{pmatrix} + \beta\begin{pmatrix} -y_2 \\ y_1 \end{pmatrix}$$

$$= \alpha L(\mathbf{x}) + \beta L(\mathbf{y})$$

The operator L has the effect of rotating each vector in R^2 by $90°$ in the counterclockwise direction (see Figure 4.1.4). □

EXAMPLE 5. Consider the mapping M defined by

$$M(\mathbf{x}) = \left(x_1^2 + x_2^2\right)^{1/2}$$

Since

$$M(\alpha\mathbf{x}) = \left(\alpha^2 x_1^2 + \alpha^2 x_2^2\right)^{1/2} = |\alpha|M(\mathbf{x})$$

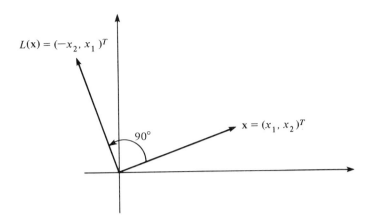

$L(\mathbf{x}) = (-x_2, x_1)^T$

$90°$

$\mathbf{x} = (x_1, x_2)^T$

FIGURE 4.1.4

it follows that

$$\alpha M(\mathbf{x}) \neq M(\alpha \mathbf{x})$$

whenever $\alpha < 0$ and $\mathbf{x} \neq \mathbf{0}$. Therefore, M is not a linear transformation. $\square$

LINEAR OPERATORS FROM R^n TO R^m

EXAMPLE 6. The mapping $L: R^2 \to R^1$ defined by

$$L(\mathbf{x}) = x_1 + x_2$$

is a linear transformation, since

$$
\begin{aligned}
L(\alpha \mathbf{x} + \beta \mathbf{y}) &= (\alpha x_1 + \beta y_1) + (\alpha x_2 + \beta y_2) \\
&= \alpha(x_1 + x_2) + \beta(y_1 + y_2) \\
&= \alpha L(\mathbf{x}) + \beta L(\mathbf{y})
\end{aligned}
$$
$\square$

EXAMPLE 7. The mapping L from R^2 to R^3 defined by

$$L(\mathbf{x}) = (x_2, x_1, x_1 + x_2)^T$$

is linear, since

$$L(\alpha \mathbf{x}) = (\alpha x_2, \alpha x_1, \alpha x_1 + \alpha x_2)^T = \alpha L(\mathbf{x})$$

and

$$
\begin{aligned}
L(\mathbf{x} + \mathbf{y}) &= (x_2 + y_2, x_1 + y_1, x_1 + y_1 + x_2 + y_2)^T \\
&= (x_2, x_1, x_1 + x_2)^T + (y_2, y_1, y_1 + y_2)^T \\
&= L(\mathbf{x}) + L(\mathbf{y})
\end{aligned}
$$

Note that if we define the matrix A by

$$A = \begin{pmatrix} 0 & 1 \\ 1 & 0 \\ 1 & 1 \end{pmatrix}$$

then

$$L(\mathbf{x}) = \begin{pmatrix} x_2 \\ x_1 \\ x_1 + x_2 \end{pmatrix} = A\mathbf{x}$$

for each $\mathbf{x} \in R^2$.   □

In general, if A is any $m \times n$ matrix, we can define a linear operator L_A from R^n to R^m by

$$L_A(\mathbf{x}) = A\mathbf{x}$$

for each $\mathbf{x} \in R^n$. The operator L_A is linear, since

$$L_A(\alpha\mathbf{x} + \beta\mathbf{y}) = A(\alpha\mathbf{x} + \beta\mathbf{y})$$
$$= \alpha A\mathbf{x} + \beta A\mathbf{y}$$
$$= \alpha L_A(\mathbf{x}) + \beta L_A(\mathbf{y})$$

Thus we can think of each $m \times n$ matrix A as a linear operator from R^n to R^m.

In Example 7 we saw that the operator L could have been defined in terms of a matrix A. In the next section we will see that this is true for all linear operators from R^n to R^m.

LINEAR TRANSFORMATIONS FROM V TO W

If L is a linear operator mapping a vector space V into a vector space W, then

(i) $L(\mathbf{0}_V) = \mathbf{0}_W$ (where $\mathbf{0}_V$ and $\mathbf{0}_W$ are the zero vectors in V and W, respectively).

(ii) If $\mathbf{v}_1, \ldots, \mathbf{v}_n$ are elements of V and $\alpha_1, \ldots, \alpha_n$ are scalars, then
$$L(\alpha_1\mathbf{v}_1 + \alpha_2\mathbf{v}_2 + \cdots + \alpha_n\mathbf{v}_n)$$
$$= \alpha_1 L(\mathbf{v}_1) + \alpha_2 L(\mathbf{v}_2) + \cdots + \alpha_n L(\mathbf{v}_n)$$

Statement (i) follows from the condition $L(\alpha\mathbf{v}) = \alpha L(\mathbf{v})$ with $\alpha = 0$. Statement (ii) can be easily proven by mathematical induction. We leave this to the reader as an exercise and conclude the section with some further examples.

EXAMPLE 8. If V is any vector space, the identity operator I is defined by

$$I(\mathbf{v}) = \mathbf{v}$$

for all $\mathbf{v} \in V$. Clearly, I is a linear transformation that maps V into itself.

$$I(\alpha \mathbf{v}_1 + \beta \mathbf{v}_2) = \alpha \mathbf{v}_1 + \beta \mathbf{v}_2 = \alpha I(\mathbf{v}_1) + \beta I(\mathbf{v}_2) \qquad \square$$

EXAMPLE 9.   Let L be the mapping from $C[a, b]$ to R^1 defined by

$$L(f) = \int_a^b f(x)\, dx$$

If f and g are any vectors in $C[a, b]$, then

$$L(\alpha f + \beta g) = \int_a^b (\alpha f + \beta g)(x)\, dx$$

$$= \alpha \int_a^b f(x)\, dx + \beta \int_a^b g(x)\, dx$$

$$= \alpha L(f) + \beta L(g)$$

Therefore, L is a linear transformation. $\qquad \square$

EXAMPLE 10.   Let D be the operator mapping $C^1[a, b]$ into $C[a, b]$ defined by

$$D(f) = f' \qquad (\text{the derivative of } f)$$

D is a linear transformation, since

$$D(\alpha f + \beta g) = \alpha f' + \beta g' = \alpha D(f) + \beta D(g) \qquad \square$$

EXERCISES

1. Show that each of the following are linear transformations from R^2 into R^2. Describe geometrically what each linear transformation accomplishes.
   (a) $L(\mathbf{x}) = (-x_1, x_2)^T$
   (b) $L(\mathbf{x}) = -\mathbf{x}$
   (c) $L(\mathbf{x}) = (x_2, x_1)^T$
   (d) $L(\mathbf{x}) = \frac{1}{2}\mathbf{x}$
   (e) $L(\mathbf{x}) = x_2 \mathbf{e}_2$

2. Let L be the linear transformation mapping R^2 into itself defined by

$$L(\mathbf{x}) = (x_1 \cos \alpha - x_2 \sin \alpha, \, x_1 \sin \alpha + x_2 \cos \alpha)^T$$

   Express x_1, x_2 and $L(\mathbf{x})$ in terms of polar coordinates. Describe geometrically the effect of the linear transformation.

3. Let $\mathbf{a}$ be a fixed nonzero vector in R^2. A mapping of the form

$$L(\mathbf{x}) = \mathbf{x} + \mathbf{a}$$

is called a *translation*. Show that a translation is not a linear transformation. Illustrate geometrically the effect of a translation.

4. Determine whether or not the following are linear transformations from R^3 into R^2.

(a) $L(\mathbf{x}) = (x_2, x_3)^T$ (b) $L(\mathbf{x}) = (0, 0)^T$
(c) $L(\mathbf{x}) = (1 + x_1, x_2)^T$ (d) $L(\mathbf{x}) = (x_3, x_1 + x_2)^T$

5. Determine whether or not the following are linear transformations from R^2 into R^3.

(a) $L(\mathbf{x}) = (x_1, x_2, 1)^T$ (b) $L(\mathbf{x}) = (x_1, x_2, x_1 + 2x_2)^T$
(c) $L(\mathbf{x}) = (x_1, 0, 0)^T$ (d) $L(\mathbf{x}) = (x_1, x_2, x_1^2 + x_2^2)^T$

6. Determine whether or not the following are linear transformations from $R^{n \times n}$ into $R^{n \times n}$.

(a) $L(A) = 2A$ (b) $L(A) = A^T$
(c) $L(A) = A + I$ (d) $L(A) = A - A^T$

7. Determine whether or not the following are linear transformations from P_2 to P_3.

(a) $L(p(x)) = xp(x)$
(b) $L(p(x)) = x^2 + p(x)$
(c) $L(p(x)) = p(x) + xp(x) + x^2 p'(x)$

8. For each $f \in C[0, 1]$ define $L(f) = F$, where

$$F(x) = \int_0^x f(t)\, dt \quad 0 \le x \le 1$$

Show that L is a linear transformation from $C[0, 1]$ to $C[0, 1]$. Determine $L(e^x)$ and $L(x^2)$.

9. Determine whether or not the following are linear transformations from $C[0, 1]$ into R^1.

(a) $L(f) = f(0)$ (b) $L(f) = |f(0)|$
(c) $L(f) = [f(0) + f(1)]/2$ (d) $L(f) = \left\{ \int_0^1 [f(x)]^2\, dx \right\}^{1/2}$

10. If L is a linear transformation from V to W, use mathematical induction to prove that

$$L(\alpha_1 \mathbf{v}_1 + \alpha_2 \mathbf{v}_2 + \cdots + \alpha_n \mathbf{v}_n)$$
$$= \alpha_1 L(\mathbf{v}_1) + \alpha_2 L(\mathbf{v}_2) + \cdots + \alpha_n L(\mathbf{v}_n)$$

11. Let $\mathbf{v}_1, \ldots, \mathbf{v}_n$ be a basis for a vector space V and let L_1 and L_2 be two linear transformations mapping V into a vector space W. Show that if

$$L_1(\mathbf{v}_i) = L_2(\mathbf{v}_i)$$

for each $i = 1, \ldots, n$ then $L_1 = L_2$, [i.e., show $L_1(\mathbf{v}) = L_2(\mathbf{v})$ for all $\mathbf{v} \in V$].

12. Let L be a linear transformation from R^1 into R^1 and let $a = L(1)$. Show that $L(x) = ax$ for all $x \in R^1$.

13. Let L be a linear operator mapping a vector space V into itself. Define $L^n, n \geq 1$, recursively by

$$L^1 = L$$
$$L^{k+1}(\mathbf{v}) = L(L^k(\mathbf{v})) \quad \text{for all } \mathbf{v} \in V$$

Show that L^n is a linear operator on V for each $n \geq 1$.

2. MATRIX REPRESENTATION OF LINEAR TRANSFORMATIONS

In Section 1 it was shown that each $m \times n$ matrix A defines a linear transformation L_A from R^n to R^m, where

$$L_A(\mathbf{x}) = A\mathbf{x}$$

for each $\mathbf{x} \in R^n$. In this section we will see that for every linear transformation L mapping R^n into R^m, there is an $m \times n$ matrix A such that

$$L(\mathbf{x}) = A\mathbf{x}$$

We will also see how any linear operator between finite-dimensional spaces can be represented by a matrix.

Theorem 4.2.1.   *If L is a linear operator mapping R^n into R^m, then there is an $m \times n$ matrix A such that*

$$L(\mathbf{x}) = A\mathbf{x}$$

for each $\mathbf{x} \in R^n$.

Proof.   For $j = 1, \ldots, n$ define

$$\mathbf{a}_j = (a_{1j}, a_{2j}, \ldots, a_{mj})^T = L(\mathbf{e}_j)$$

Let

$$A = (a_{ij}) = (\mathbf{a}_1, \mathbf{a}_2, \ldots, \mathbf{a}_n)$$

If

$$\mathbf{x} = x_1\mathbf{e}_1 + x_2\mathbf{e}_2 + \cdots + x_n\mathbf{e}_n$$

is an arbitrary element of R^n, then

$$L(\mathbf{x}) = x_1 L(\mathbf{e}_1) + x_2 L(\mathbf{e}_2) + \cdots + x_1 L(\mathbf{e}_n)$$
$$= x_1 \mathbf{a}_1 + x_2 \mathbf{a}_2 + \cdots + x_n \mathbf{a}_n$$
$$= (\mathbf{a}_1, \mathbf{a}_2, \dots, \mathbf{a}_n) \begin{pmatrix} x_1 \\ x_2 \\ \vdots \\ x_n \end{pmatrix}$$
$$= A\mathbf{x} \qquad \qquad \square$$

We have established that each linear transformation from R^n into R^m can be represented in terms of an $m \times n$ matrix. The proof of Theorem 4.2.1 tells us how to construct the matrix A corresponding to a particular linear operator L. To get the first column of A, see what L does to the first basis element $\mathbf{e}_1$ of R^n. Set $\mathbf{a}_1 = L(\mathbf{e}_1)$. To get the second column of A determine the effect of L on $\mathbf{e}_2$ and set $\mathbf{a}_2 = L(\mathbf{e}_2)$, and so on. Since the standard basis elements $\mathbf{e}_1, \mathbf{e}_2, \dots, \mathbf{e}_n$ were used for R^n, we refer to A as the *standard matrix representation* of L. Later we will see how to represent a linear operator with respect to other bases.

EXAMPLE 1.   Define the operator $L: R^3 \to R^2$ by

$$L\big((x_1, x_2, x_3)^T\big) = (x_1 + x_2, x_2 + x_3)^T$$

for each $\mathbf{x} = (x_1, x_2, x_3)^T$ in R^3. It is easily verified that L is a linear operator. We wish to find a matrix A such that $L(\mathbf{x}) = A\mathbf{x}$ for each $\mathbf{x} \in R^3$. To do this, one must determine $L(\mathbf{e}_1)$, $L(\mathbf{e}_2)$, and $L(\mathbf{e}_3)$.

$$L(\mathbf{e}_1) = L\big((1,0,0)^T\big) = \begin{pmatrix} 1 \\ 0 \end{pmatrix}$$

$$L(\mathbf{e}_2) = L\big((0,1,0)^T\big) = \begin{pmatrix} 1 \\ 1 \end{pmatrix}$$

$$L(\mathbf{e}_3) = L\big((0,0,1)^T\big) = \begin{pmatrix} 0 \\ 1 \end{pmatrix}$$

We choose these vectors to be the columns of A,

$$A = \begin{pmatrix} 1 & 1 & 0 \\ 0 & 1 & 1 \end{pmatrix}$$

To check the result, compute $A\mathbf{x}$.

$$A\mathbf{x} = \begin{pmatrix} 1 & 1 & 0 \\ 0 & 1 & 1 \end{pmatrix} \begin{pmatrix} x_1 \\ x_2 \\ x_3 \end{pmatrix} = \begin{pmatrix} x_1 + x_2 \\ x_2 + x_3 \end{pmatrix} \qquad \square$$

EXAMPLE 2.   Let L be the linear transformation mapping R^2 into R^2 which rotates each vector by an angle θ in the counter-

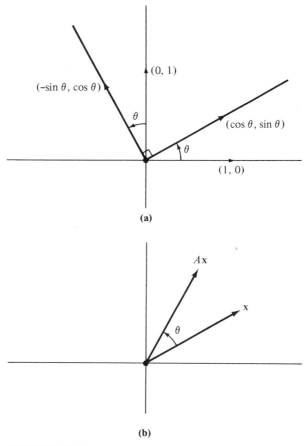

FIGURE 4.2.1

clockwise direction. We can see from Figure 4.2.1a that $\mathbf{e}_1$ will get mapped into $(\cos\theta, \sin\theta)^T$ and the image of $\mathbf{e}_2$ will be $(-\sin\theta, \cos\theta)^T$. The matrix A representing the transformation will have $(\cos\theta, \sin\theta)^T$ as its first column and $(-\sin\theta, \cos\theta)^T$ as its second column.

$$A = \begin{pmatrix} \cos\theta & -\sin\theta \\ \sin\theta & \cos\theta \end{pmatrix}$$

If $\mathbf{x}$ is any vector in R^2, then to rotate $\mathbf{x}$ counterclockwise by an angle θ, one simply multiplies by A (see Figure 4.2.1b). □

Now that we have seen how matrices are used to represent linear operators from R^n to R^m, we may ask if it is possible to find a similar representation for linear operators from V into W, where V and W are vector spaces of dimension n and m, respectively. To phrase this question properly, the following definition is necessary.

Definition. Let V be a vector space with ordered basis $[\mathbf{v}_1, \mathbf{v}_2, \ldots, \mathbf{v}_n]$. If $\mathbf{v}$ is any element of V, then $\mathbf{v}$ can be written in the form

$$\mathbf{v} = x_1\mathbf{v}_1 + x_2\mathbf{v}_2 + \cdots + x_n\mathbf{v}_n$$

where $x_1, x_2, \ldots, x_n$ are scalars. Thus we can associate with each vector $\mathbf{v}$ a unique vector $\mathbf{x} = (x_1, x_2, \ldots, x_n)^T$ in R^n. The vector $\mathbf{x}$ defined in this way is called the **coordinate vector** of $\mathbf{v}$ with respect to the ordered basis $[\mathbf{v}_1, \mathbf{v}_2, \ldots, \mathbf{v}_n]$ and the x_i's are called the **coordinates** of $\mathbf{v}$.

Let V and W be vector spaces with ordered bases $[\mathbf{v}_1, \mathbf{v}_2, \ldots, \mathbf{v}_n]$ and $[\mathbf{w}_1, \mathbf{w}_2, \ldots, \mathbf{w}_m]$, respectively. Let L be a linear operator mapping V into W. If $\mathbf{x}$ is the coordinate vector of $\mathbf{v}$ with respect to $[\mathbf{v}_1, \ldots, \mathbf{v}_n]$ and $\mathbf{y}$ is the coordinate vector of $L(\mathbf{v})$ with respect to $[\mathbf{w}_1, \ldots, \mathbf{w}_m]$, we can think of L as an operator that associates with each vector $\mathbf{x} \in R^n$ a vector $\mathbf{y} \in R^m$. We have already seen how such an operator can be represented by an $m \times n$ matrix. Thus it should be possible to find a matrix A such that $A\mathbf{x} = \mathbf{y}$. The procedure for determining A is essentially the same as before. For $j = 1, \ldots, n$ let $\mathbf{a}_j = (a_{1j}, a_{2j}, \ldots, a_{mj})^T$ be the coordinate vector of $L(\mathbf{v}_j)$ with respect to $[\mathbf{w}_1, \mathbf{w}_2, \ldots, \mathbf{w}_m]$.

$$L(\mathbf{v}_j) = a_{1j}\mathbf{w}_1 + a_{2j}\mathbf{w}_2 + \cdots + a_{mj}\mathbf{w}_m \qquad 1 \le j \le n$$

Let $A = (a_{ij}) = (\mathbf{a}_1, \ldots, \mathbf{a}_n)$. If

$$\mathbf{v} = x_1\mathbf{v}_1 + x_2\mathbf{v}_2 + \cdots + x_n\mathbf{v}_n$$

then

$$\begin{aligned}
L(\mathbf{v}) &= L\left(\sum_{j=1}^{n} x_j\mathbf{v}_j \right) \\
&= \sum_{j=1}^{n} x_j L(\mathbf{v}_j) \\
&= \sum_{j=1}^{n} x_j \left(\sum_{i=1}^{m} a_{ij}\mathbf{w}_i \right) \\
&= \sum_{i=1}^{m} \left(\sum_{j=1}^{n} a_{ij}x_j \right) \mathbf{w}_i
\end{aligned}$$

For $i = 1, \ldots, m$, let

$$y_i = \sum_{j=1}^{n} a_{ij}x_j$$

Thus

$$\mathbf{y} = (y_1, y_2, \ldots, y_m)^T = A\mathbf{x}$$

is the coordinate vector of $L(\mathbf{v})$ with respect to $[\mathbf{w}_1, \mathbf{w}_2, \ldots, \mathbf{w}_m]$.

EXAMPLE 3. Let L be a linear transformation mapping R^3 into R^2 defined by

$$L(\mathbf{x}) = x_1\mathbf{b}_1 + (x_2 + x_3)\mathbf{b}_2$$

for each $\mathbf{x} \in R^3$, where

$$\mathbf{b}_1 = \begin{pmatrix} 1 \\ 1 \end{pmatrix} \quad \text{and} \quad \mathbf{b}_2 = \begin{pmatrix} -1 \\ 1 \end{pmatrix}$$

Find the matrix A representing L with respect to the ordered bases $[\mathbf{e}_1, \mathbf{e}_2, \mathbf{e}_3]$ and $[\mathbf{b}_1, \mathbf{b}_2]$.

SOLUTION

$$L(\mathbf{e}_1) = 1\mathbf{b}_1 + 0\mathbf{b}_2$$
$$L(\mathbf{e}_2) = 0\mathbf{b}_1 + 1\mathbf{b}_2$$
$$L(\mathbf{e}_3) = 0\mathbf{b}_1 + 1\mathbf{b}_2$$

The ith column of A is determined by the coordinates of $L(\mathbf{e}_i)$ with respect to $[\mathbf{b}_1, \mathbf{b}_2]$ for $i = 1, 2, 3$. Thus

$$A = \begin{pmatrix} 1 & 0 & 0 \\ 0 & 1 & 1 \end{pmatrix}$$ □

EXAMPLE 4. Let L be a linear transformation mapping R^2 into itself defined by

$$L(\alpha\mathbf{b}_1 + \beta\mathbf{b}_2) = (\alpha + \beta)\mathbf{b}_1 + 2\beta\mathbf{b}_2$$

where $[\mathbf{b}_1, \mathbf{b}_2]$ is the ordered basis defined in Example 3. Find the matrix A representing L with respect to $[\mathbf{b}_1, \mathbf{b}_2]$.

SOLUTION

$$L(\mathbf{b}_1) = 1\mathbf{b}_1 + 0\mathbf{b}_2$$
$$L(\mathbf{b}_2) = 1\mathbf{b}_1 + 2\mathbf{b}_2$$

Thus

$$A = \begin{pmatrix} 1 & 1 \\ 0 & 2 \end{pmatrix}$$ □

EXAMPLE 5. The linear operator D defined by $D(p) = p'$ maps P_3 into P_2. Given the ordered bases $[x^2, x, 1]$ and $[x, 1]$ for P_3 and P_2, respectively, we wish to determine a matrix representation for D. To do this, we apply D to each of the basis elements of P_3.

$$D(x^2) = 2x + 0 \cdot 1$$
$$D(x) = 0x + 1 \cdot 1$$
$$D(1) = 0x + 0 \cdot 1$$

In P_2, the coordinate vectors for $D(x^2)$, $D(x)$, $D(1)$ are

$(2, 0)^T, (0, 1)^T, (0, 0)^T$, respectively. The matrix A is formed using these vectors as its columns.

$$A = \begin{pmatrix} 2 & 0 & 0 \\ 0 & 1 & 0 \end{pmatrix}$$

If $p(x) = ax^2 + bx + c$, then the coordinate vector of p with respect to the ordered basis of P_3 is $(a, b, c)^T$. To find the coordinate vector of $D(p)$ with respect to the ordered basis of P_2, we simply multiply

$$\begin{pmatrix} 2 & 0 & 0 \\ 0 & 1 & 0 \end{pmatrix} \begin{pmatrix} a \\ b \\ c \end{pmatrix} = \begin{pmatrix} 2a \\ b \end{pmatrix}$$

Thus $D(ax^2 + bx + c) = 2ax + b$.   □

EXERCISES

1. Refer back to Exercise 1 of Section 1. For each linear transformation L, find a matrix A representing L.

2. For each of the following linear transformations L mapping R^3 into R^2, find a matrix A such that $L(\mathbf{x}) = A\mathbf{x}$ for every $\mathbf{x}$ in R^3.
   (a) $L((x_1, x_2, x_3)^T) = (x_1 + x_2, 0)^T$
   (b) $L((x_1, x_2, x_3)^T) = (x_1, x_2)^T$
   (c) $L((x_1, x_2, x_3)^T) = (x_2 - x_1, x_3 - x_2)^T$

3. For each of the following linear transformations L from R^3 into R^3 find a matrix A such that $L(\mathbf{x}) = A\mathbf{x}$ for every $\mathbf{x}$ in R^3.
   (a) $L((x_1, x_2, x_3)^T) = (x_3, x_2, x_1)^T$
   (b) $L((x_1, x_2, x_3)^T) = (x_1, x_1 + x_2, x_1 + x_2 + x_3)^T$
   (c) $L((x_1, x_2, x_3)^T) = (2x_3, x_2 + 3x_1, 2x_1 - x_3)^T$

4. Let L be the linear transformation mapping R^3 into R^3 defined by

$$L(\mathbf{x}) = (2x_1 - x_2 - x_3, 2x_2 - x_1 - x_3, 2x_3 - x_1 - x_2)^T$$

Determine the standard matrix representation A of L and use A to find $L(\mathbf{x})$ for each of the following vectors $\mathbf{x}$.
   (a) $\mathbf{x} = (1, 1, 1)^T$ (b) $\mathbf{x} = (2, 1, 1)^T$ (c) $\mathbf{x} = (-5, 3, 2)^T$

5. Find the standard matrix representation for each linear operator L on R^2 described below.
   (a) L rotates each vector $\mathbf{x}$ by $45°$ in the clockwise direction.
   (b) L reflects each vector $\mathbf{x}$ about the x_1 axis and then rotates it $90°$ in the counterclockwise direction.
   (c) L doubles the length of $\mathbf{x}$ and then rotates it $30°$ in the counterclockwise direction.

(d) L reflects each vector $\mathbf{x}$ about the line $x_2 = x_1$ and then projects it onto the x_1 axis.

6. Let

$$\mathbf{b}_1 = \begin{pmatrix} 1 \\ 1 \\ 0 \end{pmatrix}, \qquad \mathbf{b}_2 = \begin{pmatrix} 1 \\ 0 \\ 1 \end{pmatrix}, \qquad \mathbf{b}_3 = \begin{pmatrix} 0 \\ 1 \\ 1 \end{pmatrix}$$

and let L be the linear transformation from R^2 into R^3 defined by

$$L(\mathbf{x}) = x_1\mathbf{b}_1 + x_2\mathbf{b}_2 + (x_1 + x_2)\mathbf{b}_3$$

Find the matrix A representing L with respect to the bases $[\mathbf{e}_1, \mathbf{e}_2]$ and $[\mathbf{b}_1, \mathbf{b}_2, \mathbf{b}_3]$.

7. Let

$$\mathbf{y}_1 = \begin{pmatrix} 1 \\ 1 \\ 1 \end{pmatrix}, \qquad \mathbf{y}_2 = \begin{pmatrix} 1 \\ 1 \\ 0 \end{pmatrix}, \qquad \mathbf{y}_3 = \begin{pmatrix} 1 \\ 0 \\ 0 \end{pmatrix}$$

and let $\mathcal{I}$ be the identity operator on R^3.
(a) Find the coordinates of $\mathcal{I}(\mathbf{e}_1)$, $\mathcal{I}(\mathbf{e}_2)$, and $\mathcal{I}(\mathbf{e}_3)$ with respect to $[\mathbf{y}_1, \mathbf{y}_2, \mathbf{y}_3]$.
(b) Find a matrix A such that $A\mathbf{x}$ is the coordinate vector of $\mathbf{x}$ with respect to $[\mathbf{y}_1, \mathbf{y}_2, \mathbf{y}_3]$.

8. Let $\mathbf{y}_1, \mathbf{y}_2, \mathbf{y}_3$ be defined as in Exercise 7 and let L be the linear transformation from R^3 into R^3 defined by

$$L(c_1\mathbf{y}_1 + c_2\mathbf{y}_2 + c_3\mathbf{y}_3)$$
$$= (c_1 + c_2 + c_3)\mathbf{y}_1 + (2c_1 + c_3)\mathbf{y}_2 - (2c_2 + c_3)\mathbf{y}_3$$

(a) Find a matrix representing L with respect to the ordered basis $[\mathbf{y}_1, \mathbf{y}_2, \mathbf{y}_3]$.
(b) For each of the following, write the vector $\mathbf{x}$ as a linear combination of $\mathbf{y}_1, \mathbf{y}_2, \mathbf{y}_3$ and use the matrix from part (a) to determine $L(\mathbf{x})$.
  (i) $\mathbf{x} = (7, 5, 2)^T$ (ii) $\mathbf{x} = (3, 2, 1)^T$ (iii) $\mathbf{x} = (1, 2, 3)^T$

9. Let L be the linear operator mapping P_2 into R^2 defined by

$$L(p(x)) = \begin{pmatrix} \int_0^1 p(x)\, dx \\ p(0) \end{pmatrix}$$

Find a matrix A such that

$$L(\alpha + \beta x) = A\begin{pmatrix} \alpha \\ \beta \end{pmatrix}$$

10. The linear operator L defined by

$$L(p(x)) = p'(x) + p(0)$$

maps P_3 into P_2. Find the matrix representation of L with respect to the ordered bases $[x^2, x, 1]$ and $[2, 1 - x]$. For each of the following vectors $p(x)$ in P_3, find the coordinates of $L(p(x))$ with respect to the ordered basis $[2, 1 - x]$.
(a) $x^2 + 2x - 3$ (b) $x^2 + 1$
(c) $3x$ (d) $4x^2 + 2x$

11. Let S be the subspace of $C[a, b]$ spanned by e^x, xe^x, and x^2e^x. Let D be the differentiation operator on S. Find the matrix representing D with respect to $[e^x, xe^x, x^2e^x]$.

12. Let L be a linear transformation from R^n into R^n. Suppose that $L(\mathbf{x}) = \mathbf{0}$ for some $\mathbf{x} \neq \mathbf{0}$. Let A be the matrix representing L with respect to the standard basis $[\mathbf{e}_1, \mathbf{e}_2, \ldots, \mathbf{e}_n]$. Show that A is singular.

13. Let L be a linear operator mapping a vector space V into itself. Let A be the matrix representing L with respect to the ordered basis $[\mathbf{v}_1, \ldots, \mathbf{v}_n]$ of V [i.e., $L(\mathbf{v}_j) = \sum_{i=1}^{n} a_{ij}\mathbf{v}_i$, $j = 1, \ldots, n$]. Show that A^m is the matrix representing L^m with respect to $[\mathbf{v}_1, \ldots, \mathbf{v}_n]$.

3. CHANGE OF BASIS

Many applied problems can be simplified by changing from one coordinate system to another. For example, in describing the motion of a particle in the plane at a particular time, it is often convenient to use a basis for R^2 consisting of a unit tangent vector $\mathbf{t}$ and a unit normal vector $\mathbf{n}$ instead of the standard basis $\mathbf{e}_1, \mathbf{e}_2$. In this section we discuss the problem of switching from one coordinate system to another. We show that this can be accomplished by multiplying a given coordinate vector $\mathbf{x}$ by a nonsingular matrix S. The product $\mathbf{y} = S\mathbf{x}$ will be the coordinate vector for the new coordinate system.

CHANGING COORDINATES IN R^2

Suppose that instead of using the standard basis $[\mathbf{e}_1, \mathbf{e}_2]$ for R^2 we wish to use a different basis, say

$$\mathbf{u}_1 = \begin{pmatrix} 3 \\ 2 \end{pmatrix}, \qquad \mathbf{u}_2 = \begin{pmatrix} 1 \\ 1 \end{pmatrix}$$

Indeed, we may want to switch back and forth between the two coordinate systems. Let us consider the following two problems.

I. Given a vector $\mathbf{x} = (x_1, x_2)^T$, find its coordinates with respect to $\mathbf{u}_1$ and $\mathbf{u}_2$.

II. Given a vector $c_1\mathbf{u}_1 + c_2\mathbf{u}_2$, find its coordinates with respect to $\mathbf{e}_1$ and $\mathbf{e}_2$.

We will solve II first, since it turns out to be the easier problem. To switch bases from $[\mathbf{u}_1, \mathbf{u}_2]$ to $[\mathbf{e}_1, \mathbf{e}_2]$, we must express the old basis elements $\mathbf{u}_1$ and $\mathbf{u}_2$ in terms of the new basis elements $\mathbf{e}_1$ and $\mathbf{e}_2$.

$$\mathbf{u}_1 = 3\mathbf{e}_1 + 2\mathbf{e}_2$$
$$\mathbf{u}_2 = \mathbf{e}_1 + \mathbf{e}_2$$

It follows then that

$$c_1\mathbf{u}_1 + c_2\mathbf{u}_2 = (3c_1\mathbf{e}_1 + 2c_1\mathbf{e}_2) + (c_2\mathbf{e}_1 + c_2\mathbf{e}_2)$$
$$= (3c_1 + c_2)\mathbf{e}_1 + (2c_1 + c_2)\mathbf{e}_2$$

Thus the coordinate vector of $c_1\mathbf{u}_1 + c_2\mathbf{u}_2$ with respect to $[\mathbf{e}_1, \mathbf{e}_2]$ is

$$\mathbf{x} = \begin{pmatrix} 3c_1 + c_2 \\ 2c_1 + c_2 \end{pmatrix} = \begin{pmatrix} 3 & 1 \\ 2 & 1 \end{pmatrix}\begin{pmatrix} c_1 \\ c_2 \end{pmatrix}$$

If we set

$$S = (\mathbf{u}_1, \mathbf{u}_2) = \begin{pmatrix} 3 & 1 \\ 2 & 1 \end{pmatrix}$$

then given any coordinate vector $\mathbf{c}$ with respect to $[\mathbf{u}_1, \mathbf{u}_2]$, to find the corresponding coordinate vector $\mathbf{x}$ with respect to $[\mathbf{e}_1, \mathbf{e}_2]$, one simply multiplies S times $\mathbf{c}$.

$$(1) \qquad\qquad\qquad \mathbf{x} = S\mathbf{c}$$

The matrix S is called the *transition matrix* from the ordered basis $[\mathbf{u}_1, \mathbf{u}_2]$ to the basis $[\mathbf{e}_1, \mathbf{e}_2]$.

To solve problem I, we must find the transition matrix from $[\mathbf{e}_1, \mathbf{e}_2]$ to $[\mathbf{u}_1, \mathbf{u}_2]$. The matrix S in (1) is nonsingular, since its column vectors, $\mathbf{u}_1$ and $\mathbf{u}_2$, are linearly independent. It follows from (1) that

$$\mathbf{c} = S^{-1}\mathbf{x}$$

Thus given a vector

$$\mathbf{x} = (x_1, x_2)^T = x_1\mathbf{e}_1 + x_2\mathbf{e}_2$$

one need only multiply by S^{-1} to find its coordinate vector with respect to $[\mathbf{u}_1, \mathbf{u}_2]$. S^{-1} is the transition matrix from $[\mathbf{e}_1, \mathbf{e}_2]$ to $[\mathbf{u}_1, \mathbf{u}_2]$.

EXAMPLE 1.   Let $\mathbf{u}_1 = (3, 2)^T$, $\mathbf{u}_2 = (1, 1)^T$, and $\mathbf{x} = (5, 4)^T$. Find the coordinates of $\mathbf{x}$ with respect to $\mathbf{u}_1$ and $\mathbf{u}_2$.

SOLUTION.   By the preceding discussion the transition matrix from $[\mathbf{e}_1, \mathbf{e}_2]$ to $[\mathbf{u}_1, \mathbf{u}_2]$ is the inverse of

$$S = (\mathbf{u}_1, \mathbf{u}_2) = \begin{pmatrix} 3 & 1 \\ 2 & 1 \end{pmatrix}$$

Thus

$$\mathbf{c} = S^{-1}\mathbf{x} = \begin{pmatrix} 1 & -1 \\ -2 & 3 \end{pmatrix}\begin{pmatrix} 5 \\ 4 \end{pmatrix} = \begin{pmatrix} 1 \\ 2 \end{pmatrix}$$

is the desired coordinate vector and

$$\mathbf{x} = 1\mathbf{u}_1 + 2\mathbf{u}_2 \qquad\qquad \square$$

EXAMPLE 2.   Let $\mathbf{b}_1 = (1, -1)^T$ and $\mathbf{b}_2 = (-2, 3)^T$. Find the transition matrix from $[\mathbf{e}_1, \mathbf{e}_2]$ to $[\mathbf{b}_1, \mathbf{b}_2]$ and determine the coordinates of $\mathbf{x} = (1, 2)^T$ with respect to $[\mathbf{b}_1, \mathbf{b}_2]$.

SOLUTION.   The transition matrix from $[\mathbf{b}_1, \mathbf{b}_2]$ to $[\mathbf{e}_1, \mathbf{e}_2]$ is

$$S = (\mathbf{b}_1, \mathbf{b}_2) = \begin{pmatrix} 1 & -2 \\ -1 & 3 \end{pmatrix}$$

and hence the transition matrix from $[\mathbf{e}_1, \mathbf{e}_2]$ to $[\mathbf{b}_1, \mathbf{b}_2]$ is

$$S^{-1} = \begin{pmatrix} 3 & 2 \\ 1 & 1 \end{pmatrix}$$

The coordinate vector of $\mathbf{x}$ with respect to $[\mathbf{b}_1, \mathbf{b}_2]$ is

$$\mathbf{c} = S^{-1}\mathbf{x} = \begin{pmatrix} 3 & 2 \\ 1 & 1 \end{pmatrix}\begin{pmatrix} 1 \\ 2 \end{pmatrix} = \begin{pmatrix} 7 \\ 3 \end{pmatrix}$$

and hence

$$\mathbf{x} = 7\mathbf{b}_1 + 3\mathbf{b}_2 \qquad\qquad \square$$

If

$$S = \begin{pmatrix} s_{11} & s_{12} \\ s_{21} & s_{22} \end{pmatrix}$$

is the transition matrix from an ordered basis $[\mathbf{v}_1, \mathbf{v}_2]$ of R^2 to an ordered basis $[\mathbf{u}_1, \mathbf{u}_2]$, then since

$$\mathbf{v}_1 = 1\mathbf{v}_1 + 0\mathbf{v}_2$$

the coordinate vector of $\mathbf{v}_1$ with respect to $[\mathbf{u}_1, \mathbf{u}_2]$ is given by

$$\mathbf{s}_1 = \begin{pmatrix} s_{11} & s_{12} \\ s_{21} & s_{22} \end{pmatrix}\begin{pmatrix} 1 \\ 0 \end{pmatrix} = \begin{pmatrix} s_{11} \\ s_{21} \end{pmatrix}$$

Similarly,

$$\mathbf{v}_2 = 0\mathbf{v}_1 + 1\mathbf{v}_2$$

and hence its coordinate vector with respect to $[\mathbf{u}_1, \mathbf{u}_2]$ is

$$\mathbf{s}_2 = \begin{pmatrix} s_{11} & s_{12} \\ s_{21} & s_{22} \end{pmatrix}\begin{pmatrix} 0 \\ 1 \end{pmatrix} = \begin{pmatrix} s_{12} \\ s_{22} \end{pmatrix}$$

Thus

(2)
$$\mathbf{v}_1 = s_{11}\mathbf{u}_1 + s_{21}\mathbf{u}_2$$
$$\mathbf{v}_2 = s_{12}\mathbf{u}_1 + s_{22}\mathbf{u}_2$$

In general, if the old basis elements $\mathbf{v}_1$ and $\mathbf{v}_2$ are written in terms of the new basis $[\mathbf{u}_1, \mathbf{u}_2]$, then the coordinate vector $\mathbf{s}_1 = (s_{11}, s_{21})^T$ corresponding to $\mathbf{v}_1$ will be the first column of the transition matrix S and the coordinate vector $\mathbf{s}_2 = (s_{12}, s_{22})^T$ corresponding to $\mathbf{v}_2$ will be the second column of S. Thus S is the transpose of the coefficient array in (2).

EXAMPLE 3. Find the transition matrix corresponding to the change of basis from $[\mathbf{v}_1, \mathbf{v}_2]$ to $[\mathbf{u}_1, \mathbf{u}_2]$, where

$$\mathbf{v}_1 = \begin{pmatrix} 5 \\ 2 \end{pmatrix}, \quad \mathbf{v}_2 = \begin{pmatrix} 7 \\ 3 \end{pmatrix} \quad \text{and} \quad \mathbf{u}_1 = \begin{pmatrix} 3 \\ 2 \end{pmatrix}, \quad \mathbf{u}_2 = \begin{pmatrix} 1 \\ 1 \end{pmatrix}$$

SOLUTION.   We must express $\mathbf{v}_1$ and $\mathbf{v}_2$ in terms of the new basis elements $\mathbf{u}_1$ and $\mathbf{u}_2$.

$$\mathbf{v}_1 = s_{11}\mathbf{u}_1 + s_{21}\mathbf{u}_2$$
$$\mathbf{v}_2 = s_{12}\mathbf{u}_1 + s_{22}\mathbf{u}_2$$

The first equation can be written

$$\begin{pmatrix} 5 \\ 2 \end{pmatrix} = \begin{pmatrix} 3s_{11} + s_{21} \\ 2s_{11} + s_{21} \end{pmatrix}$$

The solution to this system is $(s_{11}, s_{21})^T = (3, -4)^T$. Similarly, the second equation leads to the system

$$\begin{pmatrix} 7 \\ 3 \end{pmatrix} = \begin{pmatrix} 3s_{12} + s_{22} \\ 2s_{12} + s_{22} \end{pmatrix}$$

The solution to this system is $(s_{12}, s_{22})^T = (4, -5)^T$. Thus

$$S = \begin{pmatrix} 3 & 4 \\ -4 & -5 \end{pmatrix}$$

is the transition matrix from $[\mathbf{v}_1, \mathbf{v}_2]$ to $[\mathbf{u}_1, \mathbf{u}_2]$.   □

Example 3 could have been solved by finding the transition matrix S_2 from $[\mathbf{v}_1, \mathbf{v}_2]$ to $[\mathbf{e}_1, \mathbf{e}_2]$ and the transition matrix S_1 from $[\mathbf{e}_1, \mathbf{e}_2]$ to $[\mathbf{u}_1, \mathbf{u}_2]$. If

$$c_1\mathbf{v}_1 + c_2\mathbf{v}_2 = x_1\mathbf{e}_1 + x_2\mathbf{e}_2 = d_1\mathbf{u}_1 + d_2\mathbf{u}_2$$

then

$$S_2 c = \mathbf{x} \quad \text{and} \quad S_1 \mathbf{x} = \mathbf{d}$$

Thus

$$S_1 S_2 c = S_1 \mathbf{x} = \mathbf{d}$$

that is, $S_1 S_2$ is the transition matrix from $[\mathbf{v}_1, \mathbf{v}_2]$ to $[\mathbf{u}_1, \mathbf{u}_2]$ (see Exercise 3).

The examples considered so far have all dealt with changing coordinates in R^2. Similar techniques could be used for R^n. In the case of R^n the transition matrices will be $n \times n$. Indeed, if V is any n-dimensional vector space, it is possible to change from one basis to another by means of an $n \times n$ transition matrix. We will show that such a transition matrix is necessarily nonsingular.

CHANGE OF BASIS FOR A GENERAL VECTOR SPACE

Let $[\mathbf{w}_1, \ldots, \mathbf{w}_n]$ and $[\mathbf{v}_1, \ldots, \mathbf{v}_n]$ be two ordered bases for a vector space V. Each vector $\mathbf{w}_j$ can then be expressed as a linear combination of the $\mathbf{v}_i$'s,

(3)
$$
\begin{aligned}
\mathbf{w}_1 &= s_{11}\mathbf{v}_1 + s_{21}\mathbf{v}_2 + \cdots + s_{n1}\mathbf{v}_n \\
\mathbf{w}_2 &= s_{12}\mathbf{v}_1 + s_{22}\mathbf{v}_2 + \cdots + s_{n2}\mathbf{v}_n \\
&\;\;\vdots \\
\mathbf{w}_n &= s_{1n}\mathbf{v}_1 + s_{2n}\mathbf{v}_2 + \cdots + s_{nn}\mathbf{v}_n
\end{aligned}
$$

Let $\mathbf{v} \in V$. If $\mathbf{x}$ is the coordinate vector of $\mathbf{v}$ with respect to $[\mathbf{w}_1, \ldots, \mathbf{w}_n]$, it follows from (3) that

$$
\begin{aligned}
\mathbf{v} &= x_1 \mathbf{w}_1 + x_2 \mathbf{w}_2 + \cdots + x_n \mathbf{w}_n \\
&= \left(\sum_{j=1}^{n} s_{1j} x_j \right) \mathbf{v}_1 + \left(\sum_{j=1}^{n} s_{2j} x_j \right) \mathbf{v}_2 + \cdots + \left(\sum_{j=1}^{n} s_{nj} x_j \right) \mathbf{v}_n
\end{aligned}
$$

Thus, if $\mathbf{y}$ is the coordinate vector of $\mathbf{v}$ with respect to $[\mathbf{v}_1, \ldots, \mathbf{v}_n]$, then

$$
y_i = \sum_{j=1}^{n} s_{ij} x_j \quad i = 1, \ldots, n
$$

and hence

$$
\mathbf{y} = S\mathbf{x}
$$

The matrix S defined by (3) is referred to as the *transition matrix*. Once S has been determined, it is a simple matter to change coordinate systems. To find the coordinates of $\mathbf{v} = x_1 \mathbf{w}_1 + \cdots + x_n \mathbf{w}_n$ with respect to $[\mathbf{v}_1, \ldots, \mathbf{v}_n]$, one need only calculate $\mathbf{y} = S\mathbf{x}$.

The transition matrix S corresponding to the change of basis from $[\mathbf{w}_1, \ldots, \mathbf{w}_n]$ to $[\mathbf{v}_1, \ldots, \mathbf{v}_n]$ can be characterized by the condition:

(4)

$\quad S\mathbf{x} = \mathbf{y}$ if and only if $x_1 \mathbf{w}_1 + \cdots + x_n \mathbf{w}_n = y_1 \mathbf{v}_1 + \cdots + y_n \mathbf{v}_n$

Taking $\mathbf{y} = \mathbf{0}$ in (4), we see that $S\mathbf{x} = \mathbf{0}$ implies

$$
x_1 \mathbf{w}_1 + \cdots + x_n \mathbf{w}_n = \mathbf{0}
$$

Since the $\mathbf{w}_i$'s are linearly independent, it follows that $\mathbf{x} = \mathbf{0}$. Thus the equation $S\mathbf{x} = \mathbf{0}$ has only the trivial solution and hence the matrix S is nonsingular. The inverse matrix is characterized by the condition

$$S^{-1}\mathbf{y} = \mathbf{x} \qquad \text{if and only if} \qquad y_1\mathbf{v}_1 + \cdots + y_n\mathbf{v}_n = x_1\mathbf{w}_1 + \cdots + x_n\mathbf{w}_n$$

Thus S^{-1} is the transition matrix used to change basis from $[\mathbf{v}_1, \ldots, \mathbf{v}_n]$ to $[\mathbf{w}_1, \ldots, \mathbf{w}_n]$.

EXAMPLE 4. Suppose that in P_3 we want to change from the ordered basis $[1, x, x^2]$ to the ordered basis $[1, 2x, 4x^2 - 2]$. Since $[1, x, x^2]$ is the standard basis for P_3, it is easier to find the transition matrix from $[1, 2x, 4x^2 - 2]$ to $[1, x, x^2]$. Since

$$
\begin{aligned}
1 &= 1 \cdot 1 + 0x + 0x^2 \\
2x &= 0 \cdot 1 + 2x + 0x^2 \\
4x^2 - 2 &= -2 \cdot 1 + 0x + 4x^2
\end{aligned}
$$

the transition matrix is

$$S = \begin{pmatrix} 1 & 0 & -2 \\ 0 & 2 & 0 \\ 0 & 0 & 4 \end{pmatrix}$$

The inverse of S will be the transition matrix from $[1, x, x^2]$ to $[1, 2x, 4x^2 - 2]$:

$$S^{-1} = \begin{pmatrix} 1 & 0 & \frac{1}{2} \\ 0 & \frac{1}{2} & 0 \\ 0 & 0 & \frac{1}{4} \end{pmatrix}$$

Given any $p(x) = a + bx + cx^2$ in P_3, to find the coordinates of $p(x)$ with respect to $[1, 2x, 4x^2 - 2]$ one simply multiplies

$$\begin{pmatrix} 1 & 0 & \frac{1}{2} \\ 0 & \frac{1}{2} & 0 \\ 0 & 0 & \frac{1}{4} \end{pmatrix} \begin{pmatrix} a \\ b \\ c \end{pmatrix} = \begin{pmatrix} a + \frac{1}{2}c \\ \frac{1}{2}b \\ \frac{1}{4}c \end{pmatrix}$$

Thus

$$p(x) = \left(a + \tfrac{1}{2}c\right) \cdot 1 + \left(\tfrac{1}{2}b\right) \cdot 2x + \tfrac{1}{4}c \cdot (4x^2 - 2) \qquad \square$$

We have seen that each transition matrix is nonsingular. Actually, any nonsingular matrix can be thought of as a transition matrix. If S is an $n \times n$ nonsingular matrix and $[\mathbf{v}_1, \ldots, \mathbf{v}_n]$ is an ordered basis for V, then define $\mathbf{w}_1, \mathbf{w}_2, \ldots, \mathbf{w}_n$ by (3). To see that the $\mathbf{w}_j$'s are linearly independent, suppose

$$\sum_{j=1}^{n} x_j\mathbf{w}_j = \mathbf{0}$$

It follows from (3) that

$$\sum_{i=1}^{n} \left(\sum_{j=1}^{n} s_{ij} x_j \right) \mathbf{v}_i = \mathbf{0}$$

By the linear independence of the $\mathbf{v}_i$'s it follows that

$$\sum_{j=1}^{n} s_{ij} x_j = 0 \qquad i = 1, \ldots, n$$

or, equivalently,

$$S\mathbf{x} = \mathbf{0}$$

Since S is nonsingular, $\mathbf{x}$ must equal $\mathbf{0}$. Therefore, $\mathbf{w}_1, \ldots, \mathbf{w}_n$ are linearly independent and hence they form a basis for V. The matrix S is the transition matrix corresponding to the change from the ordered basis $[\mathbf{w}_1, \ldots, \mathbf{w}_n]$ to $[\mathbf{v}_1, \ldots, \mathbf{v}_n]$.

EXERCISES

1. For each of the following, find the transition matrix corresponding to the change of basis from $[\mathbf{u}_1, \mathbf{u}_2]$ to $[\mathbf{e}_1, \mathbf{e}_2]$.
   (a) $\mathbf{u}_1 = (1, 1)^T$, $\mathbf{u}_2 = (-1, 1)^T$
   (b) $\mathbf{u}_1 = (1, 2)^T$, $\mathbf{u}_2 = (2, 5)^T$
   (c) $\mathbf{u}_1 = (0, 1)^T$, $\mathbf{u}_2 = (1, 0)^T$

2. For each of the ordered bases $[\mathbf{u}_1, \mathbf{u}_2]$ in Exercise 1, find the transition matrix corresponding to the change of basis from $[\mathbf{e}_1, \mathbf{e}_2]$ to $[\mathbf{u}_1, \mathbf{u}_2]$.

3. Let $[\mathbf{v}_1, \mathbf{v}_2]$ and $[\mathbf{u}_1, \mathbf{u}_2]$ be the ordered bases defined in Example 3 and let S_1 be the transition matrix from $[\mathbf{e}_1, \mathbf{e}_2]$ to $[\mathbf{u}_1, \mathbf{u}_2]$ given in Example 1.
   (a) Determine the transition matrix S_2 from $[\mathbf{v}_1, \mathbf{v}_2]$ to $[\mathbf{e}_1, \mathbf{e}_2]$.
   (b) Compute the product $S_1 S_2$ and compare it to the transition matrix S in Example 3.

4. Let $\mathbf{v}_1 = (3, 2)^T$ and $\mathbf{v}_2 = (4, 3)^T$. For each ordered basis $[\mathbf{u}_1, \mathbf{u}_2]$ given in Exercise 1, find the transition matrix from $[\mathbf{v}_1, \mathbf{v}_2]$ to $[\mathbf{u}_1, \mathbf{u}_2]$.

5. Let $\mathbf{u}_1 = (1, 1, 1)^T$, $\mathbf{u}_2 = (1, 2, 2)^T$, $\mathbf{u}_3 = (2, 3, 4)^T$.
   (a) Find the transition matrix corresponding to the change of basis from $[\mathbf{e}_1, \mathbf{e}_2, \mathbf{e}_3]$ to $[\mathbf{u}_1, \mathbf{u}_2, \mathbf{u}_3]$.
   (b) Find the coordinates of each of the following vectors with respect to $[\mathbf{u}_1, \mathbf{u}_2, \mathbf{u}_3]$.
   (i) $(3, 2, 5)^T$,   (ii) $(1, 1, 2)^T$,   (iii) $(2, 3, 2)^T$

6. Let $\mathbf{v}_1 = (4, 6, 7)^T$, $\mathbf{v}_2 = (0, 1, 1)^T$, and $\mathbf{v}_3 = (0, 1, 2)^T$, and let $\mathbf{u}_1, \mathbf{u}_2, \mathbf{u}_3$ be the vectors given in Exercise 5. Find the transition matrix from $[\mathbf{v}_1, \mathbf{v}_2, \mathbf{v}_3]$ to $[\mathbf{u}_1, \mathbf{u}_2, \mathbf{u}_3]$. If $\mathbf{x} = 2\mathbf{v}_1 + 3\mathbf{v}_2 - 4\mathbf{v}_3$, determine the coordinates of $\mathbf{x}$ with respect to $[\mathbf{u}_1, \mathbf{u}_2, \mathbf{u}_3]$.

7. Let $[x, 1]$ and $[2x - 1, 2x + 1]$ be ordered bases for P_2.
   (a) Find the transition matrix representing the change in coordinates from $[2x - 1, 2x + 1]$ to $[x, 1]$.
   (b) Find the transition matrix representing the change in coordinates from $[x, 1]$ to $[2x - 1, 2x + 1]$.

8. Find the transition matrix representing the change of coordinates on P_3 from the ordered basis

$$[1, x, x^2]$$

to the ordered basis

$$[1, 1 + x, 1 + x + x^2]$$

4. SIMILARITY

If L is a linear transformation mapping an n-dimensional vector space V into itself, then the matrix representation of L will depend on the ordered basis chosen for V. By using different bases, it is possible to represent L by different $n \times n$ matrices. In this section we consider different matrix representations of linear operators and characterize the relationship between matrices representing the same linear operator.

Let us begin by considering an example in R^2. Let L be the linear transformation mapping R^2 into itself defined by

$$L(\mathbf{x}) = (2x_1, x_1 + x_2)^T$$

Since

$$L(\mathbf{e}_1) = \begin{pmatrix} 2 \\ 1 \end{pmatrix} \qquad \text{and} \qquad L(\mathbf{e}_2) = \begin{pmatrix} 0 \\ 1 \end{pmatrix}$$

it follows that the matrix representing L with respect to $[\mathbf{e}_1, \mathbf{e}_2]$ is

$$A = \begin{pmatrix} 2 & 0 \\ 1 & 1 \end{pmatrix}$$

If we use a different basis for R^2, then the matrix representation of L will change. If, for example, we use

$$\mathbf{u}_1 = \begin{pmatrix} 1 \\ 1 \end{pmatrix} \qquad \text{and} \qquad \mathbf{u}_2 = \begin{pmatrix} -1 \\ 1 \end{pmatrix}$$

for a basis, then to determine the matrix representation of L with respect to $[\mathbf{u}_1, \mathbf{u}_2]$ we must determine $L(\mathbf{u}_1)$, $L(\mathbf{u}_2)$ and express these vectors as linear combinations of $\mathbf{u}_1$ and $\mathbf{u}_2$. We can use the matrix A to determine $L(\mathbf{u}_1)$ and $L(\mathbf{u}_2)$.

$$L(\mathbf{u}_1) = A\mathbf{u}_1 = \begin{pmatrix} 2 & 0 \\ 1 & 1 \end{pmatrix}\begin{pmatrix} 1 \\ 1 \end{pmatrix} = \begin{pmatrix} 2 \\ 2 \end{pmatrix}$$

$$L(\mathbf{u}_2) = A\mathbf{u}_2 = \begin{pmatrix} 2 & 0 \\ 1 & 1 \end{pmatrix}\begin{pmatrix} -1 \\ 1 \end{pmatrix} = \begin{pmatrix} -2 \\ 0 \end{pmatrix}$$

To express these vectors in terms of $\mathbf{u}_1$ and $\mathbf{u}_2$, we use a transition matrix to change from the ordered basis $[\mathbf{e}_1, \mathbf{e}_2]$ to $[\mathbf{u}_1, \mathbf{u}_2]$. Let us first compute the transition matrix from $[\mathbf{u}_1, \mathbf{u}_2]$ to $[\mathbf{e}_1, \mathbf{e}_2]$. This is simply

$$S = (\mathbf{u}_1, \mathbf{u}_2) = \begin{pmatrix} 1 & -1 \\ 1 & 1 \end{pmatrix}$$

The transition matrix from $[\mathbf{e}_1, \mathbf{e}_2]$ to $[\mathbf{u}_1, \mathbf{u}_2]$ will then be

$$S^{-1} = \begin{pmatrix} \frac{1}{2} & \frac{1}{2} \\ -\frac{1}{2} & \frac{1}{2} \end{pmatrix}$$

To determine the coordinates of $L(\mathbf{u}_1)$, $L(\mathbf{u}_2)$ with respect to $[\mathbf{u}_1, \mathbf{u}_2]$, we multiply the vectors by S^{-1}.

$$S^{-1}L(\mathbf{u}_1) = S^{-1}A\mathbf{u}_1 = \begin{pmatrix} \frac{1}{2} & \frac{1}{2} \\ -\frac{1}{2} & \frac{1}{2} \end{pmatrix}\begin{pmatrix} 2 \\ 2 \end{pmatrix} = \begin{pmatrix} 2 \\ 0 \end{pmatrix}$$

$$S^{-1}L(\mathbf{u}_2) = S^{-1}A\mathbf{u}_2 = \begin{pmatrix} \frac{1}{2} & \frac{1}{2} \\ -\frac{1}{2} & \frac{1}{2} \end{pmatrix}\begin{pmatrix} -2 \\ 0 \end{pmatrix} = \begin{pmatrix} -1 \\ 1 \end{pmatrix}$$

Thus

$$L(\mathbf{u}_1) = \quad 2\mathbf{u}_1 + 0\mathbf{u}_2$$
$$L(\mathbf{u}_2) = -1\mathbf{u}_1 + 1\mathbf{u}_2$$

and the matrix representing L with respect to $[\mathbf{u}_1, \mathbf{u}_2]$ is

$$B = \begin{pmatrix} 2 & -1 \\ 0 & 1 \end{pmatrix}$$

How are A and B related? Note that the columns of B are

$$\begin{pmatrix} 2 \\ 0 \end{pmatrix} = S^{-1}A\mathbf{u}_1 \quad \text{and} \quad \begin{pmatrix} -1 \\ 1 \end{pmatrix} = S^{-1}A\mathbf{u}_2$$

Hence

$$B = (S^{-1}A\mathbf{u}_1, S^{-1}A\mathbf{u}_2) = S^{-1}A(\mathbf{u}_1, \mathbf{u}_2) = S^{-1}AS$$

Thus if

(i) B is the matrix representing L with respect to $[\mathbf{u}_1, \mathbf{u}_2]$
(ii) A is the matrix representing L with respect to $[\mathbf{e}_1, \mathbf{e}_2]$
(iii) S is the transition matrix corresponding to the change of basis from $[\mathbf{u}_1, \mathbf{u}_2]$ to $[\mathbf{e}_1, \mathbf{e}_2]$

then

(1) $$B = S^{-1}AS$$

The results that we have established for this particular linear operator on R^2 are typical of what happens in a much more general setting. We will show next that the same sort of relationship as given in (1) will hold for any two matrix representations of a linear operator which maps an n-dimensional vector space into itself.

Theorem 4.4.1. *Let* $[\mathbf{v}_1, \dots, \mathbf{v}_n]$ *and* $[\mathbf{w}_1, \dots, \mathbf{w}_n]$ *be two ordered bases for a vector space V and let L be a linear operator mapping V into itself. Let S be the transition matrix representing the change from* $[\mathbf{w}_1, \dots, \mathbf{w}_n]$ *to* $[\mathbf{v}_1, \dots, \mathbf{v}_n]$. *If A is the matrix representing L with respect to* $[\mathbf{v}_1, \dots, \mathbf{v}_n]$ *and B is the matrix representing L with respect to* $[\mathbf{w}_1, \dots, \mathbf{w}_n]$, *then* $B = S^{-1}AS$.

Proof.   Let $\mathbf{x}$ be any vector in R^n and let
$$\mathbf{v} = x_1\mathbf{w}_1 + x_2\mathbf{w}_2 + \cdots + x_n\mathbf{w}_n$$
Let

(2) $$\mathbf{y} = S\mathbf{x}, \qquad \mathbf{z} = B\mathbf{x}, \qquad \mathbf{t} = A\mathbf{y}$$

It follows from the definition of S that $\mathbf{y}$ is the coordinate vector of $\mathbf{v}$ with respect to $[\mathbf{v}_1, \dots, \mathbf{v}_n]$.
$$\mathbf{v} = y_1\mathbf{v}_1 + \cdots + y_n\mathbf{v}_n$$

Since A and B both represent L, we have
$$z_1\mathbf{w}_1 + \cdots + z_n\mathbf{w}_n = L(\mathbf{v}) = t_1\mathbf{v}_1 + \cdots + t_n\mathbf{v}_n$$

Therefore,

(3) $$S^{-1}\mathbf{t} = \mathbf{z}$$

It follows from (2) and (3) that
$$S^{-1}AS\mathbf{x} = S^{-1}A\mathbf{y} = S^{-1}\mathbf{t} = \mathbf{z} = B\mathbf{x}$$

(see Figure 4.4.1). Thus
$$S^{-1}AS\mathbf{x} = B\mathbf{x}$$

for every $\mathbf{x} \in R^n$, and hence $S^{-1}AS = B$.   □

FIGURE 4.4.1

Definition.   Let A and B be $n \times n$ matrices. B is said to be **similar** to A if there exists a nonsingular matrix S such that $B = S^{-1}AS$.

Note that if B is similar to A, then $A = SBS^{-1}$ is similar to B. Thus we may simply say that A and B are similar matrices.

It follows from Theorem 4.4.1 that if A and B are $n \times n$ matrices representing the same operator L, then A and B are similar. Conversely, suppose that A represents L with respect to the ordered basis $[\mathbf{v}_1, \dots, \mathbf{v}_n]$ and $B = S^{-1}AS$ for some nonsingular matrix S. If $\mathbf{w}_1, \dots, \mathbf{w}_n$ are defined by

$$\mathbf{w}_1 = s_{11}\mathbf{v}_1 + s_{21}\mathbf{v}_2 + \cdots + s_{n1}\mathbf{v}_n$$

$$\mathbf{w}_2 = s_{12}\mathbf{v}_1 + s_{22}\mathbf{v}_2 + \cdots + s_{n2}\mathbf{v}_n$$

$$\vdots$$

$$\mathbf{w}_n = s_{1n}\mathbf{v}_1 + s_{2n}\mathbf{v}_2 + \cdots + s_{nn}\mathbf{v}_n$$

then $[\mathbf{w}_1, \dots, \mathbf{w}_n]$ is an ordered basis for V and B is the matrix representing L with respect to $[\mathbf{w}_1, \dots, \mathbf{w}_n]$.

EXAMPLE 1.   Let D be the differentiation operator on P_3. Find the matrix B representing D with respect to $[1, x, x^2]$ and the matrix A representing D with respect to $[1, 2x, 4x^2 - 2]$.

SOLUTION

$$D(1) = 0 \cdot 1 + 0 \cdot x + 0 \cdot x^2$$

$$D(x) = 1 \cdot 1 + 0 \cdot x + 0 \cdot x^2$$

$$D(x^2) = 0 \cdot 1 + 2 \cdot x + 0 \cdot x^2$$

The matrix B is then given by

$$B = \begin{pmatrix} 0 & 1 & 0 \\ 0 & 0 & 2 \\ 0 & 0 & 0 \end{pmatrix}$$

Applying D to 1, $2x$, and $4x^2 - 2$, one obtains

$$D(1) = 0 \cdot 1 + 0 \cdot 2x + 0 \cdot (4x^2 - 2)$$

$$D(2x) = 2 \cdot 1 + 0 \cdot 2x + 0 \cdot (4x^2 - 2)$$

$$D(4x^2 - 2) = 0 \cdot 1 + 4 \cdot 2x + 0 \cdot (4x^2 - 2)$$

Thus

$$A = \begin{pmatrix} 0 & 2 & 0 \\ 0 & 0 & 4 \\ 0 & 0 & 0 \end{pmatrix}$$

The transition matrix S corresponding to the change of basis from $[1, 2x, 4x^2 - 2]$ to $[1, x, x^2]$ and its inverse were given in Example 4 of Section 3:

$$S = \begin{pmatrix} 1 & 0 & -2 \\ 0 & 2 & 0 \\ 0 & 0 & 4 \end{pmatrix} \quad \text{and} \quad S^{-1} = \begin{pmatrix} 1 & 0 & \frac{1}{2} \\ 0 & \frac{1}{2} & 0 \\ 0 & 0 & \frac{1}{4} \end{pmatrix}$$

The reader may verify that $A = S^{-1}BS$. ☐

EXAMPLE 2. Let L be the linear operator mapping R^3 into R^3 defined by $L(\mathbf{x}) = A\mathbf{x}$, where

$$A = \begin{pmatrix} 2 & 2 & 0 \\ 1 & 1 & 2 \\ 1 & 1 & 2 \end{pmatrix}$$

Thus the matrix A represents L with respect to $[\mathbf{e}_1, \mathbf{e}_2, \mathbf{e}_3]$. Find the matrix representing L with respect to $[\mathbf{y}_1, \mathbf{y}_2, \mathbf{y}_3]$, where

$$\mathbf{y}_1 = \begin{pmatrix} 1 \\ -1 \\ 0 \end{pmatrix}, \quad \mathbf{y}_2 = \begin{pmatrix} -2 \\ 1 \\ 1 \end{pmatrix}, \quad \mathbf{y}_3 = \begin{pmatrix} 1 \\ 1 \\ 1 \end{pmatrix}$$

SOLUTION

$$L(\mathbf{y}_1) = A\mathbf{y}_1 = \mathbf{0} = 0\mathbf{y}_1 + 0\mathbf{y}_2 + 0\mathbf{y}_3$$
$$L(\mathbf{y}_2) = A\mathbf{y}_2 = \mathbf{y}_2 = 0\mathbf{y}_1 + 1\mathbf{y}_2 + 0\mathbf{y}_3$$
$$L(\mathbf{y}_3) = A\mathbf{y}_3 = 4\mathbf{y}_3 = 0\mathbf{y}_1 + 0\mathbf{y}_2 + 4\mathbf{y}_3$$

Thus the matrix representing L with respect to $[\mathbf{y}_1, \mathbf{y}_2, \mathbf{y}_3]$ is

$$D = \begin{pmatrix} 0 & 0 & 0 \\ 0 & 1 & 0 \\ 0 & 0 & 4 \end{pmatrix}$$ ☐

We could have found D using the transition matrix $S = (\mathbf{y}_1, \mathbf{y}_2, \mathbf{y}_3)$ and computing

$$D = S^{-1}AS$$

This was unnecessary due to the simplicity of the action of L on the basis $[\mathbf{y}_1, \mathbf{y}_2, \mathbf{y}_3]$.

In Example 2 the linear operator L is represented by a diagonal matrix D with respect to the basis $[\mathbf{y}_1, \mathbf{y}_2, \mathbf{y}_3]$. It is much simpler to work with D than with A. For example, it is easier to compute $D\mathbf{x}$ and $D^n\mathbf{x}$ than $A\mathbf{x}$ and $A^n\mathbf{x}$. Generally, it is desirable to find as simple a representation as possible for a linear operator. In particular, if the operator can be represented by a diagonal matrix, that is usually the preferred representation. The problem of finding a diagonal representation for a linear operator will be studied in Chapter 6.

EXERCISES

1. For each of the following linear transformations L, determine the matrix A representing L with respect to $[\mathbf{e}_1, \mathbf{e}_2]$ (see Exercise 1 of Section 2) and the matrix B representing L with respect to $[\mathbf{u}_1 = (1, 1)^T, \mathbf{u}_2 = (-1, 1)^T]$.
(a) $L(\mathbf{x}) = (-x_1, x_2)^T$
(b) $L(\mathbf{x}) = -\mathbf{x}$
(c) $L(\mathbf{x}) = (x_2, x_1)^T$
(d) $L(\mathbf{x}) = \frac{1}{2}\mathbf{x}$
(e) $L(\mathbf{x}) = x_2\mathbf{e}_2$

2. Let $[\mathbf{u}_1, \mathbf{u}_2]$ and $[\mathbf{v}_1, \mathbf{v}_2]$ be ordered bases for R^2, where

$$\mathbf{u}_1 = \begin{pmatrix} 1 \\ 1 \end{pmatrix}, \quad \mathbf{u}_2 = \begin{pmatrix} -1 \\ 1 \end{pmatrix} \quad \text{and} \quad \mathbf{v}_1 = \begin{pmatrix} 2 \\ 1 \end{pmatrix}, \quad \mathbf{v}_2 = \begin{pmatrix} 1 \\ 0 \end{pmatrix}$$

Let L be the linear transformation defined by

$$L(\mathbf{x}) = (-x_1, x_2)^T$$

and let B be the matrix representing L with respect to $[\mathbf{u}_1, \mathbf{u}_2]$ [from Exercise 1(a)].
(a) Find the transition matrix S corresponding to the change of basis from $[\mathbf{u}_1, \mathbf{u}_2]$ to $[\mathbf{v}_1, \mathbf{v}_2]$.
(b) Find the matrix A representing L with respect to $[\mathbf{v}_1, \mathbf{v}_2]$ by computing SBS^{-1}.
(c) Verify that

$$L(\mathbf{v}_1) = a_{11}\mathbf{v}_1 + a_{21}\mathbf{v}_2$$
$$L(\mathbf{v}_2) = a_{12}\mathbf{v}_1 + a_{22}\mathbf{v}_2$$

3. Let L be the linear transformation on R^3 defined by

$$L(\mathbf{x}) = (2x_1 - x_2 - x_3, 2x_2 - x_1 - x_3, 2x_3 - x_1 - x_2)^T$$

and let A be the matrix representing L with respect to $[\mathbf{e}_1, \mathbf{e}_2, \mathbf{e}_3]$ (see Exercise 4 of Section 2). If $\mathbf{b}_1 = (1, 1, 0)^T$, $\mathbf{b}_2 = (1, 0, 1)^T$, and $\mathbf{b}_3 = (0, 1, 1)^T$, then $[\mathbf{b}_1, \mathbf{b}_2, \mathbf{b}_3]$ is an ordered basis for R^3.
(a) Find the transition matrix S corresponding to a change of basis from $[\mathbf{b}_1, \mathbf{b}_2, \mathbf{b}_3]$ to $[\mathbf{e}_1, \mathbf{e}_2, \mathbf{e}_3]$.
(b) Determine the matrix B representing L with respect to $[\mathbf{b}_1, \mathbf{b}_2, \mathbf{b}_3]$ by calculating $S^{-1}AS$.

4. Let L be the linear operator mapping R^3 into R^3 defined by $L(\mathbf{x}) = A\mathbf{x}$, where

$$A = \begin{pmatrix} 3 & -1 & -2 \\ 2 & 0 & -2 \\ 2 & -1 & -1 \end{pmatrix}$$

(a) Find the matrix B representing L with respect to the basis

$$\mathbf{v}_1 = \begin{pmatrix} 1 \\ 1 \\ 1 \end{pmatrix}, \qquad \mathbf{v}_2 = \begin{pmatrix} 1 \\ 2 \\ 0 \end{pmatrix}, \qquad \mathbf{v}_3 = \begin{pmatrix} 0 \\ -2 \\ 1 \end{pmatrix}$$

(b) Find the transition matrix S corresponding to a change of basis from $[\mathbf{v}_1, \mathbf{v}_2, \mathbf{v}_3]$ to $[\mathbf{e}_1, \mathbf{e}_2, \mathbf{e}_3]$ and verify that $B = S^{-1}AS$.

5. Let L be the operator on P_3 defined by

$$L(p(x)) = xp'(x) + p''(x)$$

(a) Find the matrix A representing L with respect to $[1, x, x^2]$.
(b) Find the matrix B representing L with respect to $[1, x, 1 + x^2]$.
(c) Find the matrix S such that $B = S^{-1}AS$.
(d) If $p(x) = a_0 + a_1 x + a_2(1 + x^2)$, calculate $L^n(p(x))$.

6. Let V be the subspace of $C[a, b]$ spanned by $1, e^x, e^{-x}$ and let D be the differentiation operator on V.

(a) Find the transition matrix S representing the change of coordinates from $[1, e^x, e^{-x}]$ to $[1, \cosh x, \sinh x]$. [$\cosh x = \frac{1}{2}(e^x + e^{-x})$ and $\sinh x = \frac{1}{2}(e^x - e^{-x})$.]
(b) Find the matrix A representing D with respect to $[1, \cosh x, \sinh x]$.
(c) Find the matrix B representing D with respect to $[1, e^x, e^{-x}]$.
(d) Verify that $B = S^{-1}AS$.

7. Prove that if A is similar to B and B is similar to C, then A is similar to C.

8. Suppose that $A = S\Lambda S^{-1}$, where Λ is a diagonal matrix with diagonal elements $\lambda_1, \lambda_2, \ldots, \lambda_n$. Show that:

(a) $A\mathbf{s}_i = \lambda_i \mathbf{s}_i, \ i = 1, \ldots, n$
(b) If $\mathbf{x} = \alpha_1 \mathbf{s}_1 + \alpha_2 \mathbf{s}_2 + \cdots + \alpha_n \mathbf{s}_n$, then

$$A^k \mathbf{x} = \alpha_1 \lambda_1^k \mathbf{s}_1 + \alpha_2 \lambda_2^k \mathbf{s}_2 + \cdots + \alpha_n \lambda_n^k \mathbf{s}_n$$

(c) Suppose that $|\lambda_i| < 1$ for $i = 1, \ldots, n$. What happens to $A^k \mathbf{x}$ as $k \to \infty$? Explain.

9. Suppose that $A = ST$, where S is nonsingular. Let $B = TS$. Show that B is similar to A.

10. Let A and B be $n \times n$ matrices. Show that if A is similar to B, then there exist $n \times n$ matrices S and T, with S nonsingular, such that

$$A = ST \qquad \text{and} \qquad B = TS$$

11. Let A and B be $n \times n$ matrices. Show that if A is similar to B, then A^2 is similar to B^2.

5

ORTHOGONALITY

One can add to the structure of a vector space by defining a scalar or inner product. Such a product is not a true vector multiplication, since to every pair of vectors it associates a scalar rather than a third vector. For example, in R^2 one can define the scalar product of two vectors $\mathbf{x}$ and $\mathbf{y}$ to be $\mathbf{x}^T\mathbf{y}$. One can think of vectors in R^2 as directed line segments initiating at the origin. It is not difficult to show that the angle between two line segments will be a right angle if and only if the scalar product of the corresponding vectors is zero. In general, if V is a vector space with a scalar product, then two vectors in V are said to be *orthogonal* if their scalar product is zero.

One can think of orthogonality as a generalization of the concept of *perpendicularity* to any vector space with an inner product. To see the significance of this, consider the following problem. Let l be a line passing through the origin and let Q be a point not on l. Find the point P on l that is closest to Q. The solution P to this problem is characterized by the condition that QP is perpendicular to OP (see

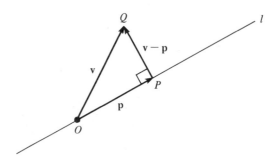

FIGURE 5.0.1

Figure 5.0.1). If we think of the line l as corresponding to a subspace of R^2 and $\mathbf{v} = OQ$ as a vector in R^2, then the problem is to find a vector in the subspace that is "closest" to $\mathbf{v}$. The solution $\mathbf{p}$ will be characterized by the property that $\mathbf{p}$ is orthogonal to $\mathbf{v} - \mathbf{p}$ (see Figure 5.0.1). In the setting of a vector space with an inner product one is able to consider general "least squares" problems. In these problems one is given a vector $\mathbf{v}$ in V and a subspace W. One wishes to find a vector in W which is "closest" to $\mathbf{v}$. A solution $\mathbf{p}$ must be orthogonal to $\mathbf{v} - \mathbf{p}$. This orthogonality condition provides the key to solving the least squares problem. Least squares problems occur in many statistical applications involving data fitting.

1. THE SCALAR PRODUCT IN R^n

Two vectors $\mathbf{x}$ and $\mathbf{y}$ in R^n may be regarded as $n \times 1$ matrices. One can then form the matrix product $\mathbf{x}^T\mathbf{y}$. This product is a 1×1 matrix which may be regarded as a vector in R^1, or more simply, as a real number. The product $\mathbf{x}^T\mathbf{y}$ is called the *scalar product* of $\mathbf{x}$ and $\mathbf{y}$. In particular, if $\mathbf{x} = (x_1, \ldots, x_n)^T$ and $\mathbf{y} = (y_1, \ldots, y_n)^T$, then

$$\mathbf{x}^T\mathbf{y} = x_1 y_1 + x_2 y_2 + \cdots + x_n y_n$$

EXAMPLE 1. If

$$\mathbf{x} = \begin{pmatrix} 3 \\ -2 \\ 1 \end{pmatrix} \quad \text{and} \quad \mathbf{y} = \begin{pmatrix} 4 \\ 3 \\ 2 \end{pmatrix}$$

then

$$\mathbf{x}^T\mathbf{y} = (3, -2, 1) \begin{pmatrix} 4 \\ 3 \\ 2 \end{pmatrix} = 3 \cdot 4 - 2 \cdot 3 + 1 \cdot 2 = 8$$

$\square$

THE SCALAR PRODUCT IN R^2 AND R^3

In order to see the geometrical significance of the scalar product, let us restrict our attention to R^2 and R^3. Vectors in R^2 and R^3 can be represented by directed line segments. Given a vector $\mathbf{x}$ in either R^2 or R^3, its Euclidean length can be defined in terms of the scalar product.

$$\|\mathbf{x}\| = \sqrt{x_1^2 + x_2^2} = (\mathbf{x}^T\mathbf{x})^{1/2} \quad \text{if} \quad \mathbf{x} \in R^2$$

$$\|\mathbf{x}\| = \sqrt{x_1^2 + x_2^2 + x_3^2} = (\mathbf{x}^T\mathbf{x})^{1/2} \quad \text{if} \quad \mathbf{x} \in R^3$$

Given two nonzero vectors $\mathbf{x}$ and $\mathbf{y}$, we can think of them as directed line segments initiating at the same point. The angle between the two vectors is then defined as the angle θ between the line segments.

Theorem 5.1.1.   *If* $\mathbf{x}$ *and* $\mathbf{y}$ *are two nonzero vectors in either* R^2 *or* R^3 *and* θ *is the angle between them, then*

(1)
$$\mathbf{x}^T\mathbf{y} = \|\mathbf{x}\| \, \|\mathbf{y}\| \cos \theta$$

Proof.   We will prove the result for R^2. The proof for R^3 is similar. The vectors $\mathbf{x}$, $\mathbf{y}$, and $\mathbf{y} - \mathbf{x}$ may be used to form a triangle as in Figure 5.1.1. By the law of cosines,

$$\|\mathbf{y} - \mathbf{x}\|^2 = \|\mathbf{x}\|^2 + \|\mathbf{y}\|^2 - 2\|\mathbf{x}\| \, \|\mathbf{y}\| \cos \theta$$

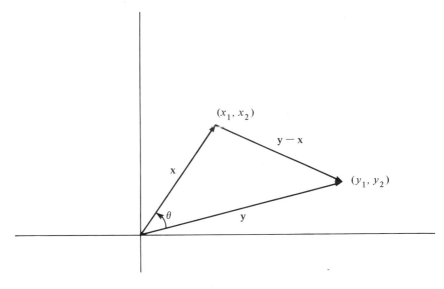

FIGURE 5.1.1

or

$$\|\mathbf{x}\|\,\|\mathbf{y}\|\cos\theta = \tfrac{1}{2}\left(\|\mathbf{x}\|^2 + \|\mathbf{y}\|^2 - \|\mathbf{x} - \mathbf{y}\|^2\right)$$
$$= \tfrac{1}{2}\left[x_1^2 + x_2^2 + y_1^2 + y_2^2 - (x_1 - y_1)^2 - (x_2 - y_2)^2\right]$$
$$= x_1 y_1 + x_2 y_2$$
$$= \mathbf{x}^T \mathbf{y} \qquad\qquad\qquad \square$$

Corollary 5.1.2 (Cauchy – Schwarz Inequality). *If* $\mathbf{x}$ *and* $\mathbf{y}$ *are vectors in either* R^2 *or* R^3, *then*

(2)
$$|\mathbf{x}^T \mathbf{y}| \le \|\mathbf{x}\|\,\|\mathbf{y}\|$$

with equality holding if and only if one of the vectors is $\mathbf{0}$ *or* $\mathbf{x} = \alpha \mathbf{y}$.

Proof.   The inequality follows from (1). If one of the vectors is $\mathbf{0}$, then both sides of (2) are 0. If $\mathbf{x} = \alpha \mathbf{y}$, then $\cos\theta = \pm 1$ in (1) and hence equality holds in (2). $\square$

If $\mathbf{x}^T \mathbf{y} = 0$, it follows from Theorem 5.1.1 that either one of the vectors is the zero vector or $\cos\theta = 0$. If $\cos\theta = 0$, the angle between the vectors is a right angle.

Definition.   The vectors $\mathbf{x}$ and $\mathbf{y}$ in R^2 (or R^3) are said to be **orthogonal** if $\mathbf{x}^T \mathbf{y} = 0$.

EXAMPLE 2
(a) The vector $\mathbf{0}$ is orthogonal to every vector in R^2.

(b) The vectors $\begin{pmatrix} 3 \\ 2 \end{pmatrix}$ and $\begin{pmatrix} -4 \\ 6 \end{pmatrix}$ are orthogonal in R^2.

(c) The vectors $\begin{pmatrix} 2 \\ -3 \\ 1 \end{pmatrix}$ and $\begin{pmatrix} 1 \\ 1 \\ 1 \end{pmatrix}$ are orthogonal in R^3. $\square$

SCALAR AND VECTOR PROJECTIONS

The scalar product can be used to find the component of one vector in the direction of another. Let $\mathbf{x}$ and $\mathbf{y}$ be nonzero vectors in either R^2 or R^3. We would like to write $\mathbf{x}$ as a sum of the form $\mathbf{p} + \mathbf{z}$, where $\mathbf{p}$ is in the direction of $\mathbf{y}$ and $\mathbf{z}$ is orthogonal to $\mathbf{p}$ (see Figure 5.1.2). To do this, let $\mathbf{u} = (1/\|\mathbf{y}\|)\mathbf{y}$. Thus $\mathbf{u}$ is a unit vector (length 1) in the direction of $\mathbf{y}$. We wish to find α such that $\mathbf{p} = \alpha\mathbf{u}$ is orthogonal to $\mathbf{z} = \mathbf{x} - \alpha\mathbf{u}$. In order for $\mathbf{p}$ and $\mathbf{z}$ to be orthogonal, the scalar α must satisfy

$$\alpha = \|\mathbf{x}\|\cos\theta$$
$$= \frac{\|\mathbf{x}\|\,\|\mathbf{y}\|\cos\theta}{\|\mathbf{y}\|}$$
$$= \frac{\mathbf{x}^T \mathbf{y}}{\|\mathbf{y}\|}$$

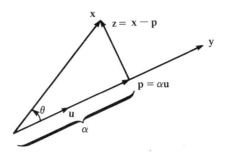

FIGURE 5.1.2

The scalar α is called the *scalar projection* of **x** onto **y** and the vector **p** is called the *vector projection* of **x** onto **y**.

Scalar projection of **x** onto **y**:

$$\alpha = \frac{\mathbf{x}^T\mathbf{y}}{\|\mathbf{y}\|}$$

Vector projection of **x** onto **y**:

$$\mathbf{p} = \alpha\mathbf{u} = \alpha\frac{1}{\|\mathbf{y}\|}\mathbf{y} = \frac{\mathbf{x}^T\mathbf{y}}{\mathbf{y}^T\mathbf{y}}\mathbf{y}$$

EXAMPLE 3.   Find the point on the line $y = \frac{1}{3}x$ that is closest to the point $(1, 4)$ (see Figure 5.1.3).

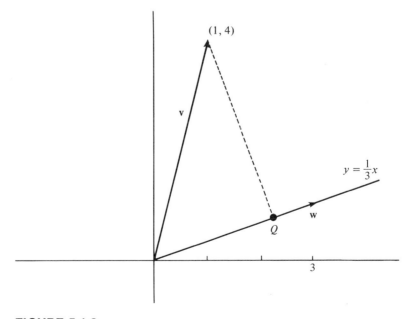

FIGURE 5.1.3

SOLUTION.   The vector $\mathbf{w} = (3, 1)^T$ is a vector in the direction of the line $y = \frac{1}{3}x$. Let $\mathbf{v} = (1, 4)^T$. If Q is the desired point, then Q^T is the vector projection of $\mathbf{v}$ onto $\mathbf{w}$.

$$Q^T = \left(\frac{\mathbf{v}^T\mathbf{w}}{\mathbf{w}^T\mathbf{w}} \right)\mathbf{w} = \frac{7}{10}\binom{3}{1} = \binom{2.1}{0.7}$$

Thus $Q = (2.1, 0.7)$ is the closest point.   □

Notation.   If P_1 and P_2 are two points in 3-space, we will denote the vector from P_1 to P_2 by $\overline{P_1P_2}$.

If $\mathbf{N}$ is a nonzero vector and P_0 is a fixed point, the set of points P such that $\overline{P_0P}$ is orthogonal to $\mathbf{N}$ forms a plane π in 3-space that passes through P_0. The vector $\mathbf{N}$ and the plane π are said to be *normal* to each other. A point $P = (x, y, z)$ will lie on π if and only if

$$\left(\overline{P_0P}\right)^T\mathbf{N} = 0$$

If $\mathbf{N} = (a, b, c)^T$ and $P_0 = (x_0, y_0, z_0)$, this equation can be written in the form

$$a(x - x_0) + b(y - y_0) + c(z - z_0) = 0$$

EXAMPLE 4.   Find the equation of the plane passing through the point $(2, -1, 3)$ and normal to the vector $\mathbf{N} = (2, 3, 4)^T$.

SOLUTION. $\overline{P_0P} = (x - 2, y + 1, z - 3)^T$. The equation is $(\overline{P_0P})^T\mathbf{N} = 0$, or

$$2(x - 2) + 3(y + 1) + 4(z - 3) = 0$$ □

EXAMPLE 5.   Find the distance from the point $(2, 0, 0)$ to the plane $x + 2y + 2z = 0$.

SOLUTION.   The vector $\mathbf{N} = (1, 2, 2)^T$ is normal to the plane and the plane passes through the origin. Let $\mathbf{v} = (2, 0, 0)^T$. The distance d from $(2, 0, 0)$ to the plane is simply the absolute value of the scalar projection of $\mathbf{v}$ onto $\mathbf{N}$. Thus

$$d = \frac{|\mathbf{v}^T\mathbf{N}|}{\|\mathbf{N}\|} = \frac{2}{3}$$ □

ORTHOGONALITY IN R^n

The definitions that have been given for R^2 and R^3 can all be generalized to R^n. Indeed, if $\mathbf{x} \in R^n$, then the Euclidean length of $\mathbf{x}$ is defined by

$$\|\mathbf{x}\| = (\mathbf{x}^T\mathbf{x})^{1/2} = \left(x_1^2 + x_2^2 + \cdots + x_n^2 \right)^{1/2}$$

and the angle θ between two nonzero vectors $\mathbf{x}$ and $\mathbf{y}$ is given by

$$\cos\theta = \frac{\mathbf{x}^T\mathbf{y}}{\|\mathbf{x}\|\,\|\mathbf{y}\|}$$

The vectors $\mathbf{x}$ and $\mathbf{y}$ are said to be *orthogonal* if $\mathbf{x}^T\mathbf{y} = 0$. Often the symbol " $\perp$ " is used to indicate orthogonality. Thus, if $\mathbf{x}$ and $\mathbf{y}$ are orthogonal, we will write $\mathbf{x} \perp \mathbf{y}$. Vector and scalar projections are defined in R^n in the same way that they were defined for R^2. One of the main applications of these concepts is in the solution of least squares problems. Least squares problems will be studied in Section 5.

EXERCISES

1. Find the angle between the vectors $\mathbf{v}$ and $\mathbf{w}$ in each of the following.
   (a) $\mathbf{v} = (2, 1, 3)^T$ $\mathbf{w} = (6, 3, 9)^T$
   (b) $\mathbf{v} = (2, -3)^T$ $\mathbf{w} = (3, 2)^T$
   (c) $\mathbf{v} = (4, 1)^T$ $\mathbf{w} = (3, 2)^T$
   (d) $\mathbf{v} = (-2, 3, 1)^T$ $\mathbf{w} = (1, 2, 4)^T$

2. For each of the pairs of vectors in Exercise 1, find the scalar projection of $\mathbf{v}$ onto $\mathbf{w}$. Also find the vector projection of $\mathbf{v}$ onto $\mathbf{w}$.

3. For each of the following pairs of vectors $\mathbf{x}$ and $\mathbf{y}$, find the vector projection $\mathbf{p}$ of $\mathbf{x}$ onto $\mathbf{y}$ and verify that $\mathbf{p}$ and $\mathbf{x} - \mathbf{p}$ are orthogonal.
   (a) $\mathbf{x} = (3, 4)^T$ $\mathbf{y} = (1, 0)^T$
   (b) $\mathbf{x} = (3, 5)^T$ $\mathbf{y} = (1, 1)^T$
   (c) $\mathbf{x} = (2, 4, 3)^T$ $\mathbf{y} = (1, 1, 1)^T$
   (d) $\mathbf{x} = (2, -5, 4)^T$ $\mathbf{y} = (1, 2, -1)^T$

4. Find the point on the line $y = 2x$ that is closest to the point $(5, 2)$.

5. Find the point on the line $y = 2x + 1$ that is closest to the point $(5, 2)$.

6. Find the distance from the point $(1, 2)$ to the line $4x - 3y = 0$.

7. In each of the following, find the equation of the plane normal to the given vector $\mathbf{N}$ and passing through the point P_0.
   (a) $\mathbf{N} = (2, 4, 3)^T$ $P_0 = (0, 0, 0)$
   (b) $\mathbf{N} = (-3, 6, 2)^T$ $P_0 = (4, 2, -5)$
   (c) $\mathbf{N} = (0, 0, 1)^T$ $P_0 = (3, 2, 4)$

8. Find the distance from the point $(1, 1, 1)$ to the plane $2x + 2y + z = 0$.

9. Find the distance from the point $(2, 1, -2)$ to the plane $6(x - 1) + 2(y - 3) + 3(z + 4) = 0$.

10. If $\mathbf{x} = (x_1, x_2)^T$, $\mathbf{y} = (y_1, y_2)^T$, and $\mathbf{z} = (z_1, z_2)^T$ are arbitrary vectors in R^2, prove:
    (a) $\mathbf{x}^T\mathbf{x} \geq 0$ (b) $\mathbf{x}^T\mathbf{y} = \mathbf{y}^T\mathbf{x}$
    (c) $\mathbf{x}^T(\mathbf{y} + \mathbf{z}) = \mathbf{x}^T\mathbf{y} + \mathbf{x}^T\mathbf{z}$

11. If $\mathbf{u}$ and $\mathbf{v}$ are any vectors in R^2, show that $\|\mathbf{u} + \mathbf{v}\|^2 \leq (\|\mathbf{u}\| + \|\mathbf{v}\|)^2$ and hence $\|\mathbf{u} + \mathbf{v}\| \leq \|\mathbf{u}\| + \|\mathbf{v}\|$. When does equality hold? Give a geometric interpretation of the inequality.

12. Let l_1 be the line $y = m_1 x + b_1$, $m_1 \neq 0$, and l be the line $y = mx + b$. Show that l is perpendicular to l_1 if and only if $m = -1/m_1$.

2. ORTHOGONAL SUBSPACES

Let A be an $m \times n$ matrix and let $\mathbf{x} \in N(A)$. Since $A\mathbf{x} = \mathbf{0}$, we have

(1) $$a_{i1}x_1 + a_{i2}x_2 + \cdots + a_{in}x_n = 0$$

for $i = 1, \ldots, m$. Equation (1) says that $\mathbf{x}$ is orthogonal to the ith column vector of A^T for $i = 1, \ldots, m$. Since $\mathbf{x}$ is orthogonal to each column vector of A^T, it is orthogonal to any linear combination of the column vectors of A^T. Thus, if $\mathbf{y}$ is any vector in the column space of A^T, then $\mathbf{x}^T\mathbf{y} = 0$. Thus each vector in $N(A)$ is orthogonal to every vector in the column space of A^T. When two subspaces of R^n have this property, we say that they are orthogonal.

Definition. Two subspaces X and Y of R^n are said to be **orthogonal** if $\mathbf{x}^T\mathbf{y} = 0$ for every $\mathbf{x} \in X$ and every $\mathbf{y} \in Y$. If X and Y are orthogonal, we write $X \perp Y$.

EXAMPLE 1. Let X be the subspace of R^3 spanned by $\mathbf{e}_1$ and let Y be the subspace spanned by $\mathbf{e}_2$. If $\mathbf{x} \in X$ and $\mathbf{y} \in Y$, these vectors must be of the form

$$\mathbf{x} = \begin{pmatrix} x_1 \\ 0 \\ 0 \end{pmatrix} \quad \text{and} \quad \mathbf{y} = \begin{pmatrix} 0 \\ y_2 \\ 0 \end{pmatrix}$$

Thus

$$\mathbf{x}^T\mathbf{y} = x_1 \cdot 0 + 0 \cdot y_2 + 0 \cdot 0 = 0$$

Therefore, $X \perp Y$.   □

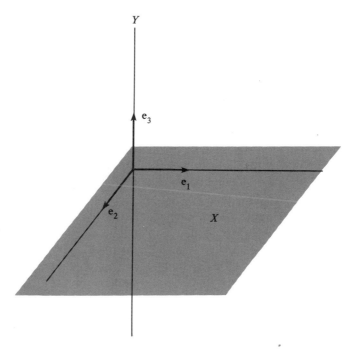

FIGURE 5.2.1

The concept of orthogonal subspaces does not always agree with our intuitive idea of perpendicularity. For example, the floor and wall of the classroom "look" orthogonal, but the xy-plane and the yz-plane are not orthogonal subspaces. Indeed, we can think of the vectors $\mathbf{x}_1 = (1, 1, 0)^T$ and $\mathbf{x}_2 = (0, 1, 1)^T$ as lying in the xy and yz planes, respectively. Since

$$\mathbf{x}_1^T \mathbf{x}_2 = 1 \cdot 0 + 1 \cdot 1 + 0 \cdot 1 = 1$$

the subspaces are not orthogonal. The next example shows that the subspace corresponding to the z axis is orthogonal to the subspace corresponding to the xy plane.

EXAMPLE 2.   Let X be the subspace of R^3 spanned by $\mathbf{e}_1$ and $\mathbf{e}_2$ and let Y be the subspace spanned by $\mathbf{e}_3$. If $\mathbf{x} \in X$ and $\mathbf{y} \in Y$, then

$$\mathbf{x}^T \mathbf{y} = x_1 \cdot 0 + x_2 \cdot 0 + 0 \cdot y_3 = 0$$

Thus $X \perp Y$. Furthermore, if $\mathbf{z}$ is any vector in R^3 that is orthogonal to every vector in Y, then $\mathbf{z} \perp \mathbf{e}_3$ and hence

$$z_3 = \mathbf{z}^T \mathbf{e}_3 = 0$$

But if $z_3 = 0$, then $\mathbf{z} \in X$. Thus X is the set of all vectors in R^3 that are orthogonal to every vector in Y (see Figure 5.2.1).   □

Definition. Let Y be a subspace of R^n. The set of all vectors in R^n that are orthogonal to every vector in Y will be denoted by $Y^\perp$. Thus

$$Y^\perp = \{ \mathbf{x} \in R^n | \mathbf{x}^T\mathbf{y} = 0 \quad \text{for every} \quad \mathbf{y} \in Y \}$$

The set $Y^\perp$ is called the **orthogonal complement** of Y.

Remarks

1. If X and Y are orthogonal subspaces of R^n, then $X \cap Y = \{\mathbf{0}\}$.
2. If Y is a subspace of R^n, then $Y^\perp$ is also a subspace of R^n.

Proof of (1). If $\mathbf{x} \in X \cap Y$ and $X \perp Y$, then $\|\mathbf{x}\|^2 = \mathbf{x}^T\mathbf{x} = 0$ and hence $\mathbf{x} = \mathbf{0}$. $\qquad \qquad \Box$

Proof of (2). If $\mathbf{x} \in Y^\perp$ and α is a scalar, then for any $\mathbf{y} \in Y$,

$$(\alpha\mathbf{x})^T\mathbf{y} = \alpha(\mathbf{x}^T\mathbf{y}) = \alpha \cdot 0 = 0$$

Therefore, $\alpha\mathbf{x} \in Y^\perp$. If $\mathbf{x}_1$ and $\mathbf{x}_2$ are elements of $Y^\perp$, then

$$(\mathbf{x}_1 + \mathbf{x}_2)^T\mathbf{y} = \mathbf{x}_1^T\mathbf{y} + \mathbf{x}_2^T\mathbf{y} = 0 + 0 = 0$$

for each $\mathbf{y} \in Y$. Hence $\mathbf{x}_1 + \mathbf{x}_2 \in Y^\perp$. Therefore, $Y^\perp$ is a subspace of R^n. $\qquad \qquad \Box$

FUNDAMENTAL SUBSPACES

Let A be an $m \times n$ matrix. We saw in Chapter 3 that a vector $\mathbf{b} \in R^m$ is in the column space of A if and only if $\mathbf{b} = A\mathbf{x}$ for some $\mathbf{x} \in R^n$. If we think of A as an operator mapping R^n into R^m, then the column space of A is the same as the range of A. Let us denote the range of A by $R(A)$. Thus

$$R(A) = \{\mathbf{b} \in R^m | \mathbf{b} = A\mathbf{x} \quad \text{for some} \quad \mathbf{x} \in R^n\}$$
$$= \text{the column space of } A$$

The column space of A^T, $R(A^T)$, is a subspace of R^n:

$$R(A^T) = \{\mathbf{y} \in R^n | \mathbf{y} = A^T\mathbf{x} \quad \text{for some} \quad \mathbf{x} \in R^m\}$$

The column space of $R(A^T)$ is essentially the same as the row space of A except that it consists of vectors in R^n ($n \times 1$ matrices) rather than n-tuples. Thus $\mathbf{y} \in R(A^T)$ if and only if $\mathbf{y}^T$ is in the row space of A. We have seen that $R(A^T) \perp N(A)$. The following theorem shows that $N(A)$ is actually the orthogonal complement of $R(A^T)$.

Theorem 5.2.1. *If A is an $m \times n$ matrix, then $N(A) = R(A^T)^\perp$ and $N(A^T) = R(A)^\perp$.*

Proof. We have already seen that $N(A) \perp R(A^T)$, and this implies that $N(A) \subset R(A^T)^\perp$. On the other hand, if $\mathbf{x}$ is any vector in $R(A^T)^\perp$, then $\mathbf{x}$ is orthogonal to each of the column vectors of A^T and consequently $A\mathbf{x} = \mathbf{0}$. Thus $\mathbf{x}$ must be an element of $N(A)$ and hence $N(A) = R(A^T)^\perp$. This proof does not depend on the dimensions of A. In particular, the result will also hold for the matrix $B = A^T$. Thus

$$N(A^T) = N(B) = R(B^T)^\perp = R(A)^\perp \qquad \square$$

EXAMPLE 3. Let

$$A = \begin{pmatrix} 1 & 0 \\ 2 & 0 \end{pmatrix}$$

The column space of A consists of all vectors of the form

$$\begin{pmatrix} \alpha \\ 2\alpha \end{pmatrix} = \alpha \begin{pmatrix} 1 \\ 2 \end{pmatrix}$$

Note that if $\mathbf{x}$ is any vector in R^n and $\mathbf{b} = A\mathbf{x}$, then

$$\mathbf{b} = \begin{pmatrix} 1 & 0 \\ 2 & 0 \end{pmatrix} \begin{pmatrix} x_1 \\ x_2 \end{pmatrix} = \begin{pmatrix} 1x_1 \\ 2x_1 \end{pmatrix} = x_1 \begin{pmatrix} 1 \\ 2 \end{pmatrix}$$

The nullspace of A^T consists of all vectors of the form $\beta(-2, 1)^T$. Since $(1, 2)^T$ and $(-2, 1)^T$ are orthogonal, it follows that every vector in $R(A)$ will be orthogonal to every vector in $N(A^T)$. The same relationship holds between $R(A^T)$ and $N(A)$. $R(A^T)$ consists of vectors of the form $\alpha \mathbf{e}_1$, and $N(A)$ consists of all vectors of the form $\beta \mathbf{e}_2$. Since $\mathbf{e}_1$ and $\mathbf{e}_2$ are orthogonal, it follows that each vector in $R(A^T)$ is orthogonal to every vector in $N(A)$. $\qquad \square$

Theorem 5.2.1 is one of the most important theorems in this chapter. In Section 5 we will see that the result $N(A^T) = R(A)^\perp$ provides a key to solving least squares problems. For the present, we will use Theorem 5.2.1 to prove the following theorem, which, in turn, will be used to establish two more important results about orthogonal subspaces.

Theorem 5.2.2. *If S is a subspace of R^n, then $\dim S + \dim S^\perp = n$. Furthermore, if $\{\mathbf{x}_1, \ldots, \mathbf{x}_r\}$ is a basis for S and $\{\mathbf{x}_{r+1}, \ldots, \mathbf{x}_n\}$ is a basis for $S^\perp$, then $\{\mathbf{x}_1, \ldots, \mathbf{x}_r, \mathbf{x}_{r+1}, \ldots, \mathbf{x}_n\}$ is a basis for R^n.*

Proof. If $S = \{\mathbf{0}\}$, then $S^\perp = R^n$ and

$$\dim S + \dim S^\perp = 0 + n = n$$

If $S \neq \{\mathbf{0}\}$, then let $\{\mathbf{x}_1, \ldots, \mathbf{x}_r\}$ be a basis for S and define X to be an $r \times n$ matrix whose ith row is $\mathbf{x}_i^T$ for each i. By construction the matrix X has rank r and $R(X^T) = S$. By Theorem 5.2.1

$$S^\perp = R(X^T)^\perp = N(X)$$

It follows from Theorem 3.5.4 that

$$\dim S^\perp = \dim N(X) = n - r$$

To show that $\{\mathbf{x}_1, \ldots, \mathbf{x}_r, \mathbf{x}_{r+1}, \ldots, \mathbf{x}_n\}$ is a basis for R^n, it suffices to show that the n vectors are linearly independent. Suppose that

$$c_1\mathbf{x}_1 + \cdots + c_r\mathbf{x}_r + c_{r+1}\mathbf{x}_{r+1} + \cdots + c_n\mathbf{x}_n = \mathbf{0}$$

Let $\mathbf{y} = c_1\mathbf{x}_1 + \cdots + c_r\mathbf{x}_r$ and $\mathbf{z} = c_{r+1}\mathbf{x}_{r+1} + \cdots + c_n\mathbf{x}_n$. We then have

$$\mathbf{y} + \mathbf{z} = \mathbf{0}$$
$$\mathbf{y} = -\mathbf{z}$$

Thus $\mathbf{y}$ and $\mathbf{z}$ are both elements of $S \cap S^\perp$. But $S \cap S^\perp = \{\mathbf{0}\}$. Therefore,

$$c_1\mathbf{x}_1 + \cdots + c_r\mathbf{x}_r = \mathbf{0}$$
$$c_{r+1}\mathbf{x}_{r+1} + \cdots + c_n\mathbf{x}_n = \mathbf{0}$$

Since $\mathbf{x}_1, \ldots, \mathbf{x}_r$ are linearly independent

$$c_1 = c_2 = \cdots = c_r = 0$$

Similarly, $\mathbf{x}_{r+1}, \ldots, \mathbf{x}_n$ are linearly independent and hence

$$c_{r+1} = c_{r+2} = \cdots = c_n = 0$$

Therefore, $\mathbf{x}_1, \mathbf{x}_2, \ldots, \mathbf{x}_n$ are linearly independent and hence form a basis for R^n. $\square$

Given a subspace S of R^n, we will use Theorem 5.2.2 to prove that each $\mathbf{x} \in R^n$ can be expressed uniquely as a sum $\mathbf{y} + \mathbf{z}$, where $\mathbf{y} \in S$ and $\mathbf{z} \in S^\perp$.

Definition. If U and V are subspaces of a vector space W and each $\mathbf{w} \in W$ can be written uniquely as a sum $\mathbf{u} + \mathbf{v}$, where $\mathbf{u} \in U$ and $\mathbf{v} \in V$, then we say that W is a **direct sum** of U and V and we write $W = U \oplus V$.

Theorem 5.2.3. *If S is a subspace of R^n, then*

$$R^n = S \oplus S^\perp$$

Proof. The result is trivial if either $S = \{\mathbf{0}\}$ or $S = R^n$. In the case where $\dim S = r$, $0 < r < n$, it follows from Theorem 5.2.2 that each vector $\mathbf{x} \in R^n$ can be represented in the form

$$\mathbf{x} = c_1\mathbf{x}_1 + \cdots + c_r\mathbf{x}_r + c_{r+1}\mathbf{x}_{r+1} + \cdots + c_n\mathbf{x}_n$$

where $\{\mathbf{x}_1, \ldots, \mathbf{x}_r\}$ is a basis for S and $\{\mathbf{x}_{r+1}, \ldots, \mathbf{x}_n\}$ is a basis for $S^\perp$. If we let

$$\mathbf{u} = c_1\mathbf{x}_1 + \cdots + c_r\mathbf{x}_r \quad \text{and} \quad \mathbf{v} = c_{r+1}\mathbf{x}_{r+1} + \cdots + c_n\mathbf{x}_n$$

then $\mathbf{u} \in S$, $\mathbf{v} \in S^\perp$, and $\mathbf{x} = \mathbf{u} + \mathbf{v}$. To show uniqueness, suppose that $\mathbf{x}$ can also be written as a sum $\mathbf{y} + \mathbf{z}$, where $\mathbf{y} \in S$ and $\mathbf{z} \in S^\perp$. Thus

$$\mathbf{u} + \mathbf{v} = \mathbf{x} = \mathbf{y} + \mathbf{z}$$
$$\mathbf{u} - \mathbf{y} = \mathbf{z} - \mathbf{v}$$

But $\mathbf{u} - \mathbf{y} \in S$ and $\mathbf{z} - \mathbf{v} \in S^\perp$, so each is in $S \cap S^\perp$. Since

$$S \cap S^\perp = \{\mathbf{0}\}$$

it follows that

$$\mathbf{u} = \mathbf{y} \qquad \text{and} \qquad \mathbf{v} = \mathbf{z} \qquad\qquad \square$$

Theorem 5.2.4. *If S is a subspace of R^n, then $(S^\perp)^\perp = S$.*

Proof. If $\mathbf{x} \in S$, then $\mathbf{x}$ is orthogonal to each $\mathbf{y}$ in $S^\perp$. Therefore, $\mathbf{x} \in (S^\perp)^\perp$ and hence $S \subset (S^\perp)^\perp$. On the other hand, suppose that $\mathbf{z}$ is an arbitrary element of $(S^\perp)^\perp$. By Theorem 5.2.3 we can write $\mathbf{z}$ as a sum $\mathbf{u} + \mathbf{v}$, where $\mathbf{u} \in S$ and $\mathbf{v} \in S^\perp$. Since $\mathbf{v} \in S^\perp$, it is orthogonal to both $\mathbf{u}$ and $\mathbf{z}$. It follows then that

$$0 = \mathbf{v}^T\mathbf{z} = \mathbf{v}^T\mathbf{u} + \mathbf{v}^T\mathbf{v} = \mathbf{v}^T\mathbf{v}$$

and consequently, $\mathbf{v} = \mathbf{0}$. Therefore, $\mathbf{z} = \mathbf{u} \in S$ and hence $S = (S^\perp)^\perp$.
$\square$

It follows from Theorem 5.2.4 that if T is the orthogonal complement of a subspace S, then S is the orthogonal complement of T and we may say simply that S and T are orthogonal complements. In particular, it follows from Theorem 5.2.1 that $N(A)$ and $R(A^T)$ are orthogonal complements of each other and $N(A^T)$ and $R(A)$ are orthogonal complements. Hence we may write

$$N(A)^\perp = R(A^T) \qquad \text{and} \qquad N(A^T)^\perp = R(A)$$

Recall that the system $A\mathbf{x} = \mathbf{b}$ is consistent if and only if $\mathbf{b} \in R(A)$. Since $R(A) = N(A^T)^\perp$, we have the following result, which may be considered a corollary to Theorem 5.2.1.

Corollary 5.2.5. *If A is an $m \times n$ matrix and $\mathbf{b} \in R^m$, then either there is a vector $\mathbf{x} \in R^n$ such that $A\mathbf{x} = \mathbf{b}$ or there is a vector $\mathbf{y} \in R^m$ such that $A^T\mathbf{y} = \mathbf{0}$ and $\mathbf{y}^T\mathbf{b} \neq 0$.*

EXAMPLE 4. Let

$$A = \begin{pmatrix} 1 & 1 & 2 \\ 0 & 1 & 1 \\ 1 & 3 & 4 \end{pmatrix}$$

Find bases for $N(A)$, $R(A^T)$, $N(A^T)$, and $R(A)$.

SOLUTION.   One can find bases for $N(A)$ and $R(A^T)$ by transforming A into reduced row echelon form.

$$\begin{pmatrix} 1 & 1 & 2 \\ 0 & 1 & 1 \\ 1 & 3 & 4 \end{pmatrix} \rightarrow \begin{pmatrix} 1 & 1 & 2 \\ 0 & 1 & 1 \\ 0 & 2 & 2 \end{pmatrix} \rightarrow \begin{pmatrix} 1 & 0 & 1 \\ 0 & 1 & 1 \\ 0 & 0 & 0 \end{pmatrix}$$

Since $(1, 0, 1)$ and $(0, 1, 1)$ form a basis for the row space of A, it follows that $\begin{pmatrix} 1 \\ 0 \\ 1 \end{pmatrix}$ and $\begin{pmatrix} 0 \\ 1 \\ 1 \end{pmatrix}$ form a basis for $R(A^T)$. If $\mathbf{x} \in N(A)$, it follows from the reduced row echelon form of A that

$$x_1 + x_3 = 0$$
$$x_2 + x_3 = 0$$

Thus

$$x_1 = x_2 = -x_3$$

Setting $x_3 = \alpha$, we see that $N(A)$ consists of all vectors of the form $\alpha(-1, -1, 1)^T$. Note that $(-1, -1, 1)^T$ is orthogonal to $(1, 0, 1)^T$ and $(0, 1, 1)^T$.

To find bases for $R(A)$ and $N(A^T)$, reduce A^T to reduced row echelon form.

$$\begin{pmatrix} 1 & 0 & 1 \\ 1 & 1 & 3 \\ 2 & 1 & 4 \end{pmatrix} \rightarrow \begin{pmatrix} 1 & 0 & 1 \\ 0 & 1 & 2 \\ 0 & 1 & 2 \end{pmatrix} \rightarrow \begin{pmatrix} 1 & 0 & 1 \\ 0 & 1 & 2 \\ 0 & 0 & 0 \end{pmatrix}$$

Thus $\begin{pmatrix} 1 \\ 0 \\ 1 \end{pmatrix}$ and $\begin{pmatrix} 0 \\ 1 \\ 2 \end{pmatrix}$ form a basis for $R(A)$. If $\mathbf{x} \in N(A^T)$, then $x_1 = -x_3$, $x_2 = -2x_3$. Thus $N(A^T)$ is the subspace of R^3 spanned by $\begin{pmatrix} -1 \\ -2 \\ 1 \end{pmatrix}$. Note that $(-1, -2, 1)^T$ is orthogonal to $(1, 0, 1)^T$ and $(0, 1, 2)^T$.   □

We saw in Chapter 3 that the row space and the column space have the same dimension. If A has rank r, then

$$\dim R(A) = \dim R(A^T) = r$$

Actually, A can be used to establish a one-to-one correspondence between $R(A^T)$ and $R(A)$.

One can think of an $m \times n$ matrix A as a linear transformation from R^n to R^m.

$$\mathbf{x} \in R^n \rightarrow A\mathbf{x} \in R^m$$

Since $R(A^T)$ and $N(A)$ are orthogonal complements in R^n,

$$R^n = R(A^T) \oplus N(A)$$

Each vector $\mathbf{x} \in R^n$ can be written as a sum

$$\mathbf{x} = \mathbf{y} + \mathbf{z}, \qquad \mathbf{y} \in R(A^T), \qquad \mathbf{z} \in N(A)$$

It follows that

$$A\mathbf{x} = A\mathbf{y} + A\mathbf{z} = A\mathbf{y} \quad \text{for each} \quad \mathbf{x} \in R^n$$

and hence

$$R(A) = \{ A\mathbf{x} | \mathbf{x} \in R^n \} = \{ A\mathbf{y} | \mathbf{y} \in R(A^T) \}$$

Thus, if one restricts the domain of A to $R(A^T)$, then A maps $R(A^T)$ onto $R(A)$. Furthermore, the mapping is one-to-one. Indeed, if $\mathbf{x}_1, \mathbf{x}_2 \in R(A^T)$ and

$$A\mathbf{x}_1 = A\mathbf{x}_2$$

then

$$A(\mathbf{x}_1 - \mathbf{x}_2) = \mathbf{0}$$

and hence

$$\mathbf{x}_1 - \mathbf{x}_2 \in R(A^T) \cap N(A)$$

Since $R(A^T) \cap N(A) = \{\mathbf{0}\}$, it follows that $\mathbf{x}_1 = \mathbf{x}_2$. Thus one can think of A as determining a one-to-one correspondence between $R(A^T)$ and $R(A)$. Since each $\mathbf{b} \in R(A)$ corresponds to exactly one $\mathbf{y} \in R(A^T)$, one can define an inverse transformation from $R(A)$ to $R(A^T)$. Indeed every $m \times n$ matrix A is invertible when viewed as a linear transformation from $R(A^T)$ to $R(A)$.

EXAMPLE 5. Let $A = \begin{pmatrix} 2 & 0 & 0 \\ 0 & 3 & 0 \end{pmatrix}$. $R(A^T)$ is spanned by $\mathbf{e}_1$ and $\mathbf{e}_2$ and $N(A)$ is spanned by $\mathbf{e}_3$. Any vector $\mathbf{x} \in R^3$ can be written as a sum

$$\mathbf{x} = \mathbf{y} + \mathbf{z}$$

where

$$\mathbf{y} = (x_1, x_2, 0)^T \in R(A^T) \quad \text{and} \quad \mathbf{z} = (0, 0, x_3)^T \in N(A)$$

If we restrict ourselves to vectors $\mathbf{y} \in R(A^T)$, then

$$\mathbf{y} = \begin{pmatrix} x_1 \\ x_2 \\ 0 \end{pmatrix} \rightarrow A\mathbf{y} = \begin{pmatrix} 2x_1 \\ 3x_2 \end{pmatrix}$$

In this case $R(A) = R^2$ and the inverse transformation from $R(A)$ to $R(A^T)$ is defined by

$$\mathbf{b} = \begin{pmatrix} b_1 \\ b_2 \end{pmatrix} \rightarrow \begin{pmatrix} \frac{1}{2}b_1 \\ \frac{1}{3}b_2 \\ 0 \end{pmatrix}$$

$\square$

EXERCISES

1. For each of the following matrices, determine a basis for each of the subspaces $R(A^T)$, $N(A)$, $R(A)$, and $N(A^T)$.

(a) $A = \begin{pmatrix} 3 & 4 \\ 6 & 8 \end{pmatrix}$ (b) $A = \begin{pmatrix} 1 & 3 & 1 \\ 2 & 4 & 0 \end{pmatrix}$

(c) $A = \begin{pmatrix} 4 & -2 \\ 1 & 3 \\ 2 & 1 \\ 3 & 4 \end{pmatrix}$ (d) $A = \begin{pmatrix} 1 & 0 & 0 & 0 \\ 0 & 1 & 1 & 1 \\ 0 & 0 & 1 & 1 \\ 1 & 1 & 2 & 2 \end{pmatrix}$

2. (a) Let S be the subspace of R^3 spanned by the vectors $\mathbf{x} = (x_1, x_2, x_3)^T$ and $\mathbf{y} = (y_1, y_2, y_3)^T$. Let

$$A = \begin{pmatrix} x_1 & x_2 & x_3 \\ y_1 & y_2 & y_3 \end{pmatrix}$$

Show that $S^\perp = N(A)$.

(b) Find the orthogonal complement of the subspace of R^3 spanned by $(1, 2, 1)^T$ and $(1, -1, 2)^T$.

3. Let S be the subspace of R^4 spanned by $\mathbf{x}_1 = (1, 0, -2, 1)^T$ and $\mathbf{x}_2 = (0, 1, 3, -2)^T$. Find a basis for $S^\perp$.

4. Let $P_1 = (1, 1, 1)$, $P_2 = (2, 4, -1)$, and $P_3 = (0, -1, 5)$

(a) Find a nonzero vector $\mathbf{N}$ that is orthogonal to $\overline{P_1 P_2}$ and $\overline{P_1 P_3}$.

(b) Find the equation of the plane determined by the three points.

5. Is it possible for a matrix to have the vector $(3, 1, 2)$ in its row space and $(2, 1, 1)^T$ in its nullspace? Explain.

6. Let $\mathbf{a}_j$ be a nonzero column vector of an $m \times n$ matrix A. Is it possible for $\mathbf{a}_j$ to be in $N(A^T)$? Explain.

7. Let S be the subspace of R^n spanned by the vectors $\mathbf{x}_1, \mathbf{x}_2, \ldots, \mathbf{x}_k$. Show that $\mathbf{y} \in S^\perp$ if and only if $\mathbf{y} \perp \mathbf{x}_i$ for $i = 1, \ldots, k$.

8. If A is an $m \times n$ matrix of rank r, what are the dimensions of $N(A)$ and $N(A^T)$? Explain.

9. Prove Corollary 5.2.5.

10. Prove: If A is an $m \times n$ matrix and $\mathbf{x} \in R^n$, then either $A\mathbf{x} = \mathbf{0}$ or there exists $\mathbf{y} \in R(A^T)$ such that $\mathbf{x}^T \mathbf{y} \neq 0$.

11. Let A be an $m \times n$ matrix. Explain why the following are true.
  (a) Any vector $\mathbf{x}$ in R^n can be uniquely written as a sum $\mathbf{y} + \mathbf{z}$, where $\mathbf{y} \in N(A)$ and $\mathbf{z} \in R(A^T)$.
  (b) Any vector $\mathbf{b} \in R^m$ can be uniquely written as a sum $\mathbf{u} + \mathbf{v}$, where $\mathbf{u} \in N(A^T)$ and $\mathbf{v} \in R(A)$.

12. Let A be an $m \times n$ matrix.
  (a) Show that if $\mathbf{x} \in N(A^T A)$, then $A\mathbf{x}$ is in both $R(A)$ and $N(A^T)$.
  (b) Show that $N(A^T A) = N(A)$.
  (c) Show that A and $A^T A$ have the same rank.
  (d) Show that if A has linearly independent columns, then $A^T A$ is nonsingular.

13. Let $A \in R^{m \times n}$, $B \in R^{n \times r}$, and $C = AB$.
  (a) Show that $N(B)$ is a subspace of $N(C)$.
  (b) Show that $N(C)^{\perp}$ is a subspace of $N(B)^{\perp}$ and consequently, $R(C^T)$ is a subspace of $R(B^T)$.

14. Let U and V be subspaces of a vector space W. If $W = U \oplus V$, show that $U \cap V = \{\mathbf{0}\}$.

15. Let A be an $m \times n$ matrix of rank r and let $\{\mathbf{x}_1, \ldots, \mathbf{x}_r\}$ be a basis for $R(A^T)$. Show that $\{A\mathbf{x}_1, \ldots, A\mathbf{x}_r\}$ is a basis for $R(A)$.

3. INNER PRODUCT SPACES

Scalar products are useful not only in R^n but in general vector spaces.

Definition. An **inner product** on a vector space V is an operation on V that assigns to each pair of vectors $\mathbf{x}$ and $\mathbf{y}$ in V a real number $\langle \mathbf{x}, \mathbf{y} \rangle$ satisfying the following conditions:
  (i) $\langle \mathbf{x}, \mathbf{x} \rangle \geq 0$ with equality if and only if $\mathbf{x} = \mathbf{0}$
  (ii) $\langle \mathbf{x}, \mathbf{y} \rangle = \langle \mathbf{y}, \mathbf{x} \rangle$ for all $\mathbf{x}$ and $\mathbf{y}$ in V
  (iii) $\langle \alpha\mathbf{x} + \beta\mathbf{y}, \mathbf{z} \rangle = \alpha\langle \mathbf{x}, \mathbf{z} \rangle + \beta\langle \mathbf{y}, \mathbf{z} \rangle$ for all $\mathbf{x}, \mathbf{y}, \mathbf{z}$ in V and all scalars α and β
A vector space V with an inner product is called an **inner product space**.

In R^n the scalar product is an example of an inner product

$$\langle \mathbf{x}, \mathbf{y} \rangle = \mathbf{x}^T \mathbf{y}$$

In $C[a, b]$ we may define an inner product by

$$(1) \qquad \langle f, g \rangle = \int_a^b f(x)g(x)\, dx$$

The reader may verify that (1) does indeed define an inner product.

If $w(x)$ is a positive continuous function on $[a, b]$, then

$$(2) \qquad \langle f, g \rangle = \int_a^b f(x)g(x)w(x)\, dx$$

defines an inner product on $C[a, b]$. The function $w(x)$ is called a *weight function*. Thus it is possible to define many different inner products on $C[a, b]$.

Let $x_1, x_2, \ldots, x_n$ be distinct real numbers. For each pair of polynomials in P_n define

$$(3) \qquad \langle p, q \rangle = \sum_{i=1}^n p(x_i)q(x_i)$$

It is easily seen that (3) satisfies conditions (ii) and (iii) of the definition of an inner product. To show that (i) holds, note that

$$\langle p, p \rangle = \sum_{i=1}^n \big(p(x_i) \big)^2 \geq 0$$

If $\langle p, p \rangle = 0$, then $x_1, x_2, \ldots, x_n$ must be roots of $p(x) = 0$. Since $p(x)$ is of degree less than n, it must be the zero polynomial.

If $w(x)$ is a positive function, then

$$(4) \qquad \langle p, q \rangle = \sum_{i=1}^n p(x_i)q(x_i)w(x_i)$$

also defines an inner product on P_n.

Two vectors $\mathbf{u}$ and $\mathbf{v}$ in an inner product space are said to be *orthogonal* if $\langle \mathbf{u}, \mathbf{v} \rangle = 0$.

EXAMPLE 1.   The functions $\cos x$ and $\sin x$ are orthogonal in $C[-\pi, \pi]$ with inner product defined by (1).

$$\langle \cos x, \sin x \rangle = \int_{-\pi}^{\pi} \cos x \sin x\, dx = 0$$

$\square$

EXAMPLE 2.   Define an inner product on $C[-1, 1]$ by

$$\langle f, g \rangle = \int_{-1}^1 f(x)g(x)e^{-x}\, dx$$

The functions $f(x) = x$ and $g(x) = e^x$ are orthogonal, since

$$\langle f, g \rangle = \int_{-1}^1 xe^x e^{-x}\, dx = 0$$

$\square$

In R^n the scalar product was used to define the length of a vector. In the same manner the length of a vector in an inner product space can be defined in terms of the inner product.

Definition.   If V is an inner product space, then for each $\mathbf{v} \in V$ the **norm** of $\mathbf{v}$ is given by

$$\|\mathbf{v}\| = \sqrt{\langle \mathbf{v}, \mathbf{v} \rangle}$$

EXAMPLE 3.   In $C[-\pi, \pi]$ with inner product defined by (1),

$$\|\sin x\| = \left(\int_{-\pi}^{\pi} \sin^2 x \, dx \right)^{1/2} = \sqrt{\pi}$$

☐

EXAMPLE 4.   In P_5 define an inner product by (3) with $x_i = (i - 1)/4$ for $i = 1, 2, \ldots, 5$. The length of the function $p(x) = x$ is given by

$$\|x\| = (\langle x, x \rangle)^{1/2}$$

$$= \left(\sum_{i=1}^{5} x_i^2 \right)^{1/2}$$

$$= \left(\sum_{i=1}^{5} \left(\frac{i - 1}{4} \right)^2 \right)^{1/2}$$

$$= \frac{\sqrt{30}}{4}$$

☐

Theorem 5.3.1 (Cauchy – Schwarz).   *If* $\mathbf{u}$ *and* $\mathbf{v}$ *are any two vectors in an inner product space* V, *then*

$$|\langle \mathbf{u}, \mathbf{v} \rangle| \le \|\mathbf{u}\| \, \|\mathbf{v}\|$$

Equality holds if and only if $\mathbf{u}$ *and* $\mathbf{v}$ *are linearly dependent.*

Proof.   Let λ be a scalar. Then

(5) $0 \le \langle \mathbf{u} - \lambda \mathbf{v}, \mathbf{u} - \lambda \mathbf{v} \rangle$ [from (i) of definition]

$$= \langle \mathbf{u}, \mathbf{u} \rangle - 2\lambda \langle \mathbf{u}, \mathbf{v} \rangle + \lambda^2 \langle \mathbf{v}, \mathbf{v} \rangle$$

$$= \|\mathbf{u}\|^2 - 2\lambda \langle \mathbf{u}, \mathbf{v} \rangle + \lambda^2 \|\mathbf{v}\|^2$$

Let $a = \|\mathbf{v}\|^2$, $b = -2\langle \mathbf{u}, \mathbf{v} \rangle$, and $c = \|\mathbf{u}\|^2$. Equation (5) can then be viewed as a quadratic inequality,

$$a\lambda^2 + b\lambda + c \ge 0$$

For this inequality to be valid, the discriminant must be less than or equal to 0.

$$b^2 - 4ac \le 0$$

Thus

$$4\langle \mathbf{u}, \mathbf{v} \rangle^2 - 4\|\mathbf{u}\|^2\|\mathbf{v}\|^2 \leq 0$$

and hence

$$|\langle \mathbf{u}, \mathbf{v} \rangle| \leq \|\mathbf{u}\|\,\|\mathbf{v}\|$$

If equality holds in (5), then $\mathbf{u} = \lambda\mathbf{v}$. $\square$

NORMS

The word *norm* in mathematics has its own meaning independent of an inner product and its use here should be justified.

Definition. A vector space V is said to be a **normed linear space** if to each vector $\mathbf{v} \in V$ there is associated a real number $\|\mathbf{v}\|$ called the **norm** of $\mathbf{v}$, satisfying:

    (i) $\|\mathbf{v}\| \geq 0$ with equality if and only if $\mathbf{v} = \mathbf{0}$
    (ii) $\|\alpha\mathbf{v}\| = |\alpha|\,\|\mathbf{v}\|$ for any scalar α
    (iii) $\|\mathbf{v} + \mathbf{w}\| \leq \|\mathbf{v}\| + \|\mathbf{w}\|$ for all $\mathbf{v}, \mathbf{w} \in V$

The third condition is called the *triangle inequality* (see Figure 5.3.1).

Theorem 5.3.2. *If V is an inner product space, then the equation*

$$\|\mathbf{v}\| = \sqrt{\langle \mathbf{v}, \mathbf{v} \rangle} \quad \textit{for all } \mathbf{v} \in V$$

defines a norm on V.

Proof. It is easily seen that conditions (i) and (ii) of the definition are satisfied. We leave this for the reader to verify and proceed to show that condition (iii) is satisfied.

$$\begin{aligned}
\|\mathbf{u} + \mathbf{v}\|^2 &= \langle \mathbf{u} + \mathbf{v}, \mathbf{u} + \mathbf{v} \rangle \\
&= \langle \mathbf{u}, \mathbf{u} \rangle + 2\langle \mathbf{u}, \mathbf{v} \rangle + \langle \mathbf{v}, \mathbf{v} \rangle \\
&\leq \|\mathbf{u}\|^2 + 2\|\mathbf{u}\|\,\|\mathbf{v}\| + \|\mathbf{v}\|^2 \quad \text{(Cauchy–Schwarz)} \\
&= \left(\|\mathbf{u}\| + \|\mathbf{v}\|\right)^2
\end{aligned}$$

Thus

$$\|\mathbf{u} + \mathbf{v}\| \leq \|\mathbf{u}\| + \|\mathbf{v}\| \qquad\qquad \square$$

FIGURE 5.3.1

FIGURE 5.3.2

As a consequence of the foregoing proof we have the *Pythagorean law*. If **u** and **v** are orthogonal, then

$$\|\mathbf{u} + \mathbf{v}\|^2 = \|\mathbf{u}\|^2 + \|\mathbf{v}\|^2$$

Interpreted in R^2, this is just the familiar Pythagorean theorem as shown in Figure 5.3.2.

It is possible to define many different norms on a given vector space. For example, in R^n one could define

$$\|\mathbf{x}\|_1 = \sum_{i=1}^{n} |x_i|$$

for every $\mathbf{x} = (x_1, x_2, \ldots, x_n)^T$. It is easily verified that $\|\cdot\|_1$ defines a norm on R^n. Another important norm on R^n is the *uniform norm*, which is defined by

$$\|\mathbf{x}\|_\infty = \max_{1 \le i \le n} |x_i|$$

More generally, one could define a norm on R^n by

$$\|\mathbf{x}\|_p = \left(\sum_{i=1}^{n} |x_i|^p \right)^{1/p}$$

for any real number $p \ge 1$. In particular, if $p = 2$, then

$$\|\mathbf{x}\|_2 = \left(\sum_{i=1}^{n} |x_i|^2 \right)^{1/2} = \sqrt{\langle \mathbf{x}, \mathbf{x} \rangle}$$

The norm $\|\cdot\|_2$ is the norm on R^n derived from the inner product. If $p \ne 2$, $\|\cdot\|_p$ does not correspond to any inner product. In the case of a norm that is not derived from an inner product, the Pythagorean law will not hold. For example,

$$\mathbf{x}_1 = \begin{pmatrix} 1 \\ 2 \end{pmatrix} \qquad \text{and} \qquad \mathbf{x}_2 = \begin{pmatrix} -4 \\ 2 \end{pmatrix}$$

are orthogonal, however,

$$\|\mathbf{x}_1\|_\infty^2 + \|\mathbf{x}_2\|_\infty^2 = 4 + 16 = 20$$

while

$$\|\mathbf{x}_1 + \mathbf{x}_2\|_\infty^2 = 16$$

On the other hand, if $\|\cdot\|_2$ is used, then

$$\|\mathbf{x}_1\|_2^2 + \|\mathbf{x}_2\|_2^2 = 5 + 20 = 25 = \|\mathbf{x}_1 + \mathbf{x}_2\|_2^2$$

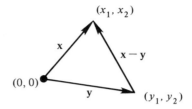

FIGURE 5.3.3

EXAMPLE 5. Let $\mathbf{x}$ be the vector $(4, -5, 3)^T$ in R^3. Compute $\|\mathbf{x}\|_1$, $\|\mathbf{x}\|_2$, and $\|\mathbf{x}\|_\infty$.

$$\|\mathbf{x}\|_1 = |4| + |-5| + |3| = 12$$

$$\|\mathbf{x}\|_2 = \sqrt{16 + 25 + 9} = 5\sqrt{2}$$

$$\|\mathbf{x}\|_\infty = \max(|4|, |-5|, |3|) = 5 \qquad \square$$

In general, a norm provides a way of measuring the distance between vectors.

Definition. Let $\mathbf{x}$ and $\mathbf{y}$ be vectors in a normed linear space. The distance between $\mathbf{x}$ and $\mathbf{y}$ is defined to be the number $\|\mathbf{x} - \mathbf{y}\|$.

EXAMPLE 6. In the vector space R^2 with norm $\|\cdot\|_2$, let $\mathbf{x} = (x_1, x_2)^T$ and $\mathbf{y} = (y_1, y_2)^T$. The distance between $\mathbf{x}$ and $\mathbf{y}$ is the length of $\mathbf{x} - \mathbf{y}$ (see Figure 5.3.3).

$$\|\mathbf{x} - \mathbf{y}\|_2 = \sqrt{(x_1 - y_1)^2 + (x_2 - y_2)^2}$$

This is the standard formula used in analytic geometry for the distance between two points in the plane. $\qquad \square$

In the next section we will define norms on the vector space $R^{m \times n}$ of all $m \times n$ matrices. The norm of an $m \times n$ matrix A is useful in determining the accuracy of calculated solutions to linear systems of the form $A\mathbf{x} = \mathbf{b}$.

EXERCISES

1. Show that

$$\langle f, g \rangle = \int_a^b f(x)g(x)\, dx$$

defines an inner product on $C[a, b]$.

2. In $C[0, 1]$ with inner product defined by (1), compute:
(a) $\langle e^x, e^{-x} \rangle$ (b) $\langle x, \sin \pi x \rangle$ (c) $\langle x^2, x^3 \rangle$

3. In $C[-\pi, \pi]$ with inner product defined by (1), show that $\cos mx$ and $\sin nx$ are orthogonal for any integers m and n.

4. Show that the functions x and x^2 are orthogonal in P_5 with inner product defined by (3), where $x_i = (i - 3)/2$ for $i = 1, \ldots, 5$.

5. If V is an inner product space, show that

$$\|\mathbf{v}\| = \sqrt{\langle \mathbf{v}, \mathbf{v} \rangle}$$

satisfies the first two conditions in the definition of a norm.

6. Show that

$$\|\mathbf{x}\|_1 = \sum_{i=1}^{n} |x_i|$$

defines a norm on R^n.

7. Show that

$$\|\mathbf{x}\|_\infty = \max_{1 \le i \le n} |x_i|$$

defines a norm on R^n.

8. Compute $\|\mathbf{x}\|_1$, $\|\mathbf{x}\|_2$, and $\|\mathbf{x}\|_\infty$ for each of the following vectors in R^3.
(a) $\mathbf{x} = (-3, 4, 0)^T$ (b) $\mathbf{x} = (-1, -1, 2)^T$
(c) $\mathbf{x} = (1, 1, 1)^T$

9. Let $\mathbf{x} = (5, 2, 4)^T$ and $\mathbf{y} = (3, 3, 2)^T$. Compute $\|\mathbf{x} - \mathbf{y}\|_1$, $\|\mathbf{x} - \mathbf{y}\|_2$, and $\|\mathbf{x} - \mathbf{y}\|_\infty$. Under which norm are the two vectors closest together? Under which norm are they farthest apart?

10. Let $\mathbf{x} = (-1, -1, 1, 1)^T$ and $\mathbf{y} = (1, 1, 5, -3)^T$. Show that $\mathbf{x} \perp \mathbf{y}$. Calculate $\|\mathbf{x}\|_2$, $\|\mathbf{y}\|_2$, $\|\mathbf{x} + \mathbf{y}\|_2$ and verify that the Pythagorean law holds.

11. Let $\mathbf{x}$ and $\mathbf{y}$ be vectors in an inner product space. Show that if $\mathbf{x} \perp \mathbf{y}$, then the distance between $\mathbf{x}$ and $\mathbf{y}$ is

$$\left(\|\mathbf{x}\|^2 + \|\mathbf{y}\|^2 \right)^{1/2}$$

12. In $C[-\pi, \pi]$ with inner product defined by (1) and norm defined by

$$\|f\| = \sqrt{\langle f, f \rangle}$$

compute:
(a) $\|\sin nx\|$ (b) $\|\cos mx\|$
(c) The distance between $\sin nx$ and $\cos mx$.

13. In P_5 with inner product as in Exercise 4 and norm defined by

$$\|p\| = \sqrt{\langle p, p \rangle} = \left\{ \sum_{i=1}^{5} [p(x_i)]^2 \right\}^{1/2}$$

compute:
(a) $\|x\|$ (b) $\|x^2\|$ (c) The distance between x and x^2

14. In R^n with inner product

$$\langle \mathbf{x}, \mathbf{y} \rangle = \mathbf{x}^T \mathbf{y}$$

derive a formula for the distance between two vectors $\mathbf{x} = (x_1, \ldots, x_n)^T$ and $\mathbf{y} = (y_1, \ldots, y_n)^T$.

15. Let $\mathbf{x} \in R^n$. Show that $\|\mathbf{x}\|_\infty \le \|\mathbf{x}\|_2$.

16. Let $\mathbf{x} \in R^2$. Show that $\|\mathbf{x}\|_2 \le \|\mathbf{x}\|_1$.

[**Hint:** Write $\mathbf{x}$ in the form $x_1 \mathbf{e}_1 + x_2 \mathbf{e}_2$ and use the triangle inequality.]

17. Give an example of a nonzero vector $\mathbf{x} \in R^2$ for which

$$\|\mathbf{x}\|_\infty = \|\mathbf{x}\|_2 = \|\mathbf{x}\|_1$$

18. Show that in any vector space with a norm

$$\|\mathbf{0}\| = 0$$

19. Show that for any $\mathbf{u}$ and $\mathbf{v}$ in a normed vector space

$$\|\mathbf{u} + \mathbf{v}\| \ge |\|\mathbf{u}\| - \|\mathbf{v}\||$$

20. Prove that for any $\mathbf{u}$ and $\mathbf{v}$ in an inner product space V

$$\|\mathbf{u} + \mathbf{v}\|^2 + \|\mathbf{u} - \mathbf{v}\|^2 = 2\|\mathbf{u}\|^2 + 2\|\mathbf{v}\|^2$$

Give a geometric interpretation of this result for the vector space R^2.

21. The result of Exercise 20 is not valid for norms other than the norm derived from the inner product. Give an example of this in R^2 using $\|\cdot\|_1$.

22. Determine whether or not the following define norms on $C[a, b]$.
(a) $\|f\| = |f(a)| + |f(b)|$
(b) $\|f\| = \int_a^b |f(x)| \, dx$
(c) $\|f\| = \max_{a \le x \le b} |f(x)|$

23. Let $\mathbf{x} \in R^n$ and show that:
(a) $\|\mathbf{x}\|_1 \leq n\|\mathbf{x}\|_\infty$　　(b) $\|\mathbf{x}\|_2 \leq \sqrt{n}\,\|\mathbf{x}\|_\infty$
Give examples of vectors in R^n for which equality holds in (a) and (b).

24. Sketch the set of points $(x_1, x_2) = \mathbf{x}^T$ in R^2 such that:
(a) $\|\mathbf{x}\|_2 = 1$　　(b) $\|\mathbf{x}\|_1 = 1$　　(c) $\|\mathbf{x}\|_\infty = 1$

25. Let $R^{2 \times 2}$ denote the vector space of all 2×2 matrices and define

$$A * B = a_{11}b_{11} + a_{12}b_{12} + a_{21}b_{21} + a_{22}b_{22}$$

Determine whether or not the operation $*$ is an inner product on $R^{2 \times 2}$.

26. Consider the vector space R^n with inner product $\langle \mathbf{x}, \mathbf{y} \rangle = \mathbf{x}^T\mathbf{y}$. For any $n \times n$ matrix A, show that:
(a) $\langle A\mathbf{x}, \mathbf{y} \rangle = \langle \mathbf{x}, A^T\mathbf{y} \rangle$　　(b) $\langle A^TA\mathbf{x}, \mathbf{x} \rangle = \|A\mathbf{x}\|^2$

4. MATRIX NORMS

In the preceding section, norms were used to measure the length of vectors. Norms can also be used to measure the size of matrices. Matrix norms are useful in a number of applications and they play an important role in numerical linear algebra. In this section we will study some of the standard matrix norms and their properties. These norms will be used throughout the last five sections of Chapter 7.

The most natural norm to use on R^n is the Euclidean norm

$$(1) \qquad \|\mathbf{x}\|_2 = \left(\sum_{i=1}^{n} x_i^2 \right)^{1/2}$$

Actually, (1) defines a whole family of norms, since it defines a norm on R^n for $n = 1, 2, \ldots$. In a similar manner we can define a family of matrix norms by

$$(2) \qquad \|A\|_F = \left(\sum_{j=1}^{n} \sum_{i=1}^{m} a_{ij}^2 \right)^{1/2}$$

for each $m \times n$ matrix. Equation (2) defines a family of norms on the vector spaces $R^{m \times n}$. Clearly, $\|A\|_F \geq 0$ and $\|A\|_F = 0$ if and only if $A = O$. It is easily seen that

$$\|\alpha A\|_F = \left(\sum_{j=1}^{n} \sum_{i=1}^{m} (\alpha a_{ij})^2 \right)^{1/2} = |\alpha| \|A\|_F$$

Thus, to show that $\| \cdot \|_F$ actually defines a norm on $R^{m \times n}$, we must show that

$$\|A + B\|_F \leq \|A\|_F + \|B\|_F$$

If we think of the matrices in $R^{m \times n}$ as being vectors in R^{mn}, this result follows immediately. If we set

$$\mathbf{a} = (a_{11}, a_{12}, \ldots, a_{1n}, a_{21}, a_{22}, \ldots, a_{2n}, \ldots, a_{m1}, a_{m2}, \ldots, a_{mn})^T$$

$$\mathbf{b} = (b_{11}, b_{12}, \ldots, b_{1n}, b_{21}, b_{22}, \ldots, b_{2n}, \ldots, b_{m1}, b_{m2}, \ldots, b_{mn})^T$$

then $\mathbf{a}$ and $\mathbf{b}$ are vectors in R^{mn} and it follows that

$$\|A + B\|_F = \|\mathbf{a} + \mathbf{b}\|_2 \leq \|\mathbf{a}\|_2 + \|\mathbf{b}\|_2 = \|A\|_F + \|B\|_F$$

Thus $\| \cdot \|_F$ defines a norm on $R^{m \times n}$ for each ordered pair of positive integers (m, n). We will refer to $\|A\|_F$ as the *Frobenius norm* of A. The $\| \cdot \|_F$ has a number of important properties, which we list below.

I. If $\mathbf{a}_j$ represents the jth column vector of A, then

$$\|A\|_F = \left(\sum_{j=1}^{n} \sum_{i=1}^{m} a_{ij}^2 \right)^{1/2} = \left(\sum_{j=1}^{n} \|\mathbf{a}_j\|_2^2 \right)^{1/2}$$

II. If $(\mathbf{a}_i)_r$ represents the ith column vector of A^T, then

$$\|A\|_F = \left(\sum_{i=1}^{m} \sum_{j=1}^{n} a_{ij}^2 \right)^{1/2} = \left(\sum_{i=1}^{m} \|(\mathbf{a}_i)_r\|_2^2 \right)^{1/2}$$

III. If $\mathbf{x} \in R^n$, then

$$\|A\mathbf{x}\|_2 = \left[\sum_{i=1}^{m} \left(\sum_{j=1}^{n} a_{ij} x_j \right)^2 \right]^{1/2} = \left[\sum_{i=1}^{m} \left((\mathbf{a}_i)_r^T \mathbf{x} \right)^2 \right]^{1/2}$$

$$\leq \left[\sum_{i=1}^{m} \|\mathbf{x}\|_2^2 \|(\mathbf{a}_i)_r\|^2 \right]^{1/2} \quad \text{(Cauchy–Schwarz)}$$

$$= \|A\|_F \|\mathbf{x}\|_2$$

IV. If $B = (\mathbf{b}_1, \ldots, \mathbf{b}_r)$ is an $n \times r$ matrix, it follows from properties I and III that

$$\|AB\|_F = \|(A\mathbf{b}_1, A\mathbf{b}_2, \ldots, A\mathbf{b}_r)\|_F$$

$$= \left(\sum_{i=1}^{r} \|A\mathbf{b}_i\|_2^2 \right)^{1/2}$$

$$\leq \|A\|_F \left(\sum_{i=1}^{r} \|\mathbf{b}_i\|_2^2 \right)^{1/2}$$

$$= \|A\|_F \|B\|_F$$

In view of property III, we say that the matrix norm $\| \cdot \|_F$ and the vector norm $\| \cdot \|_2$ are compatible. In general, a matrix norm $\| \cdot \|_M$ on $R^{m \times n}$ and a vector norm $\| \cdot \|_V$ on R^n are said to be *compatible* if

$$\|A\mathbf{x}\|_V \leq \|A\|_M \|\mathbf{x}\|_V$$

for every $\mathbf{x} \in R^n$.

There are many other norms that one could put on $R^{m \times n}$. The families of norms that turn out to be most useful generally satisfy the property
(iv) $$\|AB\| \leq \|A\| \|B\|$$

Consequently, we will only consider families of norms that have this property. One important consequence of property (iv) is that

$$\|A^n\| \leq \|A\|^n$$

In particular, if $\|A\| < 1$, then $\|A^n\| \to 0$ as $n \to \infty$.

SUBORDINATE MATRIX NORMS

We can think of each $m \times n$ matrix as a linear operator from R^n to R^m. For any family of vector norms we can define an *operator norm* by comparing $\|A\mathbf{x}\|$ and $\|\mathbf{x}\|$ for each nonzero $\mathbf{x}$ and taking

$$(3) \qquad \|A\| = \max_{\mathbf{x} \neq \mathbf{0}} \frac{\|A\mathbf{x}\|}{\|\mathbf{x}\|}$$

It can be shown that there is a particular $\mathbf{x}_0$ in R^n which maximizes $\|A\mathbf{x}\|/\|\mathbf{x}\|$, but the proof is beyond the scope of this text. Assuming that $\|A\mathbf{x}\|/\|\mathbf{x}\|$ can always be maximized, we will show that (3) actually does define a norm on $R^{m \times n}$.

(i) For each $\mathbf{x} \neq \mathbf{0}$,

$$\frac{\|A\mathbf{x}\|}{\|\mathbf{x}\|} \geq 0$$

and, consequently,

$$\|A\| = \max_{\mathbf{x} \neq \mathbf{0}} \frac{\|A\mathbf{x}\|}{\|\mathbf{x}\|} \geq 0$$

If $\|A\| = 0$, then $A\mathbf{x} = \mathbf{0}$ for every $\mathbf{x} \in R^n$. This implies that $\mathbf{a}_j = A\mathbf{e}_j = \mathbf{0}$ for $j = 1, \ldots, n$, and hence A must be the zero matrix.

(ii) $\|\alpha A\| = \displaystyle\max_{\mathbf{x} \neq \mathbf{0}} \frac{\|\alpha A\mathbf{x}\|}{\|\mathbf{x}\|} = |\alpha| \max_{\mathbf{x} \neq \mathbf{0}} \frac{\|A\mathbf{x}\|}{\|\mathbf{x}\|} = |\alpha| \|A\|$

(iii) $\|(A + B)\mathbf{x}\| \leq \|A\mathbf{x}\| + \|B\mathbf{x}\| \leq \|A\| \|\mathbf{x}\| + \|B\| \|\mathbf{x}\|$
Thus, if $\mathbf{x} \neq \mathbf{0}$, then

$$\frac{\|(A + B)\mathbf{x}\|}{\|\mathbf{x}\|} \leq \|A\| + \|B\|$$

and hence

$$\|A + B\| = \max_{\mathbf{x} \neq \mathbf{0}} \frac{\|(A + B)\mathbf{x}\|}{\|\mathbf{x}\|} \leq \|A\| + \|B\|$$

Thus (3) defines a norm on $R^{m \times n}$. For each family of vector norms $\| \cdot \|$, one can then define a family of matrix norms by (3). The matrix norms defined by (3) are said to be *subordinate* to the vector norms $\| \cdot \|$.

Theorem 5.4.1.   *If the family of matrix norms $\| \cdot \|_M$ is subordinate to the family of vector norms $\| \cdot \|$, then $\| \cdot \|_M$ and $\| \cdot \|$ are compatible and the matrix norms $\| \cdot \|_M$ satisfy property* (iv).

Proof.   If $\mathbf{x}$ is any nonzero vector in R^n, then

$$\frac{\|A\mathbf{x}\|}{\|\mathbf{x}\|} \leq \max_{\mathbf{y} \neq \mathbf{0}} \frac{\|A\mathbf{y}\|}{\|\mathbf{y}\|} = \|A\|_M$$

and hence

$$\|A\mathbf{x}\| \leq \|A\|_M \|\mathbf{x}\|$$

Since this last inequality is also valid if $\mathbf{x} = \mathbf{0}$, it follows that $\| \cdot \|_M$ and $\| \cdot \|$ are compatible. If B is an $n \times r$ matrix, then since $\| \cdot \|_M$ and $\| \cdot \|$ are compatible, we have

$$\|AB\mathbf{x}\| \leq \|A\|_M \|B\mathbf{x}\| \leq \|A\|_M \|B\|_M \|\mathbf{x}\|$$

Thus for all $\mathbf{x} \neq \mathbf{0}$,

$$\frac{\|AB\mathbf{x}\|}{\|\mathbf{x}\|} \leq \|A\|_M \|B\|_M$$

and hence

$$\|AB\|_M = \max_{\mathbf{x} \neq \mathbf{0}} \frac{\|AB\mathbf{x}\|}{\|\mathbf{x}\|} \leq \|A\|_M \|B\|_M \qquad \square$$

It is a simple matter to compute the Frobenius norm of a matrix. For example, if

$$A = \begin{pmatrix} 4 & 2 \\ 0 & 4 \end{pmatrix}$$

then

$$\|A\|_F = (4^2 + 0^2 + 2^2 + 4^2)^{1/2} = 6$$

On the other hand, it is not so obvious how to compute $\|A\|$ if $\| \cdot \|$ is a subordinate matrix norm. It turns out that the matrix norm

$$\|A\|_2 = \max_{\mathbf{x} \neq \mathbf{0}} \frac{\|A\mathbf{x}\|_2}{\|\mathbf{x}\|_2}$$

is difficult to compute; however,

$$\|A\|_1 = \max_{\mathbf{x} \neq \mathbf{0}} \frac{\|A\mathbf{x}\|_1}{\|\mathbf{x}\|_1}$$

and

$$\|A\|_\infty = \max_{\mathbf{x} \neq \mathbf{0}} \frac{\|A\mathbf{x}\|_\infty}{\|\mathbf{x}\|_\infty}$$

can be easily calculated.

Theorem 5.4.2.   *If A is an m × n matrix, then*

$$\|A\|_1 = \max_{1 \leq j \leq n} \left(\sum_{i=1}^{m} |a_{ij}| \right)$$

and

$$\|A\|_\infty = \max_{1 \leq i \leq m} \left(\sum_{j=1}^{n} |a_{ij}| \right)$$

Proof.   We will prove that

$$\|A\|_1 = \max_{1 \leq j \leq n} \left(\sum_{i=1}^{m} |a_{ij}| \right)$$

and leave the proof of the second statement as an exercise. Let

$$\alpha = \max_{1 \leq j \leq n} \sum_{i=1}^{m} |a_{ij}| = \sum_{i=1}^{m} |a_{ik}|$$

That is, k is the index of the column where the maximum occurs. Let $\mathbf{x}$ be an arbitrary vector in R^n; then

$$A\mathbf{x} = \left(\sum_{j=1}^{n} a_{1j} x_j, \sum_{j=1}^{n} a_{2j} x_j, \ldots, \sum_{j=1}^{n} a_{mj} x_j \right)^T$$

and it follows that

$$
\begin{aligned}
\|A\mathbf{x}\|_1 &= \sum_{i=1}^{m} \left| \sum_{j=1}^{n} a_{ij} x_j \right| \\
&\leq \sum_{i=1}^{m} \sum_{j=1}^{n} |a_{ij} x_j| \\
&= \sum_{j=1}^{n} \left(|x_j| \sum_{i=1}^{m} |a_{ij}| \right) \\
&\leq \alpha \sum_{j=1}^{n} |x_j| \\
&= \alpha \|\mathbf{x}\|_1
\end{aligned}
$$

Thus, for any nonzero $\mathbf{x}$ in R^n,

$$\frac{\|A\mathbf{x}\|_1}{\|\mathbf{x}\|_1} \leq \alpha$$

and hence

$$(4) \qquad \|A\|_1 = \max_{\mathbf{x} \neq \mathbf{0}} \frac{\|A\mathbf{x}\|_1}{\|\mathbf{x}\|_1} \leq \alpha$$

On the other hand,

$$\|A\mathbf{e}_k\|_1 = \|\mathbf{a}_k\|_1 = \alpha$$

Since $\|\mathbf{e}_k\|_1 = 1$, it follows that

$$(5) \qquad \|A\|_1 = \max_{\mathbf{x} \neq \mathbf{0}} \frac{\|A\mathbf{x}\|_1}{\|\mathbf{x}\|_1} \geq \frac{\|A\mathbf{e}_k\|_1}{\|\mathbf{e}_k\|_1} = \alpha$$

Together (4) and (5) imply $\|A\|_1 = \alpha$.   □

EXAMPLE 1.   Let

$$A = \begin{pmatrix} -3 & 2 & 4 & -3 \\ 5 & -2 & -3 & 5 \\ 2 & 1 & -6 & 4 \\ 1 & 1 & 1 & 1 \end{pmatrix}$$

Then

$$\|A\|_1 = |4| + |-3| + |-6| + |1| = 14$$

and

$$\|A\|_\infty = |5| + |-2| + |-3| + |5| = 15$$ □

EXERCISES

1. Determine $\|\cdot\|_F$, $\|\cdot\|_\infty$, and $\|\cdot\|_1$ for each of the following matrices.

(a) $\begin{pmatrix} 1 & 0 \\ 0 & 1 \end{pmatrix}$ (b) $\begin{pmatrix} 1 & 4 \\ -2 & 2 \end{pmatrix}$ (c) $\begin{pmatrix} \frac{1}{2} & \frac{1}{2} \\ \frac{1}{2} & \frac{1}{2} \end{pmatrix}$

(d) $\begin{pmatrix} 0 & 5 & 1 \\ 2 & 3 & 1 \\ 1 & 2 & 2 \end{pmatrix}$ (e) $\begin{pmatrix} 5 & 0 & 5 \\ 4 & 1 & 0 \\ 3 & 2 & 1 \end{pmatrix}$

2. Let

$$A = \begin{pmatrix} 2 & 0 \\ 0 & -2 \end{pmatrix} \qquad \text{and} \qquad \mathbf{x} = \begin{pmatrix} x_1 \\ x_2 \end{pmatrix}$$

Write $\|A\mathbf{x}\|_2 / \|\mathbf{x}\|_2$ in terms of x_1 and x_2. Determine the value of $\|A\|_2$.

3. Let
$$A = \begin{pmatrix} 1 & 0 \\ 0 & 0 \end{pmatrix}$$
Show that $\|A\|_2 = 1$.

4. Let I denote the $n \times n$ identity matrix. Determine the values of $\|I\|_1$, $\|I\|_\infty$, and $\|I\|_F$.

5. Let $\| \cdot \|_M$ denote a matrix norm on $R^{n \times n}$, $\| \cdot \|_V$ denote a vector norm on R^n, and let I be the $n \times n$ identity matrix.
  (a) Show that if $\| \cdot \|_M$ and $\| \cdot \|_V$ are compatible, then $\|I\|_M \geq 1$.
  (b) Show that if $\| \cdot \|_M$ is subordinate to $\| \cdot \|_V$ then $\|I\|_M = 1$.

6. Given
$$A = \begin{pmatrix} 3 & -1 & -2 \\ -1 & 2 & -7 \\ 4 & 1 & 4 \end{pmatrix}$$
  (a) Determine $\|A\|_\infty$.
  (b) Find a vector $\mathbf{x}$ whose coordinates are each ± 1 such that $\|A\mathbf{x}\|_\infty = \|A\|_\infty$. (Note that $\|\mathbf{x}\|_\infty = 1$, so that $\|A\|_\infty = \|A\mathbf{x}\|_\infty / \|\mathbf{x}\|_\infty$.)

7. Theorem 5.4.2 states that
$$\|A\|_\infty = \max_{1 \leq i \leq m} \left(\sum_{j=1}^{n} |a_{i,j}| \right)$$
Prove this in two steps.
  (a) Show first that
$$\|A\|_\infty \leq \max_{1 \leq i \leq m} \left(\sum_{j=1}^{n} |a_{i,j}| \right)$$
  (b) Construct a vector $\mathbf{x}$ whose coordinates are each ± 1 such that
$$\frac{\|A\mathbf{x}\|_\infty}{\|\mathbf{x}\|_\infty} = \|A\mathbf{x}\|_\infty = \max_{1 \leq i \leq m} \left(\sum_{j=1}^{n} |a_{i,j}| \right)$$

8. Show that $\|A\|_F = \|A^T\|_F$.

9. Let A be a symmetric $n \times n$ matrix. Show that $\|A\|_\infty = \|A\|_1$.

10. Let $\| \cdot \|$ denote a family of vector norms and let $\| \cdot \|_M$ be a subordinate matrix norm. Show that
$$\|A\|_M = \max_{\|\mathbf{x}\|=1} \|A\mathbf{x}\|$$

11. Let A be an $n \times n$ matrix and $\mathbf{x} \in R^n$. Prove:
  (a) $\|A\mathbf{x}\|_\infty \leq n^{1/2}\|A\|_2\|\mathbf{x}\|_\infty$
  (b) $\|A\mathbf{x}\|_2 \leq n^{1/2}\|A\|_\infty\|\mathbf{x}\|_2$
  (c) $n^{-1/2}\|A\|_2 \leq \|A\|_\infty \leq n^{1/2}\|A\|_2$

5. LEAST SQUARES PROBLEMS

So far we have been concerned mainly with consistent systems of linear equations. In this section we turn our attention to overdetermined systems, that is, systems involving more equations than unknowns. Such systems are usually inconsistent. Thus, given an $m \times n$ system $A\mathbf{x} = \mathbf{b}$ with $m > n$, we cannot expect in general to find a vector $\mathbf{x} \in R^n$ for which $A\mathbf{x}$ equals $\mathbf{b}$. Instead, one can look for a vector $\mathbf{x}$ for which $A\mathbf{x}$ is "closest" to $\mathbf{b}$. As you might expect, orthogonality plays an important role in finding such an $\mathbf{x}$.

Let A be an $m \times n$ matrix with $m > n$. For each $\mathbf{b} \in R^m$, define

$$\|\mathbf{b}\| = \sqrt{\langle \mathbf{b}, \mathbf{b} \rangle} = \sqrt{\mathbf{b}^T \mathbf{b}}$$

Consider the system of equations $A\mathbf{x} = \mathbf{b}$. For each $\mathbf{x} \in R^n$, we can form a *residual*

$$r(\mathbf{x}) = \mathbf{b} - A\mathbf{x}$$

The distance between $\mathbf{b}$ and $A\mathbf{x}$ is given by

$$\|\mathbf{b} - A\mathbf{x}\| = \|r(\mathbf{x})\|$$

We wish to find a vector $\mathbf{x} \in R^n$ for which $\|r(\mathbf{x})\|$ will be a minimum. Minimizing $\|r(\mathbf{x})\|$ is equivalent to minimizing $\|r(\mathbf{x})\|^2$. A vector $\hat{\mathbf{x}}$ that accomplishes this is said to be a *least squares solution* to the system $A\mathbf{x} = \mathbf{b}$.

If $\hat{\mathbf{x}}$ is a least squares solution to the system $A\mathbf{x} = \mathbf{b}$ and $\mathbf{p} = A\hat{\mathbf{x}}$, then $\mathbf{p}$ is a vector in the column space of A that is closest to $\mathbf{b}$. The following theorem guarantees that such a closest vector $\mathbf{p}$ not only exists, but is unique. Additionally, it provides an important characterization of the closest vector.

Theorem 5.5.1.   *Let S be a subspace of R^m. For each $\mathbf{b} \in R^m$ there is a unique element $\mathbf{p}$ of S that is closest to $\mathbf{b}$, that is,*

$$\|\mathbf{b} - \mathbf{y}\| > \|\mathbf{b} - \mathbf{p}\|$$

for any $\mathbf{y} \neq \mathbf{p}$ in S. Furthermore, a given vector $\mathbf{p}$ in S will be closest to a given vector $\mathbf{b} \in R^m$ if and only if $\mathbf{b} - \mathbf{p} \in S^{\perp}$.

Proof.   Since $R^m = S \oplus S^\perp$, each element **b** in R^m can be expressed uniquely as a sum

$$\mathbf{b} = \mathbf{p} + \mathbf{z}$$

where $\mathbf{p} \in S$ and $\mathbf{z} \in S^\perp$. If **y** is any other element of S, then

$$\|\mathbf{b} - \mathbf{y}\|^2 = \|(\mathbf{b} - \mathbf{p}) + (\mathbf{p} - \mathbf{y})\|^2$$

Since $\mathbf{p} - \mathbf{y} \in S$ and $\mathbf{b} - \mathbf{p} = \mathbf{z} \in S^\perp$, it follows from the Pythagorean law that

$$\|\mathbf{b} - \mathbf{y}\|^2 = \|\mathbf{b} - \mathbf{p}\|^2 + \|\mathbf{p} - \mathbf{y}\|^2$$

Therefore,

$$\|\mathbf{b} - \mathbf{y}\| > \|\mathbf{b} - \mathbf{p}\|$$

Thus, if $\mathbf{p} \in S$ and $\mathbf{b} - \mathbf{p} \in S^\perp$, then **p** is the element of S that is closest to **b**. Conversely, if $\mathbf{q} \in S$ and $\mathbf{b} - \mathbf{q} \notin S^\perp$, then $\mathbf{q} \neq \mathbf{p}$ and it follows from the preceding argument (with $\mathbf{y} = \mathbf{q}$) that

$$\|\mathbf{b} - \mathbf{q}\| > \|\mathbf{b} - \mathbf{p}\| \qquad\qquad \square$$

In the special case that **b** is in the subspace S to begin with, we have

$$\mathbf{b} = \mathbf{p} + \mathbf{z} \qquad \mathbf{p} \in S, \;\; \mathbf{z} \in S^\perp$$

and

$$\mathbf{b} = \mathbf{b} + \mathbf{0}$$

By the uniqueness of the direct sum representation,

$$\mathbf{p} = \mathbf{b} \qquad \text{and} \qquad \mathbf{z} = \mathbf{0}$$

A vector $\hat{\mathbf{x}}$ will be a solution to the least squares problem $A\mathbf{x} = \mathbf{b}$ if and only if $\mathbf{p} = A\hat{\mathbf{x}}$ is the vector in $R(A)$ that is closest to **b**. The vector **p** is said to be the *projection of* **b** *onto* $R(A)$. It follows from Theorem 5.5.1 that

$$\mathbf{b} - \mathbf{p} = \mathbf{b} - A\hat{\mathbf{x}} = r(\hat{\mathbf{x}})$$

must be an element of $R(A)^\perp$. Thus $\hat{\mathbf{x}}$ is a solution to the least squares problem if and only if

(1) $$r(\hat{\mathbf{x}}) \in R(A)^\perp$$

(See Figure 5.5.1.) How do we find a vector $\hat{\mathbf{x}}$ satisfying (1)? The key to solving the least squares problem is provided in Theorem 5.2.1, which states that

$$R(A)^\perp = N(A^T)$$

A vector $\hat{\mathbf{x}}$ will be a least squares solution to the system $A\mathbf{x} = \mathbf{b}$ if and only if

$$r(\hat{\mathbf{x}}) \in N(A^T)$$

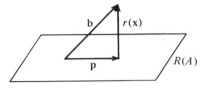

(a) $b \in R^2$ and A is a 2 × 1 matrix of rank 1.   (b) $b \in R^3$ and A is a 3 × 2 matrix of rank 2.

FIGURE 5.5.1

or, equivalently,

$$0 = A^T r(\hat{\mathbf{x}}) = A^T(\mathbf{b} - A\hat{\mathbf{x}})$$

Thus, to solve the least squares problem $A\mathbf{x} = \mathbf{b}$, we must solve

(2) $A^T A\mathbf{x} = A^T \mathbf{b}$

Equation (2) represents an $n \times n$ system of linear equations. These equations are called the *normal equations*. In general, it is possible to have more than one solution to the normal equations; however, if $\hat{\mathbf{x}}$ and $\hat{\mathbf{y}}$ are both solutions, then since the projection $\mathbf{p}$ of $\mathbf{b}$ onto $R(A)$ is unique,

$$A\hat{\mathbf{x}} = A\hat{\mathbf{y}} = \mathbf{p}$$

The following theorem characterizes the conditions under which the least squares problem $A\mathbf{x} = \mathbf{b}$ will have a unique solution.

Theorem 5.5.2. *If A is an $m \times n$ matrix of rank n, the normal equations*

$$A^T A\mathbf{x} = A^T \mathbf{b}$$

have a unique solution

$$\hat{\mathbf{x}} = (A^T A)^{-1} A^T \mathbf{b}$$

and $\hat{\mathbf{x}}$ is the unique least squares solution to the system $A\mathbf{x} = \mathbf{b}$.

Proof.   We will first show that $A^T A$ is nonsingular. To prove this, let $\mathbf{z}$ be a solution to

(3) $A^T A\mathbf{x} = \mathbf{0}$

Then $A\mathbf{z} \in N(A^T)$. Clearly, $A\mathbf{z} \in R(A) = N(A^T)^{\perp}$. Since $N(A^T) \cap N(A^T)^{\perp} = \{\mathbf{0}\}$, it follows that $A\mathbf{z} = \mathbf{0}$. If A has rank n, the column vectors of A are linearly independent and consequently $A\mathbf{x} = \mathbf{0}$ has only the trivial solution. Thus $\mathbf{z} = \mathbf{0}$ and (3) has only the trivial solution. Therefore, by Theorem 1.4.3, $A^T A$ is nonsingular. It follows that $\hat{\mathbf{x}} = (A^T A)^{-1} A^T \mathbf{b}$ is the unique solution to the normal equations and consequently the unique least squares solution to the system $A\mathbf{x} = \mathbf{b}$.   □

The projection vector

$$\mathbf{p} = A\hat{\mathbf{x}} = A(A^TA)^{-1}A^T\mathbf{b}$$

is the element of $R(A)$ that is closest to $\mathbf{b}$ in the least squares sense. The matrix $P = A(A^TA)^{-1}A^T$ is called a *projection matrix*.

APPLICATION

Hooke's law states that the force applied to a spring is proportional to the distance the spring is stretched. Thus, if F is the force applied and x is the distance the spring has been stretched, then $F = kx$. The proportionality constant k is called the *spring constant*.

Some physics students want to determine the spring constant for a given spring. They apply forces of 3, 5, and 8 pounds, which have the effect of stretching the spring 4, 7, and 11 inches, respectively. Using Hooke's law, they derive the following system of equations:

$$4k = 3$$
$$7k = 5$$
$$11k = 8$$

The system is clearly inconsistent, since each of the equations yields a different value of k. Rather than use any one of these values, the students decide to compute the least squares solution to the system.

$$(4,7,11)\begin{pmatrix} 4 \\ 7 \\ 11 \end{pmatrix}(k) = (4,7,11)\begin{pmatrix} 3 \\ 5 \\ 8 \end{pmatrix}$$

$$186k = 135$$
$$k \approx 0.726$$

EXAMPLE 1. Find the least squares solution to the system

$$x_1 + x_2 = 3$$
$$-2x_1 + 3x_2 = 1$$
$$2x_1 - x_2 = 2$$

SOLUTION. The normal equations for this system are

$$\begin{pmatrix} 1 & -2 & 2 \\ 1 & 3 & -1 \end{pmatrix}\begin{pmatrix} 1 & 1 \\ -2 & 3 \\ 2 & -1 \end{pmatrix}\begin{pmatrix} x_1 \\ x_2 \end{pmatrix} = \begin{pmatrix} 1 & -2 & 2 \\ 1 & 3 & -1 \end{pmatrix}\begin{pmatrix} 3 \\ 1 \\ 2 \end{pmatrix}$$

This simplifies to the 2×2 system

$$\begin{pmatrix} 9 & -7 \\ -7 & 11 \end{pmatrix}\begin{pmatrix} x_1 \\ x_2 \end{pmatrix} = \begin{pmatrix} 5 \\ 4 \end{pmatrix}$$

The solution to the 2×2 system is $\left(\dfrac{83}{50}, \dfrac{71}{50} \right)^T$. $\square$

Scientists often collect data and try to find a functional relationship among the variables. For example, the data may involve temperatures $T_0, T_1, \ldots, T_n$ of a liquid measured at times $t_0, t_1, \ldots, t_n$, respectively. If the temperature T can be represented as a function of the time t, this function can be used to predict the temperatures at future times. If the data consist of $n + 1$ points in the plane, it is possible to find a polynomial of degree n or less passing through all the points. Such a polynomial is called an *interpolating polynomial*. Actually, since the data usually involve experimental error, there is no reason to require

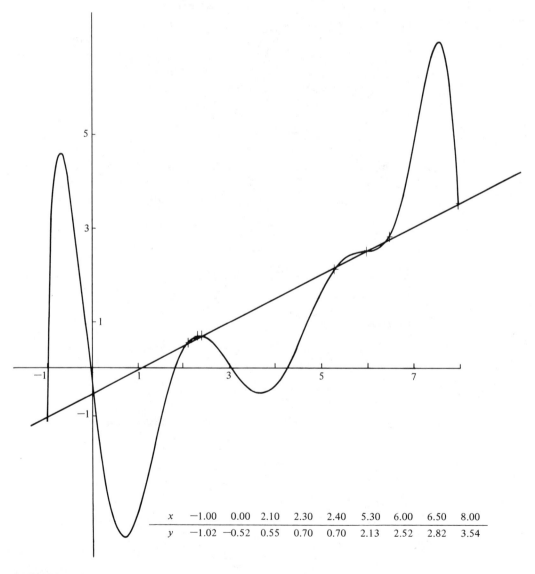

x	−1.00	0.00	2.10	2.30	2.40	5.30	6.00	6.50	8.00
y	−1.02	−0.52	0.55	0.70	0.70	2.13	2.52	2.82	3.54

FIGURE 5.5.2

that the function pass through all the points. Indeed, lower-degree polynomials that do not pass through the points exactly usually give a truer description of the relationship between the variables. If, for example, the relationship between the variables is actually linear and the data involve slight errors, it would be disastrous to use an interpolating polynomial (see Figure 5.5.2).

Given a table of data

x	x_1	x_2	$\cdots$	x_m
y	y_1	y_2	$\cdots$	y_m

we wish to find a linear function

$$y = c_0 + c_1 x$$

which best fits the data in the least squares sense. If we require that

$$y_i = c_0 + c_1 x_i \quad \text{for} \quad i = 1, \ldots, m$$

we get a system of m equations in two unknowns.

(4)
$$\begin{pmatrix} 1 & x_1 \\ 1 & x_2 \\ \vdots & \vdots \\ 1 & x_m \end{pmatrix} \begin{pmatrix} c_0 \\ c_1 \end{pmatrix} = \begin{pmatrix} y_1 \\ y_2 \\ \vdots \\ y_m \end{pmatrix}$$

The linear function whose coefficients are the least squares solution to (4) is said to be the best least squares fit to the data by a linear function.

EXAMPLE 2.   Given the data

x	0	3	6
y	1	4	5

Find the best least squares fit by a linear function.

SOLUTION.   For this example the system (4) becomes

$$A\mathbf{c} = \mathbf{y}$$

where

$$A = \begin{pmatrix} 1 & 0 \\ 1 & 3 \\ 1 & 6 \end{pmatrix}, \quad \mathbf{c} = \begin{pmatrix} c_0 \\ c_1 \end{pmatrix}, \quad \text{and} \quad \mathbf{y} = \begin{pmatrix} 1 \\ 4 \\ 5 \end{pmatrix}$$

The normal equations

$$A^T A \mathbf{c} = A^T \mathbf{y}$$

simplify to

(5)
$$\begin{pmatrix} 3 & 9 \\ 9 & 45 \end{pmatrix} \begin{pmatrix} c_0 \\ c_1 \end{pmatrix} = \begin{pmatrix} 10 \\ 42 \end{pmatrix}$$

The solution of this system is $\left(\frac{4}{3}, \frac{2}{3}\right)$. Thus the best linear least squares fit is given by

$$y = \tfrac{4}{3} + \tfrac{2}{3}x \qquad\qquad \square$$

Example 2 could also have been solved using calculus. The residual $r(\mathbf{c})$ is given by

$$r(\mathbf{c}) = \mathbf{y} - A\mathbf{c}$$

and

$$\begin{aligned}
\|r(\mathbf{c})\|^2 &= \|\mathbf{y} - A\mathbf{c}\|^2 \\
&= \left[1 - (c_0 + 0c_1)\right]^2 + \left[4 - (c_0 + 3c_1)\right]^2 \\
&\quad + \left[5 - (c_0 + 6c_1)\right]^2 \\
&= f(c_0, c_1)
\end{aligned}$$

Thus $\|r(\mathbf{c})\|^2$ can be thought of as a function of two variables, $f(c_0, c_1)$. The minimum of this function will occur when its partials are zero.

$$\frac{\partial f}{\partial c_0} = -2(10 - 3c_0 - 9c_1) = 0$$

$$\frac{\partial f}{\partial c_1} = -6(14 - 3c_0 - 15c_1) = 0$$

Dividing both equations through by 2 gives the same system as in (5) (see Figure 5.5.3).

If the data do not resemble a linear function, one could use a higher degree polynomial. To find the coefficients $c_0, c_1, \ldots, c_n$ of the best least squares fit to the data

x	x_1	x_2	$\cdots$	x_m
y	y_1	y_2	$\cdots$	y_m

by a polynomial of degree n, we must find the least squares solution to the system

(6)
$$\begin{pmatrix} 1 & x_1 & x_1^2 & \cdots & x_1^n \\ 1 & x_2 & x_2^2 & \cdots & x_2^n \\ \vdots & & & & \\ 1 & x_m & x_m^2 & \cdots & x_m^n \end{pmatrix} \begin{pmatrix} c_0 \\ c_1 \\ \vdots \\ c_n \end{pmatrix} = \begin{pmatrix} y_1 \\ y_2 \\ \vdots \\ y_m \end{pmatrix}$$

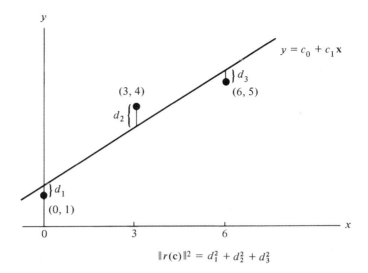

$$\|r(\mathbf{c})\|^2 = d_1^2 + d_2^2 + d_3^2$$

FIGURE 5.5.3

EXAMPLE 3. Find the best quadratic least squares fit to the data

x	0	1	2	3
y	3	2	4	4

SOLUTION.   For this example the system (6) becomes

$$\begin{pmatrix} 1 & 0 & 0 \\ 1 & 1 & 1 \\ 1 & 2 & 4 \\ 1 & 3 & 9 \end{pmatrix} \begin{pmatrix} c_0 \\ c_1 \\ c_2 \end{pmatrix} = \begin{pmatrix} 3 \\ 2 \\ 4 \\ 4 \end{pmatrix}$$

Thus the normal equations are

$$\begin{pmatrix} 1 & 1 & 1 & 1 \\ 0 & 1 & 2 & 3 \\ 0 & 1 & 4 & 9 \end{pmatrix} \begin{pmatrix} 1 & 0 & 0 \\ 1 & 1 & 1 \\ 1 & 2 & 4 \\ 1 & 3 & 9 \end{pmatrix} \begin{pmatrix} c_0 \\ c_1 \\ c_2 \end{pmatrix} = \begin{pmatrix} 1 & 1 & 1 & 1 \\ 0 & 1 & 2 & 3 \\ 0 & 1 & 4 & 9 \end{pmatrix} \begin{pmatrix} 3 \\ 2 \\ 4 \\ 4 \end{pmatrix}$$

These simplify to

$$\begin{pmatrix} 4 & 6 & 14 \\ 6 & 14 & 36 \\ 14 & 36 & 98 \end{pmatrix} \begin{pmatrix} c_0 \\ c_1 \\ c_2 \end{pmatrix} = \begin{pmatrix} 13 \\ 22 \\ 54 \end{pmatrix}$$

The solution to this system is $(2.75, -0.25, 0.25)$. The quadratic polynomial that gives the best least square fit to the data is

$$p(x) = 2.75 - 0.25x + 0.25x^2 \qquad \square$$

EXERCISES

1. Find the least squares solution to each of the following systems.

(a) $\begin{aligned} x_1 + \ x_2 &= 3 \\ 2x_1 - 3x_2 &= 1 \\ 0x_1 + 0x_2 &= 2 \end{aligned}$

(b) $\begin{aligned} -x_1 + \ x_2 &= 10 \\ 2x_1 + \ x_2 &= \ 5 \\ x_1 - 2x_2 &= 20 \end{aligned}$

(c) $\begin{aligned} x_1 + x_2 + x_3 &= 4 \\ -x_1 + x_2 + x_3 &= 0 \\ -x_2 \ \ \ \ + x_3 &= 1 \\ x_1 \ \ \ \ + x_3 &= 2 \end{aligned}$

2. For each of your solutions $\hat{\mathbf{x}}$ in Exercise 1, determine:
(i) The projection $\mathbf{p} = A\hat{\mathbf{x}}$.
(ii) The residual $r(\hat{\mathbf{x}})$.
Show that $\mathbf{p} \perp r(\hat{\mathbf{x}})$.

3. For each of the following systems $A\mathbf{x} = \mathbf{b}$, find all least squares solutions.

(a) $A = \begin{pmatrix} 1 & 2 \\ 2 & 4 \\ -1 & -2 \end{pmatrix}$ $\mathbf{b} = \begin{pmatrix} 3 \\ 2 \\ 1 \end{pmatrix}$

(b) $A = \begin{pmatrix} 1 & 1 & 3 \\ -1 & 3 & 1 \\ 1 & 2 & 4 \end{pmatrix}$ $\mathbf{b} = \begin{pmatrix} -2 \\ 0 \\ 8 \end{pmatrix}$

4. For each of the systems in Exercise 3, determine the projection $\mathbf{p}$ of $\mathbf{b}$ onto $R(A)$ and verify that $\mathbf{p} \perp \mathbf{b} - \mathbf{p}$.

5. (a) Find the best least squares fit by a linear function to the data

x	-1	0	1	2
y	0	1	3	9

(b) Plot your linear function from part (a) along with the data on a coordinate system.

6. Find the best least squares fit to the data in Exercise 5 by a quadratic polynomial. Plot the points $x = -1, 0, 1, 2$ for your function and sketch the graph.

7. Let A be an $m \times n$ matrix of rank n and let $P = A(A^TA)^{-1}A^T$.
(a) Show that $P\mathbf{b} = \mathbf{b}$ for every $\mathbf{b} \in R(A)$. Explain this in terms of projections.

(b) If $\mathbf{b} \in R(A)^{\perp}$, show that $P\mathbf{b} = \mathbf{0}$.

(c) Give a geometric illustration of parts (a) and (b) if $R(A)$ is a plane through the origin in R^3.

8. Let $P = A(A^TA)^{-1}A^T$, where A is an $m \times n$ matrix of rank n.

(a) Show that $P^2 = P$.

(b) Prove $P^k = P$ for $k = 1, 2, \ldots$.

(c) Show that P is symmetric.

[**Recall:** If B is nonsingular, then $(B^{-1})^T = (B^T)^{-1}$].

9. Show that if

$$\begin{pmatrix} A & I \\ O & A^T \end{pmatrix} \begin{pmatrix} \hat{\mathbf{x}} \\ \mathbf{r} \end{pmatrix} = \begin{pmatrix} \mathbf{b} \\ \mathbf{0} \end{pmatrix}$$

then $\hat{\mathbf{x}}$ is a least squares solution to the system $A\mathbf{x} = \mathbf{b}$ and $\mathbf{r}$ is the residual vector.

10. Let $A \in R^{m \times n}$ and let $\hat{\mathbf{x}}$ be a solution to the least squares problem $A\mathbf{x} = \mathbf{b}$. Show that a vector $\mathbf{y} \in R^n$ will also be a solution if and only if $\mathbf{y} = \hat{\mathbf{x}} + \mathbf{z}$, for some vector $\mathbf{z} \in N(A)$.

[**Hint:** $N(A^TA) = N(A)$.]

6. ORTHONORMAL SETS

In R^2 it is generally more convenient to use the standard basis $\{\mathbf{e}_1, \mathbf{e}_2\}$ than to use some other basis such as $\{(2, 1)^T, (3, 5)^T\}$. For example, it would be easier to find the coordinates of $(x_1, x_2)^T$ with respect to the standard basis. The elements of the standard basis are orthogonal unit vectors. In working with an inner product space V, it is generally desirable to have a basis of mutually orthogonal unit vectors. This is convenient not only in finding coordinates of vectors but in solving least squares problems.

Definition.   Let $\mathbf{v}_1, \mathbf{v}_2, \ldots, \mathbf{v}_n$ be vectors in an inner products space V. If $\langle \mathbf{v}_i, \mathbf{v}_j \rangle = 0$ whenever $i \neq j$, then $\{\mathbf{v}_1, \mathbf{v}_2, \ldots, \mathbf{v}_n\}$ is said to be an **orthogonal set** of vectors.

EXAMPLE 1.   The set $\{(1, 1, 1)^T, (2, 1, -3)^T, (4, -5, 1)^T\}$ is an orthogonal set in R^3, since

$$(1, 1, 1)(2, 1, -3)^T = 0$$
$$(1, 1, 1)(4, -5, 1)^T = 0$$
$$(2, 1, -3)(4, -5, 1)^T = 0 \qquad \square$$

Theorem 5.6.1.   *If* $\{\mathbf{v}_1, \ldots, \mathbf{v}_n\}$ *is an orthogonal set of nonzero vectors in an inner product space V, then $\mathbf{v}_1, \mathbf{v}_2, \ldots, \mathbf{v}_n$ are linearly independent.*

Proof.   Suppose that $\mathbf{v}_1, \mathbf{v}_2, \ldots, \mathbf{v}_n$ are mutually orthogonal vectors which are linearly dependent. Then there exist scalars $\alpha_1, \alpha_2, \ldots, \alpha_n$ not all zero for which

$$\alpha_1 \mathbf{v}_1 + \alpha_2 \mathbf{v}_2 + \cdots + \alpha_n \mathbf{v}_n = \mathbf{0}$$

We may assume without loss of generality that $\alpha_1 \neq 0$. It follows that

$$\mathbf{v}_1 = \sum_{i=2}^{n} \beta_i \mathbf{v}_i \quad \text{where} \quad \beta_i = \frac{-\alpha_i}{\alpha_1}$$

and hence

$$
\begin{aligned}
\|\mathbf{v}_1\|^2 &= \langle \mathbf{v}_1, \mathbf{v}_1 \rangle \\
&= \left\langle \mathbf{v}_1, \sum_{i=2}^{n} \beta_i \mathbf{v}_i \right\rangle \\
&= \sum_{i=2}^{n} \beta_i \langle \mathbf{v}_1, \mathbf{v}_i \rangle \\
&= 0
\end{aligned}
$$

Thus $\mathbf{v}_1$ must be the zero vector. Therefore, if $\mathbf{v}_1, \mathbf{v}_2, \ldots, \mathbf{v}_n$ are nonzero and mutually orthogonal, they must be linearly independent. $\square$

Definition.   An **orthonormal** set of vectors is an orthogonal set of unit vectors.

The set $\{\mathbf{u}_1, \mathbf{u}_2, \ldots, \mathbf{u}_n\}$ will be orthonormal if and only if

$$\langle \mathbf{u}_i, \mathbf{u}_j \rangle = \delta_{ij}$$

where

$$\delta_{ij} = \begin{cases} 1 & \text{if} \quad i = j \\ 0 & \text{if} \quad i \neq j \end{cases}$$

Given any orthogonal set of nonzero vectors $\{\mathbf{v}_1, \mathbf{v}_2, \ldots, \mathbf{v}_n\}$, it is possible to form an orthonormal set by defining

$$\mathbf{u}_i = \left(\frac{1}{\|\mathbf{v}_i\|} \right) \mathbf{v}_i \quad \text{for} \quad i = 1, 2, \ldots, n$$

The reader may verify that $\{\mathbf{u}_1, \mathbf{u}_2, \ldots, \mathbf{u}_n\}$ will be an orthonormal set.

EXAMPLE 2.   We saw in Example 1 that if $\mathbf{v}_1 = (1, 1, 1)^T$, $\mathbf{v}_2 = (2, 1, -3)^T$, and $\mathbf{v}_3 = (4, -5, 1)^T$, then $\{\mathbf{v}_1, \mathbf{v}_2, \mathbf{v}_3\}$ is an orthogonal set in R^3. To form an orthonormal set, let

$$\mathbf{u}_1 = \left(\frac{1}{\|\mathbf{v}_1\|}\right)\mathbf{v}_1 = \frac{1}{\sqrt{3}}(1,1,1)^T$$

$$\mathbf{u}_2 = \left(\frac{1}{\|\mathbf{v}_2\|}\right)\mathbf{v}_2 = \frac{1}{\sqrt{14}}(2,1,-3)^T$$

$$\mathbf{u}_3 = \left(\frac{1}{\|\mathbf{v}_3\|}\right)\mathbf{v}_3 = \frac{1}{\sqrt{42}}(4,-5,1)^T$$

□

EXAMPLE 3.   In $C[-\pi,\pi]$ with inner product

$$\langle f, g \rangle = \int_{-\pi}^{\pi} f(x)g(x)\,dx$$

the set $\{1, \cos x, \sin x\}$ is an orthogonal set of vectors, since

$$\langle 1, \cos x \rangle = \int_{-\pi}^{\pi} \cos x\,dx = 0$$

$$\langle 1, \sin x \rangle = \int_{-\pi}^{\pi} \sin x\,dx = 0$$

$$\langle \cos x, \sin x \rangle = \int_{-\pi}^{\pi} \cos x \sin x\,dx = 0$$

To form an orthonormal set, we calculate the norms of the three vectors.

$$\|1\|^2 = \langle 1, 1 \rangle = \int_{-\pi}^{\pi} 1\,dx = 2\pi$$

$$\|\cos x\|^2 = \langle \cos x, \cos x \rangle = \int_{-\pi}^{\pi} \cos^2 x\,dx = \pi$$

$$\|\sin x\|^2 = \langle \sin x, \sin x \rangle = \int_{-\pi}^{\pi} \sin^2 x\,dx = \pi$$

Thus $\{1/\sqrt{2\pi}, (1/\sqrt{\pi})\cos x, (1/\sqrt{\pi})\sin x\}$ is an orthonormal set of vectors.   □

It follows from Theorem 5.6.1 that if $B = \{\mathbf{x}_1, \mathbf{x}_2, \ldots, \mathbf{x}_k\}$ is an orthonormal set in an inner product space V, then B is a basis for a subspace S of V. We say the B is an *orthonormal basis* for S. It is generally much easier to work with an orthonormal basis than with an ordinary basis. In particular, it is usually easier to calculate the coordinates of a given vector $\mathbf{x}$ with respect to an orthonormal basis. Once these coordinates have been determined, they can be used to compute $\|\mathbf{x}\|$.

Theorem 5.6.2.   *Let* $\{\mathbf{x}_1, \mathbf{x}_2, \ldots, \mathbf{x}_n\}$ *be an orthonormal basis for an inner product space V. If*

$$\mathbf{x} = \sum_{i=1}^{n} c_i \mathbf{x}_i$$

then

$$c_i = \langle \mathbf{x}_i, \mathbf{x} \rangle$$

and

$$\|\mathbf{x}\|^2 = \sum_{i=1}^{n} c_i^2 \qquad \text{(Parseval's formula)}$$

Proof. Both of the assertions of the theorem can be proved by simple calculations.

$$\langle \mathbf{x}_i, \mathbf{x} \rangle = \left\langle \mathbf{x}_i, \sum_{j=1}^{n} c_j \mathbf{x}_j \right\rangle$$

$$= \sum_{j=1}^{n} c_j \langle \mathbf{x}_i, \mathbf{x}_j \rangle$$

$$= \sum_{j=1}^{n} c_j \delta_{ij}$$

$$= c_i$$

The second assertion follows, since

$$0 = \left\| \mathbf{x} - \sum_{i=1}^{n} c_i \mathbf{x}_i \right\|^2$$

$$= \left\langle \mathbf{x} - \sum_{i=1}^{n} c_i \mathbf{x}_i, \mathbf{x} - \sum_{i=1}^{n} c_i \mathbf{x}_i \right\rangle$$

$$= \|\mathbf{x}\|^2 - 2 \sum_{i=1}^{n} c_i \langle \mathbf{x}, \mathbf{x}_i \rangle + \|\mathbf{x}\|^2$$

$$= 2\|\mathbf{x}\|^2 - 2 \sum_{i=1}^{n} c_i^2 \qquad \square$$

EXAMPLE 4. The vectors $\mathbf{u}_1 = (1/\sqrt{2}, 1/\sqrt{2})^T$ and $\mathbf{u}_2 = (1/\sqrt{2}, -1/\sqrt{2})^T$ form an orthonormal basis for R^2. Let $\mathbf{x} = (x_1, x_2)^T$ be any vector in R^2.

$$\mathbf{x}^T \mathbf{u}_1 = \frac{x_1 + x_2}{\sqrt{2}}$$

$$\mathbf{x}^T \mathbf{u}_2 = \frac{x_1 - x_2}{\sqrt{2}}$$

It follows from the theorem that

$$\mathbf{x} = \frac{x_1 + x_2}{\sqrt{2}} \mathbf{u}_1 + \frac{x_1 - x_2}{\sqrt{2}} \mathbf{u}_2$$

and

$$\|\mathbf{x}\|^2 = \left(\frac{x_1 + x_2}{\sqrt{2}} \right)^2 + \left(\frac{x_1 - x_2}{\sqrt{2}} \right)^2 = x_1^2 + x_2^2 \qquad \square$$

EXAMPLE 5.   Given that

$$\left\{ \frac{1}{\sqrt{2\pi}}, \frac{1}{\sqrt{\pi}} \cos 2x \right\}$$

is an orthonormal set in $C[-\pi, \pi]$ (with inner product as in Example 3), determine the value of $\int_{-\pi}^{\pi} \sin^4 x \, dx$ without computing antiderivatives.

SOLUTION.   Since

$$\sin^2 x = \frac{1 - \cos 2x}{2}$$

$$= \left(\frac{\sqrt{2\pi}}{2} \right) \frac{1}{\sqrt{2\pi}} + \left(\frac{-\sqrt{\pi}}{2} \right) \frac{1}{\sqrt{\pi}} \cos 2x$$

it follows from Parseval's formula that

$$\int_{-\pi}^{\pi} \sin^4 x \, dx = \| \sin^2 x \|^2$$

$$= \left(\frac{\sqrt{2\pi}}{2} \right)^2 + \left(\frac{-\sqrt{\pi}}{2} \right)^2 = \frac{3\pi}{4} \qquad \square$$

ORTHOGONAL MATRICES

Of particular importance are $n \times n$ matrices whose column vectors form an orthonormal set in R^n. Such matrices are called *orthogonal matrices*.

Theorem 5.6.3.   *If Q is an orthogonal matrix, then Q is invertible and $Q^{-1} = Q^T$.*

Proof. Q has rank n, since its column vectors are linearly independent. Therefore, Q^{-1} exists. The (i, j) entry of $Q^T Q$ is $\mathbf{q}_i^T \mathbf{q}_j = \delta_{ij}$. Thus $Q^T Q = I$ and it follows that

$$Q^T = Q^T I = Q^T (QQ^{-1}) = (Q^T Q)Q^{-1} = IQ^{-1} = Q^{-1} \qquad \square$$

EXAMPLE 6.   For any fixed θ, the matrix

$$Q = \begin{pmatrix} \cos\theta & -\sin\theta \\ \sin\theta & \cos\theta \end{pmatrix}$$

is orthogonal and

$$Q^{-1} = \begin{pmatrix} \cos\theta & \sin\theta \\ -\sin\theta & \cos\theta \end{pmatrix} \qquad \square$$

The matrix Q in Example 6 can be thought of as a linear transformation from R^2 onto R^2 that has the effect of rotating each vector by

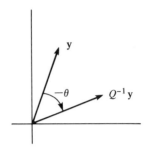

FIGURE 5.6.1

an angle θ while leaving the length of the vector unchanged (see Example 2 in Section 2 of Chapter 4). Similarly, Q^{-1} can be thought of as a rotation by the angle $-\theta$ (see Figure 5.6.1).

In general, inner products are preserved under multiplication by an orthogonal matrix [i.e., $\langle \mathbf{x}, \mathbf{y} \rangle = \langle Q\mathbf{x}, Q\mathbf{y} \rangle$]. This is true, since

$$\langle Q\mathbf{x}, Q\mathbf{y} \rangle = (Q\mathbf{y})^T Q\mathbf{x} = \mathbf{y}^T Q^T Q\mathbf{x} = \mathbf{y}^T \mathbf{x} = \langle \mathbf{x}, \mathbf{y} \rangle$$

In particular, if $\mathbf{x} = \mathbf{y}$, then $\|Q\mathbf{x}\|^2 = \|\mathbf{x}\|^2$ and hence $\|Q\mathbf{x}\| = \|\mathbf{x}\|$. Multiplication by an orthogonal matrix preserves the lengths of vectors.

Properties of Orthogonal Matrices

If Q is an $n \times n$ orthogonal matrix, then:
   1. The column vectors of Q form an orthonormal basis for R^n
   2. $Q^T Q = I$
   3. $Q^T = Q^{-1}$
   4. $\langle Q\mathbf{x}, Q\mathbf{y} \rangle = \langle \mathbf{x}, \mathbf{y} \rangle$
   5. $\|Q\mathbf{x}\|_2 = \|\mathbf{x}\|_2$
   6. $\|QA\|_2 = \|A\|_2$ for any $A \in R^{n \times r}$
   7. $\det(Q) = \pm 1$

The proofs of properties 6 and 7 are left as exercises for the reader.

PERMUTATION MATRICES

A *permutation matrix* is a matrix formed from the identity matrix by reordering its columns. Clearly, then, permutation matrices are orthogonal matrices. If P is the permutation matrix formed by reordering the columns of I in the order $(k_1, \ldots, k_n)$, then $P = (\mathbf{e}_{k_1}, \ldots, \mathbf{e}_{k_n})$. If A is an $m \times n$ matrix, then

$$AP = \left(A\mathbf{e}_{k_1}, \ldots, A\mathbf{e}_{k_n} \right) = (\mathbf{a}_{k_1}, \ldots, \mathbf{a}_{k_n})$$

Postmultiplication of A by P reorders the columns of A in the order

$(k_1, \ldots, k_n)$. For example, if

$$A = \begin{pmatrix} 1 & 2 & 3 \\ 1 & 2 & 3 \end{pmatrix} \quad \text{and} \quad P = \begin{pmatrix} 0 & 1 & 0 \\ 0 & 0 & 1 \\ 1 & 0 & 0 \end{pmatrix}$$

then

$$AP = \begin{pmatrix} 3 & 1 & 2 \\ 3 & 1 & 2 \end{pmatrix}$$

Since $P = (\mathbf{e}_{k_1}, \ldots, \mathbf{e}_{k_n})$ is orthogonal, it follows that

$$P^{-1} = P^T = \begin{pmatrix} \mathbf{e}_{k_1}^T \\ \vdots \\ \mathbf{e}_{k_n}^T \end{pmatrix}$$

The k_1 column of P^T will be $\mathbf{e}_1$, the k_2 column will be $\mathbf{e}_2$, and so on. Thus P^T is a permutation matrix. The matrix P^T can be formed directly from I by reordering its rows in the order $(k_1, k_2, \ldots, k_n)$. In general, a permutation matrix can be formed from I by either reordering its rows or its columns.

If Q is the permutation matrix formed by reordering the rows of I in the order $(k_1, k_2, \ldots, k_n)$ and B is an $n \times r$ matrix, then

$$QB = \begin{pmatrix} \mathbf{e}_{k_1}^T \\ \vdots \\ \mathbf{e}_{k_n}^T \end{pmatrix} B = \begin{pmatrix} \mathbf{e}_{k_1}^T B \\ \vdots \\ \mathbf{e}_{k_n}^T B \end{pmatrix}$$

Thus QB is the matrix formed by reordering the rows of B in the order $(k_1, k_2, \ldots, k_n)$. For example, if

$$Q = \begin{pmatrix} 0 & 0 & 1 \\ 1 & 0 & 0 \\ 0 & 1 & 0 \end{pmatrix} \quad \text{and} \quad B = \begin{pmatrix} 1 & 1 \\ 2 & 2 \\ 3 & 3 \end{pmatrix}$$

then

$$QB = \begin{pmatrix} 3 & 3 \\ 1 & 1 \\ 2 & 2 \end{pmatrix}$$

In general, if P is an $n \times n$ permutation matrix, then premultiplication of an $n \times r$ matrix B by P reorders the rows of B and postmultiplication of an $m \times n$ matrix A by P reorders the columns of A.

ORTHONORMAL SETS AND LEAST SQUARES

Orthogonality plays an important role in solving least squares problems. Recall that if A is an $m \times n$ matrix of rank n, then the least squares problem $A\mathbf{x} = \mathbf{b}$ has a unique solution $\hat{\mathbf{x}}$ that is determined by solving the normal equations $A^T A\mathbf{x} = A^T \mathbf{b}$. The projection $\mathbf{p} = A\hat{\mathbf{x}}$ is the vector in $R(A)$ that is closest to $\mathbf{b}$. The least squares problem is especially easy to solve in the case where the column vectors of A form an orthonormal set in R^m.

Theorem 5.6.4. *If the column vectors of A form an orthonormal set of vectors in R^m, then $A^T A = I$ and the solution to the least squares problem is*

$$\hat{\mathbf{x}} = A^T \mathbf{b}$$

Proof. The ijth entry of $A^T A$ is formed from the ith row of A^T and the jth column of A. Thus the ijth entry is actually the scalar product of the ith and jth columns of A. Since the column vectors of A are orthonormal, it follows that

$$A^T A = \left(\delta_{ij} \right) = I$$

Consequently, the normal equations simplify to

$$\mathbf{x} = A^T \mathbf{b} \qquad\qquad \square$$

What if the columns of A are not orthonormal? In the next section we will learn a method for finding an orthonormal basis for $R(A)$. From this method we will obtain a factorization of A into a product QR, where Q has an orthonormal set of column vectors and R is upper triangular. With this factorization, the least squares problem can be solved quickly and accurately.

If one has an orthonormal basis for $R(A)$, then the projection $\mathbf{p} = A\hat{\mathbf{x}}$ can be determined in terms of the basis elements. Indeed, this is a special case of the more general least squares problem of finding the element $\mathbf{p}$ in a subspace S of an inner product space V that is closest to a given element $\mathbf{x}$ in V. This problem is easily solved if S has an orthonormal basis. We first prove the following theorem.

Theorem 5.6.5. *Let S be a subspace of an inner product space V and let $\mathbf{x} \in V$. Let $\{\mathbf{x}_1, \mathbf{x}_2, \ldots, \mathbf{x}_n\}$ be an orthonormal basis for S. If*

(1) $$\mathbf{p} = \sum_{i=1}^{n} c_i \mathbf{x}_i$$

where

(2) $$c_i = \langle \mathbf{x}, \mathbf{x}_i \rangle \quad \text{for each } i$$

then $\mathbf{p} - \mathbf{x} \in S^{\perp}$ (see Figure 5.6.2).

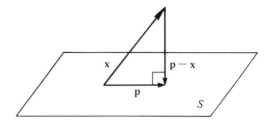

FIGURE 5.6.2

Proof.   We will show first that $(\mathbf{p} - \mathbf{x}) \perp \mathbf{x}_i$ for each i.

$$\langle \mathbf{x}_i, \mathbf{p} - \mathbf{x} \rangle = \langle \mathbf{x}_i, \mathbf{p} \rangle - \langle \mathbf{x}_i, \mathbf{x} \rangle$$

$$= \left\langle \mathbf{x}_i, \sum_{j=1}^{n} c_j \mathbf{x}_j \right\rangle - c_i$$

$$= \sum_{j=1}^{n} c_j \langle \mathbf{x}_i, \mathbf{x}_j \rangle - c_i$$

$$= 0$$

So $\mathbf{p} - \mathbf{x}$ is orthogonal to all the $\mathbf{x}_i$'s. If $\mathbf{y} \in S$, then

$$\mathbf{y} = \sum_{i=1}^{n} \alpha_i \mathbf{x}_i$$

and hence

$$\langle \mathbf{p} - \mathbf{x}, \mathbf{y} \rangle = \left\langle \mathbf{p} - \mathbf{x}, \sum_{i=1}^{n} \alpha_i \mathbf{x} \right\rangle$$

$$= \sum_{i=1}^{n} \alpha_i \langle \mathbf{p} - \mathbf{x}, \mathbf{x}_i \rangle$$

$$= 0 \qquad \qquad \square$$

If $\mathbf{x} \in S$, then the preceding result is trivial, since by Theorem 5.6.2, $\mathbf{p} - \mathbf{x} = \mathbf{0}$. If $\mathbf{x} \notin S$, then $\mathbf{p}$ is the element in S closest to $\mathbf{x}$.

Theorem 5.6.6.   *Under the hypothesis of Theorem 5.6.5, $\mathbf{p}$ is the element of S that is closest to $\mathbf{x}$, that is,*

$$\|\mathbf{y} - \mathbf{x}\| > \|\mathbf{p} - \mathbf{x}\|$$

for any $\mathbf{y} \neq \mathbf{p}$ in S.

Proof.   If $\mathbf{y} \in S$ and $\mathbf{y} \neq \mathbf{p}$, then

$$\|\mathbf{y} - \mathbf{x}\|^2 = \|(\mathbf{y} - \mathbf{p}) + (\mathbf{p} - \mathbf{x})\|^2$$

Since $\mathbf{y} - \mathbf{p} \in S$, it follows from Theorem 5.6.5 and the Pythagorean law that

$$\|\mathbf{y} - \mathbf{x}\|^2 = \|\mathbf{y} - \mathbf{p}\|^2 + \|\mathbf{p} - \mathbf{x}\|^2 > \|\mathbf{p} - \mathbf{x}\|^2$$

Therefore, $\|\mathbf{y} - \mathbf{x}\| > \|\mathbf{p} - \mathbf{x}\|$.   □

The vector $\mathbf{p}$ defined by (1) and (2) is said to be the *projection of* $\mathbf{x}$ *onto S*.

EXAMPLE 7.   Let S be the set of all vectors in R^3 of the form $(x, y, 0)^T$. Find the vector $\mathbf{p}$ in S that is closest to $\mathbf{w} = (5, 3, 4)^T$ (see Figure 5.6.3).

SOLUTION.   Let $\mathbf{u}_1 = (1, 0, 0)^T$ and $\mathbf{u}_2 = (0, 1, 0)^T$. Clearly, $\mathbf{u}_1$ and $\mathbf{u}_2$ form an orthonormal basis for S. Now

$$c_1 = \mathbf{w}^T \mathbf{u}_1 = 5$$
$$c_2 = \mathbf{w}^T \mathbf{u}_2 = 3$$

The vector $\mathbf{p}$ turns out to be exactly what we would expect

$$\mathbf{p} = 5\mathbf{u}_1 + 3\mathbf{u}_2 = (5, 3, 0)^T$$ □

EXAMPLE 8.   Find the best least squares approximation to e^x on the interval $[0, 1]$ by a linear function.

SOLUTION.   Let S be the subspace of all linear functions in $C[0, 1]$. Although the functions 1 and x span S, they are not orthogonal. We seek a function of the form $x - a$ that is orthogonal to 1.

$$\langle 1, x - a \rangle = \int_0^1 (x - a)\, dx = \tfrac{1}{2} - a$$

Thus $a = \tfrac{1}{2}$. Since $\|x - \tfrac{1}{2}\| = 1/\sqrt{12}$, it follows that

$$u_1(x) = 1 \qquad \text{and} \qquad u_2(x) = \sqrt{12}\left(x - \tfrac{1}{2}\right)$$

form an orthonormal basis for S.

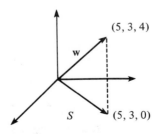

FIGURE 5.6.3

Let

$$c_1 = \int_0^1 u_1(x)\, e^x\, dx = e - 1$$

$$c_2 = \int_0^1 u_2(x)\, e^x\, dx = \sqrt{3}\,(3 - e)$$

The projection

$$
\begin{aligned}
p(x) &= c_1 u_1(x) + c_2 u_2(x) \\
&= (e - 1) \cdot 1 + \sqrt{3}\,(3 - e)\left[\sqrt{12}\left(x - \tfrac{1}{2}\right)\right] \\
&= (4e - 10) + 6(3 - e)\, x
\end{aligned}
$$

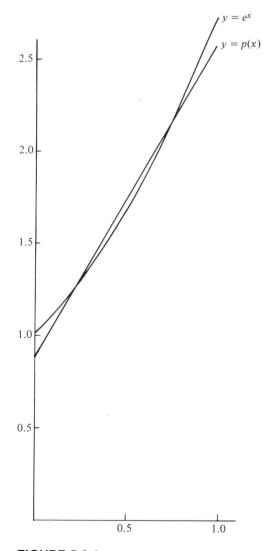

FIGURE 5.6.4

is the best linear least squares approximation to e^x on $[0, 1]$ (see Figure 5.6.4).   □

EXERCISES

1. Which of the following sets of vectors form an orthonormal basis for R^2?
   (a) $\{(1, 0)^T, (0, 1)^T\}$
   (b) $\{(\frac{3}{5}, \frac{4}{5})^T, (\frac{5}{13}, \frac{12}{13})^T\}$
   (c) $\{(1, -1)^T, (1, 1)^T\}$
   (d) $\left\{ \left(\frac{\sqrt{3}}{2}, \frac{1}{2} \right)^T, \left(-\frac{1}{2}, \frac{\sqrt{3}}{2} \right)^T \right\}$

2. Let

$$\mathbf{x}_1 = \left(\frac{1}{3\sqrt{2}}, \frac{1}{3\sqrt{2}}, -\frac{4}{3\sqrt{2}} \right)^T, \qquad \mathbf{x}_2 = \left(\frac{2}{3}, \frac{2}{3}, \frac{1}{3} \right)^T$$

$$\mathbf{x}_3 = \left(\frac{1}{\sqrt{2}}, -\frac{1}{\sqrt{2}}, 0 \right)^T$$

   (a) Show that $\{\mathbf{x}_1, \mathbf{x}_2, \mathbf{x}_3\}$ is an orthonormal basis for R^3.
   (b) Let $\mathbf{x} = (1, 1, 1)^T$. Write $\mathbf{x}$ as a linear combination of $\mathbf{x}_1$, $\mathbf{x}_2$, and $\mathbf{x}_3$ and compute $\|\mathbf{x}\|$ using Theorem 5.6.2.

3. Let S be the subspace of R^3 spanned by the vectors $\mathbf{x}_2$ and $\mathbf{x}_3$ of Exercise 2. Let $\mathbf{x} = (1, 2, 2)^T$. Find the projection $\mathbf{p}$ of $\mathbf{x}$ onto S. Show that $(\mathbf{p} - \mathbf{x}) \perp \mathbf{x}_2$ and $(\mathbf{p} - \mathbf{x}) \perp \mathbf{x}_3$.

4. Let θ be a fixed real number and let

$$\mathbf{x}_1 = \begin{pmatrix} \cos \theta \\ \sin \theta \end{pmatrix} \qquad \text{and} \qquad \mathbf{x}_2 = \begin{pmatrix} -\sin \theta \\ \cos \theta \end{pmatrix}$$

   (a) Show that $\{\mathbf{x}_1, \mathbf{x}_2\}$ is an orthonormal basis for R^2.
   (b) Given a vector $\mathbf{y}$ in R^2, write it as a linear combination $c_1 \mathbf{x}_1 + c_2 \mathbf{x}_2$.
   (c) Verify that

$$c_1^2 + c_2^2 = \|\mathbf{y}\|^2 = y_1^2 + y_2^2$$

5. Consider the set of vectors

$$S = \left\{ \frac{1}{\sqrt{2\pi}}, \frac{1}{\sqrt{\pi}} \cos x, \frac{1}{\sqrt{\pi}} \cos 2x, \frac{1}{\sqrt{\pi}} \cos 3x, \frac{1}{\sqrt{\pi}} \cos 4x \right\}$$

in the vector space $C[-\pi, \pi]$ with inner product defined by

$$\langle f, g \rangle = \int_{-\pi}^{\pi} f(x)g(x)\, dx$$

(a) Show that S is an orthonormal set in $C[-\pi, \pi]$.

(b) Use trigonométric identities to write the function $\sin^4 x$ as a linear combination of elements of S.

(c) Use part (b) and Theorem 5.6.2 to find the values of the following integrals.

(i) $\int_{-\pi}^{\pi} \sin^4 x \cos x\, dx$ (ii) $\int_{-\pi}^{\pi} \sin^4 x \cos 2x\, dx$

(iii) $\int_{-\pi}^{\pi} \sin^4 x \cos 3x\, dx$ (iv) $\int_{-\pi}^{\pi} \sin^4 x \cos 4x\, dx$

6. Let $\mathbf{u} = (\cos \theta/2, \sin \theta/2)^T$ and let $Q = I - 2\mathbf{u}\mathbf{u}^T$.
(a) Show that Q is orthogonal.
(b) Show that Q is a combination of a rotation and a reflection.

7. Prove that the transpose of an orthogonal matrix is an orthogonal matrix.

8. Let Q be an $n \times n$ orthogonal matrix. Use mathematical induction to prove each of the following:
(a) $(Q^m)^{-1} = (Q^T)^m = (Q^m)^T$ for any positive integer m
(b) $\|Q^m \mathbf{x}\| = \|\mathbf{x}\|$ for any $\mathbf{x} \in R^n$

9. Let $\mathbf{u}$ be a unit vector in R^n and let $H = I - 2\mathbf{u}\mathbf{u}^T$. Show that H is both orthogonal and symmetric and hence is its own inverse.

10. Let Q be an orthogonal matrix and let $d = \det(Q)$. Show that $|d| = 1$.

11. Show that the product of two orthogonal matrices is also an orthogonal matrix. Is the product of two permutation matrices a permutation matrix? Explain.

12. Show that if Q is an orthogonal matrix, then $\|Q\|_2 = 1$.

13. Let Q be an orthogonal $n \times n$ matrix and let $A \in R^{n \times r}$. Show that $\|QA\|_2 = \|A\|_2$.

14. Let Q and A be defined as in Exercise 13. Show that $\|A^T Q\|_2 = \|A^T\|_2$.

15. Let Q be an orthogonal $n \times n$ matrix, A be an $n \times n$ matrix, and $B = Q^T A Q$. Show that $\|B\|_2 = \|A\|_2$.

16. Let

$$A = \begin{pmatrix} \dfrac{1}{\sqrt{3}} & \dfrac{2}{\sqrt{6}} \\[2ex] \dfrac{1}{\sqrt{3}} & -\dfrac{1}{\sqrt{6}} \\[2ex] \dfrac{1}{\sqrt{3}} & -\dfrac{1}{\sqrt{6}} \end{pmatrix}$$

(a) Show that the column vectors of A form an orthonormal set in R^3.
(b) Solve the least squares problem $A\mathbf{x} = \mathbf{b}$ for each of the following choices of $\mathbf{b}$.
  (i) $\mathbf{b} = (3, 0, 0)^T$
  (ii) $\mathbf{b} = (1, 2, 3)^T$
  (iii) $\mathbf{b} = (1, 1, 2)^T$

17. Let A be an $m \times n$ matrix whose column vectors are mutually orthogonal and let $\mathbf{b} \in R^m$. Show that if $\hat{\mathbf{x}}$ is the least squares solution to the system $A\mathbf{x} = \mathbf{b}$, then

$$\hat{x}_i = \frac{\mathbf{b}^T \mathbf{a}_i}{\mathbf{a}_i^T \mathbf{a}_i} \qquad i = 1, \ldots, n$$

18. Given the vector space $C[-1, 1]$ with inner product

$$\langle f, g \rangle = \int_{-1}^{1} f(x) g(x) \, dx$$

and norm

$$\| f \| = (\langle f, f \rangle)^{1/2}$$

(a) Show that the vectors 1 and x are orthogonal.
(b) Compute $\|1\|$ and $\|x\|$.
(c) Find the best least squares approximation to $x^{1/3}$ on $[-1, 1]$ by a linear function $l(x) = c_1 1 + c_2 x$.
(d) Sketch the graphs of $x^{1/3}$ and $l(x)$ on $[-1, 1]$.

19. Let V be the subspace of $C[-\pi, \pi]$ spanned by the vectors in the set S from Exercise 5. Find the best least squares approximation to $f(x) = |x|$ on $[-\pi, \pi]$ by elements from V.

20. Let $\{\mathbf{x}_1, \mathbf{x}_2, \ldots, \mathbf{x}_k, \mathbf{x}_{k+1}, \ldots, \mathbf{x}_n\}$ be an orthonormal basis for an inner product space V. Let S_1 be the subspace of V spanned by $\mathbf{x}_1, \ldots, \mathbf{x}_k$ and let S_2 be the subspace spanned by $\mathbf{x}_{k+1}, \mathbf{x}_{k+2}, \ldots, \mathbf{x}_n$. Show that $S_1 \perp S_2$.

21. Let $\mathbf{x}$ be an element of the inner product space V in Exercise 20 and let $\mathbf{p}_1$ and $\mathbf{p}_2$ be the projections of $\mathbf{x}$ onto S_1 and S_2, respectively.
(a) Show that $\mathbf{x} = \mathbf{p}_1 + \mathbf{p}_2$.
(b) Show that if $\mathbf{x} \in S_1^{\perp}$, then $\mathbf{p}_1 = \mathbf{0}$ and hence $S_1^{\perp} = S_2$.

22. Let S be a subspace of an inner product space V. Let $\{\mathbf{x}_1, \ldots, \mathbf{x}_n\}$ be an orthogonal basis for S and let $\mathbf{x} \in V$. Show that the best least squares approximation to $\mathbf{x}$ by elements of S is given by

$$\mathbf{p} = \sum_{i=1}^{n} \frac{\langle \mathbf{x}, \mathbf{x}_i \rangle}{\langle \mathbf{x}_i, \mathbf{x}_i \rangle} \mathbf{x}_i$$

7. THE GRAM – SCHMIDT ORTHOGONALIZATION PROCESS

In this section we learn a process for constructing an orthonormal basis for an n-dimensional inner product space V. If we start with a given basis $\{\mathbf{x}_1, \mathbf{x}_2, \ldots, \mathbf{x}_n\}$, the method involves using projections to construct an orthonormal basis $\{\mathbf{u}_1, \mathbf{u}_2, \ldots, \mathbf{u}_n\}$.

We will construct the $\mathbf{u}_i$'s so that $\mathcal{S}(\mathbf{u}_1, \ldots, \mathbf{u}_k) = \mathcal{S}(\mathbf{x}_1, \ldots, \mathbf{x}_k)$ for $k = 1, \ldots, n$. To begin the process, let

$$(1) \qquad \mathbf{u}_1 = \left(\frac{1}{\|\mathbf{x}_1\|} \right) \mathbf{x}_1$$

$\mathcal{S}(\mathbf{u}_1) = \mathcal{S}(\mathbf{x}_1)$, since $\mathbf{u}_1$ is a unit vector in the direction of $\mathbf{x}_1$. Let $\mathbf{p}_1$ denote the projection of $\mathbf{x}_2$ onto $\mathcal{S}(\mathbf{x}_1) = \mathcal{S}(\mathbf{u}_1)$.

$$\mathbf{p}_1 = \langle \mathbf{x}_2, \mathbf{u}_1 \rangle \mathbf{u}_1$$

By Theorem 5.6.5,

$$(\mathbf{x}_2 - \mathbf{p}_1) \perp \mathbf{u}_1$$

Note that $\mathbf{x}_2 - \mathbf{p}_1 \neq \mathbf{0}$, since

$$(2) \qquad \mathbf{x}_2 - \mathbf{p}_1 = \frac{-\langle \mathbf{x}_2, \mathbf{u}_1 \rangle}{\|\mathbf{x}_1\|} \mathbf{x}_1 + \mathbf{x}_2$$

and $\mathbf{x}_1$ and $\mathbf{x}_2$ are linearly independent. If we set

$$(3) \qquad \mathbf{u}_2 = \frac{1}{\|\mathbf{x}_2 - \mathbf{p}_1\|} (\mathbf{x}_2 - \mathbf{p}_1)$$

then $\mathbf{u}_2$ is a unit vector orthogonal to $\mathbf{u}_1$. It follows from (1), (2), and (3) that $\mathcal{S}(\mathbf{u}_1, \mathbf{u}_2) \subset \mathcal{S}(\mathbf{x}_1, \mathbf{x}_2)$. Since $\mathbf{u}_1$ and $\mathbf{u}_2$ are linearly independent, it follows that $\{\mathbf{u}_1, \mathbf{u}_2\}$ is an orthonormal basis for $\mathcal{S}(\mathbf{x}_1, \mathbf{x}_2)$, and hence $\mathcal{S}(\mathbf{x}_1, \mathbf{x}_2) = \mathcal{S}(\mathbf{u}_1, \mathbf{u}_2)$.

FIGURE 5.7.1

To construct $\mathbf{u}_3$, continue in the same manner. Let $\mathbf{p}_2$ be the projection of $\mathbf{x}_3$ onto $\mathbb{S}(\mathbf{x}_1, \mathbf{x}_2) = \mathbb{S}(\mathbf{u}_1, \mathbf{u}_2)$,

$$\mathbf{p}_2 = \langle \mathbf{x}_3, \mathbf{u}_1 \rangle \mathbf{u}_1 + \langle \mathbf{x}_3, \mathbf{u}_2 \rangle \mathbf{u}_2$$

and set

$$\mathbf{u}_3 = \frac{1}{\|\mathbf{x}_3 - \mathbf{p}_2\|}(\mathbf{x}_3 - \mathbf{p}_2)$$

and so on (see Figure 5.7.1).

Theorem 5.7.1 (The Gram – Schmidt Process).   *Let*

$$\{\mathbf{x}_1, \mathbf{x}_2, \ldots, \mathbf{x}_n\}$$

be a basis for the inner product space V. Let

$$\mathbf{u}_1 = \left(\frac{1}{\|\mathbf{x}_1\|} \right) \mathbf{x}_1$$

and define $\mathbf{u}_2, \ldots, \mathbf{u}_n$ recursively by

$$\mathbf{u}_{k+1} = \frac{1}{\|\mathbf{x}_{k+1} - \mathbf{p}_k\|}(\mathbf{x}_{k+1} - \mathbf{p}_k) \quad \text{for } k = 1, \ldots, n-1$$

where

$$\mathbf{p}_k = \langle \mathbf{x}_{k+1}, \mathbf{u}_1 \rangle \mathbf{u}_1 + \langle \mathbf{x}_{k+1}, \mathbf{u}_2 \rangle \mathbf{u}_2 + \cdots + \langle \mathbf{x}_{k+1}, \mathbf{u}_k \rangle \mathbf{u}_k$$

is the projection of $\mathbf{x}_{k+1}$ onto $\mathbb{S}(\mathbf{u}_1, \mathbf{u}_2, \ldots, \mathbf{u}_k)$. The set $\{\mathbf{u}_1, \mathbf{u}_2, \ldots, \mathbf{u}_n\}$ is an orthonormal basis for V.

Proof.   We will argue inductively. Clearly, $\mathbb{S}(\mathbf{u}_1) = \mathbb{S}(\mathbf{x}_1)$. Suppose that $\mathbf{u}_1, \mathbf{u}_2, \ldots, \mathbf{u}_k$ have been constructed so that $\{\mathbf{u}_1, \mathbf{u}_2, \ldots, \mathbf{u}_k\}$ is an orthonormal set and

$$\mathbb{S}(\mathbf{u}_1, \mathbf{u}_2, \ldots, \mathbf{u}_k) = \mathbb{S}(\mathbf{x}_1, \mathbf{x}_2, \ldots, \mathbf{x}_k)$$

Since $\mathbf{p}_k$ is a linear combination of the $\mathbf{u}_i$'s, $1 \le i \le k$, it follows that $\mathbf{p}_k \in \mathbb{S}(\mathbf{x}_1, \ldots, \mathbf{x}_k)$ and $\mathbf{x}_{k+1} - \mathbf{p}_k \in \mathbb{S}(\mathbf{x}_1, \ldots, \mathbf{x}_{k+1})$.

$$\mathbf{x}_{k+1} - \mathbf{p}_k = \mathbf{x}_{k+1} - \sum_{i=1}^{k} c_i \mathbf{x}_i$$

Since $\mathbf{x}_1, \ldots, \mathbf{x}_{k+1}$ are linearly independent, it follows that $\mathbf{x}_{k+1} - \mathbf{p}_k$ is nonzero and by Theorem 5.6.5 it is orthogonal to each $\mathbf{u}_i$, $1 \le i \le k$. Thus $\{\mathbf{u}_1, \mathbf{u}_2, \ldots, \mathbf{u}_{k+1}\}$ is an orthonormal set of vectors in $\mathbb{S}(\mathbf{x}_1, \ldots, \mathbf{x}_{k+1})$. Since $\mathbf{u}_1, \ldots, \mathbf{u}_{k+1}$ are linearly independent, they form a basis for $\mathbb{S}(\mathbf{x}_1, \ldots, \mathbf{x}_{k+1})$, and consequently $\mathbb{S}(\mathbf{u}_1, \ldots, \mathbf{u}_{k+1}) = \mathbb{S}(\mathbf{x}_1, \ldots, \mathbf{x}_{k+1})$. It follows by induction that $\{\mathbf{u}_1, \mathbf{u}_2, \ldots, \mathbf{u}_n\}$ is an orthonormal basis for V. $\qquad\square$

EXAMPLE 1. Find an orthonormal basis for P_3 if the inner product on P_3 is defined by

$$\langle p, q \rangle = \sum_{i=1}^{3} p(x_i)q(x_i)$$

where $x_1 = -1$, $x_2 = 0$, and $x_3 = 1$.

SOLUTION. Starting with the basis $\{1, x, x^2\}$, we can use the Gram–Schmidt process to generate an orthonormal basis.

$$\|1\|^2 = \langle 1, 1 \rangle = 3$$

so

$$\mathbf{u}_1 = \left(\frac{1}{\|1\|}\right)1 = \frac{1}{\sqrt{3}}$$

Set

$$p_1 = \left\langle x, \frac{1}{\sqrt{3}} \right\rangle \frac{1}{\sqrt{3}} = \left(-1 \cdot \frac{1}{\sqrt{3}} + 0 \cdot \frac{1}{\sqrt{3}} + 1 \cdot \frac{1}{\sqrt{3}}\right)\frac{1}{\sqrt{3}} = 0$$

Therefore,

$$x - p_1 = x \quad \text{and} \quad \|x - p_1\|^2 = \langle x, x \rangle = 2$$

Hence

$$\mathbf{u}_2 = \frac{1}{\sqrt{2}}x$$

Finally,

$$p_2 = \left\langle x^2, \frac{1}{\sqrt{3}} \right\rangle \frac{1}{\sqrt{3}} + \left\langle x^2, \frac{1}{\sqrt{2}}x \right\rangle \frac{1}{\sqrt{2}}x = \frac{2}{3}$$

$$\|x^2 - p_2\|^2 = \left\langle x^2 - \tfrac{2}{3}, x^2 - \tfrac{2}{3} \right\rangle = \tfrac{2}{3}$$

and hence

$$\mathbf{u}_3 = \frac{\sqrt{6}}{2}\left(x^2 - \frac{2}{3}\right) \qquad\square$$

Orthogonal polynomials will be studied in more detail in Section 8.

EXAMPLE 2.   Let

$$A = \begin{pmatrix} 1 & 1 & 2 \\ 1 & 2 & 3 \\ 1 & 2 & 1 \\ 1 & 1 & 6 \end{pmatrix}$$

Find an orthonormal basis for the column space of A.

SOLUTION.   Let $\mathbf{x}_1$, $\mathbf{x}_2$, and $\mathbf{x}_3$ represent the respective column vectors of A. Since these vectors are linearly independent, they form a basis for a three-dimensional subspace of R^4. The Gram–Schmidt process may be used to form an orthonormal basis as follows:

(4) $\mathbf{u}_1 = \left(\dfrac{1}{\|\mathbf{x}_1\|} \right)\mathbf{x}_1 = \dfrac{1}{2}\mathbf{x}_1 = \left(\dfrac{1}{2}, \dfrac{1}{2}, \dfrac{1}{2}, \dfrac{1}{2} \right)^T$

$\mathbf{p}_1 = \langle \mathbf{x}_2, \mathbf{u}_1 \rangle \mathbf{u}_1 = 3\mathbf{u}_1$

(5) $\mathbf{x}_2 - \mathbf{p}_1 = \mathbf{x}_2 - 3\mathbf{u}_1 = \left(-\dfrac{1}{2}, \dfrac{1}{2}, \dfrac{1}{2}, -\dfrac{1}{2} \right)^T$

(6) $\mathbf{u}_2 = \left(\dfrac{1}{\|\mathbf{x}_2 - \mathbf{p}_1\|} \right)(\mathbf{x}_2 - \mathbf{p}_1)$

$= 1\left(-\dfrac{1}{2}, \dfrac{1}{2}, \dfrac{1}{2}, -\dfrac{1}{2} \right)^T$

$\mathbf{p}_2 = \langle \mathbf{x}_3, \mathbf{u}_1 \rangle \mathbf{u}_1 + \langle \mathbf{x}_3, \mathbf{u}_2 \rangle \mathbf{u}_2$

$= 6\mathbf{u}_1 - 2\mathbf{u}_2$

(7) $\mathbf{x}_3 - \mathbf{p}_2 = \mathbf{x}_3 - 6\mathbf{u}_1 + 2\mathbf{u}_2 = (-2, 1, -1, 2)^T$

(8) $\mathbf{u}_3 = \dfrac{1}{\|\mathbf{x}_3 - \mathbf{p}_2\|}(\mathbf{x}_3 - \mathbf{p}_2) = \dfrac{1}{\sqrt{10}}(-2, 1, -1, 2)^T$ □

The Gram–Schmidt process applied to the column vectors of a matrix gives rise to a useful factorization.

Theorem 5.7.2 (QR Factorization).   *If A is an $m \times n$ matrix of rank n, then A can be factored into a product QR, where Q is an $m \times n$ matrix with orthonormal columns and R is an $n \times n$ matrix that is upper triangular and invertible.*

Proof.   Let $\mathbf{x}_1, \mathbf{x}_2, \ldots, \mathbf{x}_n$ denote the respective column vectors of A. Let $\mathbf{p}_1, \ldots, \mathbf{p}_{n-1}$ be defined as in Theorem 5.7.1 and let $\{\mathbf{u}_1, \mathbf{u}_2, \ldots, \mathbf{u}_n\}$ be the orthonormal basis of $R(A)$ derived from the Gram–Schmidt process. Define

$\alpha_{11} = \|\mathbf{x}_1\|$

$\alpha_{kk} = \|\mathbf{x}_k - \mathbf{p}_{k-1}\|$ for $k = 2, \ldots, n$

and

$$\alpha_{ik} = \langle \mathbf{x}_k, \mathbf{u}_i \rangle \quad \text{for } i = 1, \ldots, k-1 \quad \text{and} \quad k = 2, \ldots, n$$

By the Gram–Schmidt process,

$$\alpha_{11}\mathbf{u}_1 = \mathbf{x}_1$$

(9)
$$\alpha_{kk}\mathbf{u}_k = \mathbf{x}_k - \alpha_{1k}\mathbf{u}_1 - \alpha_{2k}\mathbf{u}_2 - \cdots - \alpha_{k-1,k}\mathbf{u}_{k-1}$$

$$\text{for} \quad k = 2, \ldots, n$$

System (9) may be rewritten in the form

$$\mathbf{x}_1 = \alpha_{11}\mathbf{u}_1$$

$$\mathbf{x}_2 = \alpha_{12}\mathbf{u}_1 + \alpha_{22}\mathbf{u}_2$$

$$\vdots$$

$$\mathbf{x}_n = \alpha_{1n}\mathbf{u}_1 + \cdots + \alpha_{nn}\mathbf{u}_n$$

If each of these equations is transposed and the resulting system is written as a matrix equation, then

$$A^T = \begin{pmatrix} \mathbf{x}_1^T \\ \mathbf{x}_2^T \\ \vdots \\ \mathbf{x}_n^T \end{pmatrix} = \begin{pmatrix} \alpha_{11} & 0 & \cdots & 0 \\ \alpha_{12} & \alpha_{22} & \cdots & 0 \\ \vdots & & & \\ \alpha_{1n} & \alpha_{2n} & \cdots & \alpha_{nn} \end{pmatrix} \begin{pmatrix} \mathbf{u}_1^T \\ \mathbf{u}_2^T \\ \vdots \\ \mathbf{u}_n^T \end{pmatrix}$$

Transposing once more, we obtain

$$A = \begin{pmatrix} \mathbf{u}_1 & \mathbf{u}_2 & \cdots & \mathbf{u}_n \end{pmatrix} \begin{pmatrix} \alpha_{11} & \alpha_{12} & \cdots & \alpha_{1n} \\ 0 & \alpha_{22} & \cdots & \alpha_{2n} \\ \vdots & & & \\ 0 & 0 & \cdots & \alpha_{nn} \end{pmatrix} = QR$$

The columns of Q are the orthonormal vectors $\mathbf{u}_1, \mathbf{u}_2, \ldots, \mathbf{u}_n$ and R is nonsingular, since $\alpha_{kk} \neq 0$ for $k = 1, \ldots, n$.   □

EXAMPLE 3. Factor the matrix A from Example 2 into a product QR.

SOLUTION. It follows from (4) that

$$2\mathbf{u}_1 = \mathbf{x}_1$$

It follows from (5) and (6) that

$$1\mathbf{u}_2 = \mathbf{x}_2 - 3\mathbf{u}_1$$

By (7) and (8),

$$\sqrt{10}\,\mathbf{u}_3 = \mathbf{x}_3 - 6\mathbf{u}_1 + 2\mathbf{u}_2$$

Solving these equations for $\mathbf{x}_1$, $\mathbf{x}_2$, and $\mathbf{x}_3$ yields

$$\mathbf{x}_1 = 2\mathbf{u}_1$$
$$\mathbf{x}_2 = 3\mathbf{u}_1 + \mathbf{u}_2$$
$$\mathbf{x}_3 = 6\mathbf{u}_1 - 2\mathbf{u}_2 + \sqrt{10}\,\mathbf{u}_3$$

Let

$$Q = (\mathbf{u}_1 \quad \mathbf{u}_2 \quad \mathbf{u}_3) = \begin{pmatrix} \dfrac{1}{2} & -\dfrac{1}{2} & \dfrac{-2}{\sqrt{10}} \\[2mm] \dfrac{1}{2} & \dfrac{1}{2} & \dfrac{1}{\sqrt{10}} \\[2mm] \dfrac{1}{2} & \dfrac{1}{2} & \dfrac{-1}{\sqrt{10}} \\[2mm] \dfrac{1}{2} & -\dfrac{1}{2} & \dfrac{2}{\sqrt{10}} \end{pmatrix}$$

and

$$R = \begin{pmatrix} 2 & 3 & 6 \\ 0 & 1 & -2 \\ 0 & 0 & \sqrt{10} \end{pmatrix}$$

The reader may verify that $A = QR$.   □

We saw in Section 6 that if the columns of an $m \times n$ matrix A form an orthonormal set, then the least squares solution to $A\mathbf{x} = \mathbf{b}$ is simply $\hat{\mathbf{x}} = A^T\mathbf{b}$. If A has rank n but its column vectors do not form an orthonormal set in R^m, then the QR factorization can be used to solve the least squares problem.

Theorem 5.7.3.   *If A is an $m \times n$ matrix of rank n, then the solution to the least squares problem $A\mathbf{x} = \mathbf{b}$ is given by $\hat{\mathbf{x}} = R^{-1}Q^T\mathbf{b}$, where Q and R are the matrices obtained from the factorization given in Theorem 5.7.2. The solution $\hat{\mathbf{x}}$ may be obtained by using back substitution to solve $R\mathbf{x} = Q^T\mathbf{b}$.*

Proof.   Let $\hat{\mathbf{x}}$ be the solution to the least squares problem $A\mathbf{x} = \mathbf{b}$ guaranteed by Theorem 5.5.2. Thus $\hat{\mathbf{x}}$ is the solution to the normal equations

$$A^TA\mathbf{x} = A^T\mathbf{b}$$

If A is factored into a product QR, these equations become

$$(QR)^TQR\mathbf{x} = (QR)^T\mathbf{b}$$

or

$$R^T(Q^TQ)R\mathbf{x} = R^TQ^T\mathbf{b}$$

Since Q has orthonormal columns, it follows that $Q^T Q = I$ and hence

$$R^T R \mathbf{x} = R^T Q^T \mathbf{b}$$

Since R^T is invertible, this simplifies to

$$R\mathbf{x} = Q^T \mathbf{b} \quad \text{or} \quad \mathbf{x} = R^{-1} Q^T \mathbf{b} \qquad \square$$

EXAMPLE 4.   Find the least squares solution to

$$\begin{pmatrix} 1 & 1 & 2 \\ 1 & 2 & 3 \\ 1 & 2 & 1 \\ 1 & 1 & 6 \end{pmatrix} \begin{pmatrix} x_1 \\ x_2 \\ x_3 \end{pmatrix} = \begin{pmatrix} 3 \\ -4 \\ 2 \\ 1 \end{pmatrix}$$

SOLUTION.   The coefficient matrix of this system was factored in Example 3. Using that factorization, we have

$$Q^T \mathbf{b} = \begin{pmatrix} \dfrac{1}{2} & \dfrac{1}{2} & \dfrac{1}{2} & \dfrac{1}{2} \\[2mm] -\dfrac{1}{2} & \dfrac{1}{2} & \dfrac{1}{2} & -\dfrac{1}{2} \\[2mm] -\dfrac{2}{\sqrt{10}} & \dfrac{1}{\sqrt{10}} & -\dfrac{1}{\sqrt{10}} & \dfrac{2}{\sqrt{10}} \end{pmatrix} \begin{pmatrix} 3 \\ -4 \\ 2 \\ 1 \end{pmatrix} = \begin{pmatrix} 1 \\ -3 \\ -\sqrt{10} \end{pmatrix}$$

$$R\mathbf{x} = \begin{pmatrix} 2 & 3 & 6 \\ 0 & 1 & -2 \\ 0 & 0 & \sqrt{10} \end{pmatrix} \begin{pmatrix} x_1 \\ x_2 \\ x_3 \end{pmatrix} = \begin{pmatrix} 1 \\ -3 \\ -\sqrt{10} \end{pmatrix}$$

This system is easily solved by back substitution. The solution we get is $(x_1, x_2, x_3) = (11, -5, -1)$. $\square$

THE MODIFIED GRAM–SCHMIDT PROCESS

In Chapter 7 we will consider computer methods for solving least squares problems. The QR method of Example 4 does not in general produce accurate results when carried out with finite-precision arithmetic. In practice there may be a loss of orthogonality due to round off error in computing $\mathbf{u}_1, \mathbf{u}_2, \ldots, \mathbf{u}_n$. One can achieve better numerical accuracy using a modified version of the Gram–Schmidt method. In the modified version the vector $\mathbf{u}_1$ is constructed as before:

$$\mathbf{u}_1 = \frac{1}{\|\mathbf{x}_1\|} \mathbf{x}_1$$

However, the remaining vectors $\mathbf{x}_2, \ldots, \mathbf{x}_n$ are then modified so as to be orthogonal to $\mathbf{u}_1$. This can be done by subtracting from each vector $\mathbf{x}_k$ the projection of $\mathbf{x}_k$ onto $\mathbf{u}_1$.

$$\mathbf{x}_k^{(1)} = \mathbf{x}_k - \left(\mathbf{u}_1^T \mathbf{x}_k\right)\mathbf{u}_1 \qquad k = 2, \ldots, n$$

At the second step we take

$$\mathbf{u}_2 = \frac{1}{\|\mathbf{x}_2^{(1)}\|}\mathbf{x}_2^{(1)}$$

The vector $\mathbf{u}_2$ is already orthogonal to $\mathbf{u}_1$. We then modify the remaining vectors to make them orthogonal to $\mathbf{u}_2$.

$$\mathbf{x}_k^{(2)} = \mathbf{x}_k^{(1)} - \left(\mathbf{u}_2^T \mathbf{x}_k^{(1)}\right)\mathbf{u}_2 \qquad k = 3, \ldots, n$$

In a similar manner $\mathbf{u}_3, \mathbf{u}_4, \ldots, \mathbf{u}_n$ are successively determined. At the last step we need only set

$$\mathbf{u}_n = \frac{1}{\|\mathbf{x}_n^{(n-1)}\|}\mathbf{x}_n^{(n-1)}$$

to achieve an orthonormal set $\{\mathbf{u}_1, \ldots, \mathbf{u}_n\}$. The following algorithm summarizes the process.

Algorithm 5.7.4 (The Modified Gram–Schmidt Process)

For $k = 1, 2, \ldots, n$ set

$$\alpha_{kk} = \|\mathbf{x}_k\|$$

$$\mathbf{u}_k = \frac{1}{\alpha_{kk}}\mathbf{x}_k$$

For $j = k+1, k+2, \ldots, n$, set

$$\alpha_{kj} = \mathbf{u}_k^T \mathbf{x}_j$$

$$\mathbf{x}_j = \mathbf{x}_j - \alpha_{kj}\mathbf{u}_k$$

If the modified Gram–Schmidt process is applied to the column vectors of an $m \times n$ matrix A having rank n, then, as before, one can obtain a QR factorization of A. If

$$r_{ij} = \begin{cases} \alpha_{ij} & \text{if} \quad i \le j \\ 0 & \text{if} \quad i > j \end{cases}$$

and

$$Q = (\mathbf{u}_1, \ldots, \mathbf{u}_n) \qquad \text{and} \qquad R = (r_{ij})$$

then

$$A = QR$$

This factorization may then be used computationally to determine the least squares solution to $A\mathbf{x} = \mathbf{b}$.

EXERCISES

1. For each of the following, use the Gram–Schmidt process to find an orthonormal basis for $R(A)$.

   (a) $A = \begin{pmatrix} -1 & 3 \\ 1 & 5 \end{pmatrix}$ (b) $A = \begin{pmatrix} 2 & 5 \\ 1 & 10 \end{pmatrix}$

2. Factor each of the matrices in Exercise 1 into a product QR, where Q is an orthogonal matrix and R is upper triangular.

3. Given the basis $\{(1, 2, -2)^T, (4, 3, 2)^T, (1, 2, 1)^T\}$ for R^3, use the Gram–Schmidt process to obtain an orthonormal basis.

4. Consider the vector space $C[-1, 1]$ with inner product defined by

   $$\langle f, g \rangle = \int_{-1}^{1} f(x)g(x)\, dx$$

   Find an orthonormal basis for the subspace spanned by $1, x,$ and x^2.

5. Let

   $$A = \begin{pmatrix} 2 & 1 \\ 1 & 1 \\ 2 & 1 \end{pmatrix} \quad \text{and} \quad \mathbf{b} = \begin{pmatrix} 12 \\ 6 \\ 18 \end{pmatrix}$$

   (a) Use the Gram–Schmidt process to find an orthonormal basis for the column space of A.

   (b) Factor A into a product QR, where Q has an orthonormal set of column vectors and R is upper triangular.

   (c) Solve the least squares problem

   $$A\mathbf{x} = \mathbf{b}$$

6. Show that $\mathbf{u}_1 = (0, 1, 0)^T$ and $\mathbf{u}_2 = (1/\sqrt{2}, 0, 1/\sqrt{2})^T$ also form an orthonormal basis for the column space of the matrix A in Exercise 5.

7. Repeat Exercise 5 using

   $$A = \begin{pmatrix} 3 & -1 \\ 4 & 2 \\ 0 & 2 \end{pmatrix} \quad \text{and} \quad \mathbf{b} = \begin{pmatrix} 0 \\ 20 \\ 10 \end{pmatrix}$$

8. The vectors $\mathbf{x}_1 = \frac{1}{2}(1, 1, 1, -1)^T$ and $\mathbf{x}_2 = \frac{1}{6}(1, 1, 3, 5)^T$ form an orthonormal set in R^4. Extend this set to an orthonormal basis for R^4 by finding an orthonormal basis for the nullspace of

   $$\begin{pmatrix} 1 & 1 & 1 & -1 \\ 1 & 1 & 3 & 5 \end{pmatrix}$$

[**Hint:** First find a basis for the nullspace and then use the Gram–Schmidt process.]

9. Use the Gram–Schmidt process to find an orthonormal basis for the subspace of R^4 spanned by $\mathbf{x}_1 = (4, 2, 2, 1)^T$, $\mathbf{x}_2 = (2, 0, 0, 2)^T$, and $\mathbf{x}_3 = (1, 1, -1, 1)^T$.

10. Repeat Exercise 9 using the modified Gram–Schmidt process and compare answers.

11. Show that when carried out in exact arithmetic, the modified Gram–Schmidt process will produce the same orthonormal set as the classical Gram–Schmidt process.

*8. ORTHOGONAL POLYNOMIALS

We have already seen how polynomials can be used for data fitting and for approximating continuous functions. Since both of these problems are least squares problems, they can be simplified by selecting an orthogonal basis for the class of approximating polynomials. This leads us to the concept of orthogonal polynomials.

In this section we will study families of orthogonal polynomials associated with various inner products on $C[a, b]$. We will see that the polynomials in each of these classes satisfy a three-term recursion relation. This recursion relation is particularly useful in computer applications. Certain families of orthogonal polynomials have important applications in many areas of mathematics. We will refer to these polynomials as classical polynomials and examine them in more detail. In particular, the classical polynomials are solutions to certain classes of second-order linear differential equations that arise in the solution of many partial differential equations from mathematical physics.

ORTHOGONAL SEQUENCES

Since the proof of Theorem 5.7.1 was by induction, the Gram–Schmidt process is valid for a denumerable set. Thus, if $\mathbf{x}_1, \mathbf{x}_2, \ldots$ is a sequence of vectors in an inner product space V and $\mathbf{x}_1, \mathbf{x}_2, \ldots, \mathbf{x}_n$ are linearly independent for each n, then the Gram–Schmidt process may be used to form a sequence $\mathbf{u}_1, \mathbf{u}_2, \ldots$, where $\{\mathbf{u}_1, \mathbf{u}_2, \ldots\}$ is an orthonormal set and $S(\mathbf{x}_1, \mathbf{x}_2, \ldots, \mathbf{x}_n) = S(\mathbf{u}_1, \mathbf{u}_2, \ldots, \mathbf{u}_n)$ for each n. In particular, from the sequence $1, x, x^2, \ldots$ it is possible to construct an "orthonormal sequence" $p_0(x), p_1(x), \ldots$.

Let P be the vector space of all polynomials and define the inner product $\langle\,,\rangle$ on P by

$$(1) \qquad \langle p, q \rangle = \int_a^b p(x)q(x)w(x)\,dx$$

where $w(x)$ is a positive continuous function. The interval can be taken as either open or closed and may be finite or infinite. If, however,

$$\int_a^b p(x)w(x)\,dx$$

is improper, we require that it converge for every $p \in P$.

Definition.   Let $p_0(x), p_1(x), \ldots$ be a sequence of polynomials with $\deg p_i(x) = i$ for each i. If $\langle p_i(x), p_j(x) \rangle = 0$ whenever $i \neq j$, then $\{p_n(x)\}$ is said to be a **sequence of orthogonal polynomials**. If $\langle p_i, p_j \rangle = \delta_{ij}$, then $\{p_n(x)\}$ is said to be a **sequence of orthonormal polynomials**.

Theorem 5.8.1.   *If $p_0, p_1, \ldots$ is a sequence of orthogonal polynomials, then*

(i) $p_0, \ldots, p_{n-1}$ *form a basis for P_n.*

(ii) $p_n \in P_n^{\perp}$ *(i.e., p_n is orthogonal to every polynomial of degree less than n).*

Proof.   It follows from Theorem 5.6.1 that $p_0, p_1, \ldots, p_{n-1}$ are linearly independent in P_n. Since $\dim P_n = n$, these n vectors must form a basis for P_n. Let $p(x)$ be any polynomial of degree less than n. Then

$$p(x) = \sum_{i=0}^{n-1} c_i p_i(x)$$

and hence

$$\langle p_n, p \rangle = \left\langle p_n, \sum_{i=0}^{n-1} c_i p_i \right\rangle = \sum_{i=0}^{n-1} c_i \langle p_n, p_i \rangle = 0$$

Therefore, $p_n \in P_n^{\perp}$.   □

If $\{p_0, p_1, \ldots, p_{n-1}\}$ is an orthogonal set in P_n and

$$u_i = \left(\frac{1}{\|p_i\|} \right) p_i \qquad \text{for} \quad i = 0, \ldots, n-1$$

then $\{u_0, \ldots, u_{n-1}\}$ is an orthonormal basis for P_n. Hence, if $p \in P_n$,

then

$$p = \sum_{i=0}^{n-1} \langle p, u_i \rangle u_i$$

$$= \sum_{i=0}^{n-1} \left\langle p, \left(\frac{1}{\|p_i\|} \right) p_i \right\rangle \left(\frac{1}{\|p_i\|} \right) p_i$$

$$= \sum_{i=0}^{n-1} \frac{\langle p, p_i \rangle}{\langle p_i, p_i \rangle} p_i$$

Similarly, if $f \in C[a, b]$, then the best least squares approximation to f by elements of P_n is given by

$$p = \sum_{i=0}^{n-1} \frac{\langle f, p_i \rangle}{\langle p_i, p_i \rangle} p_i$$

where $p_0, p_1, \ldots, p_{n-1}$ are orthogonal polynomials.

Another nice feature of sequences of orthogonal polynomials is that they satisfy a three-term recursion relation.

Theorem 5.8.2.   *Let $p_0, p_1, \ldots$ be a sequence of orthogonal polynomials. Let a_i denote the lead coefficient of p_i for each i and define $p_{-1}(x)$ to be the zero polynomial. Then*

$$\alpha_{n+1} p_{n+1}(x) = (x - \beta_n) p_n(x) - \alpha_n \gamma_n p_{n-1}(x)$$

where

$$\alpha_n = \frac{a_{n-1}}{a_n}, \qquad \beta_n = \frac{\langle p_n, x p_n \rangle}{\langle p_n, p_n \rangle}, \qquad \gamma_n = \frac{\langle p_n, p_n \rangle}{\langle p_{n-1}, p_{n-1} \rangle}$$

Proof.   Since $p_0, p_1, \ldots, p_{n+1}$ form a basis for P_{n+2}, we can write

(2)
$$x p_n(x) = \sum_{k=0}^{n+1} c_{nk} p_k(x)$$

where

(3)
$$c_{nk} = \frac{\langle x p_n, p_k \rangle}{\langle p_k, p_k \rangle}$$

For any inner product defined by (1),

$$\langle xf, g \rangle = \langle f, xg \rangle$$

In particular,

$$\langle x p_n, p_k \rangle = \langle p_n, x p_k \rangle$$

It follows from Theorem 5.8.1 that if $k < n - 1$, then

$$c_{nk} = \frac{\langle x p_n, p_k \rangle}{\langle p_k, p_k \rangle} = \frac{\langle p_n, x p_k \rangle}{\langle p_k, p_k \rangle} = 0$$

Therefore, (2) simplifies to

$$xp_n(x) = c_{n,n-1}p_{n-1}(x) + c_{n,n}p_n(x) + c_{n,n+1}p_{n+1}(x)$$

This can be rewritten in the form

(4) $$c_{n,n+1}p_{n+1}(x) = (x - c_{n,n})p_n(x) - c_{n,n-1}p_{n-1}(x)$$

Comparing the lead coefficients of the polynomials on each side of (4), we see that

$$c_{n,n+1}a_{n+1} = a_n$$

or

(5) $$c_{n,n+1} = \frac{a_n}{a_{n+1}} = \alpha_{n+1}$$

It also follows from (4) that

$$c_{n,n+1}\langle p_n, p_{n+1}\rangle = \langle p_n, (x - c_{n,n})p_n\rangle - c_{n,n-1}\langle p_n, p_{n-1}\rangle$$
$$0 = \langle p_n, xp_n\rangle - c_{nn}\langle p_n, p_n\rangle$$

and hence

$$c_{nn} = \frac{\langle p_n, xp_n\rangle}{\langle p_n, p_n\rangle} = \beta_n$$

It follows from (3) that

$$\langle p_{n-1}, p_{n-1}\rangle c_{n,n-1} = \langle xp_n, p_{n-1}\rangle$$
$$= \langle p_n, xp_{n-1}\rangle$$
$$= \langle p_n, p_n\rangle c_{n-1,n}$$

and hence by (5) we have

$$c_{n,n-1} = \frac{\langle p_n, p_n\rangle}{\langle p_{n-1}, p_{n-1}\rangle}\alpha_n = \gamma_n\alpha_n \qquad \square$$

In generating a sequence of orthogonal polynomials by the recursion relation in Theorem 5.8.2, we are free to choose any nonzero lead coefficient a_{n+1} we want at each step. This is reasonable, since any nonzero multiple of a particular p_{n+1} will also be orthogonal to $p_0, \ldots, p_n$. If we were to choose our a_i's to be 1, for example, then the recursion relation would simplify to

$$p_{n+1}(x) = (x - \beta_n)p_n(x) - \gamma_n p_{n-1}(x)$$

CLASSICAL ORTHOGONAL POLYNOMIALS

Let us now look at some examples. Because of their importance we will consider the classical polynomials beginning with the simplest, the Legendre polynomials.

Legendre Polynomials. The Legendre polynomials are orthogonal with respect to the inner product

$$\langle p, q \rangle = \int_{-1}^{1} p(x) q(x) \, dx$$

Let $P_n(x)$ denote the Legendre polynomial of degree n. If we choose the lead coefficients so that $P_n(1) = 1$ for each n, then the recursion formula for the Legendre polynomials is

$$(n + 1) P_{n+1}(x) = (2n + 1) x P_n(x) - n P_{n-1}(x)$$

By the use of this formula, the sequence of Legendre polynomials is easily generated.

$$P_0(x) = 1$$
$$P_1(x) = x$$
$$P_2(x) = \tfrac{1}{2}(3x^2 - 1)$$
$$P_3(x) = \tfrac{1}{2}(5x^3 - 3x)$$
$$P_4(x) = \tfrac{1}{8}(35x^4 - 30x^2 + 3)$$

Tchebycheff Polynomials. These polynomials are orthogonal with respect to the inner product,

$$\langle p, q \rangle = \int_{-1}^{1} p(x) q(x) (1 - x^2)^{-1/2} \, dx$$

It is customary to normalize the lead coefficients so that $a_0 = 1$ and $a_k = 2^{k-1}$ for $k = 1, 2, \ldots$. The Tchebycheff polynomials are denoted by $T_n(x)$ and have the interesting property that

$$T_n(\cos \theta) = \cos n\theta$$

This property, together with the trigonometric identity

$$\cos(n + 1)\theta = 2 \cos \theta \cos n\theta - \cos(n - 1)\theta$$

can be used to derive the recursion relation

$$T_{n+1}(x) = 2x T_n(x) - T_{n-1}(x)$$

for $n \geq 1$.

The Legendre and the Tchebycheff polynomials are both special cases of the *Jacobi polynomials*. The Jacobi polynomials $P_n^{(\lambda, \mu)}$ are orthogonal with respect to the inner product,

$$\langle p, q \rangle = \int_{-1}^{1} p(x) q(x) (1 - x)^{\lambda} (1 + x)^{\mu} \, dx$$

where $\lambda, \mu > -1$.

TABLE 5.8.1

Tchebycheff	Hermite	Laguerre ($\lambda = 0$)
$T_{n+1} = 2xT_n - T_{n-1}, n \geq 1$	$H_{n+1} = 2xH_n - 2nH_{n-1}$	$(n+1)L_{n+1}^{(0)} = (2n+1-x)L_n^{(0)} - nL_{n-1}^{(0)}$
$T_0 = 1$	$H_0 = 1$	$L_0^{(0)} = 1$
$T_1 = x$	$H_1 = 2x$	$L_1^{(0)} = 1 - x$
$T_2 = 2x^2 - 1$	$H_2 = 4x^2 - 2$	$L_2^{(0)} = \frac{1}{2}(x^2 - 4x + 2)$
$T_3 = 4x^3 - 3x$	$H_3 = 8x^3 - 12x$	$L_3^{(0)} = \frac{1}{6}(-x^3 + 9x^2 - 18x + 6)$

The *Hermite polynomials* are defined on the interval $(-\infty, \infty)$. They are orthogonal with respect to the inner product

$$\langle p, q \rangle = \int_{-\infty}^{\infty} p(x)q(x)e^{-x^2}\,dx$$

The recursion relation for Hermite polynomials is given by

$$H_{n+1}(x) = 2xH_n(x) - 2nH_{n-1}(x)$$

The *Laguerre polynomials* are defined on the interval $(0, \infty)$ and are orthogonal with respect to the inner product,

$$\langle p, q \rangle = \int_0^{\infty} p(x)q(x)x^{\lambda}e^{-x}\,dx$$

where $\lambda > -1$. The recursion relation for the Laguerre polynomials is given by

$$(n+1)L_{n+1}^{(\lambda)}(x) = (2n + \lambda + 1 - x)L_n^{(\lambda)}(x) - (n + \lambda)L_{n-1}^{(\lambda)}(x)$$

The Tchebycheff, Hermite, and Laguerre polynomials are compared in Table 5.8.1.

APPLICATION: NUMERICAL INTEGRATION

One important application of orthogonal polynomials occurs in numerical integration. In order to approximate

(6)
$$\int_a^b f(x)w(x)\,dx$$

one first approximates $f(x)$ by an interpolating polynomial. One can determine a polynomial $P(x)$ which agrees with $f(x)$ at n points $x_1, \ldots, x_n$ in $[a, b]$ using *Lagrange's interpolation formula*,

$$P(x) = \sum_{i=1}^{n} f(x_i)\delta_i(x)$$

where

$$\delta_i(x) = \frac{\prod_{\substack{j=1 \\ j \neq i}}^{n} (x - x_j)}{\prod_{\substack{j=1 \\ j \neq i}}^{n} (x_i - x_j)}$$

The integral (6) is then approximated by

(7)
$$\int_a^b P(x)w(x)\,dx = \sum_{i=1}^{n} A_i f(x_i)$$

where

$$A_i = \int_a^b \delta_i(x)w(x)\,dx \qquad i = 1,\ldots,n$$

It can be shown that (7) will give the exact value of the integral whenever $f(x)$ is a polynomial of degree less than n. If the points $x_1,\ldots,x_n$ are chosen properly, formula (7) will be exact for higher-degree polynomials. Indeed, it can be shown that if $p_0, p_1, p_2, \ldots$ is a sequence of orthogonal polynomials with respect to the inner product (1) and $x_1,\ldots,x_n$ are the zeros of $p_n(x)$, then formula (7) will be exact for all polynomials of degree less than $2n$. The following theorem guarantees that the roots of p_n are all real and lie in the open interval (a,b).

Theorem 5.8.3. *If $p_0, p_1, p_2, \ldots$ is a sequence of orthogonal polynomials with respect to the inner product (1), then the zeros of $p_n(x)$ are all real and distinct and lie in the interval (a,b).*

Proof. Let $x_1,\ldots,x_m$ be the zeros of $p_n(x)$ that lie in (a,b) and for which $p_n(x)$ changes sign. Thus $p_n(x)$ must have a factor of $(x - x_i)^{k_i}$, where k_i is odd, for $i = 1,\ldots,m$. We may write

$$p_n(x) = a_n(x - x_1)^{k_1}(x - x_2)^{k_2} \cdots (x - x_m)^{k_m} q(x)$$

where $q(x)$ does not change sign on (a,b) and $q(x_i) \neq 0$ for $i = 1,\ldots,m$. Clearly, $m \leq n$. We will show that $m = n$. Let

$$r(x) = (x - x_1)(x - x_2) \cdots (x - x_m)$$

The product

$$p_n(x)r(x) = a_n(x - x_1)^{k_1+1}(x - x_2)^{k_2+1} \cdots (x - x_m)^{k_m+1} q(x)$$

will involve only even powers of $(x - x_i)$ for each i and hence will not change sign on (a,b). Therefore,

$$\langle p_n, r \rangle = \int_a^b p_n(x)r(x)w(x)\,dx > 0$$

Since p_n is orthogonal to all polynomials of degree less than n, it follows that $\deg(r(x)) = m \geq n$. $\qquad\square$

Numerical integration formulas of the form (7), where the x_i's are roots of orthogonal polynomials, are called *Gaussian quadrature formulas*. The proof of exactness for polynomials of degree less than $2n$ can be found in most undergraduate numerical analysis textbooks.

EXERCISES

1. Use the recursion formulas to calculate T_4, T_5 and H_4, H_5.

2. Let $p_0(x)$, $p_1(x)$, and $p_2(x)$ be orthogonal with respect to the inner product,

$$\langle p(x), q(x) \rangle = \int_{-1}^{1} \frac{p(x)q(x)}{1 + x^2}\, dx$$

Use Theorem 5.8.2 to calculate $p_1(x)$ and $p_2(x)$ if all the polynomials have lead coefficient 1.

3. Show that the Tchebycheff polynomials have the following properties:
(a) $2T_m(x)T_n(x) = T_{m+n}(x) + T_{m-n}(x)$, for $m > n$
(b) $T_m(T_n(x)) = T_{mn}(x)$

4. Find the best quadratic least squares approximation to e^x on $[-1, 1]$ with respect to the inner product,

$$\langle f, g \rangle = \int_{-1}^{1} f(x)g(x)\, dx$$

5. Let $p_0, p_1, \ldots$ be a sequence of orthogonal polynomials and let a_n denote the lead coefficient of p_n. Prove that

$$\|p_n\|^2 = a_n \langle x^n, p_n \rangle$$

6. Let $T_n(x)$ denote the Tchebycheff polynomial of degree n and define

$$U_{n-1}(x) = \frac{1}{n} T_n'(x)$$

for $n = 1, 2, \ldots$.
(a) Compute $U_0(x)$, $U_1(x)$, and $U_2(x)$
(b) If $x = \cos\theta$, show that

$$U_{n-1}(x) = \frac{\sin n\theta}{\sin \theta}$$

7. Let $U_{n-1}(x)$ be defined as in Exercise 6 for $n \geq 1$ and define $U_{-1}(x) = 0$. Show that:
(a) $T_n(x) = U_n(x) - xU_{n-1}(x)$, for $n \geq 0$
(b) $U_n(x) = 2xU_{n-1}(x) - U_{n-2}(x)$, for $n \geq 1$

8. Show the U_i's defined in Exercise 6 are orthogonal with respect to the inner product

$$\langle p, q \rangle = \int_{-1}^{1} p(x)q(x)(1 - x^2)^{1/2} \, dx$$

The U_i's are called Tchebycheff polynomials of the second kind.

9. For $n = 0, 1, 2$ show that the Legendre polynomial $P_n(x)$ satisfies the second-order equation

$$(1 - x^2)y'' - 2xy' + n(n + 1)y = 0$$

10. Use mathematical induction to prove each of the following.
(a) $H_n'(x) = 2nH_{n-1}(x)$, $n = 0, 1, \ldots$
(b) $H_n''(x) - 2xH_n'(x) + 2nH_n(x) = 0$, $n = 0, 1, \ldots$

11. Given a function $f(x)$ which passes through the points $(1, 2)$, $(2, -1)$, $(3, 4)$, use the Lagrange interpolating formula to construct a second-degree polynomial which interpolates f at the given points.

12. Show that if $f(x)$ is a polynomial of degree less than n, then $f(x)$ must equal the interpolating polynomial $P(x)$ in (7) and hence the sum in (7) gives the exact value for $\int_a^b f(x)w(x) \, dx$.

13. Use the zeros of the Legendre polynomial $P_2(x)$ to obtain a two-point quadrature formula

$$\int_{-1}^{1} f(x) \, dx \approx A_1 f(x_1) + A_2 f(x_2)$$

14. For what degree polynomials will the quadrature formula in Exercise 13 be exact? Use the formula from Exercise 13 to approximate

$$\int_{-1}^{1} (x^3 + 3x^2 + 1) \, dx \qquad \text{and} \qquad \int_{-1}^{1} \frac{1}{1 + x^2} \, dx$$

How do the approximations compare with the actual values?

6

EIGENVALUES

In Section 1 we will be concerned with the equation $A\mathbf{x} = \lambda\mathbf{x}$. This equation occurs in many applications of linear algebra. If the equation has a nonzero solution $\mathbf{x}$, then λ is said to be an *eigenvalue* of A and $\mathbf{x}$ is said to be an *eigenvector* belonging to λ. One of the main applications of eigenvalues is in the solution of systems of linear differential equations. This application is presented in Section 2.

If A is an $n \times n$ matrix, we can think of A as representing a linear transformation from R^n into itself. Eigenvalues and eigenvectors provide the key to understanding how the operator works. For example, if $\lambda > 0$, the effect of the operator on any eigenvector belonging to λ is simply a stretching or a shrinking by a constant factor. Indeed, the effect of the operator is easily determined on any linear combination of eigenvectors. In particular, if it is possible to find a basis of eigenvectors for R^n, the operator can be represented by a diagonal matrix D with respect to that basis and the matrix A can be factored into a product SDS^{-1}. In Section 3 we see how this is done and look at a number of applications.

229

In Section 4 we consider matrices with complex entries. In this setting we will be concerned with matrices whose eigenvectors can be used to form an orthonormal basis for C^n (the vector space of all n-tuples of complex numbers). Section 5 deals with the application of eigenvalues to quadratic equations in several variables and also with applications involving maximums and minimums of functions of several variables. In Section 6 we consider symmetric positive definite matrices. The eigenvalues of such matrices are all real and positive. These matrices occur in a wide variety of applications. Finally, in Section 7 we study matrices with nonnegative entries and some applications to economics.

1. EIGENVALUES AND EIGENVECTORS

In this section we are concerned with the problem of finding a scalar λ such that the $n \times n$ system

$$A\mathbf{x} = \lambda\mathbf{x}$$

has a nonzero solution.

Definition.   Let A be an $n \times n$ matrix. A scalar λ is said to be an **eigenvalue** or a **characteristic value** of A if there exists a nonzero vector $\mathbf{x}$ in R^n such that $A\mathbf{x} = \lambda\mathbf{x}$. The vector $\mathbf{x}$ is said to be an **eigenvector** or a **characteristic vector** belonging to λ.

EXAMPLE 1.   Let

$$A = \begin{pmatrix} 4 & -2 \\ 1 & 1 \end{pmatrix} \quad \text{and} \quad \mathbf{x} = \begin{pmatrix} 2 \\ 1 \end{pmatrix}$$

Since

$$A\mathbf{x} = \begin{pmatrix} 4 & -2 \\ 1 & 1 \end{pmatrix}\begin{pmatrix} 2 \\ 1 \end{pmatrix} = \begin{pmatrix} 6 \\ 3 \end{pmatrix} = 3\begin{pmatrix} 2 \\ 1 \end{pmatrix} = 3\mathbf{x}$$

it follows that $\lambda = 3$ is an eigenvalue of A and $\mathbf{x} = (2, 1)^T$ is an eigenvector belonging to λ. Actually, any nonzero multiple of $\mathbf{x}$ will be an eigenvector, since

$$A(\alpha\mathbf{x}) = \alpha A\mathbf{x} = \alpha\lambda\mathbf{x} = \lambda(\alpha\mathbf{x})$$

Thus, for example, $(4, 2)^T$ is also an eigenvector belonging to $\lambda = 3$.

$$\begin{pmatrix} 4 & -2 \\ 1 & 1 \end{pmatrix}\begin{pmatrix} 4 \\ 2 \end{pmatrix} = \begin{pmatrix} 12 \\ 6 \end{pmatrix} = 3\begin{pmatrix} 4 \\ 2 \end{pmatrix} \qquad\qquad \square$$

The equation $A\mathbf{x} = \lambda\mathbf{x}$ can be written in the form

$$(1) \qquad (A - \lambda I)\mathbf{x} = \mathbf{0}$$

Thus λ is an eigenvalue of A if and only if (1) has a nontrivial solution. The set of solutions to (1) is $N(A - \lambda I)$, which is a subspace of R^n. Thus, if λ is an eigenvalue of A, then $N(A - \lambda I) \neq \{\mathbf{0}\}$ and any nonzero vector in $N(A - \lambda I)$ is an eigenvector belonging to λ. The subspace $N(A - \lambda I)$ is called the *eigenspace* corresponding to the eigenvalue λ.

Equation (1) will have a nontrivial solution if and only if $A - \lambda I$ is singular or, equivalently,

$$(2) \qquad \det(A - \lambda I) = 0$$

If the determinant in (2) is expanded, we obtain an nth-degree polynomial in the variable λ,

$$p(\lambda) = \det(A - \lambda I)$$

This polynomial is called the *characteristic polynomial* and equation (2) is called the *characteristic equation* for the matrix A. The roots of the characteristic polynomial are the eigenvalues of A. If we count roots according to multiplicity, the characteristic polynomial will have exactly n roots. Thus A will have n eigenvalues, some of which may be repeated and some of which may be complex numbers. To take care of the latter case, it will be necessary to expand our field of scalars to the complex numbers and to allow complex entries for our vectors and matrices.

We have now established a number of equivalent conditions for λ to be an eigenvalue of A.

Let A be an $n \times n$ matrix and λ be a scalar. The following statements are equivalent.
  (a) λ is an eigenvalue of A.
  (b) $(A - \lambda I)\mathbf{x} = \mathbf{0}$ has a nontrivial solution.
  (c) $N(A - \lambda I) \neq \{\mathbf{0}\}$
  (d) $A - \lambda I$ is singular.
  (e) $\det(A - \lambda I) = 0$

We will now use statement (e) to determine the eigenvalues in a number of examples.

EXAMPLE 2. Find the eigenvalues and the corresponding eigenvectors of the matrix

$$A = \begin{pmatrix} 3 & 2 \\ 3 & -2 \end{pmatrix}$$

SOLUTION. The characteristic equation is

$$\begin{vmatrix} 3 - \lambda & 2 \\ 3 & -2 - \lambda \end{vmatrix} = 0$$

or

$$\lambda^2 - \lambda - 12 = 0$$

Thus the eigenvalues of A are $\lambda_1 = 4$ and $\lambda_2 = -3$. To find the eigenvectors belonging to $\lambda_1 = 4$, we must determine the nullspace of $A - 4I$.

$$A - 4I = \begin{pmatrix} -1 & 2 \\ 3 & -6 \end{pmatrix}$$

Solving $(A - 4I)\mathbf{x} = \mathbf{0}$, we get

$$\mathbf{x} = (2x_2, x_2)^T$$

Thus any nonzero multiple of $(2, 1)^T$ is an eigenvector belonging to λ_1 and $\{(2, 1)^T\}$ is a basis for the eigenspace corresponding to λ_1. Similarly, to find eigenvectors for λ_2, we must solve $(A + 3I)\mathbf{x} = \mathbf{0}$. In this case $\{(-1, 3)^T\}$ is a basis for $N(A + 3I)$ and any nonzero multiple of $(-1, 3)^T$ is an eigenvector belonging to λ_2.   □

EXAMPLE 3. Let

$$A = \begin{pmatrix} 2 & -3 & 1 \\ 1 & -2 & 1 \\ 1 & -3 & 2 \end{pmatrix}$$

Find the eigenvalues and the corresponding eigenspaces.

SOLUTION

$$\begin{vmatrix} 2 - \lambda & -3 & 1 \\ 1 & -2 - \lambda & 1 \\ 1 & -3 & 2 - \lambda \end{vmatrix} = -\lambda(\lambda - 1)^2$$

Thus the characteristic polynomial has roots $\lambda_1 = 0, \lambda_2 = \lambda_3 = 1$. The eigenspace corresponding to $\lambda_1 = 0$ is $N(A)$, which we determine in the usual manner.

$$\begin{pmatrix} 2 & -3 & 1 & | & 0 \\ 1 & -2 & 1 & | & 0 \\ 1 & -3 & 2 & | & 0 \end{pmatrix} \rightarrow \begin{pmatrix} 1 & 0 & -1 & | & 0 \\ 0 & 1 & -1 & | & 0 \\ 0 & 0 & 0 & | & 0 \end{pmatrix}$$

Setting $x_3 = \alpha$, we find that $x_1 = x_2 = x_3 = \alpha$. Thus the eigenspace corresponding to $\lambda_1 = 0$ consists of all vectors of the form $\alpha(1, 1, 1)^T$. To find the eigenspace corresponding to $\lambda = 1$, we must solve the system $(A - I)\mathbf{x} = \mathbf{0}$.

$$\begin{pmatrix} 1 & -3 & 1 & | & 0 \\ 1 & -3 & 1 & | & 0 \\ 1 & -3 & 1 & | & 0 \end{pmatrix} \rightarrow \begin{pmatrix} 1 & -3 & 1 & | & 0 \\ 0 & 0 & 0 & | & 0 \\ 0 & 0 & 0 & | & 0 \end{pmatrix}$$

Setting $x_2 = \alpha$ and $x_3 = \beta$, we get $x_1 = 3\alpha - \beta$. Thus the eigenspace corresponding to $\lambda = 1$ consists of all vectors of the form

$$\begin{pmatrix} 3\alpha - \beta \\ \alpha \\ \beta \end{pmatrix} = \alpha \begin{pmatrix} 3 \\ 1 \\ 0 \end{pmatrix} + \beta \begin{pmatrix} -1 \\ 0 \\ 1 \end{pmatrix}$$

□

EXAMPLE 4. Find the eigenvalues and the corresponding eigenspaces of the matrix

$$A = \begin{pmatrix} 1 & 2 \\ -2 & 1 \end{pmatrix}$$

SOLUTION

$$\begin{vmatrix} 1 - \lambda & 2 \\ -2 & 1 - \lambda \end{vmatrix} = (1 - \lambda)^2 + 4$$

The roots of the characteristic polynomial are $\lambda_1 = 1 + 2i$, $\lambda_2 = 1 - 2i$.

$$A - \lambda_1 I = \begin{pmatrix} -2i & 2 \\ -2 & -2i \end{pmatrix} = -2 \begin{pmatrix} i & -1 \\ 1 & i \end{pmatrix}$$

It follows that $\{(1, i)^T\}$ is a basis for the eigenspace corresponding to $\lambda_1 = 1 + 2i$. Similarly,

$$A - \lambda_2 I = \begin{pmatrix} 2i & 2 \\ -2 & 2i \end{pmatrix} = 2 \begin{pmatrix} i & 1 \\ -1 & i \end{pmatrix}$$

and $\{(1, -i)^T\}$ is a basis for $N(A - \lambda_2 I)$.

□

Let A be an $n \times n$ matrix and let $p(\lambda)$ be the characteristic polynomial of A.

$$p(\lambda) = \det(A - \lambda I)$$

(3)
$$= \begin{vmatrix} a_{11} - \lambda & a_{12} & \cdots & a_{1n} \\ a_{21} & a_{22} - \lambda & & a_{2n} \\ \vdots & & & \\ a_{n1} & a_{n2} & & a_{nn} - \lambda \end{vmatrix}$$

It follows from the definition of the determinant that the only term in the expansion of $\det(A - \lambda I)$ involving more than $n - 2$ of the diagonal elements will be

(4)
$$(a_{11} - \lambda)(a_{22} - \lambda) \cdots (a_{nn} - \lambda)$$

When this is expanded, the coefficient of λ^n will be $(-1)^n$. Thus, if

$\lambda_1, \ldots, \lambda_n$ are the eigenvalues of A, then

(5)
$$p(\lambda) = (-1)^n(\lambda - \lambda_1)(\lambda - \lambda_2) \cdots (\lambda - \lambda_n)$$
$$= (\lambda_1 - \lambda)(\lambda_2 - \lambda) \cdots (\lambda_n - \lambda)$$

It follows from (3) and (5) that

$$\lambda_1 \cdot \lambda_2 \cdots \lambda_n = p(0) = \det(A)$$

From (4) we also see that the coefficient of $(-\lambda)^{n-1}$ is $\sum_{i=1}^{n} a_{ii}$. If we determine this same coefficient using (5), we obtain $\sum_{i=1}^{n} \lambda_i$. It follows that

$$\sum_{i=1}^{n} \lambda_i = \sum_{i=1}^{n} a_{ii}$$

The sum of the diagonal elements of A is called the *trace* of A and is denoted by $\operatorname{tr} A$.

EXAMPLE 5.   If

$$A = \begin{pmatrix} 5 & -18 \\ 1 & -1 \end{pmatrix}$$

then

$$\det(A) = -5 + 18 = 13 \qquad \text{and} \qquad \operatorname{tr} A = 5 - 1 = 4$$

The characteristic polynomial of A is given by

$$\begin{vmatrix} 5 - \lambda & -18 \\ 1 & -1 - \lambda \end{vmatrix} = \lambda^2 - 4\lambda + 13$$

and hence the eigenvalues of A are $\lambda_1 = 2 + 3i$ and $\lambda_2 = 2 - 3i$. Note that

$$\lambda_1 + \lambda_2 = 4 = \operatorname{tr} A$$

and

$$\lambda_1 \lambda_2 = 13 = \det(A) \qquad\qquad \square$$

In the examples we have looked at so far, n has always been less than 4. For larger n it is more difficult to find the roots of the characteristic

polynomial. In Chapter 7 we will learn numerical methods for computing eigenvalues. (These methods will not involve the characteristic
polynomial at all.) If the eigenvalues of A have been computed using
some numerical method, one way to check their accuracy is to compare
their sum to the trace of A.

EXERCISES

1. Find the eigenvalues and the corresponding eigenspaces for each of
the following matrices.

(a) $\begin{pmatrix} 3 & 2 \\ 4 & 1 \end{pmatrix}$
  (b) $\begin{pmatrix} 6 & -4 \\ 3 & -1 \end{pmatrix}$

(c) $\begin{pmatrix} 3 & -1 \\ 1 & 1 \end{pmatrix}$
  (d) $\begin{pmatrix} 3 & -8 \\ 2 & 3 \end{pmatrix}$

(e) $\begin{pmatrix} 1 & 1 \\ -2 & 3 \end{pmatrix}$
  (f) $\begin{pmatrix} 0 & 1 & 0 \\ 0 & 0 & 1 \\ 0 & 0 & 0 \end{pmatrix}$

(g) $\begin{pmatrix} 1 & 1 & 1 \\ 0 & 2 & 1 \\ 0 & 0 & 1 \end{pmatrix}$
  (h) $\begin{pmatrix} 1 & 2 & 1 \\ 0 & 3 & 1 \\ 0 & 5 & -1 \end{pmatrix}$

(i) $\begin{pmatrix} 4 & -5 & 1 \\ 1 & 0 & -1 \\ 0 & 1 & -1 \end{pmatrix}$
  (j) $\begin{pmatrix} -2 & 0 & 1 \\ 1 & 0 & -1 \\ 0 & 1 & -1 \end{pmatrix}$

(k) $\begin{pmatrix} 2 & 0 & 0 & 0 \\ 0 & 2 & 0 & 0 \\ 0 & 0 & 3 & 0 \\ 0 & 0 & 0 & 4 \end{pmatrix}$
  (l) $\begin{pmatrix} 3 & 0 & 0 & 0 \\ 4 & 1 & 0 & 0 \\ 0 & 0 & 2 & 1 \\ 0 & 0 & 0 & 2 \end{pmatrix}$

2. Show that the eigenvalues of a triangular matrix are the diagonal
elements of the matrix.

3. Let A be an $n \times n$ matrix. Prove that A is singular if and only if
$\lambda = 0$ is an eigenvalue of A.

4. Let A be a nonsingular matrix and let λ be an eigenvalue of A.
Show that $1/\lambda$ is an eigenvalue of A^{-1}.

5. Let λ be an eigenvalue of A and let **x** be an eigenvector belonging
to λ. Use mathematical induction to show that λ^m is an eigenvalue
of A^m and **x** is an eigenvector of A^m belonging to λ^m for $m =$
$1, 2, \ldots$.

6. An $n \times n$ matrix A is said to be idempotent if $A^2 = A$. Show that if λ is an eigenvalue of an idempotent matrix, then λ must be either 0 or 1.

7. An $n \times n$ matrix is said to be nilpotent if $A^k = O$ for some positive integer k. Show that all the eigenvalues of a nilpotent matrix are 0.

8. Let A be an $n \times n$ matrix and let $B = A - \alpha I$ for some scalar α. How do the eigenvalues of A and B compare? Explain.

9. Show that A and A^T have the same eigenvalues. Do they necessarily have the same eigenvectors? Explain.

10. Show that the matrix

$$A = \begin{pmatrix} \cos\theta & -\sin\theta \\ \sin\theta & \cos\theta \end{pmatrix}$$

will have complex eigenvalues if θ is not a multiple of π. Give a geometric interpretation of this result.

11. Let A be a 2×2 matrix. If $\operatorname{tr} A = 8$ and $\det(A) = 12$, what are the eigenvalues of A?

12. Let $A = (a_{ij})$ be an $n \times n$ matrix with eigenvalues $\lambda_1, \ldots, \lambda_n$. Show that

$$\lambda_j = a_{jj} + \sum_{i \neq j} (a_{ii} - \lambda_i) \quad \text{for} \quad j = 1, \ldots, n$$

13. Let A be a 2×2 matrix and let $p(\lambda) = \lambda^2 + b\lambda + c$ be the characteristic polynomial of A. Show that $b = -\operatorname{tr} A$ and $c = \det(A)$.

14. Let λ be a nonzero eigenvalue of A and let $\mathbf{x}$ be an eigenvector belonging to λ. Show that $A^m \mathbf{x}$ is also an eigenvector belonging to λ for $m = 1, 2, \ldots$.

15. Let A be an $n \times n$ matrix and let λ be an eigenvalue of A. If $A - \lambda I$ has rank k, what is the dimension of the eigenspace corresponding to λ? Explain.

16. Let A be an $n \times n$ matrix. Show that a vector $\mathbf{x}$ in R^n is an eigenvector belonging to A if and only if the subspace S of R^n spanned by $\mathbf{x}$ and $A\mathbf{x}$ has dimension 1.

17. Let A be a matrix with real entries and let $\lambda = a + bi$ be a complex eigenvalue of A.

(a) Show that $\bar{\lambda} = a - bi$ is also an eigenvalue of A.

(b) If $\mathbf{z}$ is an eigenvector belonging to λ, show that $\bar{\mathbf{z}}$ is an eigenvector belonging to $\bar{\lambda}$.

18. Let $\mathbf{x}_1, \ldots, \mathbf{x}_r$ be eigenvectors of an $n \times n$ matrix A and let S be the subspace of R^n spanned by $\mathbf{x}_1, \mathbf{x}_2, \ldots, \mathbf{x}_r$. Show that S is invariant under A, (i.e., show that $A\mathbf{x} \in S$ whenever $\mathbf{x} \in S$).

19. Let A be an $n \times n$ matrix and let $\| \cdot \|$ be a matrix norm that is compatible with some vector norm on R^n. If λ is an eigenvalue of A, show that $|\lambda| \le \|A\|$.

20. Let A be an $n \times n$ matrix and let λ be a nonzero eigenvalue of A. Show that if $\mathbf{x}$ is an eigenvector belonging to λ, then $\mathbf{x}$ is in the column space of A. Hence the eigenspace corresponding to λ is a subspace of the column space of A.

21. Let $\{\mathbf{u}_1, \mathbf{u}_2, \ldots, \mathbf{u}_n\}$ be an orthonormal basis for R^n and let $\lambda_1, \ldots, \lambda_n$ be scalars. Define

$$A = \lambda_1 \mathbf{u}_1 \mathbf{u}_1^T + \lambda_2 \mathbf{u}_2 \mathbf{u}_2^T + \cdots + \lambda_n \mathbf{u}_n \mathbf{u}_n^T$$

Show that A is a symmetric matrix with eigenvalues $\lambda_1, \lambda_2, \ldots, \lambda_n$ and that $\mathbf{u}_i$ is an eigenvector belonging to λ_i for each i.

22. Let A be a matrix whose columns all add up to a fixed constant δ. Show that δ is an eigenvalue of A.

23. Let λ_1 and λ_2 be distinct real eigenvalues of A. Let $\mathbf{x}$ be an eigenvector of A belonging to λ_1 and let $\mathbf{y}$ be an eigenvector of A^T belonging to λ_2. Show that $\mathbf{x}$ and $\mathbf{y}$ are orthogonal.

24. Let A and B be $n \times n$ matrices. Show that AB and BA have the same eigenvalues.

25. Let $p(\lambda) = (-1)^n (\lambda^n - a_{n-1}\lambda^{n-1} - \cdots - a_1\lambda - a_0)$ be a polynomial of degree $n \ge 1$ and let

$$C = \begin{pmatrix} a_{n-1} & a_{n-2} & \cdots & a_1 & a_0 \\ 1 & 0 & & 0 & 0 \\ 0 & 1 & & 0 & 0 \\ \vdots & & & & \\ 0 & 0 & & 1 & 0 \end{pmatrix}$$

Show that $p(\lambda)$ is the characteristic polynomial of C. The matrix C is called the *companion matrix* of $p(\lambda)$.

2. SYSTEMS OF LINEAR DIFFERENTIAL EQUATIONS

Eigenvalues play an important role in the solution of systems of linear differential equations. In this section we see how they are used in the solution of systems of linear differential equations with constant coefficients. We begin by considering systems of first-order equations of the form

$$y_1' = a_{11}y_1 + a_{12}y_2 + \cdots + a_{1n}y_n$$
$$y_2' = a_{21}y_1 + a_{22}y_2 + \cdots + a_{2n}y_n$$
$$\vdots$$
$$y_n' = a_{n1}y_1 + a_{n2}y_2 + \cdots + a_{nn}y_n$$

where $y_i = f_i(t)$ is a function in $C^1[a, b]$ for each i. If we let $\mathbf{Y} = (y_1, y_2, \ldots, y_n)^T$ and $\mathbf{Y}' = (y_1', y_2', \ldots, y_n')^T$, the system can be written in the form

$$\mathbf{Y}' = A\mathbf{Y}$$

$\mathbf{Y}$ and $\mathbf{Y}'$ are both vector functions of t. Let us consider the simplest case first. When $n = 1$, the system is simply

$$y' = ay$$

Clearly, any function of the form

$$y(t) = ce^{at} \quad (c \text{ an arbitrary constant})$$

satisfies this equation. To generalize to the n-dimensional case, let us assume there is a solution of the form

$$\mathbf{Y} = \begin{pmatrix} x_1 e^{\lambda t} \\ x_2 e^{\lambda t} \\ \vdots \\ x_n e^{\lambda t} \end{pmatrix} = e^{\lambda t}\mathbf{x}$$

where $\mathbf{x} = (x_1, x_2, \ldots, x_n)^T$. It follows that

$$\mathbf{Y}' = \lambda e^{\lambda t}\mathbf{x} = \lambda \mathbf{Y}$$

If we choose λ to be an eigenvalue of A and $\mathbf{x}$ is an eigenvector belonging to λ, then

$$A\mathbf{Y} = e^{\lambda t}A\mathbf{x} = \lambda e^{\lambda t}\mathbf{x} = \lambda \mathbf{Y} = \mathbf{Y}'$$

and hence $\mathbf{Y}$ is a solution to the system. Thus, if λ is an eigenvalue of A and $\mathbf{x}$ is an eigenvector belonging to λ, then $e^{\lambda t}\mathbf{x}$ is a solution of the system $\mathbf{Y}' = A\mathbf{Y}$. This will be true whether λ is real or complex. Note that if $\mathbf{Y}_1$ and $\mathbf{Y}_2$ are both solutions to $\mathbf{Y}' = A\mathbf{Y}$, then $\alpha \mathbf{Y}_1 + \beta \mathbf{Y}_2$ is also a solution, since

$$
\begin{aligned}
(\alpha \mathbf{Y}_1 + \beta \mathbf{Y}_2)' &= \alpha \mathbf{Y}_1' + \beta \mathbf{Y}_2' \\
&= \alpha A\mathbf{Y}_1 + \beta A\mathbf{Y}_2 \\
&= A(\alpha \mathbf{Y}_1 + \beta \mathbf{Y}_2)
\end{aligned}
$$

It follows by induction that if $\mathbf{Y}_1, \ldots, \mathbf{Y}_n$ are solutions to $\mathbf{Y}' = A\mathbf{Y}$, then any linear combination $c_1 \mathbf{Y}_1 + \cdots + c_n \mathbf{Y}_n$ will also be a solution.

In general, the solutions to an $n \times n$ first-order system of the form

$$\mathbf{Y}' = A\mathbf{Y}$$

will form an n-dimensional subspace of the vector space of all continuous vector-valued functions. If in addition we require that $\mathbf{Y}(t)$ take on a prescribed value when $t = 0$, the problem will have a unique solution (see [27], p. 228). A problem of the form

$$\mathbf{Y}' = A\mathbf{Y} \quad \mathbf{Y}(0) = \mathbf{Y}_0$$

is called an *initial value problem*.

EXAMPLE 1. Solve the system

$$
\begin{aligned}
y_1' &= 3y_1 + 4y_2 \\
y_2' &= 3y_1 + 2y_2
\end{aligned}
$$

SOLUTION

$$
A = \begin{pmatrix} 3 & 4 \\ 3 & 2 \end{pmatrix}
$$

The eigenvalues of A are $\lambda_1 = 6$ and $\lambda_2 = -1$. Solving $(A - \lambda I)\mathbf{x} = \mathbf{0}$ with $\lambda = \lambda_1$ and $\lambda = \lambda_2$, we see that $\mathbf{x}_1 = (4, 3)^T$ is an eigenvector belonging to λ_1 and $\mathbf{x}_2 = (1, -1)^T$ is an eigenvector belonging to λ_2. Thus any vector function of the form

$$
\begin{aligned}
\mathbf{Y} &= c_1 e^{\lambda_1 t}\mathbf{x}_1 + c_2 e^{\lambda_2 t}\mathbf{x}_2 \\
&= \begin{pmatrix} 4c_1 e^{6t} + c_2 e^{-t} \\ 3c_1 e^{6t} - c_2 e^{-t} \end{pmatrix}
\end{aligned}
$$

is a solution to the system. $\qquad\qquad \square$

In Example 1 suppose that we require that $y_1 = 6$ and $y_2 = 1$ when $t = 0$. Thus

$$
\mathbf{Y}(0) = \begin{pmatrix} 4c_1 + c_2 \\ 3c_1 - c_2 \end{pmatrix} = \begin{pmatrix} 6 \\ 1 \end{pmatrix}
$$

and it follows that $c_1 = 1$ and $c_2 = 2$. Hence the solution to the initial value problem is given by

$$\mathbf{Y} = e^{6t}\mathbf{x}_1 + 2e^{-t}\mathbf{x}_2$$

$$= \begin{pmatrix} 4e^{6t} + 2e^{-t} \\ 3e^{6t} - 2e^{-t} \end{pmatrix}$$

COMPLEX EIGENVALUES

Let A be a real $n \times n$ matrix with a complex eigenvalue $\lambda = a + bi$ and let $\mathbf{x}$ be an eigenvector belonging to λ. The vector $\mathbf{x}$ can be split up into its real and imaginary parts.

$$\mathbf{x} = \begin{pmatrix} \operatorname{Re} x_1 + i \operatorname{Im} x_1 \\ \operatorname{Re} x_2 + i \operatorname{Im} x_2 \\ \vdots \\ \operatorname{Re} x_n + i \operatorname{Im} x_n \end{pmatrix} = \begin{pmatrix} \operatorname{Re} x_1 \\ \operatorname{Re} x_2 \\ \vdots \\ \operatorname{Re} x_n \end{pmatrix} + i \begin{pmatrix} \operatorname{Im} x_1 \\ \operatorname{Im} x_2 \\ \vdots \\ \operatorname{Im} x_n \end{pmatrix} = \operatorname{Re} \mathbf{x} + i \operatorname{Im} \mathbf{x}$$

$\bar{\lambda} = a - bi$ is also an eigenvalue of A and since

$$A\bar{\mathbf{x}} = \bar{A}\bar{\mathbf{x}} = \overline{A\mathbf{x}} = \overline{\lambda\mathbf{x}} = \bar{\lambda}\bar{\mathbf{x}}$$

it follows that $\bar{\mathbf{x}} = (\operatorname{Re} \mathbf{x} - i \operatorname{Im} \mathbf{x})$ is an eigenvector belonging to $\bar{\lambda}$. Thus $e^{\lambda t}\mathbf{x}$ and $e^{\bar{\lambda} t}\bar{\mathbf{x}}$ are both solutions to the first-order system $\mathbf{Y}' = A\mathbf{Y}$.

$$e^{\lambda t}\mathbf{x} = e^{(a+ib)t}\mathbf{x}$$
$$= e^{at}(\cos bt + i \sin bt)(\operatorname{Re} \mathbf{x} + i \operatorname{Im} \mathbf{x})$$
$$e^{\bar{\lambda} t}\bar{\mathbf{x}} = e^{(a-ib)t}\bar{\mathbf{x}}$$
$$= e^{at}(\cos bt - i \sin bt)(\operatorname{Re} \mathbf{x} - i \operatorname{Im} \mathbf{x})$$

Let

$$\mathbf{Y}_1 = \tfrac{1}{2}(e^{\lambda t}\mathbf{x} + e^{\bar{\lambda} t}\bar{\mathbf{x}}) = \operatorname{Re}(e^{\lambda t}\mathbf{x})$$

and

$$\mathbf{Y}_2 = \frac{1}{2i}(e^{\lambda t}\mathbf{x} - e^{\bar{\lambda} t}\bar{\mathbf{x}}) = \operatorname{Im}(e^{\lambda t}\mathbf{x})$$

The vector functions $\mathbf{Y}_1$ and $\mathbf{Y}_2$ are real-valued solutions to $\mathbf{Y}' = A\mathbf{Y}$.

$$\mathbf{Y}_1 = e^{at}[(\cos bt)\operatorname{Re} \mathbf{x} - (\sin bt)\operatorname{Im} \mathbf{x}]$$
$$\mathbf{Y}_2 = e^{at}[(\cos bt)\operatorname{Im} \mathbf{x} + (\sin bt)\operatorname{Re} \mathbf{x}]$$

EXAMPLE 2.   Solve the system

$$y_1' = \quad\ y_1 + \ y_2$$
$$y_2' = -2y_1 + 3y_2$$

SOLUTION.   Let

$$A = \begin{pmatrix} 1 & 1 \\ -2 & 3 \end{pmatrix}$$

The eigenvalues of A are $\lambda = 2 + i$ and $\bar{\lambda} = 2 - i$ with eigenvectors $\mathbf{x} = (1, 1 + i)^T$ and $\bar{\mathbf{x}} = (1, 1 - i)^T$, respectively.

$$e^{\lambda t}\mathbf{x} = \begin{pmatrix} e^{2t}(\cos t + i \sin t) \\ e^{2t}(\cos t + i \sin t)(1 + i) \end{pmatrix}$$

$$= \begin{pmatrix} e^{2t}\cos t + i e^{2t}\sin t \\ e^{2t}(\cos t - \sin t) + i e^{2t}(\cos t + \sin t) \end{pmatrix}$$

Let

$$\mathbf{Y}_1 = \mathrm{Re}(e^{\lambda t}\mathbf{x}) = \begin{pmatrix} e^{2t}\cos t \\ e^{2t}(\cos t - \sin t) \end{pmatrix}$$

and

$$\mathbf{Y}_2 = \mathrm{Im}(e^{\lambda t}\mathbf{x}) = \begin{pmatrix} e^{2t}\sin t \\ e^{2t}(\cos t + \sin t) \end{pmatrix}$$

Any linear combination

$$\mathbf{Y} = c_1\mathbf{Y}_1 + c_2\mathbf{Y}_2$$

will be a solution to the system. $\square$

If the $n \times n$ coefficient matrix A of the system $\mathbf{Y}' = A\mathbf{Y}$ has n linearly independent eigenvectors, the general solution can be obtained by the methods that have been presented. The case when A has less than n linearly independent eigenvectors is more complicated and consequently will not be covered in this book.

HIGHER-ORDER SYSTEMS

Given a second-order system of the form

$$\mathbf{Y}'' = A_1\mathbf{Y} + A_2\mathbf{Y}'$$

we may translate it into a first-order system by setting

$$y_{n+1}(t) = y_1'(t)$$

$$y_{n+2}(t) = y_2'(t)$$

$$\vdots$$

$$y_{2n}(t) = y_n'(t)$$

If we let

$$\mathbf{Y}_1 = \mathbf{Y} = (y_1, y_2, \ldots, y_n)^T$$

and

$$\mathbf{Y}_2 = \mathbf{Y}' = (y_{n+1}, \ldots, y_{2n})^T$$

then

$$\mathbf{Y}_1' = O\mathbf{Y}_1 + I\mathbf{Y}_2$$

and
$$\mathbf{Y}_2' = A_1\mathbf{Y}_1 + A_2\mathbf{Y}_2$$

The equations can be combined to give the $2n \times 2n$ first-order system

$$\begin{pmatrix} \mathbf{Y}_1' \\ \mathbf{Y}_2' \end{pmatrix} = \begin{pmatrix} O & I \\ A_1 & A_2 \end{pmatrix} \begin{pmatrix} \mathbf{Y}_1 \\ \mathbf{Y}_2 \end{pmatrix}$$

If the values of $\mathbf{Y}_1 = \mathbf{Y}$ and $\mathbf{Y}_2 = \mathbf{Y}'$ are specified when $t = 0$, then the initial value problem will have a unique solution.

EXAMPLE 3. Solve the initial value problem
$$
\begin{aligned}
y_1'' &= 2y_1 + y_2 + y_1' + y_2' \\
y_2'' &= -5y_1 + 2y_2 + 5y_1' - y_2' \\
y_1(0) &= y_2(0) = y_1'(0) = 4, \qquad y_2'(0) = -4
\end{aligned}
$$

SOLUTION. Set $y_3 = y_1'$ and $y_4 = y_2'$. This gives the first-order system

$$
\begin{aligned}
y_1' &= y_3 \\
y_2' &= y_4 \\
y_3' &= 2y_1 + y_2 + y_3 + y_4 \\
y_4' &= -5y_1 + 2y_2 + 5y_3 - y_4
\end{aligned}
$$

The coefficient matrix for this system,

$$A = \begin{pmatrix} 0 & 0 & 1 & 0 \\ 0 & 0 & 0 & 1 \\ 2 & 1 & 1 & 1 \\ -5 & 2 & 5 & -1 \end{pmatrix}$$

has eigenvalues
$$\lambda_1 = 1, \qquad \lambda_2 = -1, \qquad \lambda_3 = 3, \qquad \lambda_4 = -3$$

Corresponding to these eigenvalues are the eigenvectors
$$\mathbf{x}_1 = (1, -1, 1, -1)^T, \qquad \mathbf{x}_2 = (1, 5, -1, -5)^T,$$
$$\mathbf{x}_3 = (1, 1, 3, 3)^T, \qquad \mathbf{x}_4 = (1, -5, -3, 15)^T$$

Thus the solution will be of the form
$$c_1\mathbf{x}_1 e^t + c_2\mathbf{x}_2 e^{-t} + c_3\mathbf{x}_3 e^{3t} + c_4\mathbf{x}_4 e^{-3t}$$

We can use the initial conditions to find c_1, c_2, c_3, and c_4. For $t = 0$, we have
$$c_1\mathbf{x}_1 + c_2\mathbf{x}_2 + c_3\mathbf{x}_3 + c_4\mathbf{x}_4 = (4, 4, 4, -4)^T$$

or, equivalently,

$$\begin{pmatrix} 1 & 1 & 1 & 1 \\ -1 & 5 & 1 & -5 \\ 1 & -1 & 3 & -3 \\ -1 & -5 & 3 & 15 \end{pmatrix} \begin{pmatrix} c_1 \\ c_2 \\ c_3 \\ c_4 \end{pmatrix} = \begin{pmatrix} 4 \\ 4 \\ 4 \\ -4 \end{pmatrix}$$

The solution to this system is $\mathbf{c} = (2, 1, 1, 0)^T$ and hence the solution to the initial value problem is

$$\mathbf{Y} = 2\mathbf{x}_1 e^t + \mathbf{x}_2 e^{-t} + \mathbf{x}_3 e^{3t}$$

Therefore,

$$\begin{pmatrix} y_1 \\ y_2 \\ y_1' \\ y_2' \end{pmatrix} = \begin{pmatrix} 2e^t + e^{-t} + e^{3t} \\ -2e^t + 5e^{-t} + e^{3t} \\ 2e^t - e^{-t} + 3e^{3t} \\ -2e^t - 5e^{-t} + 3e^{3t} \end{pmatrix}$$

$\square$

In general, if we have an mth-order system of the form

$$\mathbf{Y}^{(m)} = A_1\mathbf{Y} + A_2\mathbf{Y}' + \cdots + A_m\mathbf{Y}^{(m-1)}$$

where each A_i is an $n \times n$ matrix, we can transform it into a first-order system by setting

$$\mathbf{Y}_1 = \mathbf{Y}, \mathbf{Y}_2 = \mathbf{Y}_1', \ldots, \mathbf{Y}_m = \mathbf{Y}_{m-1}'$$

We will end up with a system of the form

$$\begin{pmatrix} \mathbf{Y}_1' \\ \mathbf{Y}_2' \\ \vdots \\ \mathbf{Y}_{m-1}' \\ \mathbf{Y}_m' \end{pmatrix} = \begin{pmatrix} O & I & O & \cdots & O \\ O & O & I & \cdots & O \\ \vdots & & & & \\ O & O & O & \cdots & I \\ A_1 & A_2 & A_3 & \cdots & A_m \end{pmatrix} \begin{pmatrix} \mathbf{Y}_1 \\ \mathbf{Y}_2 \\ \vdots \\ \mathbf{Y}_{m-1} \\ \mathbf{Y}_m \end{pmatrix}$$

If in addition we require that $\mathbf{Y}, \mathbf{Y}', \ldots, \mathbf{Y}^{(m-1)}$ take on specified values when $t = 0$, there will be exactly one solution to the problem.

If the system is simply of the form $\mathbf{Y}^{(m)} = A\mathbf{Y}$, it is usually not necessary to introduce new variables. In this case one need only calculate the mth roots of the eigenvalues of A. If λ is an eigenvalue of A, $\mathbf{x}$ is an eigenvector belonging to λ, σ is an mth root of λ, and $\mathbf{Y} = e^{\sigma t}\mathbf{x}$, then

$$\mathbf{Y}^{(m)} = \sigma^m e^{\sigma t}\mathbf{x} = \lambda\mathbf{Y}$$

and

$$A\mathbf{Y} = e^{\sigma t}A\mathbf{x} = \lambda e^{\sigma t}\mathbf{x} = \lambda\mathbf{Y}$$

Therefore, $\mathbf{Y} = e^{\sigma t}\mathbf{x}$ is a solution to the system.

APPLICATION

In Figure 6.2.1 two masses are adjoined by springs and the ends A and B are fixed. The masses are free to move horizontally. We will assume that the three springs are uniform and that initially the system is in the equilibrium position. A force is exerted on the system to set the masses in motion. The horizontal displacements of the masses at time t will be denoted by $x_1(t)$ and $x_2(t)$, respectively. We will assume that there are no retarding forces such as friction. Then the only forces acting on mass m_1 at time t will be from springs 1 and 2. The force from spring 1 will

FIGURE 6.2.1

be $-kx_1$ and the force from spring 2 will be $k(x_2 - x_1)$. By Newton's second law,

$$m_1 x_1''(t) = -kx_1 + k(x_2 - x_1)$$

Similarly, the only forces acting on the second mass will be from springs 2 and 3. Using Newton's second law again, we get

$$m_2 x_2''(t) = -k(x_2 - x_1) - kx_2$$

Thus we end up with the second-order system

$$x_1'' = -\frac{k}{m_1}(2x_1 - x_2)$$

$$x_2'' = -\frac{k}{m_2}(-x_1 + 2x_2)$$

Suppose now that $m_1 = m_2 = 1$, $k = 1$, and that the initial velocity of both masses is $+2$ units per second. To determine the displacements x_1 and x_2 as functions of t, we write the system in the form

(1) $$\mathbf{X}'' = A\mathbf{X}$$

The coefficient matrix

$$A = \begin{pmatrix} -2 & 1 \\ 1 & -2 \end{pmatrix}$$

has eigenvalues $\lambda_1 = -1$ and $\lambda_2 = -3$. Corresponding to λ_1, we have the eigenvector $\mathbf{v}_1 = (1, 1)^T$ and $\sigma_1 = \pm i$. Thus $e^{it}\mathbf{v}_1$ and $e^{-it}\mathbf{v}_1$ are both solutions to (1). It follows that

$$\tfrac{1}{2}(e^{it} + e^{-it})\mathbf{v}_1 = (\mathrm{Re}\, e^{it})\mathbf{v}_1 = (\cos t)\mathbf{v}_1$$

and

$$\frac{1}{2i}(e^{it} - e^{-it})\mathbf{v}_1 = (\mathrm{Im}\, e^{it})\mathbf{v}_1 = (\sin t)\mathbf{v}_1$$

are both solutions to (1). Similarly, for $\lambda_2 = -3$, we have the eigenvector $\mathbf{v}_2 = (1, -1)^T$ and $\sigma_2 = \pm \sqrt{3}\, i$. It follows that

$$(\mathrm{Re}\, e^{\sqrt{3}\, it})\mathbf{v}_2 = (\cos \sqrt{3}\, t)\mathbf{v}_2$$

and

$$(\mathrm{Im}\, e^{\sqrt{3}\, it})\mathbf{v}_2 = (\sin \sqrt{3}\, t)\mathbf{v}_2$$

are also solutions to (1). Thus the general solution will be of the form

$$\mathbf{X}(t) = c_1(\cos t)\mathbf{v}_1 + c_2(\sin t)\mathbf{v}_1 + c_3(\cos \sqrt{3}\, t)\mathbf{v}_2 + c_4(\sin \sqrt{3}\, t)\mathbf{v}_2$$

$$= \begin{pmatrix} c_1\cos t + c_2\sin t + c_3\cos \sqrt{3}\, t + c_4\sin \sqrt{3}\, t \\ c_1\cos t + c_2\sin t - c_3\cos \sqrt{3}\, t - c_4\sin \sqrt{3}\, t \end{pmatrix}$$

At time $t = 0$, we have

$$x_1(0) = x_2(0) = 0 \quad \text{and} \quad x_1'(0) = x_2'(0) = 2$$

It follows that

$$c_1 + c_3 = 0 \quad \text{and} \quad c_2 + \sqrt{3}\, c_4 = 2$$
$$c_1 - c_3 = 0 \qquad\qquad c_2 - \sqrt{3}\, c_4 = 2$$

and hence

$$c_1 = c_3 = c_4 = 0 \quad \text{and} \quad c_2 = 2$$

Therefore, the solution to the initial value problem is simply

$$\mathbf{X}(t) = \begin{pmatrix} 2 \sin t \\ 2 \sin t \end{pmatrix}$$

The masses will oscillate with frequency 1 and amplitude 2.

EXERCISES

1. Find the general solution to each of the following systems.
   (a) $y_1' = \quad y_1 + y_2$

   $\quad y_2' = -2y_1 + 4y_2$

   (b) $y_1' = 2y_1 + 4y_2$

   $\quad y_2' = -y_1 - 3y_2$

   (c) $y_1' = \quad y_1 - 2y_2$

   $\quad y_2' = -2y_1 + 4y_2$

   (d) $y_1' = y_1 - y_2$

   $\quad y_2' = y_1 + y_2$

   (e) $y_1' = 3y_1 - 2y_2$

   $\quad y_2' = 2y_1 + 3y_2$

   (f) $y_1' = \quad y_1 + \quad y_3$

   $\quad y_2' = 2y_2 + 6y_3$

   $\quad y_3' = \quad y_2 + 3y_3$

2. Solve each of the following initial value problems.
   (a) $y_1' = -y_1 + 2y_2 \qquad y_1(0) = 3,\ y_2(0) = 1$

   $\quad y_2' = 2y_1 - \quad y_2$

   (b) $y_1' = \quad y_1 - 2y_2 \qquad y_1(0) = 1,\ y_2(0) = -2$

   $\quad y_2' = 2y_1 + \quad y_2$

   (c) $y_1' = 2y_1 - 6y_3 \qquad y_1(0) = y_2(0) - y_3(0) = 2$

   $\quad y_2' = \quad y_1 - 3y_3$

   $\quad y_3' = \quad y_2 - 2y_3$

   (d) $y_1' = \quad y_1 + 2y_3 \qquad y_1(0) = y_2(0) = 1,\ y_3(0) = 4$

   $\quad y_2' = \quad y_2 - \quad y_3$

   $\quad y_3' = \quad y_1 + \quad y_2 + y_3$

3. Given

$$\mathbf{Y} = c_1 e^{\lambda_1 t}\mathbf{x}_1 + c_2 e^{\lambda_2 t}\mathbf{x}_2 + \cdots + c_n e^{\lambda_n t}\mathbf{x}_n$$

is the solution to the initial value problem

$$\mathbf{Y}' = A\mathbf{Y}, \qquad \mathbf{Y}(0) = \mathbf{Y}_0$$

(a) Show that

$$\mathbf{Y}_0 = c_1\mathbf{x}_1 + c_2\mathbf{x}_2 + \cdots + c_n\mathbf{x}_n$$

(b) Let $S = (\mathbf{x}_1, \ldots, \mathbf{x}_n)$ and $\mathbf{c} = (c_1, \ldots, c_n)^T$. Assuming that $\mathbf{x}_1, \ldots, \mathbf{x}_n$ are linearly independent, show $\mathbf{c} = S^{-1}\mathbf{Y}_0$.

4. Find the general solution to each of the following
   (a) $y_1'' = -2y_2$ (b) $y_1'' = y_1 + y_3$
   $\quad\ y_2'' = y_1 + 3y_2$ $\quad\ y_2'' = y_2 + 2y_3$
   $\qquad\qquad\qquad\qquad\quad\ y_3'' = y_1 + y_2 + 3y_3$

5. Solve the initial value problem

$$\begin{aligned} y_1'' &= -2y_2 + \ y_1' + 2y_2' \\ y_2'' &= \quad 2y_1 + 2y_1' - \ y_2' \end{aligned}$$

$$y_1(0) = 1, \qquad y_2(0) = 0, \qquad y_1'(0) = -3, \qquad y_2'(0) = 2$$

6. In the application problem, assume that the solutions are of the form $x_1 = a_1 \sin \sigma t$, $x_2 = a_2 \sin \sigma t$. Substitute these expressions into the system and solve for the frequency σ and the amplitudes a_1 and a_2.

7. Solve the application problem with the initial conditions

$$x_1(0) = x_2(0) = 1, \qquad x_1'(0) = 4 \qquad \text{and} \qquad x_2'(0) = 2$$

8. Two masses are connected by springs as shown in the diagram. Both springs have the same spring constant and the end of the first spring is fixed. If y_1 and y_2 represent the displacements from the equilibrium position, derive a system of second-order linear differential equations that describes the motion of the system.

9. Three masses are connected by a series of springs between two fixed points as shown in the figure. Assume that the springs all have the same spring constant and let $x_1(t)$, $x_2(t)$, and $x_3(t)$ represent the displacements of the respective masses at time t.
   (a) Derive a system of second-order linear differential equations which describes the motion of this system.
   (b) Solve the system if $m_1 = m_3 = \frac{1}{3}$, $m_2 = \frac{1}{4}$, $k = 1$, and

$$x_1(0) = x_2(0) = x_3(0) = 1$$
$$x_1'(0) = x_2'(0) = x_3'(0) = 0$$

10. Transform the nth-order equation

$$y^{(n)} = a_0 y + a_1 y' + \cdots + a_{n-1} y^{(n-1)}$$

into a system of first-order equations by setting $y_1 = y$ and $y_j = y_{j-1}'$ for $j = 2, \ldots, n$. Determine the characteristic polynomial of the coefficient matrix of this system.

3. DIAGONALIZATION

In this section we consider the problem of factoring an $n \times n$ matrix A into a product of the form SDS^{-1}, where D is diagonal. We will give a necessary and sufficient condition for the existence of such a factorization and look at a number of examples. We begin by showing that eigenvectors belonging to distinct eigenvalues are linearly independent.

Theorem 6.3.1. *If $\lambda_1, \lambda_2, \ldots, \lambda_k$ are distinct eigenvalues of the $n \times n$ matrix A with corresponding eigenvectors $\mathbf{x}_1, \mathbf{x}_2, \ldots, \mathbf{x}_k$, then $\mathbf{x}_1, \ldots, \mathbf{x}_k$ are linearly independent.*

Proof. Let r be the dimension of the subspace of R^n spanned by $\mathbf{x}_1, \ldots, \mathbf{x}_k$ and suppose that $r < k$. We may assume (reordering the $\mathbf{x}_i$'s and λ_i's if necessary) that $\mathbf{x}_1, \ldots, \mathbf{x}_r$ are linearly independent. Since $\mathbf{x}_1, \mathbf{x}_2, \ldots, \mathbf{x}_r, \mathbf{x}_{r+1}$ are linearly dependent, there exist scalars $c_1, \ldots, c_r, c_{r+1}$ not all zero such that

(1) $$c_1 \mathbf{x}_1 + \cdots + c_r \mathbf{x}_r + c_{r+1} \mathbf{x}_{r+1} = \mathbf{0}$$

Note that c_{r+1} must be nonzero; otherwise, $\mathbf{x}_1, \ldots, \mathbf{x}_r$ would be dependent. So $c_{r+1}\mathbf{x}_{r+1} \neq \mathbf{0}$ and hence $c_1, \ldots, c_r$ cannot all be zero. Multiplying (1) by A, we get

$$c_1 A\mathbf{x}_1 + \cdots + c_r A\mathbf{x}_r + c_{r+1} A\mathbf{x}_{r+1} = \mathbf{0}$$

or

(2) $$c_1 \lambda_1 \mathbf{x}_1 + \cdots + c_r \lambda_r \mathbf{x}_r + c_{r+1} \lambda_{r+1} \mathbf{x}_{r+1} = \mathbf{0}$$

Subtracting λ_{r+1} times (1) from (2) gives

$$c_1(\lambda_1 - \lambda_{r+1})\mathbf{x}_1 + \cdots + c_r(\lambda_r - \lambda_{r+1})\mathbf{x}_r = \mathbf{0}$$

This contradicts the independence of $\mathbf{x}_1, \ldots, \mathbf{x}_r$. Therefore, r must equal k. $\square$

Definition.   An $n \times n$ matrix A is said to be **diagonalizable** if there exists a nonsingular matrix S and a diagonal matrix D such that

$$S^{-1}AS = D$$

We say that S *diagonalizes* A.

Theorem 6.3.2.   *An $n \times n$ matrix A is diagonalizable if and only if A has n linearly independent eigenvectors.*

Proof.   Suppose that A has n linearly independent eigenvectors $\mathbf{x}_1, \mathbf{x}_2, \ldots, \mathbf{x}_n$. Let λ_i be the eigenvalue of A corresponding to $\mathbf{x}_i$ for each i. (Some of the λ_i's may be equal.) Let S be the matrix whose jth column vector is $\mathbf{x}_j$ for $j = 1, \ldots, n$. It follows that $A\mathbf{x}_j = \lambda_j \mathbf{x}_j$ is the jth column vector of AS. Thus

$$\begin{aligned}
AS &= (A\mathbf{x}_1, A\mathbf{x}_2, \ldots, A\mathbf{x}_n) \\
&= (\lambda_1 \mathbf{x}_1, \lambda_2 \mathbf{x}_2, \ldots, \lambda_n \mathbf{x}_n) \\
&= (\mathbf{x}_1, \mathbf{x}_2, \ldots, \mathbf{x}_n) \begin{pmatrix} \lambda_1 & & & \\ & \lambda_2 & & \\ & & \ddots & \\ & & & \lambda_n \end{pmatrix} \\
&= SD
\end{aligned}$$

Since S has n linearly independent column vectors, it follows that S is nonsingular and hence

$$D = S^{-1}SD = S^{-1}AS$$

Conversely, suppose that A is diagonalizable. Then there exists a nonsingular matrix S such that $AS = SD$. If $\mathbf{x}_1, \mathbf{x}_2, \ldots, \mathbf{x}_n$ are the column vectors of S, then

$$A\mathbf{x}_j = \lambda_j \mathbf{x}_j \quad (\lambda_j = d_{jj})$$

for each j. Thus for each j, λ_j is an eigenvalue of A and $\mathbf{x}_j$ is an eigenvector belonging to λ_j. Since the column vectors of S are linearly independent, it follows that A has n linearly independent eigenvectors.

$\square$

Remarks

1. If A is diagonalizable, then the column vectors of the diagonalizing matrix S are eigenvectors of A and the diagonal elements of D are the corresponding eigenvalues of A.
2. If A has n distinct eigenvalues, then A is diagonalizable.
3. The diagonalizing matrix S is not unique. Reordering the columns of a given diagonalizing matrix S or multiplying them by nonzero scalars will produce a new diagonalizing matrix.
4. If A is diagonalizable, then A can be factored into a product SDS^{-1}.

It follows from remark 4 that

$$A^2 = (SDS^{-1})(SDS^{-1}) = SD^2S^{-1}$$

and in general

$$A^k = SD^kS^{-1}$$

$$= S \begin{pmatrix} (\lambda_1)^k & & & \\ & (\lambda_2)^k & & \\ & & \ddots & \\ & & & (\lambda_n)^k \end{pmatrix} S^{-1}$$

Once we have a factorization $A = SDS^{-1}$, it is then easy to compute powers of A.

EXAMPLE 1. Let

$$A = \begin{pmatrix} 2 & -3 \\ 2 & -5 \end{pmatrix}$$

The eigenvalues of A are $\lambda_1 = 1$ and $\lambda_2 = -4$. Corresponding to λ_1 and λ_2, we have the eigenvectors $\mathbf{x}_1 = (3, 1)^T$ and $\mathbf{x}_2 = (1, 2)^T$. Let

$$S = \begin{pmatrix} 3 & 1 \\ 1 & 2 \end{pmatrix}$$

It follows that

$$S^{-1}AS = \tfrac{1}{5}\begin{pmatrix} 2 & -1 \\ -1 & 3 \end{pmatrix}\begin{pmatrix} 2 & -3 \\ 2 & -5 \end{pmatrix}\begin{pmatrix} 3 & 1 \\ 1 & 2 \end{pmatrix}$$

$$= \begin{pmatrix} 1 & 0 \\ 0 & -4 \end{pmatrix}$$

and

$$SDS^{-1} = \begin{pmatrix} 3 & 1 \\ 1 & 2 \end{pmatrix}\begin{pmatrix} 1 & 0 \\ 0 & -4 \end{pmatrix}\begin{pmatrix} \frac{2}{5} & -\frac{1}{5} \\ -\frac{1}{5} & \frac{3}{5} \end{pmatrix} = \begin{pmatrix} 2 & -3 \\ 2 & -5 \end{pmatrix} = A$$

$\square$

EXAMPLE 2. Let

$$A = \begin{pmatrix} 3 & -1 & -2 \\ 2 & 0 & -2 \\ 2 & -1 & -1 \end{pmatrix}$$

We saw in Example 3 of Section 1 that the eigenvalues of A are $\lambda_1 = 0$, $\lambda_2 = 1$, $\lambda_3 = 1$. Corresponding to $\lambda_1 = 0$, we have the eigenvector $(1,1,1)^T$, and corresponding to $\lambda = 1$, we have the eigenvectors $(1,2,0)^T$ and $(0,-2,1)^T$. Let

$$S = \begin{pmatrix} 1 & 1 & 0 \\ 1 & 2 & -2 \\ 1 & 0 & 1 \end{pmatrix}$$

It follows that

$$S^{-1}AS = \begin{pmatrix} -2 & 1 & 2 \\ 3 & -1 & -2 \\ 2 & -1 & -1 \end{pmatrix}\begin{pmatrix} 3 & -1 & -2 \\ 2 & 0 & -2 \\ 2 & -1 & -1 \end{pmatrix}\begin{pmatrix} 1 & 1 & 0 \\ 1 & 2 & -2 \\ 1 & 0 & 1 \end{pmatrix}$$

$$= \begin{pmatrix} 0 & 0 & 0 \\ 0 & 1 & 0 \\ 0 & 0 & 1 \end{pmatrix}$$

Even though $\lambda = 1$ is a multiple eigenvalue, the matrix can still be diagonalized, since there are three linearly independent eigenvectors. Note also that

$$A^k = SD^kS^{-1} = SDS^{-1} = A$$

for any $k \geq 1$.

$\square$

If an $n \times n$ matrix A has fewer than n linearly independent eigenvectors, we say that A is *defective*. It follows from Theorem 6.3.2 that a defective matrix is not diagonalizable.

EXAMPLE 3. Let

$$A = \begin{pmatrix} 1 & 1 \\ 0 & 1 \end{pmatrix}$$

The eigenvalues of A are both equal to 1. Any eigenvector corresponding to $\lambda = 1$ must be a multiple of $\mathbf{x}_1 = (1,0)^T$. Thus A is defective and cannot be diagonalized.

$\square$

EXAMPLE 4. Let

$$A = \begin{pmatrix} 2 & 0 & 0 \\ 0 & 4 & 0 \\ 1 & 0 & 2 \end{pmatrix} \quad \text{and} \quad B = \begin{pmatrix} 2 & 0 & 0 \\ -1 & 4 & 0 \\ -3 & 6 & 2 \end{pmatrix}$$

A and B both have the same eigenvalues

$$\lambda_1 = 4, \lambda_2 = \lambda_3 = 2$$

The eigenspace of A corresponding to $\lambda_1 = 4$ is spanned by $\mathbf{e}_2$ and the eigenspace corresponding to $\lambda = 2$ is spanned by $\mathbf{e}_3$. Since A only has two linearly independent eigenvectors, it is defective. On the other hand, the matrix B has eigenvector $\mathbf{x}_1 = (0, 1, 3)^T$ corresponding to $\lambda_1 = 4$ and eigenvectors $\mathbf{x}_2 = (2, 1, 0)^T$ and $\mathbf{e}_3$ corresponding to $\lambda = 2$. Thus B has three linearly independent eigenvectors and consequently is not defective. Even though $\lambda = 2$ is an eigenvalue of multiplicity 2, the matrix B is nondefective, since the corresponding eigenspace has dimension 2. Geometrically,

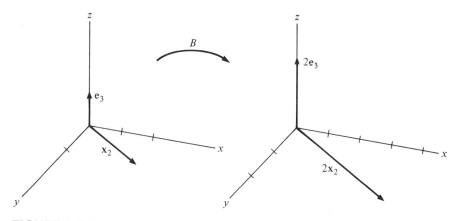

FIGURE 6.3.1

the matrix B has the effect of stretching two linearly independent vectors by a factor of 2. We can think of the eigenvalue $\lambda = 2$ as having "geometric multiplicity" 2. On the other hand, the matrix A only stretches vectors along the z-axis by a factor of 2. In this case the eigenvalue $\lambda = 2$ has algebraic multiplicity 2, but its geometric multiplicity is only 1 (see Figure 6.3.1 on page 251).   □

APPLICATION 1

Recall Application 3 from Section 3 of the first chapter. In a certain town 30 percent of the married women get divorced each year and 20 percent of the single women get married each year. There are 8000 married women and 2000 single women and the total population remains constant. Find the number of married women and single women after 5 years. What will be the long-range prospects if these percentages of marriages and divorces continue indefinitely into the future?

SOLUTION.   To find the number of married and single women after 1 year, we multiplied the vector $\mathbf{x} = (8000, 2000)^T$ by

$$A = \begin{pmatrix} 0.7 & 0.2 \\ 0.3 & 0.8 \end{pmatrix}$$

The number of married and single women after 5 years will be $A^5\mathbf{x}$. To compute $A^5\mathbf{x}$, we will factor A into a product SDS^{-1}. The eigenvalues of A are $\lambda_1 = 1$ and $\lambda_2 = \frac{1}{2}$. $\mathbf{x}_1 = (2, 3)^T$ is an eigenvector belonging to λ_1 and $\mathbf{x}_2 = (1, -1)^T$ is an eigenvector belonging to λ_2.

Let

$$S = \begin{pmatrix} 2 & 1 \\ 3 & -1 \end{pmatrix}$$

It follows that

$$A = SDS^{-1} = \tfrac{1}{5}\begin{pmatrix} 2 & 1 \\ 3 & -1 \end{pmatrix}\begin{pmatrix} 1 & 0 \\ 0 & \frac{1}{2} \end{pmatrix}\begin{pmatrix} 1 & 1 \\ 3 & -2 \end{pmatrix}$$

Thus

$$A^5\mathbf{x} = \tfrac{1}{5}\begin{pmatrix} 2 & 1 \\ 3 & -1 \end{pmatrix}\begin{pmatrix} 1 & 0 \\ 0 & \frac{1}{2} \end{pmatrix}^5\begin{pmatrix} 1 & 1 \\ 3 & -2 \end{pmatrix}\begin{pmatrix} 8000 \\ 2000 \end{pmatrix}$$

$$= \begin{pmatrix} 4125 \\ 5875 \end{pmatrix}$$

After 5 years there will be 4125 married women and 5875 single women. After n years the number of married and single women is given by $A^n\mathbf{x}$. To find the long-range trend, we take the limit as n approaches infinity.

$$\lim_{n\to\infty} A^n\mathbf{x} = \lim_{n\to\infty} SD^nS^{-1}\mathbf{x}$$

As $n \to \infty$, $D^n \to \begin{pmatrix} 1 & 0 \\ 0 & 0 \end{pmatrix}$. Thus

$$\lim_{n\to\infty} A^n\mathbf{x} = \tfrac{1}{5}\begin{pmatrix} 2 & 1 \\ 3 & -1 \end{pmatrix}\begin{pmatrix} 1 & 0 \\ 0 & 0 \end{pmatrix}\begin{pmatrix} 1 & 1 \\ 3 & -2 \end{pmatrix}\begin{pmatrix} 8000 \\ 2000 \end{pmatrix} = \begin{pmatrix} 4000 \\ 6000 \end{pmatrix}$$

In the long run, 40 percent of the women will be married and 60 percent will be single. ☐

APPLICATION 2: SEX-LINKED GENES

Sex-linked genes are genes that are located on the X chromosome. For example, the gene for blue-green color blindness is a recessive sex-linked gene. To devise a mathematical model to describe color blindness in a given population, it is necessary to divide the population into two classes, males and females. Let $x_1^{(0)}$ be the proportion of genes for color blindness in the male population and let $x_2^{(0)}$ be the proportion in the female population. [Since color blindness is recessive, the actual proportion of colorblind females will be less than $x_2^{(0)}$.] Since the male receives one X chromosome from the mother, the proportion $x_1^{(1)}$ of colorblind males in the next generation will be the same as the proportion of recessive genes in the present generation of females. Since the female receives an X chromosome from each parent, the proportion $x_2^{(1)}$ of recessive genes in the next generation of females will be the average of $x_1^{(0)}$ and $x_2^{(0)}$. Thus

$$x_2^{(0)} = x_1^{(1)}$$

$$\tfrac{1}{2}x_1^{(0)} + \tfrac{1}{2}x_2^{(0)} = x_2^{(1)}$$

If $x_1^{(0)} = x_2^{(0)}$, the proportions will not change in future generations. Let us assume that $x_1^{(0)} \neq x_2^{(0)}$ and write the system as a matrix equation.

$$\begin{pmatrix} 0 & 1 \\ \tfrac{1}{2} & \tfrac{1}{2} \end{pmatrix}\begin{pmatrix} x_1^{(0)} \\ x_2^{(0)} \end{pmatrix} = \begin{pmatrix} x_1^{(1)} \\ x_2^{(1)} \end{pmatrix}$$

Let A denote the coefficient matrix and let $\mathbf{x}^{(n)} = (x_1^{(n)}, x_2^{(n)})^T$ denote the proportion of colorblind genes in the male and female populations of the $(n + 1)$st generation. Thus

$$\mathbf{x}^{(n)} = A^n\mathbf{x}^{(0)}$$

To compute A^n, we note that A has eigenvalues 1 and $-\tfrac{1}{2}$ and consequently can be factored into a product

$$A = \begin{pmatrix} 1 & -2 \\ 1 & 1 \end{pmatrix} \begin{pmatrix} 1 & 0 \\ 0 & -\frac{1}{2} \end{pmatrix} \begin{pmatrix} \frac{1}{3} & \frac{2}{3} \\ -\frac{1}{3} & \frac{1}{3} \end{pmatrix}$$

Thus

$$\mathbf{x}^{(n)} = \begin{pmatrix} 1 & -2 \\ 1 & 1 \end{pmatrix} \begin{pmatrix} 1 & 0 \\ 0 & -\frac{1}{2} \end{pmatrix}^n \begin{pmatrix} \frac{1}{3} & \frac{2}{3} \\ -\frac{1}{3} & \frac{1}{3} \end{pmatrix} \begin{pmatrix} x_1^{(0)} \\ x_2^{(0)} \end{pmatrix}$$

$$= \frac{1}{3} \begin{pmatrix} 1 - \left(-\frac{1}{2}\right)^{n-1} & 2 + \left(-\frac{1}{2}\right)^{n-1} \\ 1 - \left(-\frac{1}{2}\right)^{n} & 2 + \left(-\frac{1}{2}\right)^{n} \end{pmatrix} \begin{pmatrix} x_1^{(0)} \\ x_2^{(0)} \end{pmatrix}$$

and hence

$$\lim_{n \to \infty} \mathbf{x}^{(n)} = \frac{1}{3} \begin{pmatrix} 1 & 2 \\ 1 & 2 \end{pmatrix} \begin{pmatrix} x_1^{(0)} \\ x_2^{(0)} \end{pmatrix}$$

$$= \begin{pmatrix} \dfrac{x_1^{(0)} + 2x_2^{(0)}}{3} \\ \dfrac{x_1^{(0)} + 2x_2^{(0)}}{3} \end{pmatrix}$$

The proportions of genes for color blindness in the male and female populations will tend to the same value as the number of generations increases. If the proportion of colorblind men is p, and over a number of generations no outsiders have entered the population, there is justification for assuming that the proportion of genes for color blindness in the female population is also p. Since color blindness is recessive, we would expect the proportion of colorblind women to be about p^2. Thus, if 1 percent of the male population is colorblind, we would expect about 0.01 percent of the female population to be colorblind.

THE EXPONENTIAL OF A MATRIX

Given a scalar a, the exponential e^a can be expressed in terms of a power series

$$e^a = 1 + a + \frac{1}{2!} a^2 + \frac{1}{3!} a^3 + \cdots$$

Similarly, for any $n \times n$ matrix A, one can define the matrix exponential e^A in terms of the convergent power series

$$(3) \qquad\qquad e^A = I + A + \frac{1}{2!} A^2 + \frac{1}{3!} A^3 + \cdots$$

The matrix exponential (3) occurs in a wide variety of applications. In the case of a diagonal matrix

$$D = \begin{pmatrix} \lambda_1 & & & \\ & \lambda_2 & & \\ & & \ddots & \\ & & & \lambda_n \end{pmatrix}$$

the matrix exponential is easy to compute:

$$e^D = \lim_{m \to \infty} \left(I + D + \frac{1}{2!}D^2 + \cdots + \frac{1}{m!}D^m \right)$$

$$= \lim_{m \to \infty} \begin{pmatrix} \sum_{k=1}^{m} \frac{1}{k!}\lambda_1^k & & \\ & \ddots & \\ & & \sum_{k=1}^{m} \frac{1}{k!}\lambda_n^k \end{pmatrix}$$

$$= \begin{pmatrix} e^{\lambda_1} & & \\ & e^{\lambda_2} & \\ & & \ddots & \\ & & & e^{\lambda_n} \end{pmatrix}$$

It is more difficult to compute the matrix exponential for a general $n \times n$ matrix A. If, however, A is diagonalizable, then

$$A^k = SD^kS^{-1} \qquad \text{for} \qquad k = 1, 2, \ldots$$

and hence

$$e^A = S\left(I + D + \frac{1}{2!}D^2 + \frac{1}{3!}D^3 + \cdots \right)S^{-1}$$

$$= Se^DS^{-1}$$

EXAMPLE 5. Compute e^A for

$$A = \begin{pmatrix} -2 & -6 \\ 1 & 3 \end{pmatrix}$$

SOLUTION. The eigenvalues of A are $\lambda_1 = 1$ and $\lambda_2 = 0$ with eigenvectors $\mathbf{x}_1 = (-2, 1)^T$ and $\mathbf{x}_2 = (-3, 1)^T$. Thus

$$A = SDS^{-1} = \begin{pmatrix} -2 & -3 \\ 1 & 1 \end{pmatrix}\begin{pmatrix} 1 & 0 \\ 0 & 0 \end{pmatrix}\begin{pmatrix} 1 & 3 \\ -1 & -2 \end{pmatrix}$$

and

$$e^A = Se^DS^{-1} = \begin{pmatrix} -2 & -3 \\ 1 & 1 \end{pmatrix}\begin{pmatrix} e^1 & 0 \\ 0 & e^0 \end{pmatrix}\begin{pmatrix} 1 & 3 \\ -1 & -2 \end{pmatrix}$$

$$= \begin{pmatrix} 3 - 2e & 6 - 6e \\ e - 1 & 3e - 2 \end{pmatrix} \qquad \square$$

The matrix exponential can be applied to the initial value problem

(4)
$$\mathbf{Y}' = A\mathbf{Y}, \qquad \mathbf{Y}(0) = \mathbf{Y}_0$$

studied in Section 2. In the case of one equation in one unknown

$$y' = ay, \qquad y(0) = y_0$$

the solution is

(5)
$$y = y_0 e^{at}$$

We can generalize this and express the solution to (4) in terms of the matrix exponential e^{At}, where $At = tA$ (i.e., t times the matrix A.) In general, a power series can be differentiated term by term within its radius of convergence. Since the expansion of e^{At} has infinite radius of convergence, we have

$$
\begin{aligned}
\frac{d}{dt} e^{At} &= \frac{d}{dt}\left(I + tA + \frac{1}{2!}t^2 A^2 + \frac{1}{3!}t^3 A^3 + \cdots \right) \\
&= \left(A + tA^2 + \frac{1}{2!}t^2 A^3 + \cdots \right) \\
&= A\left(I + tA + \frac{1}{2!}t^2 A + \cdots \right) \\
&= Ae^{At}
\end{aligned}
$$

If, as in (5), we set

$$\mathbf{Y}(t) = e^{At}\mathbf{Y}_0$$

then

$$\mathbf{Y}' = Ae^{At}\mathbf{Y}_0 = A\mathbf{Y}$$

and

$$\mathbf{Y}(0) = \mathbf{Y}_0$$

Thus the solution to

$$\mathbf{Y}' = A\mathbf{Y}, \qquad \mathbf{Y}(0) = \mathbf{Y}_0$$

is simply

(6)
$$\mathbf{Y} = e^{At}\mathbf{Y}_0$$

Although the form of this solution looks different from the solutions in Section 2, there is really no difference. In Section 2 the solution was expressed in the form

$$c_1 e^{\lambda_1 t}\mathbf{x}_1 + c_2 e^{\lambda_2 t}\mathbf{x}_2 + \cdots + c_n e^{\lambda_n t}\mathbf{x}_n$$

where $\mathbf{x}_i$ was an eigenvector belonging to λ_i for $i = 1, \ldots, n$. The c_i's that satisfied the initial conditions were determined by solving a system

$$S\mathbf{c} = \mathbf{Y}_0$$

with coefficient matrix $S = (\mathbf{x}_1, \ldots, \mathbf{x}_n)$.

If A is diagonalizable, we can write (6) in the form

$$\mathbf{Y} = Se^{Dt}S^{-1}\mathbf{Y}_0$$

Thus

$$\mathbf{Y} = Se^{Dt}\mathbf{c}$$

$$= (\mathbf{x}_1, \mathbf{x}_2, \ldots, \mathbf{x}_n)\begin{pmatrix} c_1 e^{\lambda_1 t} \\ c_2 e^{\lambda_2 t} \\ \vdots \\ c_n e^{\lambda_n t} \end{pmatrix}$$

$$= c_1 e^{\lambda_1 t}\mathbf{x}_1 + \cdots + c_n e^{\lambda_n t}\mathbf{x}_n$$

To summarize, the solution to the initial value problem (4) is given by

$$\mathbf{Y} = e^{At}\mathbf{Y}_0$$

If A is diagonalizable, this solution can be written in the form

$$\mathbf{Y} = Se^{Dt}S^{-1}\mathbf{Y}_0$$

$$= c_1 e^{\lambda_1 t}\mathbf{x}_1 + c_2 e^{\lambda_2 t}\mathbf{x}_2 + \cdots + c_n e^{\lambda_n t}\mathbf{x}_n \qquad \left(\mathbf{c} = S^{-1}\mathbf{Y}_0\right)$$

EXAMPLE 6. Use the matrix exponential to solve the initial value problem

$$\mathbf{Y}' = A\mathbf{Y}, \qquad \mathbf{Y}(0) = \mathbf{Y}_0$$

where

$$A = \begin{pmatrix} 3 & 4 \\ 3 & 2 \end{pmatrix}, \qquad \mathbf{Y}_0 = \begin{pmatrix} 6 \\ 1 \end{pmatrix}$$

(This problem was solved in Example 1 of Section 2.)

SOLUTION.   The eigenvalues of A are $\lambda_1 = 6$ and $\lambda_2 = -1$ with eigenvectors $\mathbf{x}_1 = (4, 3)^T$ and $\mathbf{x}_2 = (1, -1)^T$. Thus

$$A = SDS^{-1} = \begin{pmatrix} 4 & 1 \\ 3 & -1 \end{pmatrix}\begin{pmatrix} 6 & 0 \\ 0 & -1 \end{pmatrix}\begin{pmatrix} \frac{1}{7} & \frac{1}{7} \\ \frac{3}{7} & -\frac{4}{7} \end{pmatrix}$$

and the solution is given by

$$\mathbf{Y} = e^{At}\mathbf{Y}_0$$

$$= Se^{Dt}S^{-1}\mathbf{Y}_0$$

$$= \begin{pmatrix} 4 & 1 \\ 3 & -1 \end{pmatrix}\begin{pmatrix} e^{6t} & 0 \\ 0 & e^{-t} \end{pmatrix}\begin{pmatrix} \frac{1}{7} & \frac{1}{7} \\ \frac{3}{7} & -\frac{4}{7} \end{pmatrix}\begin{pmatrix} 6 \\ 1 \end{pmatrix}$$

$$= \begin{pmatrix} 4e^{6t} + 2e^{-t} \\ 3e^{6t} - 2e^{-t} \end{pmatrix}$$

Compare this to Example 1 in Section 2. $\square$

EXERCISES

1. In each of the following, factor the matrix A into a product SDS^{-1}, where D is diagonal.

(a) $A = \begin{pmatrix} 0 & 1 \\ 1 & 0 \end{pmatrix}$

(b) $A = \begin{pmatrix} 5 & 6 \\ -2 & -2 \end{pmatrix}$

(c) $A = \begin{pmatrix} 2 & -8 \\ 1 & -4 \end{pmatrix}$

(d) $A = \begin{pmatrix} 2 & 2 & 1 \\ 0 & 1 & 2 \\ 0 & 0 & -1 \end{pmatrix}$

(e) $A = \begin{pmatrix} 1 & 0 & 0 \\ -2 & 1 & 3 \\ 1 & 1 & -1 \end{pmatrix}$

(f) $A = \begin{pmatrix} 1 & 2 & -1 \\ 2 & 4 & -2 \\ 3 & 6 & -3 \end{pmatrix}$

2. For each of the matrices in Exercise 1, use the SDS^{-1} factorization to compute A^6.

3. For each of the nonsingular matrices in Exercise 1, use the SDS^{-1} factorization to compute A^{-1}.

4. For each of the following, find a matrix B such that $B^2 = A$.

(a) $A = \begin{pmatrix} 2 & 1 \\ -2 & -1 \end{pmatrix}$

(b) $A = \begin{pmatrix} 9 & -5 & 3 \\ 0 & 4 & 3 \\ 0 & 0 & 1 \end{pmatrix}$

5. Let $B = S^{-1}AS$ for some nonsingular matrix S. Show that B and A have the same eigenvalues.

6. Let A be a diagonalizable matrix whose eigenvalues are all either 1 or -1. Show that $A^{-1} = A$.

7. Show that any 3×3 matrix of the form

$$\begin{pmatrix} a & 1 & 0 \\ 0 & a & 1 \\ 0 & 0 & b \end{pmatrix}$$

is defective.

8. Let A be a 4×4 matrix and let λ be an eigenvalue of multiplicity 3. If $A - \lambda I$ has rank one, is A defective? Explain.

9. Let A be an $n \times n$ matrix with positive real eigenvalues $\lambda_1 > \lambda_2 > \cdots > \lambda_n$. Let x_i be an eigenvector belonging to λ_i for each i and let $x = \alpha_1 x_1 + \cdots + \alpha_n x_n$.

(a) Show that $A^m x = \displaystyle\sum_{i=1}^{n} \alpha_i \lambda_i^m x_i$.

(b) If $\lambda_1 = 1$, show that $\displaystyle\lim_{m \to \infty} A^m x = \alpha_1 x_1$.

10. Let A be an $n \times n$ matrix with an eigenvalue λ of multiplicity n. Show that A is diagonalizable if and only if $A = \lambda I$.

11. Show that a nonzero nilpotent matrix is defective.

12. Let A be a diagonalizable matrix and let S be the diagonalizing matrix. Show that the column vectors of S that correspond to nonzero eigenvalues of A form a basis for $R(A)$.

13. It follows from Exercise 12 that for a diagonalizable matrix the number of nonzero eigenvalues (counted according to multiplicity) equals the rank of the matrix. Give an example of a defective matrix whose rank is not equal to the number of nonzero eigenvalues.

14. Let A be an $n \times n$ matrix and let λ be an eigenvalue of A whose eigenspace has dimension k, where $1 < k < n$. Any basis $\{\mathbf{x}_1, \ldots, \mathbf{x}_k\}$ for the eigenspace can be extended to form a basis $\{\mathbf{x}_1, \ldots, \mathbf{x}_k, \mathbf{x}_{k+1}, \ldots, \mathbf{x}_n\}$ for R^n. Let $S = (\mathbf{x}_1, \ldots, \mathbf{x}_n)$ and $B = S^{-1}AS$.
(a) Show that B is of the form

$$\begin{pmatrix} \lambda I & B_{12} \\ O & B_{22} \end{pmatrix}$$

where I is the $k \times k$ identity matrix.
(b) Use the result from Exercise 5 to show that λ is an eigenvalue of A with multiplicity at least k.

15. Let $\mathbf{x}, \mathbf{y}$ be nonzero vectors in R^n, $n \geq 2$, and let $A = \mathbf{x}\mathbf{y}^T$.
(a) Show that 0 is an eigenvalue of A with $n - 1$ linearly independent eigenvectors and consequently has multiplicity at least $n - 1$ (see Exercise 14).
(b) Show that the remaining eigenvalue of A is

$$\lambda_n = \operatorname{tr} A = \mathbf{x}^T \mathbf{y}$$

and that $\mathbf{x}$ is an eigenvector belonging to λ_n.
(c) Show that if $\lambda_n = \mathbf{x}^T \mathbf{y} \neq 0$, then A is diagonalizable.

16. The town of Midvale maintains a constant population of 300,000 people from year to year. A political science study estimated that there were 150,000 independents, 90,000 Democrats, and 60,000 Republicans in the town. It was also estimated that each year 20 percent of the independents become Democrats and 10 percent become Republicans. Similarly, 20 percent of the Democrats become independents and 10 percent become Republicans while 10 percent of the Republicans defect to the Democrats and 10 percent become independents each year. Let

$$\mathbf{x} = \begin{pmatrix} 150{,}000 \\ 90{,}000 \\ 60{,}000 \end{pmatrix}$$

and let $\mathbf{x}^{(1)}$ be a vector representing the number of people in each group after 1 year.

(a) Find a matrix A such that $A\mathbf{x} = \mathbf{x}^{(1)}$.

(b) Show that $\lambda_1 = 1.0$, $\lambda_2 = 0.5$, and $\lambda_3 = 0.7$ are eigenvalues of A and factor A into a product SDS^{-1}, where D is diagonal.

(c) Which group will dominate in the long run? Justify your answer by computing $\lim_{n \to \infty} A^n \mathbf{x}$.

17. Compute e^A for each of the following matrices.

(a) $A = \begin{pmatrix} 1 & 1 \\ 0 & 0 \end{pmatrix}$ (b) $A = \begin{pmatrix} 3 & 4 \\ -2 & -3 \end{pmatrix}$ (c) $A = \begin{pmatrix} 1 & 1 \\ 0 & 1 \end{pmatrix}$

18. In each of the following, solve the initial value problem $\mathbf{Y}' = A\mathbf{Y}$, $\mathbf{Y}(0) = \mathbf{Y}_0$ by computing $e^{At}\mathbf{Y}_0$.

(a) $A = \begin{pmatrix} 1 & -2 \\ 0 & -1 \end{pmatrix}$, $\mathbf{Y}_0 = \begin{pmatrix} 1 \\ 1 \end{pmatrix}$

(b) $A = \begin{pmatrix} 2 & 3 \\ -1 & -2 \end{pmatrix}$, $\mathbf{Y}_0 = \begin{pmatrix} -4 \\ 2 \end{pmatrix}$

(c) $A = \begin{pmatrix} 1 & 1 & 1 \\ 0 & 0 & 1 \\ 0 & 0 & -1 \end{pmatrix}$, $\mathbf{Y}_0 = \begin{pmatrix} 1 \\ 1 \\ 1 \end{pmatrix}$

19. Let λ be an eigenvalue of an $n \times n$ matrix A and let $\mathbf{x}$ be an eigenvector belonging to λ. Show that e^λ is an eigenvalue of e^A and $\mathbf{x}$ is an eigenvector of e^A belonging to e^λ.

20. Show that e^A is nonsingular for any diagonalizable matrix A.

21. Let A be an $n \times n$ matrix and let $a_{ij}^{(k)}$ denote the ij entry of A^k. Show that

(a) $|a_{ij}^{(k)}| \leq (\|A\|_F)^k$

(b) $\sum_{k=0}^{m} \frac{1}{k!} |a_{ij}^{(k)}| \leq e^{\|A\|_F}$ for all $m \geq 0$.

4. HERMITIAN MATRICES

Let C^n denote the vector space of all n-tuples of complex numbers. The set C of all complex numbers will be taken as our field of scalars. We have already seen that a matrix A with real entries may have complex eigenvalues and eigenvectors. In this section we will study matrices with complex entries and look at the complex analogues of symmetric and orthogonal matrices.

If $\alpha = a + bi$ is a complex scalar, the length of α is given by

$$|\alpha| = \sqrt{\bar{\alpha}\alpha} = \sqrt{a^2 + b^2}$$

The length of a vector $\mathbf{z} = (z_1, z_2, \ldots, z_n)^T$ in C^n is given by

$$\|\mathbf{z}\| = \left(|z_1|^2 + |z_2|^2 + \cdots + |z_n|^2\right)^{1/2}$$
$$= \left(\bar{z}_1 z_1 + \bar{z}_2 z_2 + \cdots + \bar{z}_n z_n\right)^{1/2}$$
$$= \left(\bar{\mathbf{z}}^T \mathbf{z}\right)^{1/2}$$

As a notational convenience, we write $\mathbf{z}^H$ for the transpose of $\bar{\mathbf{z}}$. Thus

$$\bar{\mathbf{z}}^T = \mathbf{z}^H \qquad \text{and} \qquad \|\mathbf{z}\| = \left(\mathbf{z}^H \mathbf{z}\right)^{1/2}$$

Definition. Let V be a vector space over the complex numbers. An **inner product** on V is an operation that assigns to each pair of vectors $\mathbf{z}$ and $\mathbf{w}$ in V a complex number $\langle \mathbf{z}, \mathbf{w} \rangle$ satisfying the following conditions.

(i) $\langle \mathbf{z}, \mathbf{z} \rangle \geq 0$ with equality if and only if $\mathbf{z} = \mathbf{0}$
(ii) $\langle \mathbf{z}, \mathbf{w} \rangle = \overline{\langle \mathbf{w}, \mathbf{z} \rangle}$ for all $\mathbf{z}$ and $\mathbf{w}$ in V
(iii) $\langle \alpha \mathbf{z} + \beta \mathbf{w}, \mathbf{u} \rangle = \alpha \langle \mathbf{z}, \mathbf{u} \rangle + \beta \langle \mathbf{w}, \mathbf{u} \rangle$

Note that for a complex inner product space $\langle \mathbf{z}, \mathbf{w} \rangle = \overline{\langle \mathbf{w}, \mathbf{z} \rangle}$ rather than $\langle \mathbf{w}, \mathbf{z} \rangle$. If we make the proper modifications to allow for this, the theorems on real inner product spaces in Chapter 5, Section 6, will all be valid for complex inner product spaces. In particular, let us recall Theorem 5.6.2; if $\{\mathbf{u}_1, \ldots, \mathbf{u}_n\}$ is an orthonormal basis for a real inner product space V and

$$\mathbf{x} = \sum_{i=1}^{n} c_i \mathbf{u}_i$$

then

$$c_i = \langle \mathbf{u}_i, \mathbf{x} \rangle = \langle \mathbf{x}, \mathbf{u}_i \rangle \qquad \text{and} \qquad \|\mathbf{x}\|^2 = \sum_{i=1}^{n} c_i^2$$

In the case of a complex inner product space, if $\{\mathbf{w}_1, \ldots, \mathbf{w}_n\}$ is an orthonormal basis and

$$\mathbf{z} = \sum_{i=1}^{n} c_i \mathbf{w}_i$$

then

$$c_i = \langle \mathbf{z}, \mathbf{w}_i \rangle, \bar{c}_i = \langle \mathbf{w}_i, \mathbf{z} \rangle \quad, \qquad \text{and} \qquad \|\mathbf{z}\|^2 = \sum_{i=1}^{n} c_i \bar{c}_i$$

We can define an inner product on C^n by

(1) $$\langle \mathbf{z}, \mathbf{w} \rangle = \mathbf{w}^H \mathbf{z}$$

for all $\mathbf{z}$ and $\mathbf{w}$ in C^n. We leave it to the reader to verify that (1) actually does define an inner product on C^n. The complex inner product space C^n is quite similar to the real inner product space R^n. The main difference is that in the complex case it is necessary to conjugate before transposing when taking an inner product.

R^n	C^n
$\langle \mathbf{x}, \mathbf{y} \rangle = \mathbf{y}^T\mathbf{x}$	$\langle \mathbf{z}, \mathbf{w} \rangle = \mathbf{w}^H\mathbf{z}$
$\mathbf{x}^T\mathbf{y} = \mathbf{y}^T\mathbf{x}$	$\mathbf{z}^H\mathbf{w} = \overline{\mathbf{w}^H\mathbf{z}}$
$\|\mathbf{x}\|^2 = \mathbf{x}^T\mathbf{x}$	$\|\mathbf{z}\|^2 = \mathbf{z}^H\mathbf{z}$

Let $M = (m_{ij})$ be an $m \times n$ matrix with $m_{ij} = a_{ij} + ib_{ij}$ for each i and j. We may write M in the form

$$M = A + iB$$

where $A = (a_{ij})$ and $B = (b_{ij})$ have real entries. We define the conjugate of M by

$$\overline{M} = A - iB$$

Thus $\overline{M}$ is the matrix formed by conjugating each of the entries of M. The transpose of $\overline{M}$ will be denoted by M^H.

Definition. A matrix M is said to be **Hermitian** if $M = M^H$.

EXAMPLE 1. The matrix

$$M = \begin{pmatrix} 3 & 2-i \\ 2+i & 4 \end{pmatrix}$$

is Hermitian, since

$$M^H = \begin{pmatrix} \overline{3} & \overline{2-i} \\ \overline{2+i} & \overline{4} \end{pmatrix}^T = \begin{pmatrix} 3 & 2-i \\ 2+i & 4 \end{pmatrix} = M$$

$\square$

If M is a matrix with real entries, then $M^H = M^T$. In particular, if M is a real symmetric matrix, then M is Hermitian. Thus we may view Hermitian matrices as the complex analogue of real symmetric matrices. Hermitian matrices have many nice properties, as we shall see in the following theorem.

Theorem 6.4.1. *The eigenvalues of a Hermitian matrix are all real. Furthermore, eigenvectors belonging to distinct eigenvalues are orthogonal.*

Proof. Let A be a Hermitian matrix. Let λ be an eigenvalue of A and let $\mathbf{x}$ be an eigenvector belonging to λ. If $\alpha = \mathbf{x}^H A \mathbf{x}$, then

$$\bar{\alpha} = \alpha^H = (\mathbf{x}^H A \mathbf{x})^H = \mathbf{x}^H A \mathbf{x} = \alpha$$

Thus α is real. It follows that

$$\alpha = \mathbf{x}^H A \mathbf{x} = \mathbf{x}^H \lambda \mathbf{x} = \lambda \|\mathbf{x}\|^2$$

and hence

$$\lambda = \frac{\alpha}{\|\mathbf{x}\|^2}$$

is real. If $\mathbf{x}_1$ and $\mathbf{x}_2$ are eigenvectors belonging to distinct eigenvalues λ_1 and λ_2, respectively, then

$$(A\mathbf{x}_1)^H \mathbf{x}_2 = \mathbf{x}_1^H A^H \mathbf{x}_2 = \mathbf{x}_1^H A \mathbf{x}_2 = \lambda_2 \mathbf{x}_1^H \mathbf{x}_2$$

and

$$(A\mathbf{x}_1)^H \mathbf{x}_2 = (\mathbf{x}_2^H A \mathbf{x}_1)^H = (\lambda_1 \mathbf{x}_2^H \mathbf{x}_1)^H = \lambda_1 \mathbf{x}_1^H \mathbf{x}_2$$

Thus

$$\lambda_1 \mathbf{x}_1^H \mathbf{x}_2 = \lambda_2 \mathbf{x}_1^H \mathbf{x}_2$$

and since $\lambda_1 \neq \lambda_2$, it follows that

$$\langle \mathbf{x}_2, \mathbf{x}_1 \rangle = \mathbf{x}_1^H \mathbf{x}_2 = 0 \qquad \qquad \square$$

Definition. An $n \times n$ matrix U is said to be **unitary** if its column vectors form an orthonormal set in C^n.

Thus U is unitary if and only if $U^H U = I$. If U is unitary, then, since the column vectors are orthonormal, U must have rank n. It follows that

$$U^{-1} = I U^{-1} = U^H U U^{-1} = U^H$$

A real unitary matrix is an orthogonal matrix.

Corollary 6.4.2. *If the eigenvalues of a Hermitian matrix A are distinct, then there exists a unitary matrix U that diagonalizes A.*

Proof. Let $\mathbf{x}_i$ be an eigenvector belonging to λ_i for each eigenvalue λ_i of A. Let $\mathbf{u}_i = (1/\|\mathbf{x}_i\|)\mathbf{x}_i$. Thus $\mathbf{u}_i$ is a unit eigenvector belonging to λ_i for each i. It follows from Theorem 6.4.1 that $\{\mathbf{u}_1, \ldots, \mathbf{u}_n\}$ is an orthonormal set in C^n. Let U be the matrix whose ith column vector is $\mathbf{u}_i$ for each i; U is unitary and U diagonalizes A. $\qquad \square$

EXAMPLE 2. Let

$$A = \begin{pmatrix} 2 & 1-i \\ 1+i & 1 \end{pmatrix}$$

Find a unitary matrix U that diagonalizes A.

SOLUTION.   The eigenvalues of A are $\lambda_1 = 3$ and $\lambda_2 = 0$ with corresponding eigenvectors $\mathbf{x}_1 = (1 - i, 1)^T$ and $\mathbf{x}_2 = (-1, 1 + i)^T$. Let

$$\mathbf{u}_1 = \frac{1}{\|\mathbf{x}_1\|}\mathbf{x}_1 = \frac{1}{\sqrt{3}}(1 - i, 1)^T$$

and

$$\mathbf{u}_2 = \frac{1}{\|\mathbf{x}_2\|}\mathbf{x}_2 = \frac{1}{\sqrt{3}}(-1, 1 + i)^T$$

Thus

$$U = \frac{1}{\sqrt{3}}\begin{pmatrix} 1 - i & -1 \\ 1 & 1 + i \end{pmatrix}$$

and

$$U^H A U = \tfrac{1}{3}\begin{pmatrix} 1 + i & 1 \\ -1 & 1 - i \end{pmatrix}\begin{pmatrix} 2 & 1 - i \\ 1 + i & 1 \end{pmatrix}\begin{pmatrix} 1 - i & -1 \\ 1 & 1 + i \end{pmatrix}$$

$$= \begin{pmatrix} 3 & 0 \\ 0 & 0 \end{pmatrix} \qquad\qquad \square$$

Actually, Corollary 6.4.2 is valid even if the eigenvalues of A are not distinct. To show this, we will first prove the following theorem.

Theorem 6.4.3 (Schur's Theorem).   *For each $n \times n$ matrix A, there exists a unitary matrix U such that $U^H A U$ is upper triangular.*

Proof.   The proof is by induction on n. The result is obvious if $n = 1$. Assume that the hypothesis holds for all $k \times k$ matrices and let A be a $(k + 1) \times (k + 1)$ matrix. Let λ_1 be an eigenvalue of A and let $\mathbf{w}_1$ be a unit eigenvector belonging to λ_1. Using the Gram–Schmidt process, construct $\mathbf{w}_2, \ldots, \mathbf{w}_{k+1}$ such that $\{\mathbf{w}_1, \ldots, \mathbf{w}_{k+1}\}$ is an orthonormal basis for C^{k+1}. Let W be the matrix whose ith column vector is $\mathbf{w}_i$ for $i = 1, \ldots, k + 1$. Thus, by construction, W is unitary. The first column of $W^H A W$ will be $W^H A \mathbf{w}_1$.

$$W^H A \mathbf{w}_1 = \lambda_1 W^H \mathbf{w}_1 = \lambda_1 \mathbf{e}_1$$

Thus $W^H A W$ is a matrix of the form

$$\left(\begin{array}{c|cccc} \lambda_1 & \times & \times & \cdots & \times \\ \hline 0 & & & & \\ \vdots & & & M & \\ 0 & & & & \end{array}\right)$$

where M is a $k \times k$ matrix. By the induction hypothesis there exists a $k \times k$ unitary matrix V_1 such that $V_1^H M V_1 = T_1$, where T_1 is triangular.

Let

$$V = \begin{pmatrix} 1 & 0 & \cdots & 0 \\ 0 & & & \\ \vdots & & V_1 & \\ 0 & & & \end{pmatrix}$$

Here V is unitary and

$$V^H W^H A W V = \begin{pmatrix} \lambda_1 & \times & \cdots & \times \\ 0 & & & \\ \vdots & & V_1^H M V_1 & \\ 0 & & & \end{pmatrix}$$

$$= \begin{pmatrix} \lambda_1 & \times & \cdots & \times \\ 0 & & & \\ \vdots & & T_1 & \\ 0 & & & \end{pmatrix}$$

$$= T$$

Let $U = WV$. The matrix U is unitary, since

$$U^H U = (WV)^H WV = V^H W^H WV = I$$

Thus $U^H A U = T$.   □

The factorization $A = UTU^H$ is often referred to as the *Schur decomposition* of A. In the case that A is Hermitian, the matrix T will be diagonal.

Theorem 6.4.4 (Spectral Theorem).   *If A is Hermitian, then there exists a unitary matrix U that diagonalizes A.*

Proof.   By Theorem 6.4.3, there is a unitary matrix U such that $U^H A U = T$, where T is upper triangular.

$$T^H = (U^H A U)^H = U^H A^H U = U^H A U = T$$

Therefore, T is Hermitian and consequently must be diagonal.   □

In the case that A is real and symmetric, its eigenvalues and eigenvectors must be real. Thus the diagonalizing matrix U must be orthogonal.

Corollary 6.4.5.   *If A is a real symmetric matrix, then there is an orthogonal matrix U that diagonalizes A, that is, $U^T A U = D$, where D is diagonal.*

EXAMPLE 3. Given

$$A = \begin{pmatrix} 0 & 2 & -1 \\ 2 & 3 & -2 \\ -1 & -2 & 0 \end{pmatrix}$$

find an orthogonal matrix U that diagonalizes A.

SOLUTION.   The characteristic polynomial

$$p(\lambda) = -\lambda^3 + 3\lambda^2 + 9\lambda + 5 = (1 + \lambda)^2(5 - \lambda)$$

has roots $\lambda_1 = \lambda_2 = -1$, $\lambda_3 = 5$. Computing eigenvectors in the usual way we see that $x_1 = (1, 0, 1)^T$ and $x_2 = (-2, 1, 0)^T$ form a basis for the eigenspace $N(A + I)$. One can apply the Gram–Schmidt process to obtain an orthonormal basis for the eigenspace corresponding to $\lambda_1 = \lambda_2 = -1$.

$$\mathbf{u}_1 = \frac{1}{\|\mathbf{x}_1\|}\mathbf{x}_1 = \frac{1}{\sqrt{2}}(1, 0, 1)^T$$

$$\mathbf{p} = (\mathbf{x}_2^T\mathbf{u}_1)\mathbf{u}_1 = -\sqrt{2}\,\mathbf{u}_1 = (-1, 0, -1)^T$$

$$\mathbf{x}_2 - \mathbf{p} = (-1, 1, 1)^T$$

$$\mathbf{u}_2 = \frac{1}{\|\mathbf{x}_2 - \mathbf{p}\|}(\mathbf{x}_2 - \mathbf{p}) = \frac{1}{\sqrt{3}}(-1, 1, 1)^T$$

The eigenspace corresponding to $\lambda_3 = 5$ is spanned by $x_3 = (-1, -2, 1)^T$. Since $\mathbf{x}_3$ must be orthogonal to $\mathbf{u}_1$ and $\mathbf{u}_2$ (Theorem 6.4.1), we need only normalize

$$\mathbf{u}_3 = \frac{1}{\|\mathbf{x}_3\|}\mathbf{x}_3 = \frac{1}{\sqrt{6}}(-1, -2, 1)^T$$

Thus $\{\mathbf{u}_1, \mathbf{u}_2, \mathbf{u}_3\}$ is an orthonormal set and

$$U = \begin{pmatrix} \dfrac{1}{\sqrt{2}} & -\dfrac{1}{\sqrt{3}} & -\dfrac{1}{\sqrt{6}} \\[2ex] 0 & \dfrac{1}{\sqrt{3}} & -\dfrac{2}{\sqrt{6}} \\[2ex] \dfrac{1}{\sqrt{2}} & \dfrac{1}{\sqrt{3}} & \dfrac{1}{\sqrt{6}} \end{pmatrix}$$

diagonalizes A.   □

It follows from Theorem 6.4.4 that each Hermitian matrix A can be factored into a product UDU^H, where U is unitary and D is diagonal. Since U diagonalizes A, it follows that the diagonal elements of D are the eigenvalues of A and the column vectors of U are eigenvectors of A. Thus A cannot be defective. It has a complete set of eigenvectors that form an orthonormal basis for C^n. This is in a sense the ideal situation. We have seen how to express a vector as a linear combination of orthonormal basis elements (Theorem 5.6.2), and the action of A on any linear combination of eigenvectors can be easily determined. Thus, if A has an orthonormal set of eigenvectors $\{\mathbf{u}_1, \ldots, \mathbf{u}_n\}$ and $\mathbf{x} = c_1\mathbf{u}_1 + \cdots + c_n\mathbf{u}_n$, then

$$A\mathbf{x} = c_1\lambda_1\mathbf{u}_1 + \cdots + c_n\lambda_n\mathbf{u}_n$$

Furthermore,

$$c_i = \langle \mathbf{x}, \mathbf{u}_i \rangle = \mathbf{u}_i^H\mathbf{x}$$

or, equivalently, $\mathbf{c} = U^H\mathbf{x}$. Thus

$$A\mathbf{x} = \lambda_1\big(\mathbf{u}_1^H\mathbf{x}\big)\mathbf{u}_1 + \cdots + \lambda_n\big(\mathbf{u}_n^H\mathbf{x}\big)\mathbf{u}_n$$

NORMAL MATRICES

There are non-Hermitian matrices that possess complete orthonormal sets of eigenvectors. For example, skew symmetric and skew Hermitian matrices have this property. (A is *skew Hermitian* if $A^H = -A$.) In general, if A is any matrix with a complete orthonormal set of eigenvectors, then $A = UDU^H$, where U is unitary and D is a diagonal matrix (whose diagonal elements may be complex). In general, $D^H \neq D$ and, consequently, $A^H = UD^HU^H \neq A$. However,

$$AA^H = UDU^HUD^HU^H = UDD^HU^H$$

and

$$A^HA = UD^HU^HUDU^H = UD^HDU^H$$

Since

$$D^HD = DD^H = \begin{pmatrix} |\lambda_1|^2 & & & \\ & |\lambda_2|^2 & & \\ & & \ddots & \\ & & & |\lambda_n|^2 \end{pmatrix}$$

it follows that

$$AA^H = A^HA$$

Definition.   A matrix A is said to be **normal** if $AA^H = A^HA$.

We have shown that if a matrix has a complete orthonormal set of eigenvectors, it is normal. The converse is also true.

Theorem 6.4.6.   *A matrix A is normal if and only if A possesses a complete orthonormal set of eigenvectors.*

Proof.   In view of the preceding remarks, we need only show that a normal matrix A has a complete orthonormal set of eigenvectors. By Theorem 6.4.3 there exists a unitary matrix U and a triangular matrix T such that $T = U^H A U$. We claim that T is also normal.

$$T^H T = U^H A^H U U^H A U = U^H A^H A U$$

and

$$T T^H = U^H A U U^H A^H U = U^H A A^H U$$

Since $A^H A = A A^H$, it follows that $T^H T = T T^H$. Comparing the diagonal elements of $T T^H$ and $T^H T$, we see that

$$|t_{11}|^2 + |t_{12}|^2 + |t_{13}|^2 + \cdots + |t_{1n}|^2 = |t_{11}|^2$$

$$|t_{22}|^2 + |t_{23}|^2 + \cdots + |t_{2n}|^2 = |t_{12}|^2 + |t_{22}|^2$$

$$\vdots$$

$$|t_{nn}|^2 = |t_{1n}|^2 + |t_{2n}|^2 + |t_{3n}|^2 + \cdots + |t_{nn}|^2$$

It follows that $t_{ij} = 0$ whenever $i \neq j$. Thus U diagonalizes A and the column vectors of U ae eigenvectors of A. $\square$

EXERCISES

1. Which of the following matrices are Hermitian? Normal?

(a) $\begin{pmatrix} 1-i & 2 \\ 2 & 3 \end{pmatrix}$
    (b) $\begin{pmatrix} 1 & 2-i \\ 2+i & -1 \end{pmatrix}$

(c) $\begin{pmatrix} \dfrac{1}{\sqrt{2}} & -\dfrac{1}{\sqrt{2}} \\ \dfrac{1}{\sqrt{2}} & \dfrac{1}{\sqrt{2}} \end{pmatrix}$
    (d) $\begin{pmatrix} \dfrac{1}{\sqrt{2}}i & \dfrac{1}{\sqrt{2}} \\ \dfrac{1}{\sqrt{2}} & -\dfrac{1}{\sqrt{2}}i \end{pmatrix}$

(e) $\begin{pmatrix} 0 & i & 1 \\ i & 0 & -2+i \\ -1 & 2+i & 0 \end{pmatrix}$
    (f) $\begin{pmatrix} 3 & 1+i & i \\ 1-i & 1 & 3 \\ -i & 3 & 1 \end{pmatrix}$

2. Find an orthogonal or unitary diagonalizing matrix for each of the following.

(a) $\begin{pmatrix} 2 & 1 \\ 1 & 2 \end{pmatrix}$
    (b) $\begin{pmatrix} 1 & 3+i \\ 3-i & 4 \end{pmatrix}$

(c) $\begin{pmatrix} 2 & i & 0 \\ -i & 2 & 0 \\ 0 & 0 & 2 \end{pmatrix}$
    (d) $\begin{pmatrix} 2 & 1 & 1 \\ 1 & 3 & -2 \\ 1 & -2 & 3 \end{pmatrix}$

(e) $\begin{pmatrix} 0 & 0 & 1 \\ 0 & 1 & 0 \\ 1 & 0 & 0 \end{pmatrix}$
    (f) $\begin{pmatrix} 1 & 1 & 1 \\ 1 & 1 & 1 \\ 1 & 1 & 1 \end{pmatrix}$

(g) $\begin{pmatrix} 4 & 2 & -2 \\ 2 & 1 & -1 \\ -2 & -1 & 1 \end{pmatrix}$

3. Show that the diagonal entries of a Hermitian matrix must be real.

4. Let A be an $m \times n$ matrix and B be an $n \times r$ matrix. Show that $(AB)^H = B^H A^H$.

5. Show that

$$\langle \mathbf{z}, \mathbf{w} \rangle = \mathbf{w}^H \mathbf{z}$$

defines an inner product on C^n.

6. Let $\{\mathbf{u}_1, \ldots, \mathbf{u}_n\}$ be an orthonormal basis for a complex inner product space V and let $\mathbf{z}$ and $\mathbf{w}$ be elements of V. Show that

$$\langle \mathbf{z}, \mathbf{w} \rangle = \sum_{i=1}^{n} (\langle \mathbf{z}, \mathbf{u}_i \rangle \cdot \langle \mathbf{u}_i, \mathbf{w} \rangle)$$

7. Given

$$A = \begin{pmatrix} 4 & 0 & 0 \\ 0 & 1 & i \\ 0 & -i & 1 \end{pmatrix}$$

find a matrix B such that $B^H B = A$.

8. Let U be a unitary matrix. Prove:
(a) U is normal.
(b) $\|U\mathbf{x}\| = \|\mathbf{x}\|$ for all $\mathbf{x} \in C^n$.
(c) If λ is an eigenvalue of U, then $|\lambda| = 1$.

9. Let $\mathbf{u}$ be a unit vector in C^n and define $U = I - 2\mathbf{u}\mathbf{u}^H$. Show that U is both unitary and Hermitian and consequently is its own inverse.

10. Show that $M = A + iB$ (A and B real matrices) is skew Hermitian if and only if A is skew symmetric and B is symmetric.

11. Let $p(x) = -x^3 + cx^2 + (c + 3)x + 1$, where c is a real number. Let C denote the companion matrix of $p(x)$,

$$C = \begin{pmatrix} c & c+3 & 1 \\ 1 & 0 & 0 \\ 0 & 1 & 0 \end{pmatrix}$$

and let

$$A = \begin{pmatrix} -1 & 2 & -c-3 \\ 1 & -1 & c+2 \\ -1 & 1 & -c-1 \end{pmatrix}$$

(a) Compute $A^{-1}CA$.
(b) Use the result from part (a) to prove that $p(x)$ will have only real roots regardless of the value of c.

12. Let A be a Hermitian matrix with eigenvalues $\lambda_1, \ldots, \lambda_n$ and orthonormal eigenvectors $\mathbf{u}_1, \ldots, \mathbf{u}_n$. Show that

$$A = \lambda_1 \mathbf{u}_1 \mathbf{u}_1^H + \lambda_2 \mathbf{u}_2 \mathbf{u}_2^H + \cdots + \lambda_n \mathbf{u}_n \mathbf{u}_n^H$$

13. Let

$$A = \begin{pmatrix} 0 & 1 \\ 1 & 0 \end{pmatrix}$$

Write A as a sum $\lambda_1 \mathbf{u}_1 \mathbf{u}_1^T + \lambda_2 \mathbf{u}_2 \mathbf{u}_2^T$, where λ_1 and λ_2 are eigenvalues and $\mathbf{u}_1$ and $\mathbf{u}_2$ are orthonormal eigenvectors.

14. Let A be a symmetric $n \times n$ matrix with orthonormal eigenvectors $\mathbf{u}_1, \ldots, \mathbf{u}_n$. Let $\mathbf{x} \in R^n$ and let $c_i = \mathbf{u}_i^T \mathbf{x}$ for $i = 1, 2, \ldots, n$. Show that

(a) $\|A\mathbf{x}\|_2^2 = \displaystyle\sum_{i=1}^{n} (\lambda_i c_i)^2$

(b) If $\mathbf{x} \neq \mathbf{0}$, then

$$\min_{1 \leq i \leq n} |\lambda_i| \leq \frac{\|A\mathbf{x}\|_2}{\|\mathbf{x}\|_2} \leq \max_{1 \leq i \leq n} |\lambda_i|$$

(c) $\|A\|_2 = \displaystyle\max_{1 \leq i \leq n} |\lambda_i|$

15. Let A be a Hermitian matrix with eigenvalues $\lambda_1 \leq \lambda_2 \leq \cdots \leq \lambda_n$ and orthonormal eigenvectors $\mathbf{u}_1, \ldots, \mathbf{u}_n$. For any nonzero vector $\mathbf{x}$ in R^n the *Rayleigh quotient* $\rho(\mathbf{x})$ is defined by

$$\rho(\mathbf{x}) = \frac{\langle A\mathbf{x}, \mathbf{x} \rangle}{\langle \mathbf{x}, \mathbf{x} \rangle} = \frac{\mathbf{x}^H A\mathbf{x}}{\mathbf{x}^H \mathbf{x}}$$

(a) If $\mathbf{x} = c_1 \mathbf{u}_1 + \cdots + c_n \mathbf{u}_n$, show that

$$\rho(\mathbf{x}) = \frac{|c_1|^2 \lambda_1 + |c_2|^2 \lambda_2 + \cdots + |c_n|^2 \lambda_n}{\|\mathbf{c}\|^2}$$

(b) Show that

$$\lambda_1 \leq \rho(\mathbf{x}) \leq \lambda_n$$

5. QUADRATIC FORMS

By this time the reader should be well aware of the important role that matrices play in the study of linear equations. In this section we will see that matrices also play an important role in the study of quadratic equations. With each quadratic equation we can associate a vector function $f(\mathbf{x}) = \mathbf{x}^T A\mathbf{x}$. Such a vector function is called a "quadratic form." Quadratic forms arise in a wide variety of applied problems. They are particularly important in the study of optimization theory.

Definition.   A **quadratic equation** in two variables x and y is an equation of the form

(1) $$ax^2 + 2bxy + cy^2 + dx + ey + f = 0$$

Equation (1) may be rewritten in the form

(2) $$(x \quad y)\begin{pmatrix} a & b \\ b & c \end{pmatrix}\begin{pmatrix} x \\ y \end{pmatrix} + (d \quad e)\begin{pmatrix} x \\ y \end{pmatrix} + f = 0$$

Let

$$\mathbf{x} = \begin{pmatrix} x \\ y \end{pmatrix} \quad \text{and} \quad A = \begin{pmatrix} a & b \\ b & c \end{pmatrix}$$

The term

$$\mathbf{x}^T A \mathbf{x} = ax^2 + 2bxy + cy^2$$

is called the **quadratic form** associated with (1).

CONIC SECTIONS

The graph of an equation of the form (1) is called a *conic section*. [If there are no ordered pairs (x, y) that satisfy (1), we say that the equation represents an imaginary conic.] If the graph of (1) consists of a single point, a line, or a pair of lines, we say that (1) represents a degenerate conic. Of more interest are the nondegenerate conics. Graphs of nondegenerate conics turn out to be circles, ellipses, parabolas, or hyperbolas (see Figure 6.5.1). The graph of a conic is particularly easy to sketch when its equation can be put into one of the following standard forms:

(i) $$x^2 + y^2 = r^2 \qquad \text{(circle)}$$

(ii) $$\frac{x^2}{\alpha^2} + \frac{y^2}{\beta^2} = 1 \qquad \text{(ellipse)}$$

(iii) $$\frac{x^2}{\alpha^2} - \frac{y^2}{\beta^2} = 1 \quad \text{or} \quad \frac{y^2}{\alpha^2} - \frac{x^2}{\beta^2} = 1 \qquad \text{(hyperbola)}$$

(iv) $$x^2 = \alpha y \quad \text{or} \quad y^2 = \alpha x \qquad \text{(parabola)}$$

where α, β, and r are nonzero real numbers. Note that the circle is a special case of the ellipse ($\alpha = \beta = r$). A conic section is said to be in *standard position* if its equation can be put into one of these four standard forms. The graphs of (i), (ii), and (iii) will all be symmetric to both coordinate axes and the origin. We say that these curves are centered at the origin. A parabola in standard position will have its vertex at the origin and will be symmetric to one of the axes.

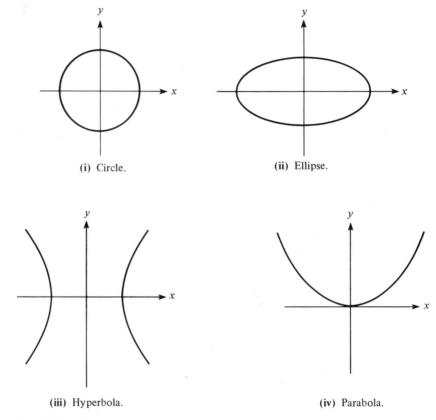

(i) Circle.   **(ii)** Ellipse.

(iii) Hyperbola.   **(iv)** Parabola.

FIGURE 6.5.1

What about conics that are not in standard position? Let us consider the following cases.

Case 1. The conic section has been translated horizontally from the standard position. This occurs when the x^2 and x terms in (1) both have nonzero coefficients.

Case 2. The conic section has been translated vertically from the standard position. This occurs when the y^2 and y terms in (1) have nonzero coefficients (i.e., $c \neq 0$ and $e \neq 0$).

Case 3. The conic section has been rotated from the standard position by an angle θ that is not a multiple of 90°. This occurs when the coefficient of the xy term is nonzero (i.e., $b \neq 0$).

In general, we may have any one or combination of these three cases. In order to graph a conic that is not in standard position, we usually find a new set of axes x' and y' such that the conic section is in standard position with respect to these new axes. This is not difficult if the conic has only been translated horizontally and vertically, in which case the new axes can be found by completing the squares. The following example illustrates how this is done.

EXAMPLE 1. Sketch the graph of the equation

$$9x^2 - 18x + 4y^2 + 16y - 11 = 0$$

SOLUTION. To see how to choose our new axis system, we complete the squares.

$$9(x^2 - 2x + 1) + 4(y^2 + 4y + 4) - 11 = 9 + 16$$

This equation can be simplified to the form

$$\frac{(x - 1)^2}{2^2} + \frac{(y + 2)^2}{3^2} = 1$$

If we let

$$x' = x - 1 \qquad \text{and} \qquad y' = y + 2$$

the equation becomes

$$\frac{(x')^2}{2^2} + \frac{(y')^2}{3^2} = 1$$

which is in standard form with respect to the variables x' and y'. Thus the graph, as shown in Figure 6.5.2, will be an ellipse that is in standard position in the $x'y'$ axis system. The center of the ellipse will be at the origin of the $x'y'$ plane [i.e., at the point $(x, y) = (1, -2)$]. The equation of the x' axis is simply $y' = 0$,

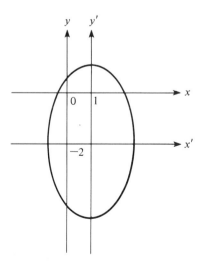

FIGURE 6.5.2

which is the equation of the line $y = -2$ in the xy plane. Similarly, the y' axis coincides with the line $x = 1$. $\square$

There is little problem if the center or vertex of the conic section has been translated. If, however, the conic section has also been rotated from the standard position, it is necessary to change coordinates so that the equation in terms of the new coordinates x' and y' involves no $x'y'$ term. Let $\mathbf{x} = (x, y)^T$ and $\mathbf{x}' = (x', y')^T$. Since the new coordinates differ from the old coordinates by a rotation, we have

$$\mathbf{x} = Q\mathbf{x}' \quad \text{or} \quad \mathbf{x}' = Q^T\mathbf{x}$$

where

$$Q = \begin{pmatrix} \cos\theta & \sin\theta \\ -\sin\theta & \cos\theta \end{pmatrix} \quad \text{and} \quad Q^T = \begin{pmatrix} \cos\theta & -\sin\theta \\ \sin\theta & \cos\theta \end{pmatrix}$$

If $0 < \theta < \pi$, the matrix Q corresponds to a rotation of θ radians in the counterclockwise direction and Q^T corresponds to a rotation of θ radians in the clockwise direction (see Example 2 in Section 2 of Chapter 4). With this change of variables, (2) becomes

$$(3) \qquad (\mathbf{x}')^T(Q^TAQ)\mathbf{x}' + (d' \quad e')\mathbf{x}' + f = 0$$

where $(d' \quad e') = (d \quad e)Q$. This equation will involve no $x'y'$ term if and only if Q^TAQ is diagonal. Since A is symmetric, it is possible to find a pair of orthonormal eigenvectors $\mathbf{q}_1 = (x_1, -y_1)^T$ and $\mathbf{q}_2 = (y_1, x_1)^T$. Thus, if we set $\cos\theta = x_1$ and $\sin\theta = y_1$, then

$$Q = (\mathbf{q}_1 \quad \mathbf{q}_2) = \begin{pmatrix} x_1 & y_1 \\ -y_1 & x_1 \end{pmatrix}$$

diagonalizes A and (3) simplifies to

$$\lambda_1(x')^2 + \lambda_2(y')^2 + d'x' + e'y' + f = 0$$

EXAMPLE 2. Consider the conic section

$$3x^2 + 2xy + 3y^2 - 8 = 0$$

This equation can be written in the form

$$(x \quad y)\begin{pmatrix} 3 & 1 \\ 1 & 3 \end{pmatrix}\begin{pmatrix} x \\ y \end{pmatrix} = 8$$

The matrix

$$\begin{pmatrix} 3 & 1 \\ 1 & 3 \end{pmatrix}$$

has eigenvalues $\lambda = 2$ and $\lambda = 4$ with corresponding unit eigenvectors

$$\left(\frac{1}{\sqrt{2}}, -\frac{1}{\sqrt{2}} \right)^T \quad \text{and} \quad \left(\frac{1}{\sqrt{2}}, \frac{1}{\sqrt{2}} \right)^T$$

Let

$$Q = \begin{pmatrix} \dfrac{1}{\sqrt{2}} & \dfrac{1}{\sqrt{2}} \\ -\dfrac{1}{\sqrt{2}} & \dfrac{1}{\sqrt{2}} \end{pmatrix} = \begin{pmatrix} \cos 45^\circ & \sin 45^\circ \\ -\sin 45^\circ & \cos 45^\circ \end{pmatrix}$$

and set

$$\begin{pmatrix} x \\ y \end{pmatrix} = \begin{pmatrix} \dfrac{1}{\sqrt{2}} & \dfrac{1}{\sqrt{2}} \\ -\dfrac{1}{\sqrt{2}} & \dfrac{1}{\sqrt{2}} \end{pmatrix} \begin{pmatrix} x' \\ y' \end{pmatrix}$$

Thus

$$Q^T A Q = \begin{pmatrix} 2 & 0 \\ 0 & 4 \end{pmatrix}$$

and the equation of the conic becomes

$$2(x')^2 + 4(y')^2 = 8$$

or

$$\frac{(x')^2}{4} + \frac{(y')^2}{2} = 1$$

In the new coordinate system the direction of the x' axis is determined by the point $x' = 1$, $y' = 0$. To translate this to the xy coordinate system, one multiplies

$$\begin{pmatrix} \dfrac{1}{\sqrt{2}} & \dfrac{1}{\sqrt{2}} \\ -\dfrac{1}{\sqrt{2}} & \dfrac{1}{\sqrt{2}} \end{pmatrix} \begin{pmatrix} 1 \\ 0 \end{pmatrix} = \begin{pmatrix} \dfrac{1}{\sqrt{2}} \\ -\dfrac{1}{\sqrt{2}} \end{pmatrix} = \mathbf{q}_1$$

The x' axis will be in the direction of $\mathbf{q}_1$. Similarly, to find the direction of the y' axis, one multiplies

$$Q\mathbf{e}_2 = \mathbf{q}_2.$$

The eigenvectors that form the columns of Q tell us the directions of the new coordinate axes (see Figure 6.5.3). ☐

EXAMPLE 3. Given the quadratic equation

$$3x^2 + 2xy + 3y^2 + 8\sqrt{2}\, y - 4 = 0$$

find a change of coordinates so that the resulting equation represents a conic in standard position.

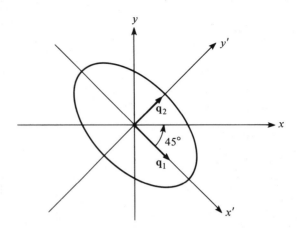

FIGURE 6.5.3

SOLUTION.   The xy term is eliminated in the same manner as in Example 2. In this case, using the rotation matrix,

$$Q = \begin{pmatrix} \dfrac{1}{\sqrt{2}} & \dfrac{1}{\sqrt{2}} \\ -\dfrac{1}{\sqrt{2}} & \dfrac{1}{\sqrt{2}} \end{pmatrix}$$

the equation becomes

$$2(x')^2 + 4(y')^2 + \begin{pmatrix} 0 & 8\sqrt{2} \end{pmatrix} Q\begin{pmatrix} x' \\ y' \end{pmatrix} = 4$$

or

$$(x')^2 - 4x' + 2(y')^2 + 4y' = 2$$

If we complete the square, we get

$$(x' - 2)^2 + 2(y' + 1)^2 = 8$$

If we set $x'' = x' - 2$ and $y'' = y' + 1$ (Figure 6.5.4), then the equation simplifies to

$$\frac{(x'')^2}{8} + \frac{(y'')^2}{4} = 1 \qquad\qquad \square$$

To summarize, a quadratic equation in the variables x and y can be written in the form

$$\mathbf{x}^T A \mathbf{x} + B\mathbf{x} + f = 0$$

where $\mathbf{x} = (x, y)^T$, A is a 2×2 symmetric matrix, B is a 1×2 matrix, and f is a scalar. If A is nonsingular, then by rotating and translating

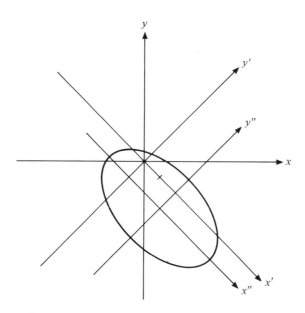

FIGURE 6.5.4

the axes, it is possible to rewrite the equation in the form

(4) $$\lambda_1(x')^2 + \lambda_2(y')^2 + f' = 0$$

where λ_1 and λ_2 are the eigenvalues of A. If (4) represents a real nondegenerate conic, it will either be an ellipse or a hyperbola, depending on whether λ_1 and λ_2 agree in sign or differ in sign. If A is singular and exactly one of its eigenvalues is zero, the quadratic equation can be reduced to either

$$\lambda_1(x')^2 + e'y' + f' = 0 \qquad \text{or} \qquad \lambda_2(y')^2 + d'x' + f' = 0$$

These equations will represent parabolas, provided that e' and d' are nonzero.

There is no reason to limit ourselves to two variables. We could just as well have quadratic equations and quadratic forms in any number of variables. Indeed, a *quadratic equation in n variables* $x_1, \ldots, x_n$ is one of the form

(5) $$\mathbf{x}^T A \mathbf{x} + B\mathbf{x} + \alpha = 0$$

where $\mathbf{x} = (x_1, \ldots, x_n)^T$, A is an $n \times n$ symmetric matrix, B is a $1 \times n$ matrix, and α is a scalar. The vector function

$$f(\mathbf{x}) = \mathbf{x}^T A \mathbf{x} = \sum_{i=1}^{n} \left(\sum_{j=1}^{n} a_{ij} x_j \right) x_i$$

is the *quadratic form in n variables* associated with the quadratic equation.

In the case of three unknowns, if

$$\mathbf{x} = \begin{pmatrix} x \\ y \\ z \end{pmatrix}, \qquad A = \begin{pmatrix} a & d & e \\ d & b & f \\ e & f & c \end{pmatrix}, \qquad B = \begin{pmatrix} g \\ h \\ i \end{pmatrix}$$

then (5) becomes

$$ax^2 + by^2 + cz^2 + 2dxy + 2exz + 2fyz + gx + hy + iz + \alpha = 0$$

The graph of a quadratic equation in three variables is called a *quadric surface*.

There are four basic types of nondegenerate quadric surfaces:

(i)   Ellipsoids
(ii) Hyperboloids (of one or two sheets)
(iii) Cones
(iv) Paraboloids (either elliptic or hyperbolic)

As in the two-dimensional case one can use translations and rotations to transform the equation into standard form,

$$\lambda_1(x')^2 + \lambda_2(y')^2 + \lambda_3(z')^2 + \alpha = 0$$

where $\lambda_1, \lambda_2, \lambda_3$ are the eigenvalues of A.

OPTIMIZATION

Let us consider the problem of maximizing and minimizing functions of several variables. In particular, we would like to determine the nature of the critical points of a real-valued vector function $w = F(\mathbf{x})$. If the function is a quadratic form, $w = \mathbf{x}^T A \mathbf{x}$, then $\mathbf{0}$ is a critical point. Whether it is a maximum, minimum, or saddle point depends upon the eigenvalues of A. More generally, if the function to be extremized is sufficiently differentiable, it behaves locally like a quadratic form. Thus each critical point can be tested by determining the signs of the eigenvalues of the matrix of an associated quadratic form.

Definition. Let $F(\mathbf{x})$ be a real-valued function on R^n. A point $\mathbf{x}_0$ in R^n is said to be a **stationary point** of F if all of the first partial derivatives of F at $\mathbf{x}_0$ exist and are zero.

If $F(\mathbf{x})$ has either a local maximum or a local minimum at a point $\mathbf{x}_0$ and the first partials of F exist at $\mathbf{x}_0$, they will all be zero. Thus, if $F(\mathbf{x})$ has first partials everywhere, its local maxima and minima will occur at stationary points.

Consider the quadratic form

$$f(x, y) = ax^2 + 2bxy + cy^2$$

The first partials of f are

$$f_x = 2ax + 2by$$
$$f_y = 2bx + 2cy$$

Setting these equal to zero, we see that $(0,0)$ is a stationary point. Moreover, if the matrix

$$A = \begin{pmatrix} a & b \\ b & c \end{pmatrix}$$

is nonsingular, this will be the only critical point. Thus, if A is nonsingular, f will have either a global minimum, a global maximum or a saddle point at $(0,0)$.

Let us write f in the form

$$f(\mathbf{x}) = \mathbf{x}^T A \mathbf{x} \qquad \text{where} \qquad \mathbf{x} = \begin{pmatrix} x \\ y \end{pmatrix}$$

Since $f(\mathbf{0}) = 0$, it follows that f will have a global minimum at $\mathbf{0}$ if and only if

$$\mathbf{x}^T A \mathbf{x} > 0 \qquad \text{for all} \qquad \mathbf{x} \neq \mathbf{0}$$

and f will have a global maximum at $\mathbf{0}$ if and only if

$$\mathbf{x}^T A \mathbf{x} < 0 \qquad \text{for all} \qquad \mathbf{x} \neq \mathbf{0}$$

If $\mathbf{x}^T A \mathbf{x}$ changes sign, then $\mathbf{0}$ is a saddle point.

In general, if f is a quadratic form in n variables, then for each $\mathbf{x} \in R^n$

$$f(\mathbf{x}) = \mathbf{x}^T A \mathbf{x}$$

where A is a symmetric $n \times n$ matrix.

Definition.   A quadratic form $f(\mathbf{x}) = \mathbf{x}^T A \mathbf{x}$ is said to be definite if it takes on only one sign as $\mathbf{x}$ varies over all nonzero vectors in R^n. The form is **positive definite** if $\mathbf{x}^T A \mathbf{x} > 0$ for all nonzero $\mathbf{x}$ in R^n and **negative definite** if $\mathbf{x}^T A \mathbf{x} < 0$ for all nonzero $\mathbf{x}$ in R^n. A quadratic form is said to be **indefinite** if it takes on values of different sign. If $f(\mathbf{x}) = \mathbf{x}^T A \mathbf{x} \geq 0$ and assumes the value 0 for some $\mathbf{x} \neq \mathbf{0}$, then $f(\mathbf{x})$ is said to be **positive semidefinite**. If $f(\mathbf{x}) \leq 0$ and assumes the value 0 for some $\mathbf{x} \neq \mathbf{0}$, then $f(\mathbf{x})$ is said to be **negative semidefinite**.

Whether the quadratic form is positive definite or negative definite depends upon the matrix A. If the quadratic form is positive definite then we say simply that A is positive definite. The previous definition can then be restated as follows.

Definition.   A real symmetric matrix A is said to be
   (i) **Positive definite** if $\mathbf{x}^T A \mathbf{x} > 0$ for all nonzero $\mathbf{x}$ in R^n
   (ii) **Negative definite** if $\mathbf{x}^T A \mathbf{x} < 0$ for all nonzero $\mathbf{x}$ in R^n
   (iii) **Positive semidefinite** if $\mathbf{x}^T A \mathbf{x} \geq 0$ for all nonzero $\mathbf{x}$ in R^n
   (iv) **Negative semidefinite** if $\mathbf{x}^T A \mathbf{x} \leq 0$ for all nonzero $\mathbf{x}$ in R^n

If A is nonsingular, then $\mathbf{0}$ will be the only stationary point of $f(\mathbf{x}) = \mathbf{x}^T A \mathbf{x}$. It will be a global minimum if A is positive definite and a global maximum if A is negative definite. If A is indefinite, then $\mathbf{0}$ is a saddle point. To classify the stationary point, we must then classify the matrix A. There are a number of ways of determining whether or not a matrix is positive definite. We will study some of these methods in the next section. The following theorem gives perhaps the most important characterization of positive definite matrices.

Theorem 6.5.1.   *Let A be a real symmetric $n \times n$ matrix. Then A is positive definite if and only if all its eigenvalues are positive.*

Proof.   If A is positive definite and λ is an eigenvalue of A, then for any eigenvector $\mathbf{x}$ belonging to λ,

$$\mathbf{x}^T A \mathbf{x} = \lambda \mathbf{x}^T \mathbf{x} = \lambda \|\mathbf{x}\|^2$$

Hence

$$\lambda = \frac{\mathbf{x}^T A \mathbf{x}}{\|\mathbf{x}\|^2} > 0$$

Conversely, suppose that all the eigenvalues of A are positive. Let $\{\mathbf{x}_1, \ldots, \mathbf{x}_n\}$ be an orthonormal set of eigenvectors of A. If $\mathbf{x}$ is any nonzero vector in R^n, then $\mathbf{x}$ can be written in the form

$$\mathbf{x} = \alpha_1 \mathbf{x}_1 + \alpha_2 \mathbf{x}_2 + \cdots + \alpha_n \mathbf{x}_n$$

where

$$\alpha_i = \mathbf{x}^T \mathbf{x}_i \quad \text{for} \quad i = 1, \ldots, n \quad \text{and} \quad \sum_{i=1}^{n} (\alpha_i)^2 = \|\mathbf{x}\|^2 > 0$$

It follows that

$$\mathbf{x}^T A \mathbf{x} = (\alpha_1 \mathbf{x}_1 + \cdots + \alpha_n \mathbf{x}_n)^T (\alpha_1 \lambda_1 \mathbf{x}_1 + \cdots + \alpha_n \lambda_n \mathbf{x}_n)$$
$$= \sum_{i=1}^{n} (\alpha_i)^2 \lambda_i$$
$$\geq (\min \lambda_i) \|\mathbf{x}\|^2 > 0$$

and hence A is positive definite. $\square$

If the eigenvalues of A are all negative, then $-A$ must be positive definite and consequently A must be negative definite. If A has eigenvalues that differ in sign, then A is indefinite. Indeed, if λ_1 is a positive eigenvalue of A and $\mathbf{x}_1$ is an eigenvector belonging to λ_1, then

$$\mathbf{x}_1^T A \mathbf{x}_1 = \lambda_1 \mathbf{x}_1^T \mathbf{x}_1 = \lambda_1 \|\mathbf{x}_1\|^2 > 0$$

and if λ_2 is a negative eigenvalue with eigenvector $\mathbf{x}_2$, then

$$\mathbf{x}_2^T A \mathbf{x}_2 = \lambda_2 \mathbf{x}_2^T \mathbf{x}_2 = \lambda_2 \|\mathbf{x}_2\|^2 < 0$$

EXAMPLE 4.   Let $f(x, y) = 2x^2 - 4xy + 5y^2$. Determine the nature of the stationary point $(0, 0)$.

SOLUTION.   The matrix A of the quadratic form is

$$\begin{pmatrix} 2 & -2 \\ -2 & 5 \end{pmatrix}$$

Its eigenvalues are $\lambda_1 = 6$ and $\lambda_2 = 1$. Since both eigenvalues are positive, A is positive definite and hence $(0, 0)$ is a global minimum.   □

Suppose now that we have a function $F(x, y)$ with a stationary point (x_0, y_0). If F has continuous third partials in a neighborhood of (x_0, y_0), it can be expanded in a Taylor series about that point.

$$F(x_0 + h, y_0 + k)$$
$$= F(x_0, y_0) + \left[hF_x(x_0, y_0) + kF_y(x_0, y_0) \right]$$
$$+ \tfrac{1}{2}\left[h^2 F_{xx}(x_0, y_0) + 2hk F_{xy}(x_0, y_0) + k^2 F_{yy}(x_0, y_0) \right] + R$$
$$= F(x_0, y_0) + \tfrac{1}{2}(ah^2 + 2bhk + ck^2) + R$$

where

$$a = F_{xx}(x_0, y_0), \qquad b = F_{xy}(x_0, y_0), \qquad c = F_{yy}(x_0, y_0)$$

and the remainder R is given by

$$R = \tfrac{1}{6}\left[h^3 F_{xxx}(\mathbf{z}) + 3h^2 k F_{xxy}(\mathbf{z}) + 3hk^2 F_{xyy}(\mathbf{z}) + k^3 F_{yyy}(\mathbf{z}) \right]$$
$$\mathbf{z} = (x_0 + \theta h, y_0 + \theta k); \qquad 0 < \theta < 1$$

If h and k are sufficiently small, $|R|$ will be less than the modulus of $\tfrac{1}{2}\{ah^2 + 2bhk + ck^2\}$ and hence $[F(x_0 + h, y_0 + k) - F(x_0, y_0)]$ will have the same sign as $(ah^2 + 2bhk + ck^2)$. The expression

$$f(h, k) = ah^2 + 2bhk + ck^2$$

is a quadratic form in the variables h and k. Thus $F(x, y)$ will have a local minimum (maximum) at (x_0, y_0) if and only if $f(h, k)$ has a minimum (maximum) at $(0, 0)$. Let

$$H = \begin{pmatrix} a & b \\ b & c \end{pmatrix} = \begin{pmatrix} F_{xx}(x_0, y_0) & F_{xy}(x_0, y_0) \\ F_{xy}(x_0, y_0) & F_{yy}(x_0, y_0) \end{pmatrix}$$

and let λ_1 and λ_2 be the eigenvalues of H. If H is nonsingular, then λ_1 and λ_2 are nonzero and we can classify the stationary point as follows:

(i) F has a minimum at (x_0, y_0) if $\lambda_1 > 0, \lambda_2 > 0$
(ii) F has a maximum at (x_0, y_0) if $\lambda_1 < 0, \lambda_2 < 0$
(iii) F has a saddle point at (x_0, y_0) if λ_1 and λ_2 differ in sign.

EXAMPLE 5. Find and describe all stationary points of the function

$$F(x, y) = \tfrac{1}{3}x^3 + xy^2 - 4xy + 1$$

SOLUTION. The first partials of F are

$$F_x = x^2 + y^2 - 4y$$
$$F_y = 2xy - 4x = 2x(y - 2)$$

Setting $F_y = 0$, we get $x = 0$ or $y = 2$. Setting $F_x = 0$, we see that if $x = 0$, then y must be either 0 or 4, and if $y = 2$, then $x = \pm 2$. Thus $(0, 0)$, $(0, 4)$, $(2, 2)$, $(-2, 2)$ are the stationary points of F. To classify the stationary points, we compute the second partials.

$$F_{xx} = 2x, \qquad F_{xy} = 2y - 4, \qquad F_{yy} = 2x$$

For each stationary point (x_0, y_0) we determine the eigenvalues of

$$\begin{pmatrix} 2x_0 & 2y_0 - 4 \\ 2y_0 - 4 & 2x_0 \end{pmatrix}$$

These values are summarized in Table 6.5.1.

TABLE 6.5.1

Stationary Point (x_0, y_0)	λ_1	λ_2	Description
$(0, 0)$	4	-4	Saddle point
$(0, 4)$	4	-4	Saddle point
$(2, 2)$	4	4	Local minimum
$(-2, 2)$	-4	-4	Local maximum

We can now generalize our method of classifying stationary points to functions of more than two variables. Let $F(\mathbf{x}) = F(x_1, \ldots, x_n)$ be a real-valued function whose third partial derivatives are all continuous. Let $\mathbf{x}_0$ be a stationary point of F and define the matrix $H = H(\mathbf{x}_0)$ by

$$h_{ij} = F_{x_i x_j}(\mathbf{x}_0)$$

$H(\mathbf{x}_0)$ is called the *Hessian* of F at $\mathbf{x}_0$.
The stationary point can be classified as follows:

(i) $\mathbf{x}_0$ is a local minimum of F if $H(\mathbf{x}_0)$ is positive definite.
(ii) $\mathbf{x}_0$ is a local maximum of F if $H(\mathbf{x}_0)$ is negative definite.
(iii) $\mathbf{x}_0$ is a saddle point of F if $H(\mathbf{x}_0)$ is indefinite.

EXAMPLE 6.   Find the local minima of the function

$$F(x, y, z) = x^2 + xz - 3\cos y + z^2$$

SOLUTION.   The first partials of F are

$$F_x = 2x + z$$
$$F_y = 3\sin y$$
$$F_z = x + 2z$$

It follows that (x, y, z) is a stationary point of F if and only if $x = z = 0$ and $y = n\pi$, where n is an integer. Let $\mathbf{x}_0 = (0, 2k\pi, 0)^T$. The Hessian of F at $\mathbf{x}_0$ is given by

$$H(\mathbf{x}_0) = \begin{pmatrix} 2 & 0 & 1 \\ 0 & 3 & 0 \\ 1 & 0 & 2 \end{pmatrix}$$

The eigenvalues of $H(\mathbf{x}_0)$ are 3, 3, and 1. Since the eigenvalues are all positive, it follows that $H(\mathbf{x}_0)$ is positive definite and hence F has a local minimum at $\mathbf{x}_0$. On the other hand, at a stationary point of the form $\mathbf{x}_1 = (0, (2k - 1)\pi, 0)^T$ the Hessian will be

$$H(\mathbf{x}_1) = \begin{pmatrix} 2 & 0 & 1 \\ 0 & -3 & 0 \\ 1 & 0 & 2 \end{pmatrix}$$

The eigenvalues of $H(\mathbf{x}_1)$ are -3, 3, and 1. It follows that $H(\mathbf{x}_1)$ is indefinite and hence $\mathbf{x}_1$ is a saddle point of F.   □

EXERCISES

1. Find the matrix associated with each of the following quadratic forms.
(a) $3x^2 - 5xy + y^2$
(b) $2x^2 + 3y^2 + z^2 + xy - 2xz + 3yz$
(c) $x^2 + 2y^2 + z^2 + xy - 2xz + 3yz$

2. Reorder the eigenvalues in Example 2 so that $\lambda_1 = 4$ and $\lambda_2 = 2$ and rework the example. In what quadrants will the positive x' and y' axes lie? Sketch the graph and compare it to Figure 6.5.3.

3. In each of the following, find a suitable change of coordinates (i.e., a rotation and/or a translation) so that the resulting conic section is in standard form; identify the curve and sketch the graph.
(a) $x^2 + xy + y^2 - 6 = 0$
(b) $3x^2 + 8xy + 3y^2 + 28 = 0$
(c) $-3x^2 + 6xy + 5y^2 - 24 = 0$
(d) $x^2 + 2xy + y^2 + 3x + y - 1 = 0$

4. Let A be a symmetric 2×2 matrix and let α be a nonzero scalar for which the equation $\mathbf{x}^T A \mathbf{x} = \alpha$ is consistent. Show that the corresponding conic section will be nondegenerate if and only if A is nonsingular.

5. Which of the following matrices are positive definite? Negative definite? Indefinite?

(a) $\begin{pmatrix} 3 & 2 \\ 2 & 2 \end{pmatrix}$ (b) $\begin{pmatrix} 3 & 4 \\ 4 & 1 \end{pmatrix}$ (c) $\begin{pmatrix} 3 & \sqrt{2} \\ \sqrt{2} & 4 \end{pmatrix}$

(d) $\begin{pmatrix} -2 & 0 & 1 \\ 0 & -1 & 0 \\ 1 & 0 & -2 \end{pmatrix}$ (e) $\begin{pmatrix} 1 & 2 & 1 \\ 2 & 1 & 1 \\ 1 & 1 & 2 \end{pmatrix}$ (f) $\begin{pmatrix} 2 & 0 & 0 \\ 0 & 5 & 3 \\ 0 & 3 & 5 \end{pmatrix}$

6. For each of the following functions, determine whether the given stationary point corresponds to a local minimum, local maximum, or saddle point.
(a) $f(x, y) = 3x^2 - xy + y^2$ $(0, 0)$
(b) $f(x, y) = \sin x + y^3 + 3xy + 2x - 3y$ $(0, -1)$
(c) $f(x, y, z) = x^3 + xyz + y^2 - 3x$ $(1, 0, 0)$
(d) $f(x, y, z) = -\frac{1}{4}(x^{-4} + y^{-4} + z^{-4}) + yz - x - 2y - 2z$
   $(1, 1, 1)$

7. Show that if A is symmetric positive definite, then $\det(A) > 0$. Give an example of a 2×2 matrix with positive determinant that is not positive definite.

8. Show that if A is a symmetric positive definite matrix, then A is nonsingular and A^{-1} is also positive definite.

9. Let A be a singular $n \times n$ matrix. Show that $A^T A$ is positive semidefinite, but not positive definite.

10. Let A be a symmetric $n \times n$ matrix with eigenvalues $\lambda_1, \ldots, \lambda_n$. Show that there exists an orthonormal set of vectors $\{\mathbf{x}_1, \ldots, \mathbf{x}_n\}$ such that

$$\mathbf{x}^T A \mathbf{x} = \sum_{i=1}^{n} \lambda_i \left(\mathbf{x}^T \mathbf{x}_i\right)^2$$

for each $\mathbf{x} \in R^n$.

11. Let A be a symmetric positive definite matrix. Show that the diagonal elements of A must all be positive.

12. Let A be a symmetric positive definite $n \times n$ matrix and let S be a nonsingular $n \times n$ matrix. Show that $S^T A S$ is positive definite.

13. Let A be a symmetric positive definite $n \times n$ matrix. Show that A can be factored into a product QQ^T, where Q is an $n \times n$ matrix whose columns are mutually orthogonal.

[**_Hint:_** See Corollary 6.4.5.]

6. POSITIVE DEFINITE MATRICES

In Section 5 we saw that a symmetric matrix is positive definite if and only if its eigenvalues are all positive. These types of matrices occur in a wide variety of applications. They frequently arise in the numerical solution of boundary value problems by finite difference methods or by finite element methods. Because of their importance in applied mathematics we devote this section to studying their properties.

Recall that a symmetric $n \times n$ matrix A is positive definite if $\mathbf{x}^T A \mathbf{x} > 0$ for all nonzero vectors $\mathbf{x}$ in R^n. In Theorem 6.5.1, symmetric positive definite matrices were characterized by the condition that all of their eigenvalues are positive. This characterization can be used to establish the following properties. If A is an $n \times n$ symmetric positive definite matrix, then

    I. A is nonsingular.
   II. $\det(A) > 0$

If A were singular, then $\lambda = 0$ would be an eigenvalue. Since all of the eigenvalues of A are positive, A must be nonsingular. The second property also follows from Theorem 6.5.1, since

$$\det(A) = \lambda_1 \cdots \lambda_n > 0$$

Given an $n \times n$ matrix A, let A_r denote the matrix formed by deleting the last $n - r$ rows and columns of A. A_r is called the _leading_ *principal submatrix* of A of order r. We can now state a third property of positive definite matrices.

III. If A is a symmetric positive definite matrix, then the leading principal submatrices $A_1, A_2, \ldots, A_n$ of A are all positive definite.

Proof. To show that A_r is positive definite, $1 \le r \le n$, let $\mathbf{x}_r = (x_1, \ldots, x_r)^T$ be any nonzero vector in R^r and set $\mathbf{x} = (x_1, \ldots, x_r, 0, \ldots, 0)^T$. Since

$$\mathbf{x}_r^T A_r \mathbf{x}_r = \mathbf{x}^T A \mathbf{x} > 0$$

it follows that A_r is positive definite. $\square$

An immediate consequence of properties **I**, **II**, and **III** is that if A_r is a leading principal submatrix of a symmetric positive definite matrix A, then A_r is nonsingular and $\det(A_r) > 0$. This has significance when related to the Gaussian elimination process. In general, if A is an $n \times n$ matrix whose leading principal submatrices are all nonsingular, then A can be reduced to upper triangular form using only row operation III; that is, the diagonal elements will never be 0 in the elimination process, so the reduction can be completed without interchanging rows.

IV. If A is a symmetric positive definite matrix, then A can be reduced to upper triangular form using only row operation III and the pivot elements will all be positive.

Let us illustrate property **IV** in the case of a 4×4 symmetric positive definite matrix A. Note first that

$$a_{11} = \det(A_1) > 0$$

so a_{11} can be used as a pivot element and row 1 is the first pivot row. Let $a_{22}^{(1)}$ denote the entry in the $(2, 2)$ position after the last three elements of column 1 have been eliminated (see Figure 6.6.1). At this step, the submatrix A_2 has been transformed into a matrix

$$\begin{pmatrix} a_{11} & a_{12} \\ 0 & a_{22}^{(1)} \end{pmatrix}$$

Since this was accomplished using only row operation III, the value of the determinant remains unchanged. Thus

$$\det(A_2) = a_{11} a_{22}^{(1)}$$

and hence

$$a_{22}^{(1)} = \frac{\det(A_2)}{a_{11}} = \frac{\det(A_2)}{\det(A_1)} > 0$$

FIGURE 6.6.1

Since $a_{22}^{(1)} \neq 0$, it can be used as a pivot in the second step of the elimination process. After step 2, the matrix A_3 has been transformed into

$$\begin{pmatrix} a_{11} & a_{12} & a_{13} \\ 0 & a_{22}^{(1)} & a_{23}^{(1)} \\ 0 & 0 & a_{33}^{(2)} \end{pmatrix}$$

Since only row operation III was used,

$$\det(A_3) = a_{11} a_{22}^{(1)} a_{33}^{(2)}$$

and hence

$$a_{33}^{(2)} = \frac{\det(A_3)}{a_{11} a_{22}^{(1)}} = \frac{\det(A_3)}{\det(A_2)} > 0$$

Thus $a_{33}^{(2)}$ can be used as a pivot in the last step. After step 3 the remaining diagonal entry will be

$$a_{44}^{(3)} = \frac{\det(A_4)}{\det(A_3)} > 0$$

In general, if an $n \times n$ matrix A can be reduced to an upper triangular form U without row interchanges, then A can be factored into a product LU where L is lower triangular with 1's on the diagonal. The (i, j) entry of L below the diagonal will be the multiple of the ith row that was subtracted from the jth row during the elimination process. We illustrate with a 3×3 example.

EXAMPLE 1. Let

$$A = \begin{pmatrix} 4 & 2 & -2 \\ 2 & 10 & 2 \\ -2 & 2 & 5 \end{pmatrix}$$

The matrix L is determined as follows. At the first step of the elimination process $\frac{1}{2}$ times the first row is subtracted from the second row and $-\frac{1}{2}$ times the first row is subtracted from the third. Corresponding to these operations, we set $l_{21} = \frac{1}{2}$ and $l_{31} = -\frac{1}{2}$. After step 1 we obtain the matrix

$$A^{(1)} = \begin{pmatrix} 4 & 2 & -2 \\ 0 & 9 & 3 \\ 0 & 3 & 4 \end{pmatrix}$$

The final elimination is carried out by subtracting $\frac{1}{3}$ times the second row from the third row. Corresponding to this step, we set $l_{32} = \frac{1}{3}$. After step 2 we end up with the upper triangular matrix,

$$U = A^{(2)} = \begin{pmatrix} 4 & 2 & -2 \\ 0 & 9 & 3 \\ 0 & 0 & 3 \end{pmatrix}$$

The matrix L is given by

$$L = \begin{pmatrix} 1 & 0 & 0 \\ \frac{1}{2} & 1 & 0 \\ -\frac{1}{2} & \frac{1}{3} & 1 \end{pmatrix}$$

and one can verify that the product $LU = A$.

$$\begin{pmatrix} 1 & 0 & 0 \\ \frac{1}{2} & 1 & 0 \\ -\frac{1}{2} & \frac{1}{3} & 1 \end{pmatrix}\begin{pmatrix} 4 & 2 & -2 \\ 0 & 9 & 3 \\ 0 & 0 & 3 \end{pmatrix} = \begin{pmatrix} 4 & 2 & -2 \\ 2 & 10 & 2 \\ -2 & 2 & 5 \end{pmatrix}$$

To see why this factorization works, let us view the process in terms of elementary matrices. Row operation III was applied three times during the process. This is equivalent to multiplying A on the left by three elementary matrices E_1, E_2, E_3. Thus $E_3 E_2 E_1 A = U$.

$$\begin{pmatrix} 1 & 0 & 0 \\ 0 & 1 & 0 \\ 0 & \frac{1}{3} & 1 \end{pmatrix}\begin{pmatrix} 1 & 0 & 0 \\ 0 & 1 & 0 \\ \frac{1}{2} & 0 & 1 \end{pmatrix}\begin{pmatrix} 1 & 0 & 0 \\ -\frac{1}{2} & 1 & 0 \\ 0 & 0 & 1 \end{pmatrix}\begin{pmatrix} 4 & 2 & -2 \\ 2 & 10 & 2 \\ -2 & 2 & 5 \end{pmatrix}$$

$$= \begin{pmatrix} 4 & 2 & -2 \\ 0 & 9 & 3 \\ 0 & 0 & 6 \end{pmatrix}$$

Since the elementary matrices are nonsingular, it follows that

$$A = \left(E_1^{-1} E_2^{-1} E_3^{-1} \right) U$$

When the inverse elementary matrices are multiplied in this order, the result is a lower triangular matrix L with 1's on the diagonal.

The entries below the diagonal of L will just be the multiples that were subtracted during the elimination process.

$$E_1^{-1}E_2^{-1}E_3^{-1} = \begin{pmatrix} 1 & 0 & 0 \\ \frac{1}{2} & 1 & 0 \\ 0 & 0 & 1 \end{pmatrix} \begin{pmatrix} 1 & 0 & 0 \\ 0 & 1 & 0 \\ -\frac{1}{2} & 0 & 1 \end{pmatrix} \begin{pmatrix} 1 & 0 & 0 \\ 0 & 1 & 0 \\ 0 & \frac{1}{3} & 1 \end{pmatrix}$$

$$= \begin{pmatrix} 1 & 0 & 0 \\ \frac{1}{2} & 1 & 0 \\ -\frac{1}{2} & \frac{1}{3} & 1 \end{pmatrix} \qquad \square$$

Given an LU factorization of a matrix A, it is possible to go one step further and factor U into a product DU_1, where D is diagonal and U_1 is upper triangular with 1's on the diagonal.

$$DU_1 = \begin{pmatrix} u_{11} & & & \\ & u_{22} & & \\ & & \ddots & \\ & & & u_{nn} \end{pmatrix} \begin{pmatrix} 1 & \frac{u_{12}}{u_{11}} & \frac{u_{13}}{u_{11}} & \cdots & \frac{u_{1n}}{u_{11}} \\ & 1 & \frac{u_{23}}{u_{22}} & \cdots & \frac{u_{2n}}{u_{22}} \\ & & & & \vdots \\ & & & & 1 \end{pmatrix}$$

It follows, then, that $A = LDU_1$. In general, if A can be factored into a product of the form LDU, where L is lower triangular, D is diagonal, U is upper triangular, and L and U both have 1's along the diagonal, then such a factorization will be unique (see Exercise 5).

If A is a symmetric positive definite matrix, then A can be factored into a product $LU = LDU_1$. The diagonal elements of D are the entries $u_{11}, \ldots, u_{nn}$ which were the pivot elements in the elimination process. By property **IV** these elements are all positive. Furthermore, since A is symmetric,

$$LDU_1 = A = A^T = (LDU_1)^T = U_1^T D^T L^T$$

It follows from the uniqueness of the LDU factorization that $L^T = U_1$. Thus

$$A = LDL^T$$

This important factorization is often used in numerical computations. There are efficient algorithms that make use of this factorization in solving symmetric positive definite linear systems.

V. If A is a symmetric positive definite matrix, then A can be factored into a product LDL^T, where L is lower triangular with 1's along the diagonal and D is a diagonal matrix whose diagonal entries are all positive.

EXAMPLE 2.   We saw in Example 1 that

$$A = \begin{pmatrix} 4 & 2 & -2 \\ 2 & 10 & 2 \\ -2 & 2 & 5 \end{pmatrix}$$

$$= \begin{pmatrix} 1 & 0 & 0 \\ \frac{1}{2} & 1 & 0 \\ -\frac{1}{2} & \frac{1}{3} & 1 \end{pmatrix} \begin{pmatrix} 4 & 2 & -2 \\ 0 & 9 & 3 \\ 0 & 0 & 3 \end{pmatrix} = LU$$

Factoring out the diagonal entries of U, we get

$$A = \begin{pmatrix} 1 & 0 & 0 \\ \frac{1}{2} & 1 & 0 \\ -\frac{1}{2} & \frac{1}{3} & 1 \end{pmatrix} \begin{pmatrix} 4 & 0 & 0 \\ 0 & 9 & 0 \\ 0 & 0 & 3 \end{pmatrix} \begin{pmatrix} 1 & \frac{1}{2} & -\frac{1}{2} \\ 0 & 1 & \frac{1}{3} \\ 0 & 0 & 1 \end{pmatrix}$$

$$= LDL^T \qquad\qquad\qquad\qquad \square$$

Since the diagonal elements $u_{11}, \ldots, u_{nn}$ are positive, it is possible to go one step further with the factorization. Let

$$D^{1/2} = \begin{pmatrix} \sqrt{u_{11}} & & & \\ & \sqrt{u_{22}} & & \\ & & \ddots & \\ & & & \sqrt{u_{nn}} \end{pmatrix}$$

and set $L_1 = LD^{1/2}$. Then

$$A = LDL^T = LD^{1/2}(D^{1/2})^T L^T = L_1 L_1^T$$

This factorization is known as the *Cholesky decomposition* of A.

VI. (Cholesky decomposition) If A is a symmetric positive definite matrix, then A can be factored into a product LL^T, where L is lower triangular with positive diagonal elements.

EXAMPLE 3.   Let A be the matrix from Examples 1 and 2. If we set

$$L_1 = LD^{1/2} = \begin{pmatrix} 1 & 0 & 0 \\ \frac{1}{2} & 1 & 0 \\ -\frac{1}{2} & \frac{1}{3} & 1 \end{pmatrix} \begin{pmatrix} 2 & 0 & 0 \\ 0 & 3 & 0 \\ 0 & 0 & \sqrt{3} \end{pmatrix} = \begin{pmatrix} 2 & 0 & 0 \\ 1 & 3 & 0 \\ -1 & 1 & \sqrt{3} \end{pmatrix}$$

then

$$L_1 L_1^T = \begin{pmatrix} 2 & 0 & 0 \\ 1 & 3 & 0 \\ -1 & 1 & \sqrt{3} \end{pmatrix} \begin{pmatrix} 2 & 1 & -1 \\ 0 & 3 & 1 \\ 0 & 0 & \sqrt{3} \end{pmatrix}$$

$$= \begin{pmatrix} 4 & 2 & -2 \\ 2 & 10 & 2 \\ -2 & 2 & 5 \end{pmatrix} = A$$

$\square$

The matrix $A = LL^T$ could also be written in terms of the upper triangular matrix $U = L^T$. Indeed, if $U = L^T$, then $A = LL^T = U^T U$. On the other hand, it is not difficult to show that any product $B^T B$ will be positive definite provided that B is nonsingular. Putting all of these results together, we have the following theorem.

Theorem 6.6.1.　*Let A be a symmetric n $\times$ n matrix. The following are equivalent.*
 (a) *A is positive definite.*
 (b) *The leading principal submatrices $A_1, \ldots, A_n$ all have positive determinants.*
 (c) *A can be reduced to upper triangular form using only row operation III and the pivot elements will all be positive.*
 (d) *A has a Cholesky factorization LL^T (where L is lower triangular with positive diagonal entries).*
 (e) *A can be factored into a product $B^T B$ for some nonsingular matrix B.*

Proof.　We have already shown that (a) implies (b), (b) implies (c), and (c) implies (d). To see that (d) implies (e), assume that $A = LL^T$. If one sets $B = L^T$, then B is nonsingular and

$$A = LL^T = B^T B$$

Finally, to show that (e) $\Rightarrow$ (a), assume that $A = B^T B$ where B is nonsingular. Let $\mathbf{x}$ be any nonzero vector in R^n and set $\mathbf{y} = B\mathbf{x}$. Since B is nonsingular, $\mathbf{y} \neq \mathbf{0}$ and it follows that

$$\mathbf{x}^T A \mathbf{x} = \mathbf{x}^T B^T B \mathbf{x} = \mathbf{y}^T \mathbf{y} = \|\mathbf{y}\|^2 > 0$$

Thus A is positive definite.　　$\square$

Analogous results to Theorem 6.6.1 are not valid for positive semi-definiteness. For example, consider the matrix

$$A = \begin{pmatrix} 1 & 1 & -3 \\ 1 & 1 & -3 \\ -3 & -3 & 5 \end{pmatrix}$$

The leading principal submatrices all have nonnegative determinants

$$\det(A_1) = 1, \qquad \det(A_2) = 0, \qquad \det(A_3) = 0$$

but A is not positive semidefinite since it has a negative eigenvalue $\lambda = -1$. Indeed, $\mathbf{x} = (1, 1, 1)^T$ is an eigenvector belonging to $\lambda = -1$ and

$$\mathbf{x}^T A \mathbf{x} = -3$$

EXERCISES

1. For each of the following, compute the determinants of all of the leading principal submatrices and use them to determine whether or not the matrix is positive definite.

   (a) $\begin{pmatrix} 2 & -1 \\ -1 & 2 \end{pmatrix}$ (b) $\begin{pmatrix} 3 & 4 \\ 4 & 2 \end{pmatrix}$

   (c) $\begin{pmatrix} 6 & 4 & -2 \\ 4 & 5 & 3 \\ -2 & 3 & 6 \end{pmatrix}$ (d) $\begin{pmatrix} 4 & 2 & 1 \\ 2 & 3 & -2 \\ 1 & -2 & 5 \end{pmatrix}$

2. Let A be a 3×3 symmetric positive definite matrix and suppose that $\det(A_1) = 3$, $\det(A_2) = 6$, and $\det(A_3) = 8$. What would the pivot elements be in the reduction of A to triangular form assuming that only row operation III is used in the reduction process?

3. For each of the following, factor the given matrix into a product LDL^T, where L is lower triangular with 1's on the diagonal and D is a diagonal matrix.

   (a) $\begin{pmatrix} 4 & 2 \\ 2 & 10 \end{pmatrix}$ (b) $\begin{pmatrix} 9 & -3 \\ -3 & 2 \end{pmatrix}$

   (c) $\begin{pmatrix} 16 & 8 & 4 \\ 8 & 6 & 0 \\ 4 & 0 & 7 \end{pmatrix}$ (d) $\begin{pmatrix} 9 & 3 & -6 \\ 3 & 4 & 1 \\ -6 & 1 & 9 \end{pmatrix}$

4. Find the Cholesky decomposition LL^T for each of the matrices in Exercise 3.

5. Let A be a nonsingular $n \times n$ matrix and suppose that $A = L_1 D_1 U_1 = L_2 D_2 U_2$, where L_1 and L_2 are lower triangular, D_1 and D_2 are diagonal, U_1 and U_2 are upper triangular, and L_1, L_2, U_1, U_2 all have 1's along the diagonal. Show that $L_1 = L_2$, $D_1 = D_2$ and $U_1 = U_2$. [**Hint:** L_2^{-1} is lower triangular and U_1^{-1} is upper triangular. Compare both sides of the equation $D_2^{-1} L_2^{-1} L_1 D_1 = U_2 U_1^{-1}$.]

6. Let A be an $n \times n$ symmetric positive definite matrix. For each $\mathbf{x}, \mathbf{y} \in R^n$ define

$$\langle \mathbf{x}, \mathbf{y} \rangle = \mathbf{x}^T A \mathbf{y}$$

Show that $\langle \, , \, \rangle$ defines an inner product on R^n.

7. Let A be a symmetric positive definite matrix and let Q be an orthogonal diagonalizing matrix. Use the factorization $A = QDQ^T$ to find a nonsingular matrix B such that $B^TB = A$.

8. Let B be an $m \times n$ matrix of rank n. Show that B^TB is positive definite.

9. Let A be a symmetric $n \times n$ matrix. Show that e^A is positive definite.

*7. NONNEGATIVE MATRICES

In many of the types of linear systems that occur in applications, the entries of the coefficient matrix represent nonnegative quantities. This section deals with the study of such matrices and some of their properties.

Definition. An $n \times n$ matrix A with real entries is said to be **nonnegative** if $a_{ij} \geq 0$ for each i and j and **positive** if $a_{ij} > 0$ for each i and j.

Similarly, a vector $\mathbf{x} = (x_1, \ldots, x_n)^T$ is said to be **nonnegative** if each $x_i \geq 0$ and **positive** if each $x_i > 0$.

For an example of one of the applications of nonnegative matrices, we consider the Leontief input–output models.

APPLICATION 1: THE OPEN MODEL

Suppose that there are n industries producing n different products. Each industry requires input of the products from the other industries and possibly even of its own product. In the open model it is assumed that there is an additional demand for each of the products from an outside sector. The problem is to determine the output of each of the industries that is necessary to meet the total demand.

We will show that this problem can be represented by a linear system of equations and that the system has a unique nonnegative solution. Let a_{ij} denote the amount of input from the ith industry necessary to produce one unit of output in the jth industry. By a unit of input or output we mean one dollar's worth of the product. Thus producing one dollar's worth of the jth product will involve a cost of $\sum\limits_{i=1}^{n} a_{ij}$ dollars. Clearly, production of the jth product will not be profitable unless $\sum\limits_{i=1}^{n} a_{ij} < 1$. Let d_i denote the demand of the open sector for the ith product. Finally, let x_i represent the amount of output of the ith product necessary to meet the total demand. If the jth industry is to have an output of x_j, it will need an input of $a_{ij}x_j$ units from the ith industry. Thus the total demand for the ith product will be

$$a_{i1}x_1 + a_{i2}x_2 + \cdots + a_{in}x_n + d_i$$

and hence we require that

$$x_i = a_{i1}x_1 + a_{i2}x_2 + \cdots + a_{in}x_n + d_i$$

for $i = 1, \ldots, n$. This leads to the system

$$
\begin{aligned}
(1 - a_{11})x_1 &\quad - a_{12}x_2 - \cdots &\quad - a_{1n}x_n = d_1 \\
-a_{21}x_1 &+ (1 - a_{22})x_2 - \cdots &\quad - a_{2n}x_n = d_2 \\
&\;\;\vdots \\
-a_{n1}x_1 &\quad - a_{n2}x_2 - \cdots - (1 - a_{nn})x_n = d_n
\end{aligned}
$$

which may be written in the form

(1) $(I - A)\mathbf{x} = \mathbf{d}$

The entries of A have two important properties:

(i) $a_{ij} \geq 0$ for each i and j

(ii) $\sum\limits_{i=1}^{n} a_{ij} < 1$

The vector $\mathbf{x}$ must not only be a solution to (1), it must be nonnegative. (It would not make any sense to have a negative output.) It follows from condition (ii) that $\|A\|_1 < 1$. Thus, if λ is any eigenvalue of A and $\mathbf{x}$ is an eigenvector belonging to λ, then

$$|\lambda| \|\mathbf{x}\|_1 = \|\lambda \mathbf{x}\|_1 = \|A\mathbf{x}\|_1 \leq \|A\|_1 \|\mathbf{x}\|_1$$

and hence

$$|\lambda| \leq \|A\|_1 < 1$$

Thus 1 is not an eigenvalue of A. It follows that $I - A$ is nonsingular and hence the system (1) has a unique solution.

$$\mathbf{x} = (I - A)^{-1}\mathbf{d}$$

We would like to show that this solution must be nonnegative. To do this, we will show that $(I - A)^{-1}$ is nonnegative. First note that

$$\|A^m\|_1 \leq \|A\|_1^m$$

Since $\|A\|_1 < 1$, it follows that

$$\|A^m\|_1 \to 0 \qquad \text{as} \quad m \to \infty$$

and hence A^m approaches the zero matrix as $m \to \infty$.

Since

$$(I - A)(I + A + \cdots + A^m) = I - A^{m+1}$$

it follows that

$$I + A + \cdots + A^m = (I - A)^{-1} - (I - A)^{-1} A^{m+1}$$

and hence $I + A + \cdots + A^m$ approaches $(I - A)^{-1}$ as $m \to \infty$. By condition (i), $I + A + \cdots + A^m$ is nonnegative for each m and therefore $(I - A)^{-1}$ must be nonnegative. Since $\mathbf{d}$ is nonnegative, it follows that the solution $\mathbf{x}$ must be nonnegative. We see, then, that conditions (i) and (ii) guarantee that the system (1) will have a unique nonnegative solution $\mathbf{x}$.

As you have probably guessed, there is also a closed version of the Leontief input–output model. In the closed version it is assumed that each industry must produce enough output to meet the input needs of only the other industries and itself. The open sector is ignored. Thus, in place of the system (1), we have

$$(I - A)\mathbf{x} = \mathbf{0}$$

and we require that $\mathbf{x}$ be a positive solution. The existence of such an $\mathbf{x}$ in this case is a much deeper result than in the open version and requires some more advanced theorems.

Theorem 6.7.1 (Perron).   *If A is a positive $n \times n$ matrix, A has a positive real eigenvalue r with the following properties*:
  (i) *r is a simple root of the characteristic equation.*
  (ii) *r has a positive eigenvector $\mathbf{x}$.*
  (iii) *If λ is any other eigenvalue of A, then $|\lambda| < r$.*

The Perron theorem may be thought of as a special case of a more general theorem due to Frobenius. The Frobenius theorem applies to "irreducible" nonnegative matrices.

Definition.   A nonnegative matrix A is said to be **reducible** if there exists a partition of the index set $\{1, 2, \ldots, n\}$ into nonempty disjoint sets I_1 and I_2 such that $a_{ij} = 0$ whenever $i \in I_1$ and $j \in I_2$. Otherwise, A is said to be **irreducible**.

EXAMPLE 1.   Let A be a matrix of the form

$$\begin{pmatrix} \times & \times & 0 & 0 & \times \\ \times & \times & 0 & 0 & \times \\ \times & \times & \times & \times & \times \\ \times & \times & \times & \times & \times \\ \times & \times & 0 & 0 & \times \end{pmatrix}$$

Let $I_1 = \{1, 2, 5\}$ and $I_2 = \{3, 4\}$. Then $I_1 \cup I_2 = \{1, 2, 3, 4, 5\}$ and $a_{ij} = 0$ whenever $i \in I_1$ and $j \in I_2$. Therefore, A is reducible. Notice that if P is the permutation matrix formed by interchanging the third and fifth rows of the identity matrix I, then

$$PA = \begin{pmatrix} \times & \times & 0 & 0 & \times \\ \times & \times & 0 & 0 & \times \\ \times & \times & 0 & 0 & \times \\ \times & \times & \times & \times & \times \\ \times & \times & \times & \times & \times \end{pmatrix}$$

and

$$PAP^T = \left(\begin{array}{ccc|cc} \times & \times & \times & 0 & 0 \\ \times & \times & \times & 0 & 0 \\ \times & \times & \times & 0 & 0 \\ \hline \times & \times & \times & \times & \times \\ \times & \times & \times & \times & \times \end{array} \right)$$

In general, it can be shown that an $n \times n$ matrix A is reducible if and only if there exists a permutation matrix P such that PAP^T is a matrix of the form

$$\left(\begin{array}{c|c} B & O \\ \hline X & C \end{array} \right)$$

where B and C are square matrices. □

Theorem 6.7.2 (Frobenius). *If A is an irreducible nonnegative matrix, then A has a positive real eigenvalue r with the following properties:*
  (i) *r has a positive eigenvector $\mathbf{x}$.*
  (ii) *If λ is any other eigenvalue of A, then $|\lambda| \leq r$. The eigenvalues of modulus r are all simple roots of the characteristic equation. Indeed, if there are m eigenvalues of modulus r, they must be of the form*

$$\lambda_k = r\left[\exp\left(\frac{2k\pi i}{m} \right) \right] \qquad k = 0, 1, \ldots, m - 1$$

The proof of this theorem is beyond the scope of the text. We refer the reader to Gantmacher [2, Vol. 2]. Perron's theorem follows as a special case of the Frobenius theorem.

APPLICATION 2: THE CLOSED MODEL

In the closed Leontief input–output model, we assume that there is no demand from the open sector, and we wish to find outputs to satisfy the demands of all n industries. Thus, defining the x_i's and a_{ij}'s as in the open model, we have

$$x_i = a_{i1}x_1 + a_{i2}x_2 + \cdots + a_{in}x_n$$

for $i = 1, \ldots, n$. The resulting system may be written in the form

(2) $(A - I)\mathbf{x} = \mathbf{0}$

As before, we have the condition

(i) $a_{ij} \geq 0$

Since there is no open sector, the amount of output for the jth industry should be the same as the total input for that industry. Thus

$$x_j = \sum_{i=1}^{n} a_{ij}x_j$$

and hence we have as our second condition

(ii) $\sum_{i=1}^{n} a_{ij} = 1 \qquad j = 1, \ldots, n$

Condition (ii) implies that $A - I$ is singular since the sum of its row vectors is $\mathbf{0}$. Therefore, 1 is an eigenvalue of A, and since $\|A\|_1 = 1$, it follows that all the eigenvalues of A have moduli less than or equal to 1. Let us assume that enough of the coefficients of A are nonzero so that A is irreducible. Then by Theorem 6.7.2, $\lambda = 1$ has a positive eigenvector $\mathbf{x}$. Thus any positive multiple of $\mathbf{x}$ will be a positive solution to (2).

EXERCISES

1. Find the eigenvalues of each of the following matrices and verify that conditions (i), (ii), and (iii) of Theorem 6.7.1 hold.

(a) $\begin{pmatrix} 2 & 3 \\ 2 & 1 \end{pmatrix}$ (b) $\begin{pmatrix} 4 & 2 \\ 2 & 7 \end{pmatrix}$ (c) $\begin{pmatrix} 1 & 2 & 4 \\ 2 & 4 & 1 \\ 1 & 2 & 4 \end{pmatrix}$

2. Find the eigenvalues of each of the following matrices and verify that conditions (i) and (ii) of Theorem 6.7.2 hold.

(a) $\begin{pmatrix} 2 & 3 \\ 1 & 0 \end{pmatrix}$ (b) $\begin{pmatrix} 0 & 2 \\ 2 & 0 \end{pmatrix}$ (c) $\begin{pmatrix} 0 & 0 & 8 \\ 1 & 0 & 0 \\ 0 & 1 & 0 \end{pmatrix}$

3. Find the output vector $\mathbf{x}$ in the open version of the Leontief input–output model if

$$A = \begin{pmatrix} 0.2 & 0.4 & 0.4 \\ 0.4 & 0.2 & 0.2 \\ 0.0 & 0.2 & 0.2 \end{pmatrix} \qquad \text{and} \qquad \mathbf{d} = \begin{pmatrix} 16,000 \\ 8,000 \\ 24,000 \end{pmatrix}$$

4. Consider the closed version of the Leontief input–output model with input matrix

$$A = \begin{pmatrix} 0.5 & 0.4 & 0.1 \\ 0.5 & 0.0 & 0.5 \\ 0.0 & 0.6 & 0.4 \end{pmatrix}$$

If $\mathbf{x} = (x_1, x_2, x_3)^T$ is any output vector for this model, how are the coordinates x_1, x_2, and x_3 related?

5. Prove: If $A^m = O$ for some positive integer m, then $I - A$ is nonsingular.

6. Let

$$A = \begin{pmatrix} 0 & 1 & 1 \\ 0 & -1 & 1 \\ 0 & -1 & 1 \end{pmatrix}$$

(a) Compute $(I - A)^{-1}$.
(b) Compute A^2 and A^3. Verify that $(I - A)^{-1} = I + A + A^2$.

7. Which of the following matrices are reducible? For each reducible matrix, find a permutation matrix P such that PAP^T is of the form

$$\left(\begin{array}{c|c} B & O \\ \hline X & C \end{array} \right)$$

where B and C are square matrices.

(a) $\begin{pmatrix} 1 & 1 & 1 & 0 \\ 1 & 1 & 1 & 0 \\ 1 & 1 & 1 & 1 \\ 1 & 1 & 1 & 1 \end{pmatrix}$ (b) $\begin{pmatrix} 1 & 0 & 1 & 1 \\ 1 & 1 & 1 & 1 \\ 1 & 0 & 1 & 1 \\ 1 & 0 & 1 & 1 \end{pmatrix}$

(c) $\begin{pmatrix} 1 & 0 & 1 & 0 & 0 \\ 0 & 1 & 1 & 1 & 1 \\ 1 & 0 & 1 & 0 & 0 \\ 1 & 1 & 0 & 1 & 1 \\ 1 & 1 & 1 & 1 & 1 \end{pmatrix}$ (d) $\begin{pmatrix} 1 & 1 & 1 & 1 & 1 \\ 1 & 1 & 0 & 0 & 1 \\ 1 & 1 & 1 & 1 & 1 \\ 1 & 1 & 0 & 0 & 1 \\ 1 & 1 & 0 & 0 & 1 \end{pmatrix}$

8. Let A be a nonnegative irreducible 3×3 matrix whose eigenvalues satisfy $\lambda_1 = 2 = |\lambda_2| = |\lambda_3|$. Determine λ_2 and λ_3.

9. Let

$$A = \left(\begin{array}{c|c} B & O \\ \hline O & C \end{array} \right)$$

where B and C are square matrices.
(a) If λ is an eigenvalue of B with eigenvector $\mathbf{x} = (x_1, \ldots, x_k)^T$, show that λ is also an eigenvalue of A with eigenvector $\tilde{\mathbf{x}} = (x_1, \ldots, x_k, 0, \ldots, 0)^T$.

(b) If B and C are positive matrices, show that A has a positive real eigenvalue r with the property that $|\lambda| < r$ for any eigenvalue $\lambda \neq r$. Show also that the multiplicity of r is at most 2 and that r has a nonnegative eigenvector.

(c) If $B = C$, show that the eigenvalue r in part (b) has multiplicity 2 and possesses a positive eigenvector.

10. Prove that a 2×2 matrix A is reducible if and only if $a_{12}a_{21} = 0$.

11. Prove the Frobenius theorem in the case where A is a 2×2 matrix.

7

NUMERICAL LINEAR ALGEBRA

In this chapter we consider computer methods for solving linear algebra problems. To understand these methods, one should be familiar with the type of number system used by the computer. When data are read into the computer, they are translated into the finite number system of the computer. This translation will usually involve some roundoff error. Further roundoff errors will occur when the algebraic operations of the algorithm are carried out. Because of this, we cannot expect to get the exact solution to the original problem. The best we can hope for is a good approximation to a slightly perturbed problem. Suppose, for example, that one wanted to solve $A\mathbf{x} = \mathbf{b}$. When the entries of A and $\mathbf{b}$ are read into the computer, roundoff errors will generally occur. Thus the program will actually be attempting to compute a good approximation to the solution of a system of the form $(A + E)\mathbf{x} = \hat{\mathbf{b}}$. An algorithm is said to be stable if it will produce a good approximation to a slightly perturbed problem. Algorithms that ordinarily would converge to the solution using exact arithmetic could very well fail to be stable, owing to the growth of error in the algebraic processes. Even using a stable algorithm, we may encounter problems that are very sensitive to perturbations. For example, if A is "nearly singular," the exact solutions of $A\mathbf{x} = \mathbf{b}$ and $(A + E)\mathbf{x} = \mathbf{b}$ may vary greatly even though all the entries of E are small. The major part of this chapter is devoted to numerical methods for solving linear systems. We will pay particular attention to the growth of error and to the sensitivity of systems to small changes.

Another problem that is very important in numerical applications is the problem of finding the eigenvalues of a matrix. Two iterative methods for computing eigenvalues are presented in Section 8. The second of these methods is the powerful QR algorithm, which makes use of the special types of orthogonal transformations presented in Section 5.

In many matrix computations it is important to know whether or not a matrix is close to being singular. In Section 7 the concept of the singular values of a matrix is introduced. The smallest nonzero singular value can be used as a measure of how close the matrix is to a matrix of lower rank. In Section 9 we show how the singular values can be used in the solution of least squares problems. The Golub–Reinsch algorithm for computing the singular values is also presented in this section.

1. FLOATING-POINT NUMBERS

In solving a numerical problem on the computer, we do not usually expect to get the exact answer. Some amount of error is inevitable. Roundoff error may occur initially when the data are represented in the finite number system of the computer. Further roundoff error may occur whenever arithmetic operations are used. Like a cancer, these errors may grow to such an extent that the computed solution may be completely unreliable. To avoid this, one must understand how computational errors occur. To do this, one must be familiar with the type of numbers used by the computer.

Definition. A **floating-point number** in base b is a number of the form

$$\pm\left(\frac{d_1}{b} + \frac{d_2}{b^2} + \cdots + \frac{d_t}{b^t}\right) \times b^e$$

where $t, d_1, d_2, \ldots, d_t, b, e$ are all integers and

$$0 \le d_i \le b - 1 \qquad i = 1, \ldots, t$$

The integer t refers to the number of digits and this depends on the word length of the computer. The exponent e is restricted to be within certain bounds, $L \leq e \leq U$, which also depend on the particular computer. Most computers use base 2, although some use other bases such as 8 or 16. Hand calculators generally use base 10.

EXAMPLE 1. The following are five-digit decimal (base 10) floating-point numbers:

$$0.53216 \times 10^{-4}$$

$$-0.81724 \times 10^{21}$$

$$0.00112 \times 10^{8}$$

$$0.11200 \times 10^{6}$$

Note that the numbers 0.00112×10^{8} and 0.11200×10^{6} are equal. Thus the floating-point representation of a number need not be unique. Floating-point numbers that are written with no leading zeros are said to be normalized. ☐

EXAMPLE 2. $(0.236)_8 \times 8^2$ and $(0.132)_8 \times 8^4$ are normalized three-digit base 8 floating-point numbers. Here $(0.236)_8$ represents

$$\frac{2}{8} + \frac{3}{8^2} + \frac{6}{8^3}$$

Thus $(0.236)_8 \times 8^2$ is the base 8 floating-point representation of the decimal number 19.75. Similarly,

$$(0.132)_8 \times 8^4 = \left(\frac{1}{8} + \frac{3}{8^2} + \frac{2}{8^3} \right) \times 8^4 = 720$$

☐

To better understand the type of number system we are working with, it may help to look at a very simple example.

EXAMPLE 3. Suppose that $t = 1$, $L = -1$, $U = 1$, and $b = 10$. There are altogether 55 one-digit decimal floating-point numbers in this system. These are

$$0, \pm 0.1 \times 10^{-1}, \pm 0.2 \times 10^{-1}, \ldots, \pm 0.9 \times 10^{-1}$$

$$\pm 0.1 \times 10^{0}, \pm 0.2 \times 10^{0}, \ldots, \pm 0.9 \times 10^{0}$$

$$\pm 0.1 \times 10^{1}, \pm 0.2 \times 10^{1}, \ldots, \pm 0.9 \times 10^{1}$$

Although all these numbers lie in the interval $[-9, 9]$, over one third of the numbers have absolute value less than 0.1 and over two thirds of the numbers have absolute value less than 1. Figure 7.1.1 illustrates how the floating-point numbers in the interval $[0, 2]$ are distributed. ☐

FIGURE 7.1.1

Most real numbers have to be rounded off in order to be represented as t-digit floating-point numbers. The difference between the floating-point number x' and the original number x is called the *roundoff error*. The size of the roundoff error is perhaps more meaningful when compared to the size of the original number.

Definition. If x is a real number and x' is its floating-point approximation, then the difference $x' - x$ is called the **absolute error**. Also, the quotient $(x' - x)/x$ is called the **relative error**.

Real Number, x	Four-Digit Decimal Floating-Point Number, x'	Absolute Error, $x' - x$	Relative Error, $(x' - x)/x$
62,133	0.6213×10^5	-3	$\dfrac{-3}{62,133} \approx -4.8 \times 10^{-5}$
0.12658	0.1266×10^0	2×10^{-5}	$\dfrac{1}{6329} \approx 1.6 \times 10^{-4}$
47.213	0.4721×10^2	-3.0×10^{-3}	$\dfrac{-0.003}{47.213} \approx -6.4 \times 10^{-5}$
π	0.3142×10^1	$3.142 - \pi \approx 4 \times 10^{-4}$	$\dfrac{3.142 - \pi}{\pi} \approx 1.3 \times 10^{-4}$

When arithmetic operations are applied to floating-point numbers, additional roundoff errors may occur.

EXAMPLE 4. Let $a' = 0.263 \times 10^4$ and $b' = 0.446 \times 10^1$ be three-digit decimal floating-point numbers. If these numbers are added, the exact sum will be

$$a' + b' = 0.263446 \times 10^4$$

However, the floating-point representation of this sum is 0.263×10^4. This then should be the computed sum. We will denote the floating-point sum by $fl(a' + b')$. The absolute error in the sum is

$$fl(a' + b') - (a' + b') = -4.46$$

and the relative error is

$$\frac{-4.46}{0.26344 \times 10^4} \approx -0.17 \times 10^{-2}$$

The actual value of $a'b'$ is 11,729.8; however, $fl(a'b')$ is 0.117×10^5. The absolute error in the product is -29.8 and the relative error is approximately -0.25×10^{-2}. Floating-point subtraction and division can be done in a similar manner. $\qquad\qquad\square$

The relative error in approximating a number x by its floating-point representation x' is usually denoted by the symbol δ. Thus

(1) $$\delta = \frac{x' - x}{x} \qquad \text{or} \qquad x' = x(1 + \delta)$$

$|\delta|$ can be bounded by a positive constant ϵ, called the *machine precision* or the *machine epsilon*. The machine epsilon is defined to be the smallest floating-point number ϵ for which

$$fl(1 + \epsilon) > 1$$

For example, if the computer uses three-digit decimal floating-point numbers, then

$$fl(1 + 0.499 \times 10^{-2}) = 1$$

while

$$fl(1 + 0.500 \times 10^{-2}) = 1.01$$

In this case the machine epsilon would be 0.500×10^{-2}.

It follows from (1) that if a' and b' are two floating-point numbers, then

$$fl(a' + b') = (a' + b')(1 + \delta_1)$$
$$fl(a'b') = (a'b')(1 + \delta_2)$$
$$fl(a' - b') = (a' - b')(1 + \delta_3)$$
$$fl(a' \div b') = (a' \div b')(1 + \delta_4)$$

The δ_i's are the relative errors and will all have absolute values less than ϵ. Note that in Example 4, $\delta_1 \approx -0.17 \times 10^{-2}$, $\delta_2 \approx -0.25 \times 10^{-2}$, and $\epsilon = 0.5 \times 10^{-2}$.

If the numbers you are working with involve some slight errors, arithmetic operations may compound these errors. If two numbers agree to k decimal places and one number is subtracted from the other, there will be a loss of significant digits in your answer. In this case the relative error in the difference will be many times as great as the relative error in either of the numbers.

EXAMPLE 5. Let $c = 3.4215298$ and $d = 3.4213851$. Calculate $c - d$ and $1/(c - d)$ using six-digit decimal floating-point arithmetic.

SOLUTION
  (i) The first step is to represent c and d by six-digit decimal floating-point numbers.

$$c' = 0.342153 \times 10^1$$
$$d' = 0.342139 \times 10^1$$

The relative errors in c and d are

$$\frac{c' - c}{c} \approx 0.6 \times 10^{-7} \quad \text{and} \quad \frac{d' - d}{d} \approx 1.4 \times 10^{-6}$$

(ii) $fl(c' - d') = c' - d' = 0.140000 \times 10^{-3}$.

The actual value of $c - d$ is 0.1447×10^{-3}. The absolute and relative errors in approximating $c - d$ by $fl(c' - d')$ are

$$fl(c' - d') - (c - d) = -0.47 \times 10^{-5}$$

and

$$\frac{fl(c' - d') - (c - d)}{c - d} \approx -3.2 \times 10^{-2}$$

Note that the magnitude of the relative error in the difference is more than 10^4 times the relative error in either c or d.

(iii) $fl[1/(c' - d')] = 0.714286 \times 10^4$, and the correct answer to six significant figures is

$$\frac{1}{c - d} \approx 0.691085 \times 10^4$$

The absolute and relative errors are approximately 232 and 0.03.   □

EXERCISES

1. Find the three-digit decimal floating-point representation of each of the following numbers.

   (a) 2312   (b) 32.56   (c) 0.01277   (d) 82,431

2. Find the absolute error and the relative error when each of the real numbers in Exercise 1 is approximated by a three-digit decimal floating-point number.

3. Represent each of the following as five-digit base 2 floating-point numbers.

   (a) 21   (b) $\frac{3}{8}$ (c) 9.872   (d) -0.1

4. Do each of the following using four-digit decimal floating-point arithmetic and calculate the absolute and relative errors in your answers.

   (a) $10{,}420 + 0.0018$ (b) $10{,}424 - 10{,}416$

   (c) $0.12347 - 0.12342$ (d) $(3626.6) \cdot (22.656)$

5. Let $x_1 = 94{,}210$, $x_2 = 8631$, $x_3 = 1440$, $x_4 = 133$, and $x_5 = 34$. Calculate each of the following using four-digit decimal floating-point arithmetic.

   (a) $(((x_1 + x_2) + x_3) + x_4) + x_5$

   (b) $x_1 + ((x_2 + x_3) + (x_4 + x_5))$

   (c) $(((x_5 + x_4) + x_3) + x_2) + x_1$

6. What would the machine epsilon be for a computer that uses five-digit base 2 floating-point arithmetic?

7. How many floating-point numbers are there in the system if $t = 2$, $L = -2$, $U = 2$, and $b = 2$?

2. GAUSSIAN ELIMINATION

In this section we will discuss the problem of solving a system of n linear equations in n unknowns using Gaussian elimination. Gaussian elimination is generally considered to be the most efficient computational method, since it involves the least amount of arithmetic operations.

GAUSSIAN ELIMINATION WITHOUT INTERCHANGES

Let $A = A^{(1)} = (a_{ij}^{(1)})$ be a nonsingular matrix. Then A can be reduced to triangular form using row operations I and III. For simplicity, let us assume that the reduction can be done using only row operation III.

$$A^{(1)} = \begin{pmatrix} a_{11}^{(1)} & a_{12}^{(1)} & \cdots & a_{1n}^{(1)} \\ a_{21}^{(1)} & a_{22}^{(1)} & \cdots & a_{2n}^{(1)} \\ \vdots & & & \\ a_{n1}^{(1)} & a_{n2}^{(1)} & \cdots & a_{nn}^{(1)} \end{pmatrix}$$

Step 1.   Let $m_{k1} = a_{k1}^{(1)}/a_{11}^{(1)}$ for $k = 2, \ldots, n$ [by our assumption, $a_{11}^{(1)} \neq 0$]. The first step of the elimination process is to apply row operation III $n - 1$ times to eliminate the entries below the diagonal in the first column of A. Note that m_{k1} is the multiple of the first row that is to be subtracted from the kth row. The new matrix obtained will be

$$A^{(2)} = \begin{pmatrix} a_{11}^{(1)} & a_{12}^{(1)} & \cdots & a_{1n}^{(1)} \\ 0 & a_{22}^{(2)} & \cdots & a_{2n}^{(2)} \\ \vdots & & & \\ 0 & a_{n2}^{(2)} & \cdots & a_{nn}^{(2)} \end{pmatrix}$$

where

$$a_{kj}^{(2)} = a_{kj}^{(1)} - m_{k1}a_{1j}^{(1)} \qquad (2 \leq k \leq n, 2 \leq j \leq n)$$

The first step of the elimination process requires $n - 1$ divisions, $(n - 1)^2$ multiplications, and $(n - 1)^2$ additions/subtractions.

Step 2.   If $a_{22}^{(2)} \neq 0$, then it can be used as a pivot element to eliminate $a_{32}^{(2)}, a_{42}^{(2)}, \ldots, a_{n2}^{(2)}$. For $k = 3, \ldots, n$, set

$$m_{k2} = \frac{a_{k2}^{(2)}}{a_{22}^{(2)}}$$

and subtract m_{k2} times the second row of $A^{(2)}$ from the kth row. The new matrix obtained will be

$$A^{(3)} = \begin{pmatrix} a_{11}^{(1)} & a_{12}^{(1)} & a_{13}^{(1)} & \cdots & a_{1n}^{(1)} \\ 0 & a_{22}^{(2)} & a_{23}^{(2)} & \cdots & a_{2n}^{(2)} \\ 0 & 0 & a_{33}^{(3)} & \cdots & a_{3n}^{(3)} \\ \vdots & \vdots & \vdots & & \vdots \\ 0 & 0 & a_{n3}^{(3)} & \cdots & a_{nn}^{(3)} \end{pmatrix}$$

The second step requires $n - 2$ divisions, $(n - 2)^2$ multiplications, and $(n - 2)^2$ additions/subtractions.

After $n - 1$ steps we will end up with a triangular matrix $U = A^{(n)}$. The operation count for the entire process can be determined as follows:

Divisions:

$$(n - 1) + (n - 2) + \cdots + 1 = \frac{n(n - 1)}{2}$$

Multiplications:

$$(n - 1)^2 + (n - 2)^2 + \cdots + 1^2 = \frac{n(2n - 1)(n - 1)}{6}$$

Additions and/or subtractions:

$$(n - 1)^2 + \cdots + 1^2 = \frac{n(2n - 1)(n - 1)}{6}$$

The elimination process is summarized in the following algorithm.

Algorithm 7.2.1 (Gaussian Elimination Without Interchanges)

For $i = 1, 2, \ldots, n - 1$

    For $k = i + 1, \ldots, n$

        Set $m_{ki} = \dfrac{a_{ki}^{(i)}}{a_{ii}^{(i)}}$ (provided that $a_{ii}^{(i)} \neq 0$)

        For $j = i + 1, \ldots, n$

           Set $a_{kj}^{(i+1)} = a_{kj}^{(i)} - m_{ki} a_{ij}^{(i)}$

To solve the system $A\mathbf{x} = \mathbf{b}$, we could augment A by $\mathbf{b}$. Thus $\mathbf{b}$ would be stored in an extra column of A. The reduction process could then be done using Algorithm 7.2.1 and letting j run from $i + 1$ to

$n + 1$ instead of from $i + 1$ to n. The triangular system could then be solved using back substitution.

Most of the work involved in solving a system $A\mathbf{x} = \mathbf{b}$ occurs in the reduction of A to triangular form. Suppose that after having solved $A\mathbf{x} = \mathbf{b}$, we want to solve another system, $A\mathbf{x} = \mathbf{b}_1$. We know the triangular form U from the first system, and consequently we would like to be able to solve the new system without having to go through the entire reduction process again. We can do this if we make use of the LU factorization discussed in Section 6 of Chapter 6 (see Example 1 of that section). To compute the LU factorization, we must keep track of the numbers m_{ki} used in Algorithm 7.2.1. These numbers are called *multipliers*. The multiplier m_{ki} is the multiple of the ith row that is subtracted from the kth row during the ith step of the reduction process. To see how the multipliers can be used to solve $A\mathbf{x} = \mathbf{b}_1$, it is helpful to view the reduction process in terms of matrix multiplications.

TRIANGULAR FACTORIZATION

The first step of the reduction process involves multiplying A by $n - 1$ elementary matrices,

$$A^{(2)} = E_{n1} \cdots E_{31} E_{21} A^{(1)}$$

where

$$E_{k1} = \begin{pmatrix} 1 & & & & & \\ 0 & 1 & & & & \\ \vdots & & \ddots & & & \\ -m_{k1} & 0 & \cdots & 1 & & \\ \vdots & & & & \ddots & \\ 0 & 0 & \cdots & 0 & \cdots & 1 \end{pmatrix}$$

Each E_{k1} is nonsingular with

$$E_{k1}^{-1} = \begin{pmatrix} 1 & & & & & \\ 0 & 1 & & & & \\ \vdots & & \ddots & & & \\ m_{k1} & 0 & \cdots & 1 & & \\ \vdots & & & & \ddots & \\ 0 & 0 & \cdots & 0 & \cdots & 1 \end{pmatrix}$$

Let

$$M_1 = E_{n1} \cdots E_{31} E_{21} = \begin{pmatrix} 1 & & & & \\ -m_{21} & 1 & & & \\ -m_{31} & 0 & 1 & & \\ \vdots & & & \ddots & \\ -m_{n1} & 0 & 0 & \cdots & 1 \end{pmatrix}$$

Thus $A^{(2)} = M_1 A$. The matrix M_1 is nonsingular and

$$M_1^{-1} = E_{21}^{-1} E_{31}^{-1} \cdots E_{n1}^{-1} = \begin{pmatrix} 1 & & & & \\ m_{21} & 1 & & & \\ m_{31} & 0 & 1 & & \\ \vdots & & & \ddots & \\ m_{n1} & 0 & 0 & \cdots & 1 \end{pmatrix}$$

Similarly,

$$\begin{aligned} A^{(3)} &= E_{n2} \cdots E_{42} E_{32} A^{(2)} \\ &= M_2 A^{(2)} \\ &= M_2 M_1 A \end{aligned}$$

where

$$M_2 = E_{n2} \cdots E_{32} = \begin{pmatrix} 1 & & & & \\ 0 & 1 & & & \\ 0 & -m_{32} & 1 & & \\ \vdots & \vdots & \vdots & \ddots & \\ 0 & -m_{n2} & 0 & \cdots & 1 \end{pmatrix}$$

and

$$M_2^{-1} = \begin{pmatrix} 1 & & & & \\ 0 & 1 & & & \\ 0 & m_{32} & 1 & & \\ \vdots & \vdots & \vdots & \ddots & \\ 0 & m_{n2} & 0 & \cdots & 1 \end{pmatrix}$$

At the end of the reduction process, we have

$$U = A^{(n)} = M_{n-1} \cdots M_2 M_1 A$$

It follows that

$$A = M_1^{-1} M_2^{-1} \cdots M_{n-1}^{-1} U$$

The M_j^{-1}'s multiply out to give the following lower triangular matrix when they are taken in this order:

$$L = M_1^{-1} M_2^{-1} \cdots M_{n-1}^{-1} = \begin{pmatrix} 1 & 0 & 0 & \cdots & 0 \\ m_{21} & 1 & 0 & \cdots & 0 \\ m_{31} & m_{32} & 1 & \cdots & 0 \\ \vdots & & & & \\ m_{n1} & m_{n2} & m_{n3} & \cdots & 1 \end{pmatrix}$$

Thus $A = LU$, where L is lower triangular and U is upper triangular.

EXAMPLE 1. Let

$$A = \begin{pmatrix} 2 & 3 & 1 \\ 4 & 1 & 4 \\ 3 & 4 & 6 \end{pmatrix}$$

The elimination can be carried out in two steps:

$$\begin{pmatrix} 2 & 3 & 1 \\ 4 & 1 & 4 \\ 3 & 4 & 6 \end{pmatrix} \overset{1}{\rightarrow} \begin{pmatrix} 2 & 3 & 1 \\ 0 & -5 & 2 \\ 0 & -\frac{1}{2} & \frac{9}{2} \end{pmatrix} \overset{2}{\rightarrow} \begin{pmatrix} 2 & 3 & .1 \\ 0 & -5 & 2 \\ 0 & 0 & 4.3 \end{pmatrix}$$

The multipliers used were $m_{21} = 2$, $m_{31} = \frac{3}{2}$ (step 1), and $m_{32} = \frac{1}{10}$ (step 2). Let

$$L = \begin{pmatrix} 1 & 0 & 0 \\ m_{21} & 1 & 0 \\ m_{31} & m_{32} & 1 \end{pmatrix} = \begin{pmatrix} 1 & 0 & 0 \\ 2 & 1 & 0 \\ \frac{3}{2} & \frac{1}{10} & 1 \end{pmatrix}$$

and

$$U = \begin{pmatrix} 2 & 3 & 1 \\ 0 & -5 & 2 \\ 0 & 0 & 4.3 \end{pmatrix}$$

The reader may verify that $LU = A$.   □

Once A has been reduced to triangular form and the factorization LU has been determined, the system $A\mathbf{x} = \mathbf{b}$ can be solved in two steps.

Step 1.   Forward Substitution.   The system $A\mathbf{x} = \mathbf{b}$ can be written in the form

$$LU\mathbf{x} = \mathbf{b}$$

Let $\mathbf{y} = U\mathbf{x}$. It follows that

$$L\mathbf{y} = LU\mathbf{x} = \mathbf{b}$$

Thus one can find $\mathbf{y}$ by solving the lower triangular system

$$\begin{aligned} y_1 &= b_1 \\ m_{21}y_1 + y_2 &= b_2 \\ m_{31}y_1 + m_{32}y_2 + y_3 &= b_3 \\ &\vdots \\ m_{n1}y_1 + m_{n2}y_2 + m_{n3}y_3 + \cdots + y_n &= b_n \end{aligned}$$

It follows from the first equation that $y_1 = b_1$. This value can be used in the second equation to solve for y_2. The values of y_1 and y_2 can be used in the third equation to solve for y_3, and so on. This method of solving a lower triangular system is called *forward substitution*.

Step 2.   Back Substitution.   Once **y** has been determined, we need only solve the upper triangular system $U\mathbf{x} = \mathbf{y}$ to find the solution **x** to the system. The upper triangular system is solved by back substitution.

EXAMPLE 2.   Solve the system

$$2x_1 + 3x_2 + \ x_3 = -4$$

$$4x_1 + \ x_2 + 4x_3 = \ \ \ 9$$

$$3x_1 + 4x_2 + 6x_3 = \ \ \ 0$$

SOLUTION.   The coefficient matrix for this system is the matrix A in Example 1. Since L and U have been determined, the system can be solved using forward and back substitution.

$$\left(\begin{array}{ccc|c} 1 & 0 & 0 & -4 \\ 2 & 1 & 0 & 9 \\ \frac{3}{2} & \frac{1}{10} & 1 & 0 \end{array}\right) \qquad \begin{array}{l} y_1 = -4 \\ y_2 = 9 - 2y_1 = 17 \\ y_3 = 0 - \frac{3}{2}y_1 - \frac{1}{10}y_2 = 4.3 \end{array}$$

$$\left(\begin{array}{ccc|c} 2 & 3 & 1 & -4 \\ 0 & -5 & 2 & 17 \\ 0 & 0 & 4.3 & 4.3 \end{array}\right) \qquad \begin{array}{rcrcr} 2x_1 + 3x_2 + & x_3 & = -4 & x_1 = & 2 \\ -5x_2 + & 2x_3 & = 17 & x_2 = & -3 \\ & 4.3x_3 & = 4.3 & x_3 = & 1 \end{array}$$

The solution to the system is $\mathbf{x} = (2, -3, 1)^T$.   □

Algorithm 7.2.2 (Forward and Back Substitution)

$$\boxed{\begin{array}{l} \text{For } k = 1, \ldots, n \\[2mm] \qquad \text{Set } y_k = b_k - \sum_{i=1}^{k-1} m_{ki} y_i \end{array}}$$

$$\boxed{\begin{array}{l} \text{For } k = n, n-1, \ldots, 1 \\[2mm] \qquad \text{Set } x_k = \dfrac{y_k - \sum_{j=k+1}^{n} u_{kj} x_j}{u_{kk}} \end{array}}$$

Operation Count. Algorithm 7.2.2 requires n divisions, $n(n-1)$ multiplications, and $n(n-1)$ additions/subtractions. The total operation count for solving a system $A\mathbf{x} = \mathbf{b}$ using Algorithms 7.2.1 and 7.2.2 is then

Multiplications/divisions: $\frac{1}{3}n^3 + n^2 - \frac{1}{3}n$

Additions/subtractions: $\frac{1}{3}n^3 + \frac{1}{2}n^2 - \frac{5}{6}n$

In both cases the $\frac{1}{3}n^3$ is the dominant term. We will say that solving a system by Gaussian elimination involves roughly $\frac{1}{3}n^3$ multiplications/divisions and $\frac{1}{3}n^3$ additions/subtractions.

Storage.   It is not necessary to store the multipliers in a separate matrix L. Each multiplier m_{ki} can be stored in the matrix A in place of the entry $a_{ki}^{(i)}$ eliminated. At the end of the reduction process, A is being used to store the m_{ki}'s and the u_{ij}'s.

$$\begin{pmatrix} u_{11} & u_{12} & \cdots & u_{1,n-1} & u_{1n} \\ m_{21} & u_{22} & \cdots & u_{2,n-1} & u_{2n} \\ \vdots & & & & \\ m_{n1} & m_{n2} & \cdots & m_{n,n-1} & u_{nn} \end{pmatrix}$$

Algorithm 7.2.1 breaks down if at any step $a_{kk}^{(k)}$ is 0. If this happens, it is necessary to perform row interchanges. In the next section we will see how to incorporate interchanges into our elimination algorithm.

EXERCISES

1. Let

$$A = \begin{pmatrix} 1 & 1 & 1 \\ 2 & 4 & 1 \\ -3 & 1 & -2 \end{pmatrix}$$

Factor A into a product LU, where L is lower triangular with 1's along the diagonal and U is upper triangular.

2. Let A be the matrix in Exercise 1. Use the LU factorization of A to solve $A\mathbf{x} = \mathbf{b}$ for each of the following choices of $\mathbf{b}$.
(a) $(4, 3, -13)^T$ (b) $(3, 1, -10)^T$ (c) $(7, 23, 0)^T$

3. Let A and B be $n \times n$ matrices and let $\mathbf{x} \in R^n$.
(a) How many scalar additions and multiplications are necessary to compute the product $A\mathbf{x}$?
(b) How many scalar additions and multiplications are necessary to compute the product AB?

4. Let E_{ki} be the elementary matrix formed by subtracting α times the ith row of the identity matrix from the kth row.
(a) Show that $E_{ki} = I - \alpha\mathbf{e}_k\mathbf{e}_i^T$.
(b) Let $E_{ji} = I - \beta\mathbf{e}_j\mathbf{e}_i^T$. Show that $E_{ji}E_{ki} = I - (\alpha\mathbf{e}_k + \beta\mathbf{e}_j)\mathbf{e}_i^T$.

5. Let A be an $n \times n$ matrix with triangular factorization LU. Show that

$$\det(A) = u_{11}u_{22} \cdots u_{nn}$$

6. If A is a symmetric $n \times n$ matrix with a triangular factorization LU, then A can be factored further into a product LDL^T (where D is diagonal). Devise an algorithm, similar to Algorithm 7.2.2, for solving $LDL^T\mathbf{x} = \mathbf{b}$.

7. Write an algorithm for solving the tridiagonal system

$$\begin{pmatrix} a_1 & b_1 & & & \\ c_1 & a_2 & \ddots & & \\ & \ddots & \ddots & & \\ & & \ddots & a_{n-1} & b_{n-1} \\ & & & c_{n-1} & a_n \end{pmatrix} \begin{pmatrix} x_1 \\ x_2 \\ \vdots \\ x_{n-1} \\ x_n \end{pmatrix} = \begin{pmatrix} d_1 \\ d_2 \\ \vdots \\ d_{n-1} \\ d_n \end{pmatrix}$$

using Gaussian elimination with the diagonal elements as pivots. How many additions/subtractions and multiplications/divisions are necessary?

8. Let $A = LU$, where L is lower triangular with 1's on the diagonal and U is upper triangular.
(a) How many scalar additions and multiplications are necessary to solve $L\mathbf{y} = \mathbf{e}_j$ using forward substitution?
(b) How many additions/subtractions and multiplications/divisions are necessary to solve $A\mathbf{x} = \mathbf{e}_j$? The solution $\mathbf{x}_j$ to $A\mathbf{x} = \mathbf{e}_j$ will be the jth column of A^{-1}.
(c) Given the factorization $A = LU$, how many additional multiplications/divisions and additions/subtractions are needed to compute A^{-1}?

9. Suppose that A^{-1} and the LU factorization of A have already been determined. How many scalar additions and multiplications are necessary to compute $A^{-1}\mathbf{b}$? Compare this with the number of operations required to solve $LU\mathbf{x} = \mathbf{b}$ using Algorithm 7.2.2. Suppose that we have a number of systems to solve with the same coefficient matrix A. Is it worthwhile to compute A^{-1}? Explain.

3. PIVOTING STRATEGIES

In this section we will present an algorithm for Gaussian elimination with row interchanges. At each step of the algorithm it will be necessary to choose a pivotal row. One can often avoid unnecessarily large error accumulations by choosing the pivotal rows in a reasonable manner.

GAUSSIAN ELIMINATION WITH INTERCHANGES

Consider the following example.

EXAMPLE 1. Let

$$A = \begin{pmatrix} 6 & -4 & 2 \\ 4 & 2 & 1 \\ 2 & -1 & 1 \end{pmatrix}$$

We wish to reduce A to triangular form using row operations I and III. To keep track of the interchanges, we will use a row vector **p**. The coordinates of **p** will be denoted by $p(1)$, $p(2)$, $p(3)$. Initially, we set $\mathbf{p} = (1, 2, 3)$. Suppose that at the first step of the reduction process, the third row is chosen as the pivotal row. Instead of interchanging the first and third rows, we will interchange the first and third entries of **p**. Setting $p(1) = 3$ and $p(3) = 1$, the vector **p** becomes $(3, 2, 1)$. The vector **p** is used to keep track of the reordering of the rows. We can think of **p** as a renumbering of the rows. The actual physical reordering of the rows can be deferred until the end of the reduction process.

$$\begin{array}{c} \text{row} \\ \begin{array}{c} p(3) \\ p(2) \\ p(1) \end{array} \end{array} \begin{pmatrix} \boxed{6} & -4 & 2 \\ \boxed{4} & 2 & 1 \\ \textcircled{2} & -1 & 1 \end{pmatrix} \rightarrow \begin{pmatrix} 0 & -1 & -1 \\ 0 & 4 & -1 \\ 2 & -1 & 1 \end{pmatrix}$$

If at the second step row $p(3)$ is chosen as the pivotal row, the entries of $p(3)$ and $p(2)$ are switched. The final step of the elimination process is then carried out.

$$\begin{array}{c} p(2) \\ p(3) \\ p(1) \end{array} \begin{pmatrix} 0 & \textcircled{-1} & -1 \\ 0 & \boxed{4} & -1 \\ 2 & -1 & 1 \end{pmatrix} \rightarrow \begin{pmatrix} 0 & -1 & -1 \\ 0 & 0 & -5 \\ 2 & -1 & 1 \end{pmatrix}$$

If the rows of this last matrix are reordered in the order $(p(1), p(2), p(3)) = (3, 1, 2)$, then the resulting matrix will be in triangular form.

$$\begin{matrix} p(1) = 3 \\ p(2) = 1 \\ p(3) = 2 \end{matrix} \quad \begin{pmatrix} 2 & -1 & 1 \\ 0 & -1 & -1 \\ 0 & 0 & -5 \end{pmatrix}$$

Had the rows been written in the order $(3, 1, 2)$ to begin with, then the reduction would have been exactly the same except there would have been no need for interchanges. Reordering the rows of A in the order $(3, 1, 2)$ is the same as premultiplying A by the permutation matrix

$$P = \begin{pmatrix} 0 & 0 & 1 \\ 1 & 0 & 0 \\ 0 & 1 & 0 \end{pmatrix}$$

Let us perform the reduction on A and PA simultaneously and compare the results. The multipliers used in the reduction process were 3, 2, and -4. These will be stored in the place of the terms eliminated and underlined to distinguish them from the other entries of the matrix.

$$A = \begin{pmatrix} 6 & -4 & 2 \\ 4 & 2 & 1 \\ 2 & -1 & 1 \end{pmatrix} \qquad PA = \begin{pmatrix} 2 & -1 & 1 \\ 6 & -4 & 2 \\ 4 & 2 & 1 \end{pmatrix}$$

$$\downarrow \qquad\qquad\qquad\qquad \downarrow$$

$$\begin{pmatrix} \underline{3} & -1 & -1 \\ \underline{2} & 4 & -1 \\ 2 & -1 & 1 \end{pmatrix} \qquad \begin{pmatrix} 2 & -1 & 1 \\ \underline{3} & -1 & -1 \\ \underline{2} & 4 & -1 \end{pmatrix}$$

$$\downarrow \qquad\qquad\qquad\qquad \downarrow$$

$$\begin{pmatrix} \underline{3} & -1 & -1 \\ \underline{2} & -\underline{4} & -5 \\ 2 & -1 & 1 \end{pmatrix} \qquad \begin{pmatrix} 2 & -1 & 1 \\ \underline{3} & -1 & -1 \\ \underline{2} & -\underline{4} & -5 \end{pmatrix}$$

After reordering,

$$\begin{pmatrix} 2 & -1 & 1 \\ \underline{3} & -1 & -1 \\ \underline{2} & -\underline{4} & -5 \end{pmatrix}$$

the results are the same. The matrix PA can be factored into a product LU, where

$$L = \begin{pmatrix} 1 & 0 & 0 \\ 3 & 1 & 0 \\ 2 & -4 & 1 \end{pmatrix} \quad \text{and} \quad U = \begin{pmatrix} 2 & -1 & 1 \\ 0 & -1 & -1 \\ 0 & 0 & -5 \end{pmatrix} \qquad \square$$

On the computer it is not necessary to actually interchange the rows of A. We simply treat row $p(k)$ as the kth row and use $a_{p(k),j}$ in place of $a_{k,j}$.

Algorithm 7.3.1 (Gaussian Elimination with Interchanges)

$\ulcorner$*For $i = 1, \ldots, n$*
$\quad \llcorner\rightarrow$ *Set $p(i) = i$*

$\ulcorner$*For $i = 1, \ldots, n$*
$\quad$ (1) *Choose a pivot element $a_{p(j),i}$ from the elements*

$$a_{p(i),i}, a_{p(i+1),i}, \ldots, a_{p(n),i}$$

$\quad$ (*Strategies for doing this will be discussed later in the section.*)
$\quad$ (2) *Switch the ith and jth entries of p.*
$\quad$ (3) $\ulcorner$*For $k = i + 1, \ldots, n$*
$\qquad$ *Set $m_{p(k),i} = a_{p(k),i}/a_{p(i),i}$*
$\qquad \ulcorner$*For $j = i + 1, \ldots, n$*
$\qquad \quad \llcorner\rightarrow$ *Set $a_{p(k),j} = a_{p(k),j} - m_{p(k),i}a_{p(i),j}$*

Remarks

1. The multiplier $m_{p(k),i}$ is stored in the position of the element $a_{p(k),i}$ being eliminated.
2. The vector **p** can be used to form a permutation matrix P whose ith row is the $p(i)$th row of the identity matrix.
3. The matrix PA can be factored into a product LU, where

$$l_{ki} = \begin{cases} m_{p(k),i} & \text{if} \quad k > i \\ 1 & \text{if} \quad k = i \\ 0 & \text{if} \quad k < i \end{cases} \quad \text{and} \quad u_{ki} = \begin{cases} a_{p(k),i} & \text{if} \quad k \le i \\ 0 & \text{if} \quad k > i \end{cases}$$

4. Since P is nonsingular, the system $A\mathbf{x} = \mathbf{b}$ is equivalent to the system $PA\mathbf{x} = P\mathbf{b}$. Let $\hat{\mathbf{b}} = P\mathbf{b}$. Since $PA = LU$, it follows that the system is equivalent to

$$LU\mathbf{x} - \hat{\mathbf{b}}$$

5. If $PA = LU$, then $A = P^{-1}LU = P^TLU$.

It follows from Remarks 4 and 5 that if $A = P^TLU$, the system $A\mathbf{x} = \mathbf{b}$ can be solved in three steps:

Step 1. Reordering.   Reorder the entries of **b** to form $\hat{\mathbf{b}} = P\mathbf{b}$.

Step 2. Forward Substitution.   Solve the system $L\mathbf{y} = \hat{\mathbf{b}}$ for **y**.

Step 3. Back Substitution.   Solve $U\mathbf{x} = \mathbf{y}$.

EXAMPLE 2. Solve the system

$$6x_1 - 4x_2 + 2x_3 = -2$$
$$4x_1 + 2x_2 + \ x_3 = \ \ 4$$
$$2x_1 - \ x_2 + \ x_3 = -1$$

SOLUTION. The coefficient matrix of this system is the matrix A from Example 1. P, L, and U have already been determined and they can be used to solve the system as follows:

Step 1. $\hat{\mathbf{b}} = P\mathbf{b} = (-1, -2, 4)^T$

Step 2. $\begin{aligned} y_1 \quad\quad\quad\quad &= -1 \quad & y_1 &= -1 \\ 3y_1 + y_2 \quad\quad &= -2 \quad & y_2 &= -2 + 3 = 1 \\ 2y_1 - 4y_2 + y_3 &= \ \ 4 \quad & y_3 &= 4 + 2 + 4 = 10 \end{aligned}$

Step 3. $\begin{aligned} 2x_1 - x_2 + x_3 &= -1 \quad & x_1 &= \ \ 1 \\ -x_2 - x_3 &= \ \ 1 \quad & x_2 &= \ \ 1 \\ -5x_3 &= \ \ 10 \quad & x_3 &= -2 \end{aligned}$

The solution to the system is $\mathbf{x} = (1, 1, -2)^T$.   □

It is possible to do Gaussian elimination without row interchanges if the diagonal entries $a_{ii}^{(i)}$ are nonzero at each step. However, in finite-precision arithmetic, pivots $a_{ii}^{(i)}$ that are near 0 can cause problems.

EXAMPLE 3. Consider the system

$$0.0001x_1 + 2x_2 = 4$$
$$x_1 + \ x_2 = 3$$

The exact solution to the system is

$$\mathbf{x} = \left(\frac{2}{1.9999}, \frac{3.9997}{1.9997} \right)^T$$

When rounded off to four decimal places, the solution is $(1.0001, 1.9999)^T$. Let us solve the system using three-digit decimal floating-point arithmetic.

$$\begin{pmatrix} 0.0001 & 2 & | & 4 \\ 1 & 1 & | & 3 \end{pmatrix} \rightarrow \begin{pmatrix} 0.0001 & 2 & | & 4 \\ 0 & -0.200 \times 10^5 & | & -0.400 \times 10^5 \end{pmatrix}$$

The computed solution is $\mathbf{x}' = (0, 2)^T$. There is a 100 percent error in the x_1 coordinate. On the other hand, if we interchange rows to avoid the small pivot, then three-digit decimal arithmetic gives

$$\begin{pmatrix} 1 & 1 \,\big|\, 3 \\ 0.0001 & 2 \,\big|\, 4 \end{pmatrix} \rightarrow \begin{pmatrix} 1 & 1 \,\big|\, 3 \\ 0 & 2.00 \,\big|\, 4.00 \end{pmatrix}$$

In this case the computed solution is $\mathbf{x}' = (1, 2)^T$.   □

If the pivot $a_{ii}^{(i)}$ is small in absolute value, the multipliers $m_{ki} = a_{ki}^{(i)}/a_{ii}^{(i)}$ will be large in absolute value. If there is an error in the computed value of $a_{ij}^{(i)}$, it will be multiplied by m_{ki}. In general, large multipliers contribute to the propagation of error. On the other hand, multipliers that are less than 1 in absolute value generally retard the growth of error. By careful selection of the pivot elements, one can try to avoid small pivots and at the same time keep the multipliers less than 1 in modulus. The most commonly used strategy for doing this is called *partial pivoting*.

PARTIAL PIVOTING

At the ith step of the reduction process, there are $n - i + 1$ candidates for the pivot element:

$$a_{p(i),i}, a_{p(i+1),i}, \ldots, a_{p(n),i}$$

Choose the candidate $a_{p(j),i}$ with maximum modulus

$$|a_{p(j),i}| = \max_{i \leq k \leq n} |a_{p(k),i}|$$

and interchange the ith and jth entries of $\mathbf{p}$. The pivot element $a_{p(i),i}$ has the property

$$|a_{p(i),i}| \geq |a_{p(k),i}|$$

for $k = i + 1, \ldots, n$. Thus the multipliers will all satisfy

$$|m_{p(k),i}| = \left| \frac{a_{p(k),i}}{a_{p(i),i}} \right| \leq 1$$

One could always carry things one step further and do *complete pivoting*. In complete pivoting the pivot element is chosen to be the element of maximum modulus among all the elements in the remaining rows and columns. In this case one must keep track of both the rows and the columns. At the ith step the element $a_{p(j)q(k)}$ is chosen so that

$$|a_{p(j)q(k)}| = \max_{\substack{i \leq s \leq n \\ i \leq t \leq n}} |a_{p(s)q(t)}|$$

The ith and jth entries of $\mathbf{p}$ are interchanged and the ith and kth entries of $\mathbf{q}$ are interchanged. The new pivot element is $a_{p(i)q(i)}$. The major drawback to complete pivoting is that at each step one must search for a pivot element among $(n - i + 1)^2$ elements of A. This may be too costly in terms of computer time.

EXERCISES

1. Let

$$A = \begin{pmatrix} 0 & 3 & 1 \\ 1 & 2 & -2 \\ 2 & 5 & 4 \end{pmatrix} \quad \text{and} \quad b = \begin{pmatrix} 1 \\ 7 \\ -1 \end{pmatrix}$$

  (a) Reorder the rows of $(A|\mathbf{b})$ in the order $(2, 3, 1)$ and then solve this reordered system.
  (b) Factor A into a product $P^T LU$, where P is the permutation matrix corresponding to the reordering in part (a).

2. Let A be the matrix in Exercise 1. Use the factorization $P^T LU$ to solve $A\mathbf{x} = \mathbf{c}$ for each of the following choices of $\mathbf{c}$.
  (a) $(8, 1, 20)^T$ (b) $(-9, -2, -7)^T$ (c) $(4, 1, 11)^T$

3. Let

$$A = \begin{pmatrix} 1 & 8 & 6 \\ -1 & -4 & 5 \\ 2 & 4 & -6 \end{pmatrix} \quad \text{and} \quad b = \begin{pmatrix} 8 \\ 1 \\ 4 \end{pmatrix}$$

  Solve the system $A\mathbf{x} = \mathbf{b}$ using partial pivoting. If P is the permutation matrix corresponding to the pivoting strategy, factor PA into a product LU.

4. Let

$$A = \begin{pmatrix} 3 & 2 \\ 2 & 4 \end{pmatrix} \quad \text{and} \quad b = \begin{pmatrix} 5 \\ -2 \end{pmatrix}$$

  Solve the system $A\mathbf{x} = \mathbf{b}$ using complete pivoting. Let P be the permutation matrix determined by the pivot rows and Q the permutation matrix determined by the pivot columns. Factor PAQ into a product LU.

5. Let A be the matrix in Exercise 4 and let $\mathbf{c} = (6, -4)^T$. Solve the system $A\mathbf{x} = \mathbf{c}$ in two steps:
  1. Set $\mathbf{z} = Q\mathbf{x}$ and solve $LU\mathbf{z} = P\mathbf{c}$ for $\mathbf{z}$.
  2. Calculate $\mathbf{x} = Q^T\mathbf{z}$.

6. The exact solution to the system

$$0.6000x_1 + 2000x_2 = 2003$$

$$0.3076x_1 - 0.4010x_2 = 1.137$$

  is $\mathbf{x} = (5, 1)^T$. Suppose that the calculated value of x_2 is $x_2' = 1 + \epsilon$.

Use this value in the first equation and solve for x_1. What will the error be? Calculate the relative error in x_1 if $\epsilon = 0.001$.

7. Solve the system in Exercise 6 using four-digit decimal floating-point arithmetic and Gaussian elimination with partial pivoting.

8. Solve the system in Exercise 6 using four-digit decimal floating-point arithmetic and Gaussian elimination with complete pivoting.

9. Use four-digit decimal floating-point arithmetic and scale the system in Exercise 6 by multiplying the first equation through by $1/2000$ and the second equation through by $1/0.4010$. Solve the scaled system using partial pivoting.

4. THE CONDITION NUMBER OF A MATRIX

In this section we will be concerned with the accuracy of computed solutions to linear systems. How accurate can we expect the computed solutions to be, and how can one test their accuracy? The answer to these questions depends largely upon how sensitive the coefficient matrix of the system is to small changes. Consider the following example.

EXAMPLE 1. Solve the following system.

(1)
$$2.0000x_1 + 2.0000x_2 = 6.0000$$
$$2.0000x_1 + 2.0005x_2 = 6.0010$$

If one uses five-digit decimal floating-point arithmetic, the computed solution will be the exact solution $\mathbf{x} = (1, 2)^T$. Suppose, however, that one is forced to use four-digit decimal floating-point numbers. Thus, in place of (1), we have

(2)
$$2.000x_1 + 2.000x_2 - 6.000$$
$$2.000x_1 + 2.001x_2 = 6.001$$

The computed solution to system (2) is the exact solution $\mathbf{x}' = (2, 1)^T$.

The systems (1) and (2) agree except for the coefficient a_{22}. The relative error in this coefficient is

$$\frac{a'_{22} - a_{22}}{a_{22}} \approx 0.00025$$

However, the relative errors in the coordinates of the solutions $\mathbf{x}$

and $\mathbf{x}'$ are

$$\frac{x_1' - x_1}{x_1} = 1.0 \qquad \text{and} \qquad \frac{x_2' - x_2}{x_2} = -0.5 \qquad \square$$

Definition. A matrix A is said to be **ill conditioned** if relatively small changes in the entries of A can cause relatively large changes in the solutions to $A\mathbf{x} = \mathbf{b}$. A is said to be **well conditioned** if relatively small changes in the entries of A result in relatively small changes in the solutions to $A\mathbf{x} = \mathbf{b}$.

If the matrix A is ill conditioned, the computed solution to $A\mathbf{x} = \mathbf{b}$ generally will not be very accurate. Even if the entries of A can be represented exactly as floating-point numbers, small roundoff errors occurring in the reduction process may have a drastic effect on the computed solution. On the other hand, if the matrix is well conditioned and the proper pivoting strategy is used, one should be able to compute solutions quite accurately. In general, the accuracy of the solution depends upon the conditioning of the matrix. If we could measure the conditioning of A, this measure could be used to derive a bound for the relative error in the computed solution.

Let A be an $n \times n$ nonsingular matrix and consider the system $A\mathbf{x} = \mathbf{b}$. If $\mathbf{x}$ is the exact solution to the system and $\mathbf{x}'$ is the calculated solution, then the error can be represented by the vector $\mathbf{e} = \mathbf{x} - \mathbf{x}'$. If $\| \cdot \|$ is a norm on R^n, then $\|\mathbf{e}\|$ is a measure of the absolute error and $\|\mathbf{e}\|/\|\mathbf{x}\|$ is a measure of the relative error. In general, we have no way of determining the exact values of $\|\mathbf{e}\|$ and $\|\mathbf{e}\|/\|\mathbf{x}\|$. One possible way of testing the accuracy of $\mathbf{x}'$ is to put it back into the original system and to see how close $\mathbf{b}' = A\mathbf{x}'$ comes to $\mathbf{b}$. The vector

$$\mathbf{r} = \mathbf{b} - \mathbf{b}' = \mathbf{b} - A\mathbf{x}'$$

is called the *residual* and can be easily calculated. The quantity

$$\frac{\|\mathbf{b} - A\mathbf{x}'\|}{\|\mathbf{b}\|} = \frac{\|\mathbf{r}\|}{\|\mathbf{b}\|}$$

is called the *relative residual*. Is the relative residual a good estimate of the relative error? The answer to this depends upon the conditioning of A. In Example 1 the residual for the computed solution $\mathbf{x}' = (2, 1)^T$ is

$$\mathbf{r} = \mathbf{b} - A\mathbf{x}' = (0, 0.0005)^T$$

The relative residual in terms of the ∞ norm is

$$\frac{\|\mathbf{r}\|_\infty}{\|\mathbf{b}\|_\infty} = \frac{0.0005}{6.0010} \approx 0.000083$$

and the relative error is given by

$$\frac{\|\mathbf{e}\|_\infty}{\|\mathbf{x}\|_\infty} = 0.5$$

The relative error is more than 6000 times the relative residual. In general, we will show that if A is ill conditioned, the relative residual may be much smaller than the relative error. On the other hand, for well-conditioned matrices, the relative residual and the relative error are quite close. To show this, we need to make use of matrix norms. Recall that if $\| \cdot \|$ is a compatible matrix norm on $R^{n \times n}$, then for any $n \times n$ matrix C and any vector $\mathbf{y} \in R^n$, we have

$$(3) \qquad \|C\mathbf{y}\| \le \|C\| \|\mathbf{y}\|$$

Now

$$\mathbf{r} = \mathbf{b} - A\mathbf{x}' = A\mathbf{x} - A\mathbf{x}' = A\mathbf{e}$$

and consequently,

$$\mathbf{e} = A^{-1}\mathbf{r}$$

It follows from property (3) that

$$\|\mathbf{e}\| \le \|A^{-1}\| \|\mathbf{r}\|$$

and

$$\|\mathbf{r}\| = \|A\mathbf{e}\| \le \|A\| \|\mathbf{e}\|$$

Therefore,

$$(4) \qquad \frac{\|\mathbf{r}\|}{\|A\|} \le \|\mathbf{e}\| \le \|A^{-1}\| \|\mathbf{r}\|$$

Now $\mathbf{x}$ is the exact solution to $A\mathbf{x} = \mathbf{b}$, and hence $\mathbf{x} = A^{-1}\mathbf{b}$. By the same reasoning used to derive (4), we have

$$(5) \qquad \frac{\|\mathbf{b}\|}{\|A\|} \le \|\mathbf{x}\| \le \|A^{-1}\| \|\mathbf{b}\|$$

It follows from (4) and (5) that

$$\frac{1}{\|A\| \|A^{-1}\|} \frac{\|\mathbf{r}\|}{\|\mathbf{b}\|} \le \frac{\|\mathbf{e}\|}{\|\mathbf{x}\|} \le \|A\| \|A^{-1}\| \frac{\|\mathbf{r}\|}{\|\mathbf{b}\|}$$

The number $\|A\| \|A^{-1}\|$ is called the *condition number* of A and will be denoted by cond (A). Thus

$$(6) \qquad \frac{1}{\text{cond}(A)} \frac{\|\mathbf{r}\|}{\|\mathbf{b}\|} \le \frac{\|\mathbf{e}\|}{\|\mathbf{x}\|} \le \text{cond}(A) \frac{\|\mathbf{r}\|}{\|\mathbf{b}\|}$$

Inequality (6) relates the size of the relative error $\|\mathbf{e}\|/\|\mathbf{x}\|$ to the relative residual $\|\mathbf{r}\|/\|\mathbf{b}\|$. If the condition number is close to 1, the relative error and the relative residual will be close. If the condition number is large, the relative error could be many times as large as the relative residual.

EXAMPLE 2.   Let

$$A = \begin{pmatrix} 3 & 3 \\ 4 & 5 \end{pmatrix}$$

Then

$$A^{-1} = \tfrac{1}{3}\begin{pmatrix} 5 & -3 \\ -4 & 3 \end{pmatrix}$$

$\|A\|_\infty = 9$ and $\|A^{-1}\|_\infty = \tfrac{8}{3}$. (We use $\|\cdot\|_\infty$ because it is easy to calculate.) Thus

$$\mathrm{cond}_\infty(A) = 9 \cdot \tfrac{8}{3} = 24$$

Theoretically, the relative error in the calculated solution to a system $A\mathbf{x} = \mathbf{b}$ could be as much as 24 times the relative residual.   □

EXAMPLE 3.   Suppose that $\mathbf{x}' = (2.0, 0.1)^T$ is the calculated solution to

$$3x_1 + 3x_2 = 6$$
$$4x_1 + 5x_2 = 9$$

Determine the residual $\mathbf{r}$ and the relative residual $\|\mathbf{r}\|_\infty / \|\mathbf{b}\|_\infty$.

SOLUTION

$$\mathbf{r} = \begin{pmatrix} 6 \\ 9 \end{pmatrix} - \begin{pmatrix} 3 & 3 \\ 4 & 5 \end{pmatrix}\begin{pmatrix} 2.0 \\ 0.1 \end{pmatrix} = \begin{pmatrix} -0.3 \\ 0.5 \end{pmatrix}$$

$$\frac{\|\mathbf{r}\|_\infty}{\|\mathbf{b}\|_\infty} = \frac{0.5}{9} = \frac{1}{18}$$

□

We can see by inspection that the actual solution to the system in the example above is $\mathbf{x} = \begin{pmatrix} 1 \\ 1 \end{pmatrix}$. The error $\mathbf{e}$ is given by

$$\mathbf{e} = \mathbf{x} - \mathbf{x}' = \begin{pmatrix} -1.0 \\ 0.9 \end{pmatrix}$$

The relative error is given by

$$\frac{\|\mathbf{e}\|_\infty}{\|\mathbf{x}\|_\infty} = \frac{1.0}{1} = 1$$

The relative error is 18 times the relative residual. This is not surprising, since $\mathrm{cond}(A) = 24$. The results are similar using $\|\cdot\|_1$. In this case

$$\frac{\|\mathbf{r}\|_1}{\|\mathbf{b}\|_1} = \frac{0.8}{15} = \frac{4}{75} \quad \text{and} \quad \frac{\|\mathbf{e}\|_1}{\|\mathbf{x}\|_1} = \frac{1.9}{2} = \frac{19}{20}$$

The condition number of a nonsingular matrix actually gives us valuable information about the conditioning of A. Let A' be a new matrix formed by altering the entries of A slightly. Let $E = A' - A$. Thus $A' = A + E$, where the entries of E are small relative to the entries of A. A will be ill conditioned if for some such E the solutions to $A'\mathbf{x} = \mathbf{b}$ and $A\mathbf{x} = \mathbf{b}$ vary greatly. Let $\mathbf{x}'$ be the solution to $A'\mathbf{x} = \mathbf{b}$ and $\mathbf{x}$ be the solution to $A\mathbf{x} = \mathbf{b}$. The condition number allows us to compare the change in solution relative to $\mathbf{x}'$, to the relative change in the matrix A.

$$\mathbf{x} = A^{-1}\mathbf{b} = A^{-1}A'\mathbf{x}' = A^{-1}(A + E)\mathbf{x}' = \mathbf{x}' + A^{-1}E\mathbf{x}'$$

Hence

$$\mathbf{x} - \mathbf{x}' = A^{-1}E\mathbf{x}'$$

Using inequality (3), we see that

$$\|\mathbf{x} - \mathbf{x}'\| \leq \|A^{-1}\| \, \|E\| \, \|\mathbf{x}'\|$$

or

(7) $$\frac{\|\mathbf{x} - \mathbf{x}'\|}{\|\mathbf{x}'\|} \leq \|A^{-1}\| \, \|E\| = \text{cond}(A)\frac{\|E\|}{\|A\|}$$

Let us return to Example 1 and see how the inequality (7) applies. Let A and A' be the two coefficient matrices in Example 1.

$$E = A' - A = \begin{pmatrix} 0 & 0 \\ 0 & 0.0005 \end{pmatrix}$$

and

$$A^{-1} = \begin{pmatrix} 2000.5 & -2000 \\ -2000 & 2000 \end{pmatrix}$$

In terms of the ∞-norm, the relative error in A is

$$\frac{\|E\|_\infty}{\|A\|_\infty} = \frac{0.0005}{4.0005} \approx 0.0001$$

and the condition number is

$$\text{cond}(A) = \|A\|_\infty \|A^{-1}\|_\infty = (4.0005)(4000.5) \approx 16{,}004$$

The bound on the relative error given in (7) is then

$$\text{cond}(A)\frac{\|E\|}{\|A\|} = \|A^{-1}\| \, \|E\| = (4000.5)(0.0005) \approx 2$$

The actual relative error for the systems in Example 1 is

$$\frac{\|\mathbf{x} - \mathbf{x}'\|_\infty}{\|\mathbf{x}'\|_\infty} = \frac{1}{2}$$

EXERCISES

1. Let

$$A = \begin{pmatrix} 1 & -0.99 \\ -1 & 1 \end{pmatrix}$$

Find A^{-1} and $\text{cond}_\infty(A)$.

2. Solve the two systems below and compare the two solutions. Are the coefficient matrices well conditioned? Ill conditioned? Explain.

$$1.0x_1 + 2.0x_2 = 1.12 \qquad 1.000x_1 + 2.011x_2 = 1.120$$
$$2.0x_1 + 3.9x_2 = 2.16 \qquad 2.000x_1 + 3.982x_2 = 2.160$$

3. Let

$$A_n = \begin{pmatrix} 1 & 1 \\ 1 & 1 - \dfrac{1}{n} \end{pmatrix}$$

for each positive integer n. Calculate:
(a) A_n^{-1}
(b) $\text{cond}_\infty(A_n)$
(c) $\lim\limits_{n \to \infty} \text{cond}_\infty(A_n)$

4. Let A be a nonsingular $n \times n$ matrix. Show that $\text{cond}(A) \geq 1$.

5. Let

$$A = \begin{pmatrix} 1 & 0 & 1 \\ 2 & 2 & 3 \\ 1 & 1 & 2 \end{pmatrix}$$

Calculate $\text{cond}_\infty(A) = \|A\|_\infty \|A^{-1}\|_\infty$.

6. Let A be the matrix in Exercise 5 and let $\mathbf{b} = (4, 7, 5)^T$. Suppose that $\mathbf{x}' = (0.8, -1.8, 3.3)^T$ is the calculated solution to $A\mathbf{x} = \mathbf{b}$.
(a) Calculate the residual $\mathbf{r}$ and $\|\mathbf{r}\|_\infty$.
(b) Without computing the exact solution $\mathbf{x}$, find a bound for the relative error $\|\mathbf{e}\|_\infty / \|\mathbf{x}\|_\infty$.
(c) Compute the exact solution $\mathbf{x}$ and the relative error $\|\mathbf{e}\|_\infty / \|\mathbf{x}\|_\infty$.

7. Let

$$A = \begin{pmatrix} -0.50 & 0.75 & -0.25 \\ -0.50 & 0.25 & 0.25 \\ 1.00 & -0.50 & 0.50 \end{pmatrix}$$

Calculate $\text{cond}_1(A) = \|A\|_1 \|A^{-1}\|_1$.

8. Let A be the matrix in Exercise 7 and let

$$A' = \begin{pmatrix} -0.5 & 0.8 & -0.3 \\ -0.5 & 0.3 & 0.3 \\ 1.0 & -0.5 & 0.5 \end{pmatrix}$$

Let $\mathbf{x}$ and $\mathbf{x}'$ be the solutions to $A\mathbf{x} = \mathbf{b}$ and $A'\mathbf{x} = \mathbf{b}$, respectively, for some $\mathbf{b} \in R^3$. Find a bound for the relative error $(\|\mathbf{x} - \mathbf{x}'\|_1)/\|\mathbf{x}'\|_1$.

9. Let A and B be nonsingular $n \times n$ matrices. Show that

$$\text{cond}(AB) \leq \text{cond}(A)\text{cond}(B)$$

10. Let D be a nonsingular $n \times n$ diagonal matrix. Show that

$$\text{cond}_1(D) = \text{cond}_\infty(D) = \frac{d_{max}}{d_{min}}$$

where

$$d_{max} = \max_{1 \leq i \leq n} |d_{ii}| \quad \text{and} \quad d_{min} = \min_{1 \leq i \leq n} |d_{ii}|$$

11. Let D, d_{max}, and d_{min} be defined as in Exercise 10. Show that
(a) $\|D\|_2 = d_{max}$
(b) $\text{cond}_2(D) = \dfrac{d_{max}}{d_{min}}$

12. Let Q be an $n \times n$ orthogonal matrix. Show that
(a) $\text{cond}_2(Q) = 1$
(b) For any $\mathbf{b} \in R^n$, the relative error in the solution to $Q\mathbf{x} = \mathbf{b}$ is equal to the relative residual, that is,

$$\frac{\|\mathbf{e}\|_2}{\|\mathbf{x}\|_2} = \frac{\|\mathbf{r}\|_2}{\|\mathbf{b}\|_2}$$

13. Let A be a nonsingular $n \times n$ matrix and let Q be an $n \times n$ orthogonal matrix. Show that:
(a) $\text{cond}_2(QA) = \text{cond}_2(AQ) = \text{cond}_2(A)$.
(b) If $B = Q^T A Q$, then $\text{cond}_2(B) = \text{cond}_2(A)$.

14. Let A be a symmetric nonsingular $n \times n$ matrix with eigenvalues $\lambda_1, \ldots, \lambda_n$. Show that

$$\text{cond}_2(A) = \frac{\max\limits_{1 \leq i \leq n} |\lambda_i|}{\min\limits_{1 \leq i \leq n} |\lambda_i|}$$

5. ORTHOGONAL TRANSFORMATIONS

Orthogonal transformations are one of the most important tools in numerical linear algebra. The types of orthogonal transformations that will be introduced in this section are easy to work with and do not require much storage. Most important, processes that involve orthogonal transformations are inherently stable. For example, let $\mathbf{x} \in R^n$ and $\mathbf{x}' = \mathbf{x} + \mathbf{e}$ be an approximation to $\mathbf{x}$; if Q is an orthogonal matrix,

then

$$Q\mathbf{x}' = Q\mathbf{x} + Q\mathbf{e}$$

The error in $Q\mathbf{x}'$ is $Q\mathbf{e}$. With respect to the 2-norm, the vector $Q\mathbf{e}$ is the same size as $\mathbf{e}$.

$$\|Q\mathbf{e}\|_2 = \|\mathbf{e}\|_2$$

Similarly, if $A' = A + E$, then

$$QA' = QA + QE$$

and

$$\|QE\|_2 = \|E\|_2$$

When an orthogonal transformation is applied to a vector or a matrix, the error will not grow with respect to the 2-norm.

ELEMENTARY ORTHOGONAL TRANSFORMATIONS

By an *elementary orthogonal matrix*, we mean a matrix of the form

$$Q = I - 2\mathbf{u}\mathbf{u}^T$$

where $\mathbf{u} \in R^n$ and $\|\mathbf{u}\|_2 = 1$. To see that Q is orthogonal, note that

$$Q^T = (I - 2\mathbf{u}\mathbf{u}^T)^T = I - 2\mathbf{u}\mathbf{u}^T = Q$$

and

$$\begin{aligned}
Q^T Q = Q^2 &= (I - 2\mathbf{u}\mathbf{u}^T)(I - 2\mathbf{u}\mathbf{u}^T) \\
&= I - 4\mathbf{u}\mathbf{u}^T + 4\mathbf{u}(\mathbf{u}^T\mathbf{u})\mathbf{u}^T \\
&= I
\end{aligned}$$

Thus, if Q is an elementary orthogonal matrix, then

$$Q^T = Q^{-1} = Q$$

The matrix $Q = I - 2\mathbf{u}\mathbf{u}^T$ is completely determined by the unit vector $\mathbf{u}$. Rather than store all n^2 entries of Q, we need store only the vector $\mathbf{u}$. To compute $Q\mathbf{x}$, note that

$$\begin{aligned}
Q\mathbf{x} &= (I - 2\mathbf{u}\mathbf{u}^T)\mathbf{x} \\
&= \mathbf{x} - 2\alpha\mathbf{u} \quad \text{where} \quad \alpha = \mathbf{u}^T\mathbf{x}
\end{aligned}$$

The matrix product QA is computed as follows:

$$QA = (Q\mathbf{a}_1, Q\mathbf{a}_2, \ldots, Q\mathbf{a}_n)$$

where

$$Q\mathbf{a}_i = \mathbf{a}_i - 2\alpha_i\mathbf{u} \qquad \alpha_i = \mathbf{u}^T\mathbf{a}_i$$

Elementary orthogonal transformations can be used to obtain a QR factorization of A, and this in turn can be used to solve a linear system $A\mathbf{x} = \mathbf{b}$. As with Gaussian elimination, the elementary matrices are chosen so as to produce zeros in the coefficient matrix. To see how this is done, let us consider the problem of finding a unit vector **u** such that

$$(I - 2\mathbf{u}\mathbf{u}^T)\mathbf{x} = (\alpha, 0, \ldots, 0)^T = \alpha\mathbf{e}_1$$

for a given vector $\mathbf{x} \in R^n$.

HOUSEHOLDER TRANSFORMATIONS

Let $H = I - 2\mathbf{u}\mathbf{u}^T$. If $H\mathbf{x} = \alpha\mathbf{e}_1$, then, since H is orthogonal, it follows that

$$|\alpha| = \|\alpha\mathbf{e}_1\|_2 = \|H\mathbf{x}\|_2 = \|\mathbf{x}\|_2$$

If we take $\alpha = \|\mathbf{x}\|_2$ and $H\mathbf{x} = \alpha\mathbf{e}_1$, then, since H is its own inverse, it follows that

(1) $$\mathbf{x} = H(\alpha\mathbf{e}_1) = \alpha(\mathbf{e}_1 - (2u_1)\mathbf{u})$$

Thus

$$x_1 = \alpha(1 - 2u_1^2)$$

$$x_2 = -2\alpha u_1 u_2$$

$$\vdots$$

$$x_n = -2\alpha u_1 u_n$$

Solving for the u_i's, we get

$$u_1 = \pm\left(\frac{\alpha - x_1}{2\alpha}\right)^{1/2}$$

$$u_i = \frac{-x_i}{2\alpha u_1} \qquad \text{for} \qquad i = 2, \ldots, n$$

If we let

$$u_1 = -\left(\frac{\alpha - x_1}{2\alpha}\right)^{1/2} \qquad \text{and set} \qquad \beta = \alpha(\alpha - x_1)$$

then

$$-2\alpha u_1 = [2\alpha(\alpha - x_1)]^{1/2} = (2\beta)^{1/2}$$

It follows that

$$\mathbf{u} = \left(-\frac{1}{2\alpha u_1}\right)(-2\alpha u_1^2, x_2, \ldots, x_n)^T$$

$$= \frac{1}{\sqrt{2\beta}}(x_1 - \alpha, x_2, \ldots, x_n)^T$$

If we set $\mathbf{v} = (x_1 - \alpha, x_2, \ldots, x_n)^T$, then

$$\|\mathbf{v}\|_2^2 = (x_1 - \alpha)^2 + \sum_{i=2}^{n} x_i^2 = 2\alpha(\alpha - x_1)$$

and hence

$$\|\mathbf{v}\|_2 = \sqrt{2\beta}$$

Thus

$$\mathbf{u} = \frac{1}{\sqrt{2\beta}}\mathbf{v} = \frac{1}{\|\mathbf{v}\|_2}\mathbf{v}$$

In summation, given a vector $\mathbf{x} \in R^n$, if we set

$$\alpha = \|\mathbf{x}\|_2, \qquad \beta = \alpha(\alpha - x_1)$$

$$\mathbf{v} = (x_1 - \alpha, x_2, \ldots, x_n)^T$$

$$\mathbf{u} = \frac{1}{\|\mathbf{v}\|_2}\mathbf{v} = \frac{1}{\sqrt{2\beta}}\mathbf{v}$$

and

$$H = I - 2\mathbf{u}\mathbf{u}^T = I - \frac{1}{\beta}\mathbf{v}\mathbf{v}^T$$

then

$$H\mathbf{x} = \alpha\mathbf{e}_1$$

The matrix H formed in this way is called a *Householder transformation*. The matrix H is determined by the vector $\mathbf{v}$ and the scalar β. For any vector $\mathbf{y} \in R^n$,

$$H\mathbf{y} = \left(I - \frac{1}{\beta}\mathbf{v}\mathbf{v}^T\right)\mathbf{y} = \mathbf{y} - \left(\frac{1}{\beta}\mathbf{v}^T\mathbf{y}\right)\mathbf{v}$$

Rather than store all n^2 entries of H, we need store only $\mathbf{v}$ and β.

Suppose now that we wish to zero out only the last $n - k$ components of a vector $\mathbf{x} = (x_1, \ldots, x_k, x_{k+1}, \ldots, x_n)^T$. To do this, we let $\mathbf{x}^{(1)} = (x_1, \ldots, x_{k-1})^T$ and $\mathbf{x}^{(2)} = (x_k, x_{k+1}, \ldots, x_n)^T$. Let $I^{(1)}$ and $I^{(2)}$ denote the $(k - 1) \times (k - 1)$ and $(n - k + 1) \times (n - k + 1)$ identity matrices, respectively. By the methods just described, we can construct a Householder matrix $H_k^{(2)} = I^{(2)} - (1/\beta_k)\mathbf{v}_k\mathbf{v}_k^T$ such that

$$H_k^{(2)}\mathbf{x}^{(2)} = \|\mathbf{x}^{(2)}\|_2\mathbf{e}_1$$

Let

$$H_k = \begin{pmatrix} I^{(1)} & O \\ O & H_k^{(2)} \end{pmatrix}$$

It follows that

$$H_k\mathbf{x} = \begin{pmatrix} I^{(1)} & O \\ O & H_k^{(2)} \end{pmatrix}\begin{pmatrix} \mathbf{x}^{(1)} \\ \mathbf{x}^{(2)} \end{pmatrix}$$

$$= \begin{pmatrix} I^{(1)}\mathbf{x}^{(1)} \\ H_k^{(2)}\mathbf{x}^{(2)} \end{pmatrix}$$

$$= \left(x_1, \ldots, x_{k-1}, \left(\sum_{i=k}^{n} x_i^2\right)^{1/2}, 0, \ldots, 0\right)^T$$

Remarks

1. The Householder matrix H_k defined above is an elementary orthogonal matrix. If we let $\mathbf{v} = \begin{pmatrix} \mathbf{0} \\ \mathbf{v}_k \end{pmatrix}$ and $\mathbf{u} = (1/\|\mathbf{v}\|)\mathbf{v}$, then

$$H_k = I - \frac{1}{\beta_k}\mathbf{v}\mathbf{v}^T = I - 2\mathbf{u}\mathbf{u}^T$$

2. H_k acts like the identity on the first $k - 1$ coordinates of any vector $\mathbf{y} \in R^n$. If $\mathbf{y} = (y_1, \ldots, y_{k-1}, y_k, \ldots, y_n)^T$, $\mathbf{y}^{(1)} = (y_1, \ldots, y_{k-1})^T$ and $\mathbf{y}^{(2)} = (y_k, \ldots, y_n)^T$, then

$$H_k\mathbf{y} = \begin{pmatrix} I^{(1)} & O \\ O & H_k^{(2)} \end{pmatrix}\begin{pmatrix} \mathbf{y}^{(1)} \\ \mathbf{y}^{(2)} \end{pmatrix} = \begin{pmatrix} \mathbf{y}^{(1)} \\ H_k^{(2)}\mathbf{y}^{(2)} \end{pmatrix}$$

In particular, if $\mathbf{y}^{(2)} = \mathbf{0}$, then $H_k\mathbf{y} = \mathbf{y}$.

3. It is generally not necessary to store the entire matrix H_k. It suffices to store the $n - k + 1$ vector $\mathbf{v}_k$ and the scalar β_k.

We are now ready to apply Householder transformations to solve linear systems. If A is a nonsingular $n \times n$ matrix, we can use Householder transformations to reduce A to triangular form. To begin with, we can find a Householder transformation $H_1 = I - (1/\beta_1)\mathbf{v}_1\mathbf{v}_1^T$, which when applied to the first column of A will give a multiple of $\mathbf{e}_1$. Thus H_1A will be of the form

$$\begin{pmatrix} \times & \times & \cdots & \times \\ 0 & \times & \cdots & \times \\ 0 & \times & \cdots & \times \\ \vdots & & & \\ 0 & \times & \cdots & \times \end{pmatrix}$$

We can then find a Householder transformation H_2 that will zero out the last $n - 2$ elements in the second column of H_1A while leaving the first element in that column unchanged. It follows from remark (2) that H_2 will have no effect on the first column of H_1A.

$$H_2H_1A = \begin{pmatrix} \times & \times & \times & \cdots & \times \\ 0 & \times & \times & \cdots & \times \\ 0 & 0 & \times & \cdots & \times \\ \vdots & & & & \\ 0 & 0 & \times & \cdots & \times \end{pmatrix}$$

We can continue applying Householder transformations in this fashion until we end up with an upper triangular matrix, which we will denote by R. Thus

$$H_{n-1} \cdots H_2H_1A = R$$

It follows that

$$A = H_1^{-1} H_2^{-1} \cdots H_{n-1}^{-1} R$$
$$= H_1 H_2 \cdots H_{n-1} R$$

Let $Q = H_1 H_2 \cdots H_{n-1}$. The matrix Q is orthogonal and A can be factored into an orthogonal matrix times an upper triangular matrix:

$$A = QR$$

Once A has been factored into a product QR, the system $A\mathbf{x} = \mathbf{b}$ is easily solved.

$$A\mathbf{x} = \mathbf{b}$$
$$QR\mathbf{x} = \mathbf{b}$$
$$R\mathbf{x} = Q^T\mathbf{b}$$

(2) $$R\mathbf{x} = H_{n-1} \cdots H_2 H_1 \mathbf{b}$$

Once $H_{n-1} \cdots H_2 H_1 \mathbf{b}$ has been calculated, the system (2) can be solved using back substitution.

Storage. The vector $\mathbf{v}_k$ can be stored in the kth column of A. Since $\mathbf{v}_k$ has $n - k + 1$ nonzero entries and there are only $n - k$ zeros in the kth column of the reduced matrix, it is necessary to store r_{kk} elsewhere. The diagonal elements of R can either be stored in an n vector or in an additional row added to A. The β_k's can also be stored in an additional row of A.

$$\begin{pmatrix} v_{11} & r_{12} & r_{13} & r_{14} \\ v_{12} & v_{22} & r_{23} & r_{24} \\ v_{13} & v_{23} & v_{33} & r_{34} \\ v_{14} & v_{24} & v_{34} & 0 \\ r_{11} & r_{22} & r_{33} & r_{44} \\ \beta_1 & \beta_2 & \beta_3 & 0 \end{pmatrix}$$

Operation Count. In solving an $n \times n$ system using Householder transformations, most of the work is done in reducing A to triangular form. The number of operations required is approximately $\frac{2}{3}n^3$ multiplications, $\frac{2}{3}n^3$ additions, and $n - 1$ square roots.

ROTATIONS AND REFLECTIONS

Often it will be desirable to have a transformation that annihilates only a single entry of a vector. In this case it is convenient to use either a rotation or a reflection. Let us consider first the two-dimensional case. Let

$$R = \begin{pmatrix} \cos\theta & -\sin\theta \\ \sin\theta & \cos\theta \end{pmatrix} \quad \text{and} \quad G = \begin{pmatrix} \cos\theta & \sin\theta \\ \sin\theta & -\cos\theta \end{pmatrix}$$

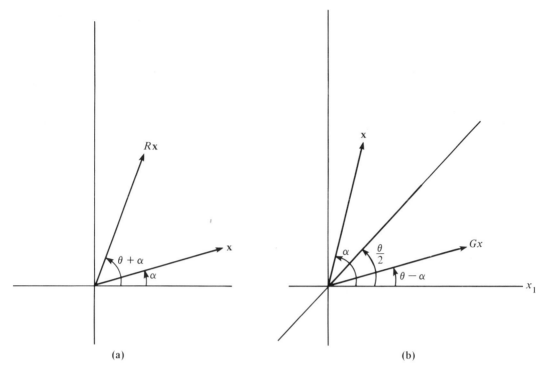

FIGURE 7.5.1

and let

$$\mathbf{x} = \begin{pmatrix} x_1 \\ x_2 \end{pmatrix} = \begin{pmatrix} r\cos\alpha \\ r\sin\alpha \end{pmatrix}$$

be a vector in R^2.

$$R\mathbf{x} = \begin{pmatrix} r\cos(\theta + \alpha) \\ r\sin(\theta + \alpha) \end{pmatrix} \quad \text{and} \quad G\mathbf{x} = \begin{pmatrix} r\cos(\theta - \alpha) \\ r\sin(\theta - \alpha) \end{pmatrix}$$

R represents a rotation in the plane by an angle θ. The matrix G has the effect of reflecting $\mathbf{x}$ about the line $x_2 = [\tan(\theta/2)]x_1$ (see Figure 7.5.1).

If we set $\cos\theta = x_1/r$ and $\sin\theta = -x_2/r$, then

$$R\mathbf{x} = \begin{pmatrix} x_1\cos\theta - x_2\sin\theta \\ x_1\sin\theta + x_2\cos\theta \end{pmatrix} = \begin{pmatrix} r \\ 0 \end{pmatrix}$$

If we set $\cos\theta = x_1/r$ and $\sin\theta = x_2/r$, then

$$G\mathbf{x} = \begin{pmatrix} x_1\cos\theta + x_2\sin\theta \\ x_1\sin\theta - x_2\cos\theta \end{pmatrix} = \begin{pmatrix} r \\ 0 \end{pmatrix}$$

Both R and G are orthogonal. The matrix G is also symmetric. Indeed, G is an elementary orthogonal matrix. If we let $\mathbf{u} = (\sin\theta/2, -\cos\theta/2)^T$, then $G = I - 2\mathbf{u}\mathbf{u}^T$.

Let us now consider the n-dimensional case. Let R and G be $n \times n$ matrices with

$$r_{ii} = r_{jj} = \cos\theta \qquad\qquad g_{ii} = \cos\theta,\ g_{jj} = -\cos\theta$$

$$r_{ji} = \sin\theta,\ r_{ij} = -\sin\theta \qquad g_{ji} = g_{ij} = \sin\theta$$

and $r_{st} = g_{st} = \delta_{st}$ for all other entries of R and G. Thus R and G resemble the identity matrix except for the ii, ij, jj, and ji positions. Let $c = \cos\theta$ and $s = \sin\theta$. If $\mathbf{x} \in R^n$, then

$$R\mathbf{x} = (x_1, \ldots, x_{i-1}, x_i c - x_j s, x_{i+1}, \ldots, x_{j-1}, x_i s + x_j c, x_{j+1}, \ldots, x_n)^T$$

and

$$G\mathbf{x} = (x_1, \ldots, x_{i-1}, x_i c + x_j s, x_{i+1}, \ldots, x_{j-1}, x_i s - x_j c, x_{j+1}, \ldots, x_n)^T$$

The transformations R and G only alter the ith and jth components of a vector. They have no effect on the other coordinates. We will refer to R as a *plane rotation* and to G as a *Givens transformation* or a *Givens reflection*. If we set $c = x_i/r$ and $s = -x_j/r$, $(r = \sqrt{x_i^2 + x_j^2})$, the jth component of $R\mathbf{x}$ will be 0. If we set $c = x_i/r$ and $s = x_j/r$, the jth component of $G\mathbf{x}$ will be 0.

Given a nonsingular $n \times n$ matrix A, we can use either plane rotations or Givens transformations to obtain a QR factorization of A. Let G_{21} be the Givens transformation acting on the first and second coordinates, which when applied to A results in a zero in the $(2, 1)$ position. We can apply another Givens transformation, G_{31}, to $G_{21}A$ to obtain a zero in the $(3, 1)$ position. This process can be continued until the last $n - 1$ entries in the first column have been eliminated.

$$G_{n1} \cdots G_{31} G_{21} A = \begin{pmatrix} \times & \times & \cdots & \times \\ 0 & \times & \cdots & \times \\ 0 & \times & \cdots & \times \\ \vdots & & & \\ 0 & \times & \cdots & \times \end{pmatrix}$$

At the next step, Givens transformations $G_{32}, G_{42}, \ldots, G_{n2}$ are used to eliminate the last $n - 2$ entries in the second column. This process is continued until all the elements below the diagonal have been eliminated.

$$(G_{n,\,n-1}) \cdots (G_{n2} \cdots G_{32})(G_{n1} \cdots G_{21}) A = R \qquad (R \text{ upper triangular})$$

If we let $Q^T = (G_{n,\,n-1}) \cdots (G_{n2} \cdots G_{32})(G_{n1} \cdots G_{21})$, then $A = QR$ and the system $A\mathbf{x} = \mathbf{b}$ is equivalent to the system

$$R\mathbf{x} = Q^T\mathbf{b}$$

This system can be solved by back substitution.

Operation Count. The QR factorization of A using Givens transformations or plane rotations requires roughly $\frac{4}{3}n^3$ multiplications, $\frac{2}{3}n^3$ additions, and $\frac{1}{2}n^2$ square roots.

EXERCISES

1. For each of the following vectors $\mathbf{x}$, find a rotation matrix R such that $R\mathbf{x} = \|\mathbf{x}\|_2\mathbf{e}_1$.
  (a) $\mathbf{x} = (1, 1)^T$ (b) $\mathbf{x} = (\sqrt{3}, -1)^T$ (c) $\mathbf{x} = (-4, 3)^T$

2. Given $\mathbf{x} \in R^3$, define

$$r_{ij} = \left(x_i^2 + x_j^2\right)^{1/2} \qquad i, j = 1, 2, 3$$

For each of the following, determine a Givens transformation G_{ij} such that the ith and jth coordinates of $G_{ij}\mathbf{x}$ are r_{ij} and 0, respectively.
  (a) $\mathbf{x} = (3, 1, 4)^T$, $i = 1, \; j = 3$
  (b) $\mathbf{x} = (1, -1, 2)^T$, $i = 1, \; j = 2$
  (c) $\mathbf{x} = (4, 1, \sqrt{3})^T$, $i = 2, \; j = 3$
  (d) $\mathbf{x} = (4, 1, \sqrt{3})^T$, $i = 3, \; j = 2$

3. For each of the given vectors $\mathbf{x}$, find a Householder transformation H such that $H\mathbf{x} = \alpha\mathbf{e}_1$, where $\alpha = \|\mathbf{x}\|_2$.
  (a) $\mathbf{x} = (8, -1, -4)^T$ (b) $\mathbf{x} = (6, 2, 3)^T$ (c) $\mathbf{x} = (7, 4, -4)^T$

4. For each of the following, find a Householder transformation that zeros out the last two coordinates of the vector.
  (a) $\mathbf{x} = (5, 8, 4, 1)^T$ (b) $\mathbf{x} = (4, -3, -2, -1, 2)^T$

5. Given

$$A = \begin{pmatrix} 1 & 2 & -4 \\ 2 & 6 & 7 \\ -2 & 1 & 8 \end{pmatrix} \qquad \text{and} \qquad \mathbf{b} = \begin{pmatrix} 9 \\ 9 \\ -3 \end{pmatrix}$$

  (a) Use Householder transformations to transform A into an upper triangular matrix R. Also transform the vector $\mathbf{b}$, that is, compute $\mathbf{b}^{(1)} = H_2 H_1 \mathbf{b}$.
  (b) Solve $R\mathbf{x} = \mathbf{b}^{(1)}$ for $\mathbf{x}$ and check your answer by computing the residual $\mathbf{b} - A\mathbf{x}$.

6. Suppose that you wish to eliminate the last coordinate of a vector $\mathbf{x}$ and leave the first $n - 2$ coordinates unchanged. How many operations are necessary if this is to be done by a Givens transformation G? A Householder transformation H? If A is an $n \times n$ matrix, how many operations are required to compute GA and HA?

7. Let $H_k = I - 2\mathbf{u}\mathbf{u}^T$ be a Householder transformation with $\mathbf{u} = (0, \ldots, 0, u_k, u_{k+1}, \ldots, u_n)^T$. Let $\mathbf{b} \in R^n$ and let A be an $n \times n$

matrix. How many additions and multiplications are necessary to compute $H_k\mathbf{b}$? $H_k A$?

8. Let $Q^T = G_{n-k} \cdots G_2 G_1$, where each G_i is a Givens transformation. Let $\mathbf{b} \in R^n$ and let A be an $n \times n$ matrix. How many additions and multiplications are necessary to compute $Q^T\mathbf{b}$? $Q^T A$?

9. Let R_1 and R_2 be two 2×2 rotation matrices and let G_1 and G_2 be two 2×2 Givens transformations. What type of transformations are each of the following?
(a) $R_1 R_2$ (b) $G_1 G_2$ (c) $R_1 G_1$ (d) $G_1 R_1$

10. Let $\mathbf{x}$ and $\mathbf{y}$ be distinct vectors in R^n with $\|\mathbf{x}\|_2 = \|\mathbf{y}\|_2$. Define

$$\mathbf{u} = \frac{1}{\|\mathbf{x} - \mathbf{y}\|_2}(\mathbf{x} - \mathbf{y}) \qquad \text{and} \qquad Q = I - 2\mathbf{u}\mathbf{u}^T$$

Show that:
(a) $\|\mathbf{x} - \mathbf{y}\|_2^2 = 2(\mathbf{x} - \mathbf{y})^T\mathbf{x}$ (b) $Q\mathbf{x} = \mathbf{y}$

11. Let $A = Q_1 R_1 = Q_2 R_2$, where Q_1 and Q_2 are orthogonal and R_1 and R_2 are upper triangular.
(a) Show that $Q_1^T Q_2$ is diagonal.
(b) How do R_1 and R_2 compare? Explain.

*6. ITERATIVE METHODS FOR SOLVING LINEAR SYSTEMS

In this section we study iterative methods for solving a linear system $A\mathbf{x} = \mathbf{b}$. Iterative methods start out with an initial approximation $\mathbf{x}^{(0)}$ to the solution and go through a fixed procedure to obtain a better approximation, $\mathbf{x}^{(1)}$. The same procedure is then repeated on $\mathbf{x}^{(1)}$ to obtain an improved approximation, $\mathbf{x}^{(2)}$; and so on. The iterations terminate when a desired accuracy has been achieved.

Iterative methods are most useful in solving large sparse systems. Such systems occur, for example, in the solution of boundary value problems for partial differential equations. The amount of storage required for a sparse coefficient matrix A is proportional to n. Gaussian elimination and the other direct methods studied in Section 5 usually tend to fill in the zeros of A and hence require an amount of storage proportional to n^2. This can present quite a problem when n is very large, say $n \geq 1000$. The iterative methods we will describe will not require any extra storage and hence are more practical. One disadvantage they do have is that after solving $A\mathbf{x} = \mathbf{b}_1$, one must start all over again from the beginning in order to solve $A\mathbf{x} = \mathbf{b}_2$.

Given a system $A\mathbf{x} = \mathbf{b}$, we write the coefficient matrix A in the form $A = C - M$, where C is a nonsingular matrix which is in some form that is easily invertible (e.g., diagonal or triangular). The system can then be rewritten in the form

$$C\mathbf{x} = M\mathbf{x} + \mathbf{b}$$
$$\mathbf{x} = C^{-1}M\mathbf{x} + C^{-1}\mathbf{b}$$

If we set

$$B = C^{-1}M - I \quad C^{-1}A \quad \text{and} \quad \mathbf{c} = C^{-1}\mathbf{b}$$

then

(1)
$$\mathbf{x} = B\mathbf{x} + \mathbf{c}$$

To solve (1), we start out with an initial guess $\mathbf{x}^{(0)}$, which may be any vector in R^n. We then set

$$\mathbf{x}^{(1)} = B\mathbf{x}^{(0)} + \mathbf{c}$$
$$\mathbf{x}^{(2)} = B\mathbf{x}^{(1)} + \mathbf{c}$$

and in general

$$\mathbf{x}^{(k+1)} = B\mathbf{x}^{(k)} + \mathbf{c}$$

Let $\mathbf{x}$ be a solution to (1). If $\| \cdot \|$ denotes some vector norm on R^n and the corresponding matrix norm of B is less than 1, we claim that $\|\mathbf{x}^{(k)} - \mathbf{x}\| \to 0$ as $k \to \infty$. Indeed,

$$\mathbf{x}^{(1)} - \mathbf{x} = (B\mathbf{x}^{(0)} + \mathbf{c}) - (B\mathbf{x} + \mathbf{c}) = B(\mathbf{x}^{(0)} - \mathbf{x})$$
$$\mathbf{x}^{(2)} - \mathbf{x} = (B\mathbf{x}^{(1)} + \mathbf{c}) - (B\mathbf{x} + \mathbf{c}) = B(\mathbf{x}^{(1)} - \mathbf{x}) = B^2(\mathbf{x}^{(0)} - \mathbf{x})$$

and so on. In general,

$$\mathbf{x}^{(k)} - \mathbf{x} = B^k(\mathbf{x}^{(0)} - \mathbf{x})$$

and hence

$$\|\mathbf{x}^{(k)} - \mathbf{x}\| = \|B^k(\mathbf{x}^{(0)} - \mathbf{x})\|$$
$$\leq \|B^k\| \|\mathbf{x}^{(0)} - \mathbf{x}\|$$
$$\leq \|B\|^k \|\mathbf{x}^{(0)} - \mathbf{x}\|$$

Thus, if $\|B\| < 1$, then $\|\mathbf{x}^{(k)} - \mathbf{x}\| \to 0$ as $k \to \infty$.

The foregoing result holds for any norm on R^n, although in practice it is simplest to use the $\| \cdot \|_\infty$ or the $\| \cdot \|_1$. Essentially, then, we require that the matrix C be easily invertible and that C^{-1} be a good enough approximation to A^{-1} so that

$$\|I - C^{-1}A\| = \|B\| < 1$$

This last condition implies that all the eigenvalues of B are less than 1 in modulus.

Definition. Let $\lambda_1, \ldots, \lambda_n$ be the eigenvalues of B and let $\rho(B)$ $= \max_{1 \le i \le n} |\lambda_i|$. The constant $\rho(B)$ is called the **spectral radius** of B.

Theorem 7.6.1. *Let $\mathbf{x}^{(0)}$ be an arbitrary vector in R^n and define $\mathbf{x}^{(i+1)} = B\mathbf{x}^{(i)} + \mathbf{c}$ for $i = 0, 1, \ldots$. If $\mathbf{x}$ is the solution to (1), then a necessary and sufficient condition for $\mathbf{x}^{(k)} \to \mathbf{x}$ is that $\rho(B) < 1$.*

Proof. We will prove the theorem only in the case where B has n linearly independent eigenvectors. The case where B is not diagonalizable is beyond the scope of this book. If $\mathbf{x}_1, \ldots, \mathbf{x}_n$ are n linearly independent eigenvectors of B, we can write

$$\mathbf{x}^{(0)} - \mathbf{x} = \alpha_1 \mathbf{x}_1 + \cdots + \alpha_n \mathbf{x}_n$$

and

$$\mathbf{x}^{(k)} - \mathbf{x} = B^k(\alpha_1 \mathbf{x}_1 + \cdots + \alpha_n \mathbf{x}_n)$$

$$= \alpha_1 \lambda_1^k \mathbf{x}_1 + \cdots + \alpha_n \lambda_n^k \mathbf{x}_n$$

It follows that

$$(\mathbf{x}^{(k)} - \mathbf{x}) \to \mathbf{0}$$

if and only if $|\lambda_i| < 1$ for $i = 1, \ldots, n$. Thus $\mathbf{x}^{(k)} \to \mathbf{x}$ if and only if $\rho(B) < 1$.   □

The simplest choice of C is to let C be a diagonal matrix whose diagonal elements are the diagonal elements of A. The iteration scheme with this choice of C is called *Jacobi iteration*.

JACOBI ITERATION

Let

$$C = \begin{pmatrix} a_{11} & 0 & \cdots & 0 \\ 0 & a_{22} & & \\ \vdots & & \ddots & \\ 0 & 0 & & a_{nn} \end{pmatrix}$$

and

$$M = -\begin{pmatrix} 0 & a_{12} & \cdots & a_{1n} \\ a_{21} & 0 & & a_{2n} \\ \vdots & & \ddots & \\ a_{n1} & a_{n2} & & 0 \end{pmatrix}$$

and set $B = C^{-1}M$ and $\mathbf{c} = C^{-1}\mathbf{b}$. Thus

$$B = \begin{pmatrix} 0 & \dfrac{-a_{12}}{a_{11}} & \cdots & \dfrac{-a_{1n}}{a_{11}} \\ \dfrac{-a_{21}}{a_{22}} & 0 & \cdots & \dfrac{-a_{2n}}{a_{22}} \\ \vdots & & & \\ \dfrac{-a_{n1}}{a_{nn}} & \dfrac{-a_{n2}}{a_{nn}} & \cdots & 0 \end{pmatrix} \quad \text{and} \quad \mathbf{c} = \begin{pmatrix} \dfrac{b_1}{a_{11}} \\ \dfrac{b_2}{a_{22}} \\ \vdots \\ \dfrac{b_n}{a_{nn}} \end{pmatrix}$$

At the $(i + 1)$st iteration, the vector $\mathbf{x}^{(i+1)}$ is calculated by

$$(2) \qquad x_j^{(i+1)} = \frac{1}{a_{jj}} \left(-\sum_{\substack{k=1 \\ k \neq j}}^{n} a_{jk} x_k^{(i)} + b_j \right) \qquad j = 1, \ldots, n$$

The vector $\mathbf{x}^{(i)}$ is used in calculating $\mathbf{x}^{(i+1)}$. Consequently, these two vectors must be stored separately.

If the diagonal elements of A are much larger than the off-diagonal elements, the entries of B should all be small and the Jacobi iteration should converge. We say that A is *diagonally dominant* if

$$|a_{ii}| > \sum_{\substack{j=1 \\ j \neq i}}^{n} |a_{ij}| \qquad \text{for} \qquad i = 1, \ldots, n$$

If A is diagonally dominant, the matrix B of the Jacobi iteration will have the property

$$\sum_{j=1}^{n} |b_{ij}| = \sum_{\substack{j=1 \\ j \neq i}}^{n} \frac{|a_{ij}|}{|a_{ii}|} < 1 \qquad \text{for} \qquad i = 1, \ldots, n$$

Thus

$$\|B\|_\infty = \max_{1 \leq i \leq n} \left(\sum_{j=1}^{n} |b_{ij}| \right) < 1$$

It follows, then, that if A is diagonally dominant, the Jacobi iteration will converge to the solution of $A\mathbf{x} = \mathbf{b}$.

An alternative to the Jacobi iteration is to take C to be the lower triangular part of A (i.e., $c_{ij} = a_{ij}$ if $i \geq j$ and $c_{ij} = 0$ if $i < j$). Since C is a better approximation to A than the diagonal matrix in the Jacobi iteration, we would expect that C^{-1} is a better approximation to A^{-1}, and hopefully B will have a smaller norm. The iteration scheme with this choice of C is called *Gauss–Seidel iteration*. It usually converges faster than Jacobi iteration.

GAUSS–SEIDEL ITERATION

Let

$$
L = - \begin{pmatrix}
0 & 0 & \cdots & 0 & 0 \\
a_{21} & 0 & \cdots & 0 & 0 \\
\vdots & & & & \\
a_{n-1,1} & a_{n-1,2} & & 0 & 0 \\
a_{n1} & a_{n2} & \cdots & a_{n,n-1} & 0
\end{pmatrix}
$$

$$
D = \begin{pmatrix}
a_{11} & 0 & \cdots & 0 \\
0 & a_{22} & \cdots & 0 \\
\vdots & & & \\
0 & 0 & \cdots & a_{nn}
\end{pmatrix}
$$

and

$$
U = - \begin{pmatrix}
0 & a_{12} & \cdots & a_{1,n-1} & a_{1n} \\
0 & 0 & \cdots & a_{2,n-1} & a_{2n} \\
\vdots & & & & \\
0 & 0 & & 0 & a_{n-1,n} \\
0 & 0 & & 0 & 0
\end{pmatrix}
$$

Set $C = D - L$ and $M = U$. Let $\mathbf{x}^{(0)}$ be an arbitrary nonzero vector in R^n. We have

$$
C\mathbf{x}^{(i+1)} = M\mathbf{x}^{(i)} + \mathbf{b}
$$

$$
(D - L)\mathbf{x}^{(i+1)} = U\mathbf{x}^{(i)} + \mathbf{b}
$$

$$
D\mathbf{x}^{(i+1)} = L\mathbf{x}^{(i+1)} + U\mathbf{x}^{(i)} + \mathbf{b}
$$

We can solve this last equation for $\mathbf{x}^{(i+1)}$ one coordinate at a time. The first coordinate of $\mathbf{x}^{(i+1)}$ is given by

$$
x_1^{(i+1)} = \frac{1}{a_{11}} \left(- \sum_{k=2}^{n} a_{1k} x_k^{(i)} + b_1 \right)
$$

The second coordinate of $\mathbf{x}^{(i+1)}$ can be solved for in terms of the first coordinate and the last $n - 2$ coordinates of $x^{(i)}$.

$$
x_2^{(i+1)} = \frac{1}{a_{22}} \left(- a_{21} x_1^{(i+1)} - \sum_{k=3}^{n} a_{2k} x_k^{(i)} + b_2 \right)
$$

In general,

$$
(3) \qquad x_j^{(i+1)} = \frac{1}{a_{jj}} \left(- \sum_{k=1}^{j-1} a_{jk} x_k^{(i+1)} - \sum_{k=j+1}^{n} a_{jk} x_k^{(i)} + b_j \right)
$$

It is interesting to compare (2) and (3). The difference between the

Jacobi and Gauss–Seidel iterations is that in the latter case, one is using the coordinates of $\mathbf{x}^{(i+1)}$ as soon as they are calculated rather than in the next iteration. The program for the Gauss–Seidel iteration is actually simpler than the program for the Jacobi iteration. The vectors $\mathbf{x}^{(i)}$ and $\mathbf{x}^{(i+1)}$ are both stored in the same vector, $\mathbf{x}$. As a coordinate of $\mathbf{x}^{(i+1)}$ is calculated, it replaces the corresponding coordinate of $\mathbf{x}^{(i)}$.

Theorem 7.6.2. *If A is diagonally dominant, then the Gauss–Seidel iteration converges to a solution of $A\mathbf{x} = \mathbf{b}$.*

Proof.   For $j = 1, \ldots, n$, let

$$\alpha_j = \sum_{i=1}^{j-1} |a_{ji}|, \qquad \beta_j = \sum_{i=j+1}^{n} |a_{ji}|, \qquad \text{and} \qquad M_j = \frac{\beta_j}{(|a_{jj}| - \alpha_j)}$$

Since A is diagonally dominant, it follows that

$$|a_{jj}| > \alpha_j + \beta_j$$

and consequently $M_j < 1$ for $j = 1, \ldots, n$. Thus

$$M = \max_{1 \le j \le n} M_j < 1$$

We will show that

$$\|B\|_\infty = \max_{\mathbf{x} \ne \mathbf{0}} \frac{\|B\mathbf{x}\|_\infty}{\|\mathbf{x}\|_\infty} \le M < 1$$

Let $\mathbf{x}$ be a nonzero vector in R^n and let $\mathbf{y} = B\mathbf{x}$. Choose k so that

$$\|\mathbf{y}\|_\infty = \max_{1 \le i \le n} |y_i| = |y_k|$$

It follows from the definition of B that

$$\mathbf{y} = B\mathbf{x} = (D - L)^{-1} U\mathbf{x}$$

and hence

$$\mathbf{y} = D^{-1}(L\mathbf{y} + U\mathbf{x})$$

Comparing the kth coordinates of each side, we see that

$$y_k = \frac{1}{a_{kk}} \left(-\sum_{i=1}^{k-1} a_{ki} y_i - \sum_{i=k+1}^{n} a_{ki} x_i \right)$$

and hence

(4) $$\|\mathbf{y}\|_\infty = |y_k| \le \frac{1}{|a_{kk}|} \left(\alpha_k \|\mathbf{y}\|_\infty + \beta_k \|\mathbf{x}\|_\infty \right)$$

It follows from (4) that

$$\frac{\|B\mathbf{x}\|_\infty}{\|\mathbf{x}\|_\infty} = \frac{\|\mathbf{y}\|_\infty}{\|\mathbf{x}\|_\infty} \le M_k \le M$$

Thus

$$\|B\|_\infty = \max_{\mathbf{x} \neq \mathbf{0}} \frac{\|B\mathbf{x}\|_\infty}{\|\mathbf{x}\|_\infty} \leq M < 1$$

and hence the iteration will converge to the solution of $A\mathbf{x} = \mathbf{b}$.   □

EXERCISES

1. Let

$$A = \begin{pmatrix} 10 & 1 \\ 2 & 10 \end{pmatrix}, \qquad \mathbf{b} = \begin{pmatrix} 11 \\ 12 \end{pmatrix}, \qquad \text{and} \qquad \mathbf{x}^{(0)} = \begin{pmatrix} 0 \\ 0 \end{pmatrix}$$

Use Jacobi iteration to compute $\mathbf{x}^{(1)}$ and $\mathbf{x}^{(2)}$. [The exact solution is $\mathbf{x} = (1, 1)^T$.]

2. Let

$$A = \begin{pmatrix} 1 & 1 & 1 \\ 0 & 1 & 1 \\ 0 & 0 & 1 \end{pmatrix}, \qquad \mathbf{b} = \begin{pmatrix} 3 \\ 2 \\ 1 \end{pmatrix}, \qquad \text{and} \qquad \mathbf{x}^{(0)} = \begin{pmatrix} 1 \\ 0 \\ 0 \end{pmatrix}$$

Use Jacobi iteration to compute $\mathbf{x}^{(1)}$, $\mathbf{x}^{(2)}$, $\mathbf{x}^{(3)}$, and $\mathbf{x}^{(4)}$.

3. Repeat Exercise 1 using Gauss–Seidel iteration.

4. Let

$$A = \begin{pmatrix} 10 & 1 & 1 \\ 1 & 10 & 1 \\ 1 & 1 & 10 \end{pmatrix}, \qquad \mathbf{b} = \begin{pmatrix} 12 \\ 12 \\ 12 \end{pmatrix}, \qquad \text{and} \qquad \mathbf{x}^{(0)} = \begin{pmatrix} 1 \\ 0 \\ 0 \end{pmatrix}$$

(a) Calculate $\mathbf{x}^{(1)}$ using Jacobi iteration.
(b) Calculate $\mathbf{x}^{(1)}$ using Gauss–Seidel iteration.
(c) Compare your answers to (a) and (b) with the correct solution $\mathbf{x} = (1, 1, 1)^T$. Which is closer?

5. For which of the following matrices, will the iteration scheme

$$\mathbf{x}^{(k+1)} = B\mathbf{x}^{(k)} + \mathbf{c}$$

converge to a solution of $\mathbf{x} = B\mathbf{x} + \mathbf{c}$? Explain.

(a) $B = \begin{pmatrix} 1 & 1 & 1 \\ 0 & 1 & 1 \\ 0 & 0 & 1 \end{pmatrix}$ (b) $B = \begin{pmatrix} 0.9 & 1 & 1 \\ 0 & 0.9 & 1 \\ 0 & 0 & 0.9 \end{pmatrix}$

(c) $B = \begin{pmatrix} \frac{1}{2} & 10 & 100 \\ 0 & \frac{1}{2} & 10 \\ 0 & 0 & \frac{1}{2} \end{pmatrix}$ (d) $B = \begin{pmatrix} \frac{1}{4} & \frac{1}{4} & \frac{1}{4} \\ \frac{1}{4} & \frac{1}{2} & \frac{1}{8} \\ \frac{1}{2} & \frac{1}{4} & \frac{1}{8} \end{pmatrix}$

(e) $B = \begin{pmatrix} \frac{1}{3} & \frac{1}{3} & \frac{1}{3} \\ \frac{1}{2} & \frac{1}{3} & \frac{1}{6} \\ 0 & \frac{1}{6} & \frac{1}{3} \end{pmatrix}$

6. Let $\mathbf{x}$ be the solution to $\mathbf{x} = B\mathbf{x} + \mathbf{c}$. Let $\mathbf{x}^{(0)}$ be an arbitrary vector in R^n and define

$$\mathbf{x}^{(k+1)} = B\mathbf{x}^{(k)} + \mathbf{c}$$

for $k = 0, 1, \ldots$. Prove that if B^m is the zero matrix, then $\mathbf{x}^{(m)} = \mathbf{x}$.

7. Let A be a nonsingular upper triangular matrix. Show that the Jacobi iteration will give the exact solution (assuming no roundoff errors) to $A\mathbf{x} = \mathbf{b}$ after n iterations.

8. For an iterative method based on the splitting $A = C - M, C$ nonsingular, show that

$$\mathbf{x}^{(k+1)} = \mathbf{x}^{(k)} + C^{-1}\mathbf{r}^{(k)}$$

where $\mathbf{r}^{(k)}$ denotes the residual $\mathbf{b} - A\mathbf{x}^{(k)}$.

9. Let $A = D - L - U$, where D, L, and U are defined as in Gauss–Seidel iteration and let ω be a nonzero scalar. The system $\omega A\mathbf{x} = \omega\mathbf{b}$ can be solved iteratively by splitting ωA into $C - M$, where $C = D - \omega L$. Determine the B and $\mathbf{c}$ corresponding to this splitting. (The constant ω is called a *relaxation parameter*. The case $\omega = 1$ corresponds to Gauss–Seidel iteration.)

10. Let $\mathbf{x}$ be the solution to $\mathbf{x} = B\mathbf{x} + \mathbf{c}$. Let $\mathbf{x}^{(0)}$ be an arbitrary vector in R^n and define

$$\mathbf{x}^{(i+1)} = B\mathbf{x}^{(i)} + \mathbf{c}$$

for $i = 0, 1, \ldots$. If $\|B\| = \alpha < 1$, show that

$$\|\mathbf{x}^{(k)} - \mathbf{x}\| \le \frac{\alpha}{1 - \alpha}\|\mathbf{x}^{(k)} - \mathbf{x}^{(k-1)}\|$$

7. THE SINGULAR VALUE DECOMPOSITION _____

In many applications it is necessary to either determine the rank of a matrix or to determine whether or not the matrix is deficient in rank. Theoretically, one can use Gaussian elimination to reduce the matrix to row echelon form and then count the number of nonzero rows. However, this approach is not practical when working in finite precision arithmetic. If A is rank deficient and U is the computed echelon form, then because of roundoff errors in the elimination process, it is unlikely that U will have the proper number of nonzero rows. In practice, the coefficient matrix A usually involves some error. This may be due to errors in the data or to the finite number system. Thus it is generally more practical to ask whether A is "close" to a rank deficient matrix. However, it may well turn out that A is close to being rank deficient and the computed echelon form U is not.

In this section we will assume throughout that A is an $m \times n$ matrix with $m \geq n$. (This assumption is made for convenience only; all the results will also hold if $m < n$.) We will present a method for determining how close A is to a matrix of smaller rank. The method involves factoring A into a product $U\Sigma V^T$, where U is an $m \times m$ orthogonal matrix, V is an $n \times n$ orthogonal matrix, and Σ is an $m \times n$ matrix whose off diagonal entries are all 0's and whose diagonal elements satisfy

$$\sigma_1 \geq \sigma_2 \geq \cdots \geq \sigma_n \geq 0$$

$$\Sigma = \begin{pmatrix} \sigma_1 & & & \\ & \sigma_2 & & \\ & & \ddots & \\ & & & \sigma_n \\ & & & \end{pmatrix}$$

The σ_i's determined by this factorization are unique and are called the *singular values* of A. The factorization $U\Sigma V^T$ is called the *singular value decomposition* of A. We will show that the rank of A equals the number of nonzero singular values and that the magnitudes of the nonzero singular values provide a measure of how close A is to a matrix of lower rank.

We begin by showing that such a decomposition is always possible.

Theorem 7.7.1. *If A is an $m \times n$ matrix, then A has a singular value decomposition.*

Proof. The matrix A^TA is positive semidefinite, since

$$\mathbf{x}^T A^T A \mathbf{x} = \|A\mathbf{x}\|_2^2 \geq 0$$

for all $\mathbf{x} \in R^n$. Thus A^TA has nonnegative eigenvalues. The eigenvalues may be ordered so that

$$\lambda_1 \geq \lambda_2 \geq \cdots \geq \lambda_n \geq 0$$

Define

$$\sigma_i = \sqrt{\lambda_i} \qquad i = 1, \ldots, n$$

Without loss of generality, one may assume that exactly r of the σ_i's are nonzero, so that

$$\sigma_1 \ge \sigma_2 \ge \cdots \ge \sigma_r > 0 \quad \text{and} \quad \sigma_{r+1} = \sigma_{r+2} = \cdots = \sigma_n = 0$$

Let

$$\Sigma = \begin{pmatrix} \sigma_1 & & & & & & \\ & \sigma_2 & & & & & \\ & & \ddots & & & & \\ & & & \sigma_r & & & \\ & & & & 0 & & \\ & & & & & \ddots & \\ & & & & & & 0 \end{pmatrix} = \begin{pmatrix} \Sigma_r & O \\ O & O \end{pmatrix}$$

where Σ_r is an $r \times r$ diagonal matrix whose diagonal elements are $\sigma_1, \ldots, \sigma_r$. Since $A^T A$ is symmetric, there is an orthogonal matrix V that diagonalizes $A^T A$.

$$V^T A^T A V = \Sigma^T \Sigma$$

Let $\mathbf{v}_1, \mathbf{v}_2, \ldots, \mathbf{v}_n$ be the column vectors of V. Thus $\mathbf{v}_i$ is an eigenvector belonging to σ_i^2 for $i = 1, \ldots, r$ and $\mathbf{v}_{r+1}, \ldots, \mathbf{v}_n$ are eigenvectors belonging to 0. Let $V_1 = (\mathbf{v}_1, \mathbf{v}_2, \ldots, \mathbf{v}_r)$ and let $V_2 = (\mathbf{v}_{r+1}, \ldots, \mathbf{v}_n)$. Since

$$A^T A \mathbf{v}_i = \mathbf{0}$$

for $i = r + 1, \ldots, n$, it follows that

$$(A V_2)^T A V_2 = V_2^T A^T A V_2 = O$$

and hence

$$A V_2 = O$$

Since

$$A^T A V_1 = V_1 \Sigma_r^2$$

it follows that

$$\Sigma_r^{-1} V_1^T A^T A V_1 \Sigma_r^{-1} = I_r$$

where I_r denotes the $r \times r$ identity matrix. If we set

$$U_1 = A V_1 \Sigma_r^{-1}$$

then

$$U_1^T U_1 = I_r$$

and hence U_1 is an $m \times r$ matrix with orthonormal column vectors $\mathbf{u}_1, \ldots, \mathbf{u}_r$. The set $\{\mathbf{u}_1, \ldots, \mathbf{u}_r\}$ can be extended to form an orthonormal basis for R^m. Let $U_2 = (\mathbf{u}_{r+1}, \ldots, \mathbf{u}_m)$ and let $U = (U_1 \quad U_2)$. It follows that

$$U^T A V = \begin{pmatrix} U_1^T \\ U_2^T \end{pmatrix} A(V_1 \quad V_2)$$

$$= \begin{pmatrix} U_1^T \\ U_2^T \end{pmatrix} (AV_1 \quad O)$$

$$= \begin{pmatrix} U_1^T A V_1 & O \\ U_2^T A V_1 & O \end{pmatrix}$$

But

$$U_1^T A V_1 = \Sigma_r^{-1} V_1^T A^T A V_1 = \Sigma_r$$

and

$$U_2^T A V_1 = U_2^T U_1 \Sigma_r = O$$

Therefore, $U^T A V = \Sigma$ and hence $A = U \Sigma V^T$. $\square$

Observations.   Let A be an $m \times n$ matrix with a singular value decomposition $U \Sigma V^T$.

1. Since

$$A^T A = V \Sigma^T \Sigma V^T$$

and

$$\Sigma^T \Sigma = \begin{pmatrix} \sigma_1^2 & & & \\ & \sigma_2^2 & & \\ & & \ddots & \\ & & & \sigma_n^2 \end{pmatrix}$$

it follows that the eigenvalues of $A^T A$ are $\sigma_1^2, \sigma_2^2, \ldots, \sigma_n^2$. Since the σ_i's are the nonnegative square roots of the eigenvalues of $A^T A$, they are unique. However, the orthogonal matrices U and V are not unique.

2. Since V diagonalizes $A^T A$, it follows that the $\mathbf{v}_j$'s are eigenvectors of $A^T A$.

3. Since $AA^T = U \Sigma \Sigma^T U^T$, it follows that U diagonalizes AA^T and that the $\mathbf{u}_j$'s are eigenvectors of AA^T.

4. Comparing the jth columns of each side of the equation

$$AV = U\Sigma$$

we get

$$A\mathbf{v}_j = \sigma_j \mathbf{u}_j \qquad j = 1, \ldots, n$$

Similarly,

$$A^T U = V \Sigma^T$$

and hence

$$A^T \mathbf{u}_j = \sigma_j \mathbf{v}_j \qquad \text{for} \quad j = 1, \ldots, n$$

$$A^T \mathbf{u}_j = \mathbf{0} \qquad \text{for} \quad j = n + 1, \ldots, m$$

The $\mathbf{v}_j$'s are called the *right singular vectors* of A and the $\mathbf{u}_j$'s are called the *left singular vectors* of A.

5. Since multiplying a matrix on either side by a nonsingular matrix does not alter its rank, it follows that A and Σ have the same rank. Thus, if $\sigma_1 \geq \sigma_2 \geq \cdots \geq \sigma_r > 0$ and $\sigma_{r+1} = \cdots = \sigma_n = 0$, then A has rank r. The rank of a matrix is the number of nonzero singular values (where singular values are counted according to multiplicity). The reader should be careful not to make a similar assumption about eigenvalues. The matrix

$$M = \begin{pmatrix} 0 & 1 & 0 & 0 \\ 0 & 0 & 1 & 0 \\ 0 & 0 & 0 & 1 \\ 0 & 0 & 0 & 0 \end{pmatrix}$$

for example, has rank 3 even though all its eigenvalues are 0.

EXAMPLE 1.   Let

$$A = \begin{pmatrix} 1 & 1 \\ 1 & 1 \\ 0 & 0 \end{pmatrix}$$

Compute the singular values and the singular value decomposition of A.

SOLUTION.   The matrix

$$A^T A = \begin{pmatrix} 2 & 2 \\ 2 & 2 \end{pmatrix}$$

has eigenvalues $\lambda_1 = 4$ and $\lambda_2 = 0$. Consequently, the singular values of A are $\sigma_1 = \sqrt{4} = 2$ and $\sigma_2 = 0$. The eigenvalue λ_1 has eigenvectors of the form $\alpha(1, 1)^T$ and λ_2 has eigenvectors $\beta(1, -1)^T$. Therefore, the orthogonal matrix

$$V = \frac{1}{\sqrt{2}} \begin{pmatrix} 1 & 1 \\ 1 & -1 \end{pmatrix}$$

diagonalizes $A^T A$. From observation 4 it follows that

$$\mathbf{u}_1 = \frac{1}{\sigma_1} A \mathbf{v}_1 = \frac{1}{2} \begin{pmatrix} 1 & 1 \\ 1 & 1 \\ 0 & 0 \end{pmatrix} \begin{pmatrix} \dfrac{1}{\sqrt{2}} \\ \dfrac{1}{\sqrt{2}} \end{pmatrix} = \begin{pmatrix} \dfrac{1}{\sqrt{2}} \\ \dfrac{1}{\sqrt{2}} \\ 0 \end{pmatrix}$$

The remaining column vectors of U must be eigenvectors of AA^T belonging to the eigenvalue 0. Thus we can determine $\mathbf{u}_2$ and $\mathbf{u}_3$ by finding an orthonormal basis for $N(AA^T)$. Since

$$AA^T = \begin{pmatrix} 2 & 2 & 0 \\ 2 & 2 & 0 \\ 0 & 0 & 0 \end{pmatrix}$$

we can choose $\mathbf{u}_2 = \left(\dfrac{1}{\sqrt{2}}, -\dfrac{1}{\sqrt{2}}, 0\right)^T$ and $\mathbf{u}_3 = (0, 0, 1)^T$. It follows then that

$$A = U\Sigma V^T = \begin{pmatrix} \dfrac{1}{\sqrt{2}} & \dfrac{1}{\sqrt{2}} & 0 \\ \dfrac{1}{\sqrt{2}} & -\dfrac{1}{\sqrt{2}} & 0 \\ 0 & 0 & 1 \end{pmatrix} \begin{pmatrix} 2 & 0 \\ 0 & 0 \\ 0 & 0 \end{pmatrix} \begin{pmatrix} \dfrac{1}{\sqrt{2}} & \dfrac{1}{\sqrt{2}} \\ \dfrac{1}{\sqrt{2}} & -\dfrac{1}{\sqrt{2}} \end{pmatrix} \qquad \square$$

If A is an $m \times n$ matrix of rank r and $0 < k < r$, we can use the singular value decomposition to find a matrix in $R^{m \times n}$ of rank k that is closest to A with respect to the Frobenius norm. Let $\mathbb{S}$ be the set of all $m \times n$ matrices of rank k or less. It can be shown that there is a matrix X in $\mathbb{S}$ such that

$$(1) \qquad \|A - X\|_F = \min_{S \in \mathbb{S}} \|A - S\|_F$$

We will not prove this, since the proof is beyond the scope of this book. Assuming that the minimum is achieved, we will show how such a matrix X can be derived from the singular value decomposition of A. The following lemma will be useful.

Lemma 7.7.2. *If A is an $m \times n$ matrix and Q is an $m \times m$ orthogonal matrix, then*

$$\|QA\|_F = \|A\|_F$$

Proof

$$\begin{aligned}
\|QA\|_F^2 &= \|(Q\mathbf{a}_1, Q\mathbf{a}_2, \ldots, Q\mathbf{a}_n)\|_F^2 \\
&= \sum_{i=1}^{n} \|Q\mathbf{a}_i\|_2^2 \\
&= \sum_{i=1}^{n} \|\mathbf{a}_i\|_2^2 \\
&= \|A\|_F^2 \qquad\qquad\qquad\qquad\qquad \square
\end{aligned}$$

It follows from the lemma that if A has singular value decomposition $U\Sigma V^T$, then

$$\|A\|_F = \|\Sigma V^T\|_F$$

Since

$$\|\Sigma V^T\|_F = \|(\Sigma V^T)^T\|_F = \|V\Sigma^T\|_F = \|\Sigma^T\|_F$$

it follows that

$$\|A\|_F = \left(\sigma_1^2 + \sigma_2^2 + \cdots + \sigma_n^2\right)^{1/2}$$

Theorem 7.7.3. *Let $A = U\Sigma V^T$ be an $m \times n$ matrix and let $\mathbb{S}$ denote the set of all $m \times n$ matrices of rank k or less, where $0 < k < \text{rank}\,(A)$. If X is a matrix in $\mathbb{S}$ satisfying (1), then*

$$\|A - X\|_F = \left(\sigma_{k+1}^2 + \sigma_{k+2}^2 + \cdots + \sigma_n^2\right)^{1/2}$$

In particular, if $A' = U\Sigma' V^T$, where

$$\Sigma' = \begin{pmatrix} \sigma_1 & & & & \\ & \ddots & & & O \\ & & \sigma_k & & \\ \hline & O & & & O \end{pmatrix} = \begin{pmatrix} \Sigma_k & O \\ O & O \end{pmatrix}$$

then

$$\|A - A'\|_F = \left(\sigma_{k+1}^2 + \cdots + \sigma_n^2\right)^{1/2} = \min_{S \in \mathbb{S}}\|A - S\|_F$$

Proof.   Let X be a matrix in $\mathbb{S}$ satisfying (1). Since $A' \in \mathbb{S}$, it follows that

$$(2) \qquad \|A - X\|_F \le \|A - A'\|_F = \left(\sigma_{k+1}^2 + \cdots + \sigma_n^2\right)^{1/2}$$

We will show that

$$\|A - X\|_F \ge \left(\sigma_{k+1}^2 + \cdots + \sigma_n^2\right)^{1/2}$$

and hence that equality holds in (2). Let $Q\Omega P^T$ be the singular value decomposition of X.

$$\Omega = \begin{pmatrix} \omega_1 & & & & \\ & \omega_2 & & & \\ & & \ddots & & O \\ & & & \omega_k & \\ \hline & & O & & O \end{pmatrix} = \begin{pmatrix} \Omega_k & O \\ O & O \end{pmatrix}$$

If we set $B = Q^T A P$, then $A = QBP^T$ and it follows that

$$\|A - X\|_F = \|Q(B - \Omega)P^T\|_F = \|B - \Omega\|_F$$

Let us partition B in the same manner as Ω.

$$B = \begin{pmatrix} \overbrace{B_{11}}^{k \times k} & \Big| & \overbrace{B_{12}}^{k \times (n-k)} \\ \hline B_{21} & \Big| & B_{22} \\ \underbrace{}_{(m-k) \times k} & & \underbrace{}_{(m-k) \times (n-k)} \end{pmatrix}$$

It follows that

$$\|A - X\|_F^2 = \|B_{11} - \Omega_k\|_F^2 + \|B_{12}\|_F^2 + \|B_{21}\|_F^2 + \|B_{22}\|_F^2$$

We claim $B_{12} = O$. If not, then define

$$Y = Q\begin{pmatrix} B_{11} & B_{12} \\ O & O \end{pmatrix} P^T$$

$Y \in \mathbb{S}$ and

$$\|A - Y\|_F^2 = \|B_{21}\|_F^2 + \|B_{22}\|_F^2 < \|A - X\|_F^2$$

This contradicts the definition of X. Therefore, $B_{12} = O$. In a similar manner it can be shown that B_{21} must equal O. If we set

$$Z = Q\begin{pmatrix} B_{11} & O \\ O & O \end{pmatrix} P^T$$

then $Z \in \mathbb{S}$ and

$$\|A - Z\|_F^2 = \|B_{22}\|_F^2 \leq \|B_{11} - \Omega_k\|_F^2 + \|B_{22}\|_F^2 = \|A - X\|_F^2$$

It follows from the definition of X that B_{11} must equal Ω_k. If B_{22} has singular value decomposition $U_1 \Lambda V_1^T$, then

$$\|A - X\|_F = \|B_{22}\|_F = \|\Lambda\|_F$$

Let

$$U_2 = \begin{pmatrix} I_k & O \\ O & U_1 \end{pmatrix} \qquad \text{and} \qquad V_2 = \begin{pmatrix} I_k & O \\ O & V_1 \end{pmatrix}$$

Now

$$U_2^T Q^T A P V_2 = \begin{pmatrix} \Omega_k & O \\ O & \Lambda \end{pmatrix}$$

$$A = (QU_2)\begin{pmatrix} \Omega_k & O \\ O & \Lambda \end{pmatrix}(PV_2)^T$$

and hence it follows that the diagonal elements of Λ are singular values of A. Thus

$$\|A - X\|_F = \|\Lambda\|_F \geq \left(\sigma_{k+1}^2 + \cdots + \sigma_n^2\right)^{1/2}$$

It follows from (2) that

$$\|A - X\|_F = \left(\sigma_{k+1}^2 + \cdots + \sigma_n^2\right)^{1/2} = \|A - A'\|_F \qquad \square$$

If $A = U\Sigma V^T$ and we define $E_j = \mathbf{u}_j \mathbf{v}_j^T$ for $j = 1, \ldots, n$, then each E_j is of rank 1 and

(3) $$A = \sigma_1 E_1 + \sigma_2 E_2 + \cdots + \sigma_n E_n$$

To prove (3), note that

$$\left(\mathbf{v}_1 \mathbf{v}_1^T + \cdots + \mathbf{v}_n \mathbf{v}_n^T\right)\mathbf{x} = \left(\mathbf{v}_1^T \mathbf{x}\right)\mathbf{v}_1 + \cdots + \left(\mathbf{v}_n^T \mathbf{x}\right)\mathbf{v}_n = \mathbf{x}$$

for every $\mathbf{x} \in R^n$, since the column vectors of V form an orthonormal basis for R^n. Thus

$$N\left(\mathbf{v}_1\mathbf{v}_1^T + \cdots + \mathbf{v}_n\mathbf{v}_n^T - I\right) = R^n$$

and hence

$$\mathbf{v}_1\mathbf{v}_1^T + \cdots + \mathbf{v}_n\mathbf{v}_{nn}^T = I$$

It follows that

$$A = A\left(\mathbf{v}_1\mathbf{v}_1^T + \cdots + \mathbf{v}_n\mathbf{v}_n^T\right)$$
$$= \sigma_1\mathbf{u}_1\mathbf{v}_1^T + \cdots + \sigma_n\mathbf{u}_n\mathbf{v}_n^T$$
$$= \sigma_1 E_1 + \cdots + \sigma_n E_n$$

If A is of rank n, then

$$A' = U\begin{pmatrix} \sigma_1 & & & & \\ & \sigma_2 & & & \\ & & \ddots & & \\ & & & \sigma_{n-1} & \\ & & & & 0 \end{pmatrix} V^T = \sigma_1 E_1 + \cdots + \sigma_{n-1}E_{n-1}$$

will be a matrix of rank $n - 1$ which is nearest to A with respect to the Frobenius norm. Similarly, $A'' = \sigma_1 E_1 + \cdots + \sigma_{n-2}E_{n-2}$ will be the nearest matrix of rank $n - 2$, and so on. In particular, if A is a nonsingular $n \times n$ matrix, then A' is singular and $\|A - A'\|_F = \sigma_n$. Thus σ_n may be taken as a measure of how close a matrix is to being singular.

The reader should be careful not to use the value of $\det(A)$ as a measure of how close A is to being singular. If, for example, A is the 100×100 diagonal matrix whose diagonal entries are all $\frac{1}{2}$, then $\det(A) = 2^{-100}$; however, $\sigma_{100} = \frac{1}{2}$. On the other hand, the matrix in the following example is very close to being singular even though its determinant is 1 and all of its eigenvalues are equal to 1.

EXAMPLE 2. Let A be an $n \times n$ upper triangular matrix whose diagonal elements are all 1 and whose entries above the main diagonal are all -1.

$$A = \begin{pmatrix} 1 & -1 & -1 & \cdots & -1 & -1 \\ 0 & 1 & -1 & \cdots & -1 & -1 \\ 0 & 0 & 1 & \cdots & -1 & -1 \\ \vdots & & & & & \\ 0 & 0 & 0 & \cdots & 1 & -1 \\ 0 & 0 & 0 & \cdots & 0 & 1 \end{pmatrix}$$

Notice that $\det(A) = \det(A^{-1}) = 1$ and all the eigenvalues of A are 1. However, if n is large, then A is close to being singular. To see this, let

$$B = \begin{pmatrix} 1 & -1 & -1 & \cdots & -1 & -1 \\ 0 & 1 & -1 & \cdots & -1 & -1 \\ 0 & 0 & 1 & \cdots & -1 & -1 \\ \vdots & & & & & \\ 0 & 0 & 0 & \cdots & 1 & -1 \\ \dfrac{-1}{2^{n-2}} & 0 & 0 & \cdots & 0 & 1 \end{pmatrix}$$

B is singular, since the system $B\mathbf{x} = \mathbf{0}$ has a nontrivial solution $\mathbf{x} = (2^{n-2}, 2^{n-3}, \ldots, 2^0, 1)^T$. The matrices A and B only differ in the $(n, 1)$ position.

$$\|A - B\|_F = \frac{1}{2^{n-2}}$$

It follows from Theorem 7.7.3 that

$$\sigma_n = \min_{X \text{ singular}} \|A - X\|_F \leq \|A - B\|_F = \frac{1}{2^{n-2}}$$

Thus, if $n = 100$, then $\sigma_n \leq 1/2^{98}$, and consequently A is very close to singular. $\square$

The inverse of the matrix A in Example 2 is given by

$$A^{-1} = \begin{pmatrix} 1 & 1 & 2 & 4 & \cdots & 2^{n-2} \\ 0 & 1 & 1 & 2 & \cdots & 2^{n-3} \\ \vdots & & & & & \\ 0 & 0 & 0 & 0 & \cdots & 2^0 \\ 0 & 0 & 0 & 0 & \cdots & 1 \end{pmatrix}$$

Thus

$$\text{cond}_\infty A = \|A\|_\infty \|A^{-1}\|_\infty = n 2^{n-1}$$

If n is large, A will be extremely ill conditioned. This is not surprising. We have already seen that A is very close to a singular matrix. Consequently, we would expect that a small change in the entries of A would have a drastic effect. It seems reasonable that there should be some relation between the condition number and the singular values of a matrix. This is indeed the case when the condition number is defined in terms of the 2-norm of the matrix. Recall that

$$\|A\|_2 = \max_{\mathbf{x} \neq \mathbf{0}} \frac{\|A\mathbf{x}\|_2}{\|\mathbf{x}\|_2}$$

$$\text{cond}_2(A) = \|A\|_2 \|A^{-1}\|_2$$

Theorem 7.7.4. *If A is an $m \times n$ matrix with singular value decomposition $U\Sigma V^T$, then*

$$\|A\|_2 = \sigma_1 \qquad (\textit{the largest singular value})$$

Proof. Since U and V are orthogonal,

$$\|A\|_2 = \|U\Sigma V^T\|_2 = \|\Sigma\|_2$$

Now

$$\|\Sigma\|_2 = \max_{\mathbf{x} \neq \mathbf{0}} \frac{\|\Sigma\mathbf{x}\|_2}{\|\mathbf{x}\|_2}$$

$$= \max_{\mathbf{x} \neq \mathbf{0}} \frac{\left(\sum_{i=1}^{n} (\sigma_i x_i)^2 \right)^{1/2}}{\left(\sum_{i=1}^{n} x_i^2 \right)^{1/2}}$$

$$\leq \sigma_1$$

However, if we choose $\mathbf{x} = \mathbf{e}_1$, then

$$\frac{\|\Sigma\mathbf{x}\|_2}{\|\mathbf{x}\|_2} = \sigma_1$$

and hence it follows that

$$\|A\|_2 = \|\Sigma\|_2 = \sigma_1 \qquad \qquad \square$$

Corollary 7.7.5. *If $A = U\Sigma V^T$ is nonsingular, then*

$$\operatorname{cond}_2(A) = \frac{\sigma_1}{\sigma_n}$$

Proof. The singular values of $A^{-1} = V\Sigma^{-1}U^T$ arranged in decreasing order are

$$\frac{1}{\sigma_n} \geq \frac{1}{\sigma_{n-1}} \geq \cdots \geq \frac{1}{\sigma_1}$$

Therefore,

$$\|A^{-1}\|_2 = \frac{1}{\sigma_n} \qquad \text{and} \qquad \operatorname{cond}_2(A) = \frac{\sigma_1}{\sigma_n} \qquad \qquad \square$$

If two matrices A and B are close, their singular values must also be close. More precisely, if A has the singular values $\sigma_1 \geq \sigma_2 \geq \cdots \geq \sigma_n$ and B has the singular values $\omega_1 \geq \omega_2 \geq \cdots \geq \omega_n$, then

$$|\sigma_i - \omega_i| \leq \|A - B\|_2 \qquad i = 1, \ldots, n$$

(see [20], p. 321). Thus, in computing the singular values of a matrix A, we need not worry that small changes in the entries of A will cause drastic changes in the computed singular values. In Section 9 we will learn an algorithm for computing the singular values of a matrix. We will also see how the singular value decomposition can be used to solve least squares problems.

APPLICATION: DIGITAL IMAGE PROCESSING

A video image or photograph can be digitized by breaking it up into a rectangular array of cells (or pixels) and measuring the gray level of each cell. This information can be stored and transmitted using an $n \times n$ matrix A. The entries of A are nonnegative numbers corresponding to the measures of the gray levels. Because the gray levels of any one cell generally turn out to be close to the gray levels of its neighboring cells, it is possible to reduce the amount of storage necessary from n^2 to a multiple of n. Generally, the matrix A will have many small singular values. Consequently, A can be approximated by a matrix of much lower rank.

If A has singular value decomposition $U\Sigma V^T$, then A can be represented by the outer product expansion

$$A = \sigma_1 \mathbf{u}_1 \mathbf{v}_1^T + \sigma_2 \mathbf{u}_2 \mathbf{v}_2^T + \cdots + \sigma_n \mathbf{u}_n \mathbf{v}_n^T$$

The closest matrix of rank k is obtained by truncating this sum after the first k terms:

$$A_k = \sigma_1 \mathbf{u}_1 \mathbf{v}_1^T + \sigma_2 \mathbf{u}_2 \mathbf{v}_2^T + \cdots + \sigma_k \mathbf{u}_k \mathbf{v}_k^T$$

The total storage for A_k is $k(2n + 1)$. One can choose k to be considerably less than n and still have the digital image corresponding to A_k very close to the original. For typical choices of k, the storage required for A_k will be less than 20 percent of the amount of storage necessary for the entire matrix A.

A list of references is given below for further reading. References 4 and 5 use the same approach presented here, but use different outer product expansions to approximate A.

REFERENCES

1. Andrews, H. C., and B. R. Hunt, *Digital Image Processing*. Englewood Cliffs, N.J.: Prentice-Hall, Inc., 1977.
2. Andrews, H. C., and C. L. Patterson, "Outer Product Expansions and Their Uses in Digital Image Processing," *American Mathematical Monthly*, *82*, 1975.
3. Moler, Cleve, "Numerical Linear Algebra," in *Numerical Analysis*, Proceedings of Symposia in Applied Mathematics XXII, American Mathematical Society, Providence, R.I., 1978.
4. Moler, C. B., and G. N. Stewart, "An Efficient Matrix Factorization for Digital Image Processing," manuscript, Computer Science Department, University of Maryland, 1979.

5. O'Leary, Diane P., and Shmuel Peleg, "Digital Image Compression by Outer Product Expansion," manuscript, Computer Science Department, University of Maryland, 1981.

EXERCISES

1. Show that A and A^T have the same nonzero singular values. How are their singular value decompositions related?

2. Use the method of Example 1 to find the singular value decomposition of each of the following matrices.

(a) $\begin{pmatrix} 1 & 1 \\ 2 & 2 \end{pmatrix}$ (b) $\begin{pmatrix} 2 & -2 \\ 1 & 2 \end{pmatrix}$

(c) $\begin{pmatrix} 1 & 3 \\ 3 & 1 \\ 0 & 0 \\ 0 & 0 \end{pmatrix}$ (d) $\begin{pmatrix} 2 & 0 & 0 \\ 0 & 2 & 1 \\ 0 & 1 & 2 \\ 0 & 0 & 0 \end{pmatrix}$

3. For each of the matrices in Exercise 2:
   (i) Determine the rank.
   (ii) Determine the 2-norm.
   (iii) Find the closest (with respect to the Frobenius norm) matrix of rank 1.

4. Prove that if A is a symmetric matrix with eigenvalues $\lambda_1, \lambda_2, \ldots, \lambda_n$, then the singular values of A are $|\lambda_1|, |\lambda_2|, \ldots, |\lambda_n|$.

5. Let A be an $m \times n$ matrix with singular value decomposition $U \Sigma V^T$ and suppose that A has rank r, where $r < n$.
   (a) Show that $\{\mathbf{v}_{r+1}, \ldots, \mathbf{v}_n\}$ is an orthonormal basis for $N(A)$.
   (b) Show that $\{\mathbf{u}_{r+1}, \ldots, \mathbf{u}_m\}$ is an orthonormal basis for $N(A^T)$.
   (c) Use Theorem 5.2.1 to show that $\{\mathbf{v}_1, \ldots, \mathbf{v}_r\}$ is an orthonormal basis for $R(A^T)$ and $\{\mathbf{u}_1, \ldots, \mathbf{u}_r\}$ is an orthonormal basis for $R(A)$.

6. Let A be an $m \times n$ matrix with singular value decomposition $U \Sigma V^T$. Show that

$$\min_{\mathbf{x} \neq \mathbf{0}} \frac{\|A\mathbf{x}\|_2}{\|\mathbf{x}\|_2} = \sigma_n$$

7. Let A be an $m \times n$ matrix with singular value decomposition $U \Sigma V^T$. Show that for any vector $\mathbf{x} \in R^n$,

$$\sigma_n \|\mathbf{x}\|_2 \leq \|A\mathbf{x}\|_2 \leq \sigma_1 \|\mathbf{x}\|_2$$

8. Show that if σ is a singular value of A, then there exists a nonzero vector $\mathbf{x}$ such that

$$\sigma = \frac{\|A\mathbf{x}\|_2}{\|\mathbf{x}\|_2}$$

9. Let A be an $m \times n$ matrix with singular value decomposition $U\Sigma V^T$. Show that if A has rank r, $r < n$, then

$$A = U_r \Sigma_r V_r^T$$

where $V_r = (\mathbf{v}_1, \mathbf{v}_2, \ldots, \mathbf{v}_r)$, $U_r = (\mathbf{u}_1, \mathbf{u}_2, \ldots, \mathbf{u}_r)$ and Σ_r is an $r \times r$ diagonal matrix with diagonal entries $\sigma_1, \ldots, \sigma_r$.

10. Let A be an $m \times n$ matrix of rank n with singular value decomposition $U\Sigma V^T$. Let Σ^+ denote the $n \times m$ matrix

$$\begin{pmatrix} \frac{1}{\sigma_1} & & & & \\ & \frac{1}{\sigma_2} & & & O \\ & & \ddots & & \\ & & & \frac{1}{\sigma_n} & \end{pmatrix}$$

and define $A^+ = V\Sigma^+ U^T$. Show that $\hat{\mathbf{x}} = A^+\mathbf{b}$ satisfies the normal equations $A^T A\mathbf{x} = A^T\mathbf{b}$.

11. Let A^+ be defined as in Exercise 10 and let $P = AA^+$. Show that $P^2 = P$ and $P^T = P$.

8. THE EIGENVALUE PROBLEM

In this section we will be concerned with numerical methods for computing the eigenvalues and eigenvectors of an $n \times n$ matrix A. The first method we will study is called the power method. The power method is an iterative method for finding the dominant eigenvalue of a matrix and a corresponding eigenvector. By the dominant eigenvalue we mean an eigenvalue λ_1 satisfying $|\lambda_1| > |\lambda_i|$ for $i = 2, \ldots, n$. If the eigenvalues of A satisfy

$$|\lambda_1| > |\lambda_2| > \cdots > |\lambda_n|$$

then the power method can be used to compute the eigenvalues one at a time. The second method is called the QR algorithm. The QR algorithm is an iterative method involving orthogonal similarity transformations. It has many advantages over the power method. It will converge whether or not A has a dominant eigenvalue and it calculates all of the eigenvalues at the same time.

In the examples in Chapter 6 the eigenvalues were determined by forming the characteristic polynomial and finding its roots. However, this procedure is generally not recommended for numerical computations. The difficulty with this procedure is that often a small change in one or more of the coefficients of the characteristic polynomial can result in a relatively large change in the computed zeros of the polynomial. For example, consider the polynomial $p(x) = x^{10}$. The lead coefficient is 1 and the remaining coefficients are all 0. If the constant term is altered by adding -10^{-10}, we obtain the polynomial $q(x) = x^{10} - 10^{-10}$. Although the coefficients of $p(x)$ and $q(x)$ only differ by 10^{-10}, the roots of $q(x)$ all have absolute value $\frac{1}{10}$, whereas the roots of $p(x)$ are all 0. Thus even when the coefficients of the characteristic polynomial have been determined quite accurately, the computed eigenvalues may involve significant error. For this reason, the methods presented in this section will not involve the characteristic polynomial. To see that there is some advantage to working directly with the matrix A, we must determine the effect that small changes in the entries of A have upon the eigenvalues. This is done in the following theorem.

Theorem 7.8.1. *Let A be an $n \times n$ matrix with a complete set of eigenvectors and let X be a matrix that diagonalizes A.*

$$X^{-1}AX = D = \begin{pmatrix} \lambda_1 & & & \\ & \lambda_2 & & \\ & & \ddots & \\ & & & \lambda_n \end{pmatrix}$$

If $A' = A + E$ and λ' is an eigenvalue of A', then

(1) $$\min_{1 \le i \le n} |\lambda' - \lambda_i| \le \text{cond}_2(X)\|E\|_2$$

Proof. We may assume that λ' is unequal to any of the λ_i's (otherwise there is nothing to prove). Thus, if we set $D_1 = D - \lambda'I$, then D_1 is a nonsingular diagonal matrix. Since λ' is an eigenvalue of A', it is also an eigenvalue of $X^{-1}A'X$. Therefore, $X^{-1}A'X - \lambda'I$ is singular and hence $D_1^{-1}(X^{-1}A'X - \lambda'I)$ is also singular. But

$$D_1^{-1}(X^{-1}A'X - \lambda'I) = D_1^{-1}X^{-1}(A + E - \lambda'I)X$$
$$= D_1^{-1}X^{-1}EX + I$$

Therefore, -1 is an eigenvalue of $D_1^{-1}X^{-1}EX$. It follows that

$$|-1| \le \|D_1^{-1}X^{-1}EX\|_2 \le \|D_1^{-1}\|_2 \, \text{cond}_2(X)\|E\|_2$$

The 2-norm of D_1^{-1} is given by

$$\|D_1^{-1}\|_2 = \max_{1 \le i \le n} |\lambda' - \lambda_i|^{-1}$$

The index i that maximizes $|\lambda' - \lambda_i|^{-1}$ is the same index that minimizes $|\lambda' - \lambda_i|$. Thus

$$\min_{1 \le i \le n} |\lambda' - \lambda_i| \le \text{cond}_2(X)\|E\|_2 \qquad \square$$

If the matrix A is symmetric, we can choose an orthogonal diagonalizing matrix. In general, if Q is any orthogonal matrix, then

$$\text{cond}_2(Q) = \|Q\|_2\|Q^{-1}\|_2 = 1$$

Hence (1) simplifies to

$$\min_{1 \le i \le n} |\lambda' - \lambda_i| \le \|E\|_2$$

Thus, if A is symmetric and $\|E\|_2$ is small, the eigenvalues of A' will be close to the eigenvalues of A.

We are now ready to talk about some methods for calculating the eigenvalues and eigenvectors of an $n \times n$ matrix A. The first method we will present computes an eigenvector $\mathbf{x}$ of A by successively applying A to a given vector in R^n. To see the idea behind the method, let us assume that A has n linearly independent eigenvectors $\mathbf{x}_1, \ldots, \mathbf{x}_n$ and that the corresponding eigenvalues satisfy

$$(2) \qquad |\lambda_1| > |\lambda_2| \ge \cdots \ge |\lambda_n|$$

Given an arbitrary vector $\mathbf{v}_0$ in R^n, we can write

$$\mathbf{v}_0 = \alpha_1 \mathbf{x}_1 + \alpha_2 \mathbf{x}_2 + \cdots + \alpha_n \mathbf{x}_n$$
$$A\mathbf{v}_0 = \alpha_1 \lambda_1 \mathbf{x}_1 + \alpha_2 \lambda_2 \mathbf{x}_2 + \cdots + \alpha_n \lambda_n \mathbf{x}_n$$
$$A^2\mathbf{v}_0 = \alpha_1 \lambda_1^2 \mathbf{x}_1 + \alpha_2 \lambda_2^2 \mathbf{x}_2 + \cdots + \alpha_n \lambda_n^2 \mathbf{x}_n$$

and in general

$$A^k\mathbf{v}_0 = \alpha_1 \lambda_1^k \mathbf{x}_1 + \alpha_2 \lambda_2^k \mathbf{x}_2 + \cdots + \alpha_n \lambda_n^k \mathbf{x}_n$$

If we define

$$\mathbf{v}_k = A^k\mathbf{v}_0 \qquad k = 1, 2, \ldots$$

then

$$(3) \qquad \frac{1}{\lambda_1^k}\mathbf{v}_k = \alpha_1 \mathbf{x}_1 + \alpha_2\left(\frac{\lambda_2}{\lambda_1}\right)^k \mathbf{x}_2 + \cdots + \alpha_n\left(\frac{\lambda_n}{\lambda_1}\right)^k \mathbf{x}_n$$

Since

$$\left|\frac{\lambda_i}{\lambda_1}\right| < 1 \qquad \text{for} \qquad i = 2, 3, \ldots, n$$

it follows that

$$\frac{1}{\lambda_1^k} \mathbf{v}_k \to \alpha_1 \mathbf{x}_1 \qquad \text{as} \qquad k \to \infty$$

Thus, if $\alpha_1 \neq 0$, the sequence $\{(1/\lambda_1^k)\mathbf{v}_k\}$ converges to an eigenvector $\alpha_1 \mathbf{x}_1$ of A. There are some obvious difficulties with the method as it has been presented so far. We cannot compute $(1/\lambda_1^k)\mathbf{v}_k$, since λ_1 is unknown. Even if λ_1 were known, there would be difficulties because of λ_1^k approaching 0 or $\pm\infty$. Fortunately, however, we do not have to scale the sequence $\{\mathbf{v}_k\}$ using $1/\lambda_1^k$. If the $\mathbf{v}_k$'s are scaled so that one obtains unit vectors at each step, the sequence will converge to a unit vector in the direction of $\mathbf{x}_1$. The eigenvalue λ_1 can be computed at the same time. This method of computing the eigenvalue of largest magnitude and a corresponding eigenvector is called the *power method*. It may be summarized as follows.

THE POWER METHOD

Two sequences $\{\mathbf{v}_k\}$ and $\{\mathbf{u}_k\}$ are defined recursively. To start, $\mathbf{u}_0$ can be any nonzero vector in R^n. Once $\mathbf{u}_k$ has been determined, the vectors $\mathbf{v}_{k+1}$ and $\mathbf{u}_{k+1}$ are calculated as follows:

  (i) Set $\mathbf{v}_{k+1} = A\mathbf{u}_k$.
  (ii) Find the coordinate j_{k+1} of $\mathbf{v}_{k+1}$ of maximum modulus.
  (iii) Set $\mathbf{u}_{k+1} = (1/v_{j_{k+1}})\mathbf{v}_{k+1}$.

The sequence $\{\mathbf{u}_k\}$ has the property that for $k \geq 1$, $\|\mathbf{u}_k\|_\infty = u_{j_k} = 1$. If the eigenvalues of A satisfy (2) and $\mathbf{u}_0$ can be written as a linear combination of eigenvectors $\alpha_1 \mathbf{x}_1 + \cdots + \alpha_n \mathbf{x}_n$ with $\alpha_1 \neq 0$, the sequence $\{\mathbf{u}_k\}$ will converge to an eigenvector $\mathbf{y}$ of λ_1. If k is large, then $\mathbf{u}_k$ will be a good approximation to $\mathbf{y}$ and $\mathbf{v}_{k+1} = A\mathbf{u}_k$ will be a good approximation to $\lambda_1\mathbf{y}$. Since the j_kth coordinate of $\mathbf{u}_k$ is 1, it follows that the j_kth coordinate of $\mathbf{v}_{k+1}$ will be a good approximation to λ_1.

In view of (3), we can expect that the $\mathbf{u}_k$'s will converge to $\mathbf{y}$ at the same rate at which $(\lambda_2/\lambda_1)^k$ is converging to 0. Thus, if $|\lambda_2|$ is nearly as large as $|\lambda_1|$, the convergence will be slow.

EXAMPLE 1. Let

$$A = \begin{pmatrix} 2 & 1 \\ 1 & 2 \end{pmatrix}$$

It is an easy matter to determine the exact eigenvalues of A. These turn out to be $\lambda_1 = 3$ and $\lambda_2 = 1$, with corresponding eigenvectors $\mathbf{x}_1 = (1, 1)^T$ and $\mathbf{x}_2 = (1, -1)^T$. To illustrate how the vectors generated by the power method converge, we will apply the method with

$\mathbf{u}_0 = (2, 1)^T.$

$$\mathbf{v}_1 = A\mathbf{u}_0 = \begin{pmatrix} 5 \\ 4 \end{pmatrix}, \qquad \mathbf{u}_1 = \frac{1}{5}\mathbf{v}_1 = \begin{pmatrix} 1.0 \\ 0.8 \end{pmatrix}$$

$$\mathbf{v}_2 = A\mathbf{u}_1 = \begin{pmatrix} 2.8 \\ 2.6 \end{pmatrix}, \qquad \mathbf{u}_2 = \frac{1}{2.8}\mathbf{v}_2 = \begin{pmatrix} 1 \\ \frac{13}{14} \end{pmatrix} \approx \begin{pmatrix} 1.00 \\ 0.93 \end{pmatrix}$$

$$\mathbf{v}_3 = A\mathbf{u}_2 = \frac{1}{14}\begin{pmatrix} 41 \\ 40 \end{pmatrix}, \quad \mathbf{u}_3 = \frac{14}{41}\mathbf{v}_3 = \begin{pmatrix} 1 \\ \frac{40}{41} \end{pmatrix} \approx \begin{pmatrix} 1.00 \\ 0.98 \end{pmatrix}$$

$$\mathbf{v}_4 = A\mathbf{u}_3 \approx \begin{pmatrix} 2.98 \\ 2.95 \end{pmatrix}$$

If $\mathbf{u}_3 = (1.00, 0.98)^T$ is taken as an approximate eigenvector, then 2.98 is the approximate value of λ_1. Thus, with only a few iterations, the approximation for λ_1 involves an error of only 0.02.

$\square$

The power method can be used to compute the eigenvalue λ_1 of largest magnitude and a corresponding eigenvector $\mathbf{y}_1$. What about the remaining eigenvalues and eigenvectors? If we could reduce the problem of finding the remaining eigenvalues of A to that of finding the eigenvalues of some $(n - 1) \times (n - 1)$ matrix A_1, then the power method could be applied to A_1. This can actually be done by a process called *deflation*.

DEFLATION

The idea behind deflation is to find a nonsingular matrix H such that HAH^{-1} is a matrix of the form

(4)
$$\begin{pmatrix} \lambda_1 & \times & \cdots & \times \\ \hline 0 & & & \\ \vdots & & A_1 & \\ 0 & & & \end{pmatrix}$$

Since A and HAH^{-1} are similar, they have the same characteristic polynomials. Thus, if HAH^{-1} is of the form (4), then

$$\det(A - \lambda I) = \det(HAH^{-1} - \lambda I) = (\lambda_1 - \lambda)\det(A_1 - \lambda I)$$

and it follows that the remaining $n - 1$ eigenvalues of A are the eigenvalues of A_1. The question remains: How do we find such a matrix H? Note that the form (4) requires that the first column of HAH^{-1} be $\lambda_1\mathbf{e}_1$. The first column of HAH^{-1} is $HAH^{-1}\mathbf{e}_1$. Thus

$$HAH^{-1}\mathbf{e}_1 = \lambda_1\mathbf{e}_1$$

or, equivalently,

$$A\left(H^{-1}\mathbf{e}_1\right) = \lambda_1\left(H^{-1}\mathbf{e}_1\right)$$

So $H^{-1}\mathbf{e}_1$ is in the eigenspace corresponding to λ_1. Thus, for some eigenvector $\mathbf{x}_1$ belonging to λ_1,

$$H^{-1}\mathbf{e}_1 = \mathbf{x}_1 \qquad \text{or} \qquad H\mathbf{x}_1 = \mathbf{e}_1$$

We must find a matrix H such that $H\mathbf{x}_1 = \mathbf{e}_1$ for some eigenvector $\mathbf{x}_1$ belonging to λ_1. This can be done by means of a Householder transformation. If $\mathbf{y}_1$ is the computed eigenvector belonging to λ_1, set

$$\mathbf{x}_1 = \frac{1}{\|\mathbf{y}_1\|_2}\mathbf{y}_1$$

Since $\|\mathbf{x}_1\|_2 = 1$, one can find a Householder transformation H such that

$$H\mathbf{x}_1 = \mathbf{e}_1$$

Because H is a Householder transformation, it follows that $H^{-1} = H$ and hence HAH is the desired similarity transformation.

REDUCTION TO HESSENBERG FORM

The standard methods for finding eigenvalues are all iterative. The amount of work required in each iteration is often prohibitively high unless initially A is in some special form that is easier to work with. If this is not the case, the standard procedure is to reduce A to a simpler form by means of similarity transformations. Generally, Householder matrices are used to transform A into a matrix of the form

$$\begin{pmatrix} \times & \times & \cdots & \times & \times & \times \\ \times & \times & \cdots & \times & \times & \times \\ 0 & \times & \cdots & \times & \times & \times \\ 0 & 0 & \cdots & \times & \times & \times \\ \vdots & & & & & \\ 0 & 0 & \cdots & \times & \times & \times \\ 0 & 0 & \cdots & 0 & \times & \times \end{pmatrix}$$

A matrix in this form is said to be in *upper Hessenberg form*. Thus B is in upper Hessenberg form if and only if $b_{ij} = 0$ whenever $i \geq j + 2$.

A matrix A can be transformed into upper Hessenberg form in the following manner. First, choose a Householder matrix H_1 so that $H_1 A$ is of the form

$$\begin{pmatrix} a_{11} & a_{12} & \cdots & a_{1n} \\ \times & \times & \cdots & \times \\ 0 & \times & \cdots & \times \\ \vdots & & & \\ 0 & \times & \cdots & \times \end{pmatrix}$$

The matrix H_1 will be of the form

$$\begin{pmatrix} 1 & 0 & \cdots & 0 \\ 0 & \times & \cdots & \times \\ \vdots & & & \\ 0 & \times & \cdots & \times \end{pmatrix}$$

and hence postmultiplication of $H_1 A_1$ by H_1 will leave the first column unchanged. If $A^{(1)} = H_1 A H_1$, then $A^{(1)}$ is a matrix of the form

$$\begin{pmatrix} a_{11}^{(1)} & a_{12}^{(1)} & \cdots & a_{1n}^{(1)} \\ a_{21}^{(1)} & a_{22}^{(1)} & \cdots & a_{2n}^{(1)} \\ 0 & a_{32}^{(1)} & \cdots & a_{3n}^{(1)} \\ \vdots & & & \\ 0 & a_{n2}^{(1)} & \cdots & a_{nn}^{(1)} \end{pmatrix}$$

Since H_1 is a Householder matrix, it follows that $H_1^{-1} = H_1$, and hence $A^{(1)}$ is similar to A. Next a Householder matrix H_2 is chosen so that

$$H_2\left(a_{12}^{(1)}, a_{22}^{(1)}, \ldots, a_{n2}^{(1)}\right)^T = \left(a_{12}^{(1)}, a_{22}^{(1)}, \times, 0, \ldots, 0\right)^T$$

The matrix H_2 will be of the form

$$\begin{pmatrix} 1 & 0 & 0 & \cdots & 0 \\ 0 & 1 & 0 & \cdots & 0 \\ 0 & 0 & \times & \cdots & \times \\ \vdots & & & & \\ 0 & 0 & \times & \cdots & \times \end{pmatrix} = \left(\begin{array}{c|c} I_2 & O \\ \hline O & X \end{array}\right)$$

Multiplication of $A^{(1)}$ on the left by H_2 will leave the first two rows and the first column unchanged.

$$H_2 A^{(1)} = \begin{pmatrix} a_{11}^{(1)} & a_{12}^{(1)} & a_{13}^{(1)} & \cdots & a_{1n}^{(1)} \\ a_{21}^{(1)} & a_{22}^{(1)} & a_{23}^{(1)} & \cdots & a_{2n}^{(1)} \\ 0 & \times & \times & \cdots & \times \\ 0 & 0 & \times & \cdots & \times \\ \vdots & & & & \\ 0 & 0 & \times & \cdots & \times \end{pmatrix}$$

Postmultiplication of $H_2 A^{(1)}$ by H_2 will leave the first two columns unchanged. Thus $A^{(2)} = H_2 A^{(1)} H_2$ is of the form

$$\begin{pmatrix} \times & \times & \times & \cdots & \times \\ \times & \times & \times & \cdots & \times \\ 0 & \times & \times & \cdots & \times \\ 0 & 0 & \times & \cdots & \times \\ \vdots & & & & \\ 0 & 0 & \times & \cdots & \times \end{pmatrix}$$

This process may be continued until one ends up with an upper Hessenberg matrix

$$H = A^{(n-2)} = H_{n-2} \cdots H_2 H_1 A H_1 H_2 \cdots H_{n-2}$$

which is similar to A.

If, in particular, A is symmetric, then since

$$H^T = H_{n-2}^T \cdots H_2^T H_1^T A^T H_1^T H_2^T \cdots H_{n-2}^T$$
$$= H_{n-2} \cdots H_2 H_1 A H_1 H_2 \cdots H_{n-2}$$
$$= H$$

it follows that H is tridiagonal. Thus any $n \times n$ matrix A can be reduced by similarity transformations to upper Hessenberg form. If A is symmetric, the reduction will yield a tridiagonal matrix.

We close this section by outlining one of the best methods available for computing the eigenvalues of a matrix. The method is called the *QR* algorithm and was presented by K. G. F. Francis in 1961.

QR ALGORITHM

Given an $n \times n$ matrix A, factor it into a product $Q_1 R_1$ where Q_1 is orthogonal and R_1 is upper triangular. Define

$$A_1 = A = Q_1 R_1$$

and

$$A_2 = Q_1^T A Q_1 = R_1 Q_1$$

Factor A_2 into a product $Q_2 R_2$, where Q_2 is orthogonal and R_2 is upper triangular. Define

$$A_3 = Q_2^T A_2 Q_2 = R_2 Q_2$$

Note that $A_2 = Q_1^T A Q_1$ and $A_3 = (Q_1 Q_2)^T A (Q_1 Q_2)$ are both similar to A. We can continue in this manner to obtain a sequence of similar matrices. In general, if

$$A_k = Q_k R_k$$

then A_{k+1} is defined to be $R_k Q_k$. It can be shown that under very general conditions the sequence of matrices defined in this way converges to a matrix of the form

$$\begin{pmatrix} B_1 & \times & \cdots & \times \\ & B_2 & & \times \\ & & \ddots & \\ O & & & B_s \end{pmatrix}$$

where the B_i's are either 1×1 or 2×2 diagonal blocks. Each 2×2 block will correspond to a pair of complex conjugate eigenvalues of A.

The eigenvalues of A will be the eigenvalues of the B_i's. In the case where A is symmetric, each of the A_k's will also be symmetric and the sequence will converge to a diagonal matrix.

EXAMPLE 2. Let A_1 be the matrix from Example 1. The QR factorization of A_1 requires only a single Givens transformation,

$$G_1 = \frac{1}{\sqrt{5}} \begin{pmatrix} 2 & 1 \\ 1 & -2 \end{pmatrix}$$

Thus

$$A_2 = G_1 A G_1 = \tfrac{1}{5} \begin{pmatrix} 2 & 1 \\ 1 & -2 \end{pmatrix} \begin{pmatrix} 2 & 1 \\ 1 & 2 \end{pmatrix} \begin{pmatrix} 2 & 1 \\ 1 & -2 \end{pmatrix} = \begin{pmatrix} 2.8 & -0.6 \\ -0.6 & 1.2 \end{pmatrix}$$

The QR factorization of A_2 can be accomplished using the Givens transformation

$$G_2 = \frac{1}{\sqrt{8.2}} \begin{pmatrix} 2.8 & -0.6 \\ -0.6 & -2.8 \end{pmatrix}$$

It follows that

$$A_3 = G_2 A_2 G_2 \approx \begin{pmatrix} 2.98 & 0.22 \\ 0.22 & 1.02 \end{pmatrix}$$

The off-diagonal elements are getting closer to 0 after each iteration, and the diagonal elements are approaching the eigenvalues $\lambda_1 = 3$ and $\lambda_2 = 1$.   □

Remarks

1. Because of the amount of work required at each iteration of the QR algorithm, it is important that the starting matrix A be in either Hessenberg or symmetric tridiagonal form. If this is not the case, we should perform similarity transformations on A to obtain a matrix A_1 that is in one of these forms.
2. If A_k is in upper Hessenberg form, the QR factorization can be carried out using $n - 1$ Givens transformations.

$$G_{n,\,n-1} \cdots G_{32} G_{21} A_k = R_k$$

Setting

$$Q_k^T = G_{n,\,n-1} \cdots G_{32} G_{21}$$

we have

$$A_k = Q_k R_k$$

and

$$A_{k+1} = Q_k^T A_k Q_k$$

To compute A_{k+1}, it is not necessary to determine Q_k explicitly. One need only keep track of the $n - 1$ Givens transformations. When R_k

is postmultiplied by G_{21}, the resulting matrix will have the $(2, 1)$ entry filled in. The other entries below the diagonals will all still be zero. Postmultiplying $R_k G_{21}$ by G_{32} will have the effect of filling in the $(3, 2)$ position. Postmultiplication of $R_k G_{21} G_{32}$ by G_{43} will fill in the $(4, 3)$ position, and so on. Thus the resulting matrix $A_{k+1} = R_k G_{21} G_{32} \cdots G_{n, n-1}$ will be in upper Hessenberg form. If A_1 is a symmetric tridiagonal matrix, each of the succeeding A_i's will be upper Hessenberg and symmetric. Thus $A_2, A_3, \ldots$ will all be tridiagonal.

3. As in the power method, convergence may be slow when some of the eigenvalues are close together. To speed up convergence, it is customary to introduce *origin shifts*. At the kth step a scalar α_k is chosen and $A_k - \alpha_k I$ (rather than A_k) is decomposed into a product $Q_k R_k$. The matrix A_{k+1} is defined by

$$A_{k+1} = R_k Q_k + \alpha_k I$$

Note that

$$Q_k^T A_k Q_k = Q_k^T (Q_k R_k + \alpha_k I) Q_k = R_k Q_k + \alpha_k I = A_{k+1}$$

so that A_k and A_{k+1} are similar. With the proper choice of shifts α_k, the convergence can be greatly accelerated.

4. In our brief discussion we have presented only an outline of the method. Many of the details, such as how to choose the origin shifts, have been omitted. For a more thorough discussion and a proof of convergence, see Wilkinson ([22], Chapter 8).

EXERCISES

1. Let

$$A = \begin{pmatrix} 1 & 1 \\ 1 & 1 \end{pmatrix}$$

   (a) Apply one iteration of the power method with any nonzero starting vector.
   (b) Apply one iteration of the QR algorithm to A.
   (c) Determine the exact eigenvalues of A by solving the characteristic equation and determine the eigenspace corresponding to the largest eigenvalue. Compare your answers with those to parts (a) and (b).

2. Let

$$A = \begin{pmatrix} 2 & 1 & 0 \\ 1 & 3 & 1 \\ 0 & 1 & 2 \end{pmatrix} \quad \text{and} \quad \mathbf{u}_0 = \begin{pmatrix} 1 \\ 1 \\ 1 \end{pmatrix}$$

   (a) Apply the power method to compute $\mathbf{v}_1, \mathbf{u}_1, \mathbf{v}_2, \mathbf{u}_2,$ and $\mathbf{v}_3$. (Round off to two decimal places.)

(b) Determine an approximation λ_1' to the largest eigenvalue of A from the coordinates of $\mathbf{v}_3$. Determine the exact value of λ_1 and compare it to λ_1'. What is the relative error?

3. Let

$$A = \begin{pmatrix} 1. & 2 \\ -1 & -1 \end{pmatrix} \quad \text{and} \quad \mathbf{u}_0 = \begin{pmatrix} 1 \\ 1 \end{pmatrix}$$

(a) Compute $\mathbf{u}_1$, $\mathbf{u}_2$, $\mathbf{u}_3$, and $\mathbf{u}_4$ using the power method.
(b) Explain why the power method will fail to converge in this case.

4. Let

$$A = A_1 = \begin{pmatrix} 1 & 1 \\ 1 & 3 \end{pmatrix}$$

Compute A_2 and A_3 using the QR algorithm. Compute the exact eigenvalues of A and compare them to the diagonal elements of A_3. To how many decimal places do they agree?

5. Given

$$A = \begin{pmatrix} 5 & 2 & 2 \\ -2 & 1 & -2 \\ -3 & -4 & 2 \end{pmatrix}$$

(a) Verify that $\lambda_1 = 4$ is an eigenvalue of A and $\mathbf{y}_1 = (2, -2, 1)^T$ is an eigenvector belonging to λ_1.
(b) Find a Householder transformation H such that HAH is of the form

$$\begin{pmatrix} 4 & \times & \times \\ 0 & \times & \times \\ 0 & \times & \times \end{pmatrix}$$

(c) Compute HAH and find the remaining eigenvalues of A.

6. Let A be an $n \times n$ matrix with distinct real eigenvalues $\lambda_1, \lambda_2, \ldots, \lambda_n$. Let λ be a scalar that is not an eigenvalue of A and let $B = (A - \lambda I)^{-1}$.
(a) Show that the scalars $\mu_j = 1/(\lambda_j - \lambda)$, $j = 1, \ldots, n$ are the eigenvalues of B.
(b) Show that if $\mathbf{x}_j$ is an eigenvector of B belonging to μ_j, then $\mathbf{x}_j$ is an eigenvector of A belonging to λ_j.
(c) Show that if the power method is applied to B, then the sequence of vectors will converge to an eigenvector of A belonging to the eigenvalue that is closest to λ. [The convergence will be quite rapid if λ is much closer to one λ_i than to any of the others. This method of computing eigenvectors using powers of $(A - \lambda I)^{-1}$ is called the *inverse power method*.]

7. Let $\mathbf{x} = (x_1, \ldots, x_n)^T$ be an eigenvector of A belonging to λ. If $|x_i| = \|\mathbf{x}\|_\infty$, show that:

(a) $\displaystyle\sum_{j=1}^{n} a_{ij}x_j = \lambda x_i$

(b) $\displaystyle|\lambda - a_{ii}| \le \sum_{\substack{j=1 \\ j \ne i}}^{n} |a_{ij}|$ (Gerschgorin's theorem)

8. Let A be a matrix with eigenvalues $\lambda_1, \ldots, \lambda_n$, and let λ be an eigenvalue of $A + E$. Let X be a matrix that diagonalizes A, and let $C = X^{-1}EX$. Prove:

(a) For some i

$$|\lambda - \lambda_i| \le \sum_{j=1}^{n} |c_{ij}|$$

[**Hint:** λ is an eigenvalue of $X^{-1}(A + E)X$. Apply Gerschgorin's theorem from Exercise 7.]

(b) $\displaystyle\min_{1 \le j \le n} |\lambda - \lambda_j| \le \text{cond}_\infty(X)\|E\|_\infty$

9. Let $A_k = Q_k R_k$, $k = 1, 2, \ldots$ be the sequence of matrices derived from $A = A_1$ by applying the QR algorithm. For each positive integer k, define

$$P_k = Q_1 Q_2 \cdots Q_k \qquad \text{and} \qquad U_k = R_k \cdots R_2 R_1$$

Show that

$$P_k A_{k+1} = A P_k$$

for all $k \ge 1$.

10. Let P_k and U_k be defined as in Exercise 9.

(a) Show that

$$P_{k+1}U_{k+1} = P_k A_{k+1} U_k = A P_k U_k$$

(b) Show that $P_k U_k = A^k$ and hence that

$$(Q_1 Q_2 \cdots Q_k)(R_k \cdots R_2 R_1)$$

is the QR factorization of A^k.

9. LEAST SQUARES PROBLEMS

In this section we study computational methods for finding least squares solutions to overdetermined systems. Let A be an $m \times n$ matrix with $m \ge n$ and let $\mathbf{b} \in R^m$. We consider some methods for computing a vector $\hat{\mathbf{x}}$ that minimizes $\|\mathbf{b} - A\mathbf{x}\|_2^2$.

THE NORMAL EQUATIONS

We saw in Chapter 5 that if $\hat{\mathbf{x}}$ satisfies the normal equations

$$A^T A \mathbf{x} = A^T \mathbf{b}$$

then $\hat{\mathbf{x}}$ is a solution to the least squares problem. If A is of full rank (rank n), then $A^T A$ is nonsingular and hence the system will have a unique solution. Thus, if $A^T A$ is invertible, one possible method for solving the least squares problem is to form the normal equations and then solve them using Gaussian elimination. An algorithm for doing this would have two main parts.

1. Compute $B = A^T A$ and $\mathbf{c} = A^T \mathbf{b}$.
2. Solve $B\mathbf{x} = \mathbf{c}$.

Note that forming the normal equations requires roughly $mn^2/2$ multiplications. Since $A^T A$ is nonsingular, the matrix B is positive definite. For positive definite matrices there are reduction algorithms that require only half the usual number of multiplications. Thus the solution of $B\mathbf{x} = \mathbf{c}$ requires roughly $n^3/6$ multiplications. Most of the work then occurs in forming the normal equations rather than solving them. However, the main difficulty with this method is that in forming the normal equations, we may well end up transforming the problem into an ill-conditioned one. Recall from Section 4 that if $\mathbf{x}'$ is the computed solution to $B\mathbf{x} = \mathbf{c}$ and $\mathbf{x}$ is the exact solution, then the inequality

$$\frac{1}{\text{cond}(B)} \frac{\|\mathbf{r}\|}{\|\mathbf{c}\|} \le \frac{\|\mathbf{x} - \mathbf{x}'\|}{\|\mathbf{x}\|} \le \text{cond}(B) \frac{\|\mathbf{r}\|}{\|\mathbf{c}\|}$$

shows how the relative error compares to the relative residual. If A has singular values $\sigma_1 \ge \sigma_2 \ge \cdots \ge \sigma_n > 0$, then $\text{cond}_2(A) = \sigma_1/\sigma_n$. The singular values of B are $\sigma_1^2, \sigma_2^2, \ldots, \sigma_n^2$. Thus

$$\text{cond}_2(B) = \frac{\sigma_1^2}{\sigma_n^2} = \left[\text{cond}_2(A)\right]^2$$

If, for example, $\text{cond}_2(A) = 100$, the relative error in the computed solution to the normal equations could be 10^4 times as large as the relative residual. For this reason one should be very careful about using the normal equations to solve least squares problems.

In Chapter 5 we saw how to use the Gram–Schmidt process to obtain a QR factorization of a matrix A with full rank. In that case the matrix Q was an $m \times n$ matrix with orthonormal columns and R was an $n \times n$ upper triangular matrix. In the following numerical method, we will use Householder transformations to obtain a QR factorization of A. In this case Q will be an $m \times m$ orthogonal matrix and R will be an $m \times n$ matrix whose subdiagonal entries are all 0.

THE *QR* FACTORIZATION

Given an $m \times n$ matrix A of full rank, we can apply n Householder transformations to zero out all the elements below the diagonal. Thus

$$H_n H_{n-1} \cdots H_1 A = R$$

where R is of the form

$$\binom{R_1}{O} = \begin{pmatrix} \times & \times & \times & \cdots & \times \\ & \times & \times & \cdots & \times \\ & & \times & \cdots & \times \\ & & & \ddots & \vdots \\ & & & & \times \\ & & & & \\ & & & & \end{pmatrix}$$

with nonzero diagonal. Let

$$Q^T = H_n \cdots H_1 = \begin{pmatrix} Q_1^T \\ Q_2^T \end{pmatrix}$$

where Q_1^T is an $n \times m$ matrix consisting of the first n rows of Q^T. Since

$$Q^T A = \begin{pmatrix} Q_1^T A \\ Q_2^T A \end{pmatrix} = \begin{pmatrix} R_1 \\ O \end{pmatrix}$$

it follows that $A = Q_1 R_1$. Let

$$\mathbf{c} = Q^T \mathbf{b} = \begin{pmatrix} Q_1^T \mathbf{b} \\ Q_2^T \mathbf{b} \end{pmatrix} = \begin{pmatrix} \mathbf{c}_1 \\ \mathbf{c}_2 \end{pmatrix}$$

The normal equations can be written in the form

$$R_1^T Q_1^T Q_1 R_1 \mathbf{x} = R_1^T Q_1^T \mathbf{b}$$

Since $Q_1^T Q_1 = I$ and R_1^T is nonsingular, this simplifies to

$$R_1 \mathbf{x} = \mathbf{c}_1$$

This system can be solved by back substitution. The solution $\mathbf{x} = R_1^{-1} \mathbf{c}_1$ will be the unique solution to the least squares problem. To compute the residual, note that

$$Q^T \mathbf{r} = \begin{pmatrix} \mathbf{c}_1 \\ \mathbf{c}_2 \end{pmatrix} - \begin{pmatrix} R_1 \\ O \end{pmatrix} \mathbf{x} = \begin{pmatrix} \mathbf{0} \\ \mathbf{c}_2 \end{pmatrix}$$

so that

$$\mathbf{r} = Q \begin{pmatrix} \mathbf{0} \\ \mathbf{c}_2 \end{pmatrix} \qquad \text{and} \qquad \|\mathbf{r}\|_2 = \|\mathbf{c}_2\|_2$$

In summation, when A has full rank, the least squares problem can be solved as follows:

1. Use Householder transformations to compute

$$R = H_n \cdots H_2 H_1 A \quad \text{and} \quad \mathbf{c} = H_n \cdots H_2 H_1 \mathbf{b}.$$

2. Use back substitution to solve $R_1 \mathbf{x} = \mathbf{c}_1$.

Now consider the case where the matrix A has rank $r < n$. The singular value decomposition provides the key to solving the least squares problem in this case. It can be used to construct a generalized inverse of A. In the case where A is a nonsingular $n \times n$ matrix with singular value decomposition $U\Sigma V^T$, the inverse is given by

$$A^{-1} = V\Sigma^{-1}U^T$$

More generally, if $A = U\Sigma V^T$ is an $m \times n$ matrix of rank r, the matrix Σ will be an $m \times n$ matrix of the form

$$\Sigma = \left(\begin{array}{c|c} \Sigma_1 & O \\ \hline O & O \end{array} \right) = \left(\begin{array}{cccc|c} \sigma_1 & & & & \\ & \sigma_2 & & & O \\ & & \ddots & & \\ & & & \sigma_r & \\ \hline & O & & & O \end{array} \right)$$

and one can define

$$(1) \qquad A^+ = V\Sigma^+ U^T$$

where Σ^+ is the $n \times m$ matrix

$$\Sigma^+ = \left(\begin{array}{c|c} \Sigma_1^{-1} & O \\ \hline O & O \end{array} \right) = \left(\begin{array}{cccc|c} \dfrac{1}{\sigma_1} & & & & \\ & \ddots & & & O \\ & & & \dfrac{1}{\sigma_r} & \\ \hline & O & & & O \end{array} \right)$$

Equation (1) gives a natural generalization of the inverse of a matrix. The matrix A^+ defined by (1) is called the *pseudoinverse* of A.

It is also possible to define A^+ by its algebraic properties. These properties are given in the following four conditions.

THE PENROSE CONDITIONS

1. $AXA = A$
2. $XAX = X$
3. $(AX)^T = AX$
4. $(XA)^T = XA$

If A is an $m \times n$ matrix, we claim that there is a unique $n \times m$ matrix X that satisfies these conditions. Indeed, if we choose $X = A^+ = V\Sigma^+U^T$, then it is easily verified that X satisfies all four conditions. We leave this as an exercise for the reader. To show uniqueness, suppose that Y also satisfies the Penrose conditions. By successively applying these conditions, one can argue as follows:

$$\begin{array}{llll} X = XAX & (2) & Y = YAY & (2) \\ \quad = A^TX^TX & (4) & \quad = YY^TA^T & (3) \\ \quad = (AYA)^TX^TX & (1) & \quad = YY^T(AXA)^T & (1) \\ \quad = (A^TY^T)(A^TX^T)X & & \quad = Y(Y^TA^T)(X^TA^T) & \\ \quad = YAXAX & (4) & \quad = YAYAX & (3) \\ \quad = YAX & (1) & \quad = YAX & (1) \end{array}$$

Therefore, $X = Y$. Thus A^+ is the unique matrix satisfying the four Penrose conditions. These conditions are often used to define the pseudoinverse and A^+ is often referred to as the *Moore–Penrose pseudoinverse*.

To see how the pseudoinverse can be used in solving least squares problems, let us first consider the case where A is an $m \times n$ matrix of rank n. Thus Σ is of the form

$$\Sigma = \begin{pmatrix} \Sigma_1 \\ O \end{pmatrix}$$

where Σ_1 is a nonsingular $n \times n$ diagonal matrix. The matrix A^TA is nonsingular and

$$(A^TA)^{-1} = V(\Sigma^T\Sigma)^{-1}V^T$$

The solution to the normal equations is given by

$$\begin{aligned} \mathbf{x} &= (A^TA)^{-1}A^T\mathbf{b} \\ &= V(\Sigma^T\Sigma)^{-1}V^TV\Sigma^TU^T\mathbf{b} \\ &= V(\Sigma^T\Sigma)^{-1}\Sigma^TU^T\mathbf{b} \\ &= V\Sigma^+U^T\mathbf{b} \\ &= A^+\mathbf{b} \end{aligned}$$

Thus, if A has full rank, $A^+\mathbf{b}$ is the solution to the least squares problem. What about the case where A has rank $r < n$? In this case there are infinitely many solutions to the least squares problem. The following theorem shows that not only is $A^+\mathbf{b}$ a solution, but it is the minimal solution with respect to the 2-norm.

Theorem 7.9.1.   *If A is an $m \times n$ matrix of rank $r < n$ with singular value decomposition $U\Sigma V^T$, then the vector*

$$\mathbf{x} = A^+\mathbf{b} = V\Sigma^+U^T\mathbf{b}$$

minimizes $\|\mathbf{b} - A\mathbf{x}\|_2^2$. *Moreover, if* $\mathbf{z}$ *is any other vector that minimizes* $\|\mathbf{b} - A\mathbf{x}\|_2^2$, *then* $\|\mathbf{z}\|_2 > \|\mathbf{x}\|_2$.

Proof.   Let $\mathbf{x}$ be a vector in R^n and define

$$\mathbf{c} = U^T\mathbf{b} = \begin{pmatrix} \mathbf{c}_1 \\ \mathbf{c}_2 \end{pmatrix} \quad \text{and} \quad \mathbf{y} = V^T\mathbf{x} = \begin{pmatrix} \mathbf{y}_1 \\ \mathbf{y}_2 \end{pmatrix}$$

where $\mathbf{c}_1$ and $\mathbf{y}_1$ are vectors in R^r. Since U^T is orthogonal, it follows that

$$\|\mathbf{b} - A\mathbf{x}\|_2^2 = \|U^T\mathbf{b} - \Sigma(V^T\mathbf{x})\|_2^2$$

$$= \|\mathbf{c} - \Sigma\mathbf{y}\|_2^2$$

$$= \left\| \begin{pmatrix} \mathbf{c}_1 \\ \mathbf{c}_2 \end{pmatrix} - \begin{pmatrix} \Sigma_1 & O \\ O & O \end{pmatrix} \begin{pmatrix} \mathbf{y}_1 \\ \mathbf{y}_2 \end{pmatrix} \right\|_2^2$$

$$= \left\| \begin{pmatrix} \mathbf{c}_1 - \Sigma_1\mathbf{y}_1 \\ \mathbf{c}_2 \end{pmatrix} \right\|_2^2$$

$$= \|\mathbf{c}_1 - \Sigma_1\mathbf{y}_1\|_2^2 + \|\mathbf{c}_2\|_2^2$$

Since $\mathbf{c}_2$ is independent of $\mathbf{x}$, it follows that $\|\mathbf{b} - A\mathbf{x}\|^2$ will be minimal if and only if

$$\|\mathbf{c}_1 - \Sigma_1\mathbf{y}_1\| = 0$$

Thus $\mathbf{x}$ is a solution to the least squares problem if and only if $\mathbf{x} = V\mathbf{y}$, where $\mathbf{y}$ is a vector of the form

$$\begin{pmatrix} \Sigma_1^{-1}\mathbf{c}_1 \\ \mathbf{y}_2 \end{pmatrix}$$

In particular,

$$\mathbf{x} = V\begin{pmatrix} \Sigma_1^{-1}\mathbf{c}_1 \\ \mathbf{0} \end{pmatrix}$$

$$= V\begin{pmatrix} \Sigma_1^{-1} & O \\ O & O \end{pmatrix} \begin{pmatrix} \mathbf{c}_1 \\ \mathbf{c}_2 \end{pmatrix}$$

$$= V\Sigma^+ U^T\mathbf{b}$$

$$= A^+\mathbf{b}$$

is a solution. If $\mathbf{z}$ is any other solution, $\mathbf{z}$ must be of the form

$$V\mathbf{y} = V\begin{pmatrix} \Sigma_1^{-1}\mathbf{c}_1 \\ \mathbf{y}_2 \end{pmatrix}$$

with $\mathbf{y}_2 \neq \mathbf{0}$. It follows then that

$$\|\mathbf{z}\|^2 = \|\mathbf{y}\|^2 = \|\Sigma_1^{-1}\mathbf{c}_1\|^2 + \|\mathbf{y}_2\|^2 > \|\Sigma_1^{-1}\mathbf{c}_1\|^2 = \|\mathbf{x}\|^2 \qquad \square$$

If the singular value decomposition $U\Sigma V^T$ of A is known, it is a simple matter to compute the solution to the least squares problem. If $U = (\mathbf{u}_1, \ldots, \mathbf{u}_m)$ and $V = (\mathbf{v}_1, \ldots, \mathbf{v}_n)$, then defining $\mathbf{y} = \Sigma^+ U^T \mathbf{b}$, we have

$$y_i = \frac{1}{\sigma_i}\mathbf{u}_i^T\mathbf{b} \qquad i = 1, \ldots, r \qquad (r = \text{rank of } A)$$

$$y_i = 0 \qquad i = r + 1, \ldots, n$$

and hence

$$A^+\mathbf{b} = V\mathbf{y} = \begin{pmatrix} v_{11}y_1 + v_{12}y_2 + \cdots + v_{1r}y_r \\ v_{21}y_1 + v_{22}y_2 + \cdots + v_{2r}y_r \\ \vdots \\ v_{n1}y_1 + v_{n2}y_2 + \cdots + v_{nr}y_r \end{pmatrix}$$

$$= y_1\mathbf{v}_1 + y_2\mathbf{v}_2 + \cdots + y_r\mathbf{v}_r$$

Thus the solution $\mathbf{x} = A^+\mathbf{b}$ can be computed in two steps:

1. Set $y_i = (1/\sigma_i)\mathbf{u}_i^T\mathbf{b}$ for $i = 1, \ldots, r$.
2. Let $\mathbf{x} = y_1\mathbf{v}_1 + \cdots + y_r\mathbf{v}_r$.

We conclude this section by outlining a method for computing the singular values of a matrix. Observe first that if A has singular value decomposition $U\Sigma V^T$ and $B = HAP^T$, where H is an $m \times m$ orthogonal matrix and P is an $n \times n$ orthogonal matrix, then B has singular value decomposition $(HU)\Sigma(PV)^T$. Thus the problem of finding the singular values of A can be simplified by applying orthogonal transformations to A to obtain a simpler matrix B with the same singular values. Golub and Kahan have shown that A can be reduced to upper bidiagonal form using Householder transformations.

BIDIAGONALIZATION

Let H_1 be a Householder transformation that annihilates all the elements below the diagonal in the first column of A. Let P_1 be a Householder transformation such that postmultiplication of H_1A by P_1 annihilates the last $n - 2$ entries of the first row of H_1A while leaving the first column unchanged.

$$H_1AP_1 = \begin{pmatrix} \times & \times & 0 & \cdots & 0 \\ 0 & \times & \times & \cdots & \times \\ \vdots & & & & \\ 0 & \times & \times & \cdots & \times \end{pmatrix}$$

The next step is to apply a Householder transformation H_2 that annihilates the elements below the diagonal in the second column of H_1AP_1 while leaving the first row and column unchanged.

$$H_2H_1AP_1 = \begin{pmatrix} \times & \times & 0 & \cdots & 0 \\ 0 & \times & \times & \cdots & \times \\ 0 & 0 & \times & \cdots & \times \\ \vdots & & & & \\ 0 & 0 & \times & \cdots & \times \end{pmatrix}$$

$H_2H_1AP_1$ is then postmultiplied by a Householder transformation P_2 that annihilates the last $n-3$ elements in the second row while leaving the first two columns and the first row unchanged.

$$H_2H_1AP_1P_2 = \begin{pmatrix} \times & \times & 0 & 0 & \cdots & 0 \\ 0 & \times & \times & 0 & \cdots & 0 \\ 0 & 0 & \times & \times & \cdots & \times \\ \vdots & & & & & \\ 0 & 0 & \times & \times & \cdots & \times \end{pmatrix}$$

We continue in this manner until we obtain a matrix

$$B = H_n \cdots H_1AP_1 \cdots P_{n-2}$$

of the form

$$\begin{pmatrix} \times & \times & & & \\ & \times & \times & & \\ & & \ddots & \ddots & \\ & & & \times & \times \\ & & & & \times \end{pmatrix}$$

Since $H = H_n \cdots H_1$ and $P^T = P_1 \cdots P_{n-2}$ are orthogonal, it follows that B has the same singular values as A.

The problem has now been simplified to that of finding the singular values of an upper bidiagonal matrix B. One could at this point form the symmetric tridiagonal matrix B^TB and then compute its eigenvalues using the QR algorithm. The problem with this approach is that in forming B^TB, we would still be squaring the condition number, and consequently our computed solution would be much less reliable. The method we will outline produces a sequence of bidiagonal matrices $B_1, B_2, \ldots$ that converges to a diagonal matrix Σ. The method involves applying a sequence of Givens transformations to B alternately on the right- and left-hand sides.

THE GOLUB–REINSCH ALGORITHM

Let

$$R_k = \begin{pmatrix} I_{k-1} & O & O \\ O & G(\theta_k) & O \\ O & O & I_{n-k-1} \end{pmatrix}$$

and

$$L_k = \begin{pmatrix} I_{k-1} & O & O \\ O & G(\varphi_k) & O \\ O & O & I_{n-k-1} \end{pmatrix}$$

where the 2×2 matrices $G(\theta_k)$ and $G(\varphi_k)$ are given by

$$G(\theta_k) = \begin{pmatrix} \cos\theta_k & \sin\theta_k \\ \sin\theta_k & -\cos\theta_k \end{pmatrix} \quad \text{and} \quad G(\varphi_k) = \begin{pmatrix} \cos\varphi_k & \sin\varphi_k \\ \sin\varphi_k & -\cos\varphi_k \end{pmatrix}$$

for some angles θ_k and φ_k. The matrix $B = B_1$ is first multiplied on the right by R_1. This will have the effect of filling in the $(2, 1)$ position.

$$B_1 R_1 = \begin{pmatrix} \times & \times & & & & \\ \times & \times & \times & & & \\ & & \times & & & \\ & & & \ddots & & \\ & & & & \ddots & \times \\ & & & & & \times \end{pmatrix}$$

Next L_1 is chosen so as to annihilate the element filled in by R_1. It will also have the effect of filling in the $(1, 3)$ position. Thus

$$L_1 B_1 R_1 = \begin{pmatrix} \times & \times & \times & & \\ & \times & \times & & \\ & & & \ddots & \\ & & & \ddots & \times \\ & & & & \times \end{pmatrix}$$

R_2 is chosen so as to annihilate the $(1, 3)$ entry. It will fill in the $(3, 2)$ entry of $L_1 B_1 R_1$. Next L_2 annihilates the $(3, 2)$ entry and fills in the

(2, 4) entry, and so on.

$$\begin{pmatrix} \times & \times & & & & \\ & \times & \times & & & \\ & \times & \times & \times & & \\ & & & \ddots & \ddots & \\ & & & & \ddots & \times \\ & & & & & \times \end{pmatrix} \qquad \begin{pmatrix} \times & \times & & & & \\ & \times & \times & \times & & \\ & \times & \times & & & \\ & & & \ddots & \ddots & \\ & & & & & \times \\ & & & & & \times \end{pmatrix}$$

$$L_1 B_1 R_1 R_2 \qquad\qquad\qquad L_2 L_1 B_1 R_1 R_2$$

We continue this process until we end up with a new bidiagonal matrix,

$$B_2 = L_{n-1} \cdots L_1 B_1 R_1 \cdots R_{n-1}$$

Why should we be any better off with B_2 than with B_1? It can be shown that if the first transformation R_1 is chosen correctly, $B_2^T B_2$ will be the matrix obtained from $B_1^T B_1$ by applying one iteration of the QR algorithm with shift. The same process can now be applied to B_2 to obtain a new bidiagonal matrix B_3 such that $B_3^T B_3$ would be the matrix obtained by applying two iterations of the QR algorithm with shifts to $B_1^T B_1$. Even though the $B_i^T B_i$'s are never computed, we know that with the proper choice of shifts, these matrices will converge rapidly to a diagonal matrix. The B_i's then must also converge to a diagonal matrix Σ. Since each of the B_i's has the same singular values as B, the diagonal elements of Σ will be the singular values of B. The matrices U and V^T can be determined by keeping track of all the orthogonal transformations.

Only a brief sketch of the algorithm has been given. To include more would have been beyond the scope of this book. For complete details of the algorithm and an ALGOL program, we refer the reader to the paper by Golub and Reinsch in ([23], p. 135).

EXERCISES

1. Find the solution x to the least squares problem given that $A = QR$ in each of the following.

$$\text{(a)} \quad Q = \begin{pmatrix} \dfrac{1}{\sqrt{2}} & \dfrac{1}{\sqrt{2}} \\ \dfrac{1}{\sqrt{2}} & -\dfrac{1}{\sqrt{2}} \\ 0 & 0 \end{pmatrix}, \qquad R = \begin{pmatrix} 1 & 1 \\ 0 & 1 \end{pmatrix}, \qquad \mathbf{b} = \begin{pmatrix} 1 \\ 1 \\ 1 \end{pmatrix}$$

(b) $Q = \begin{pmatrix} 1 & 0 & 0 \\ 0 & \dfrac{1}{\sqrt{2}} & -\dfrac{1}{\sqrt{2}} \\ 0 & \dfrac{1}{\sqrt{2}} & \dfrac{1}{\sqrt{2}} \\ 0 & 0 & 0 \end{pmatrix}$, $R = \begin{pmatrix} 1 & 1 & 0 \\ 0 & 1 & 1 \\ 0 & 0 & 1 \end{pmatrix}$, $\mathbf{b} = \begin{pmatrix} 1 \\ 3 \\ 1 \\ 2 \end{pmatrix}$

(c) $Q = \begin{pmatrix} 1 & 0 & 0 \\ 0 & \dfrac{1}{\sqrt{2}} & -\dfrac{1}{\sqrt{2}} \\ 0 & \dfrac{1}{\sqrt{2}} & \dfrac{1}{\sqrt{2}} \end{pmatrix}$, $R = \begin{pmatrix} 1 & 1 \\ 0 & 1 \\ 0 & 0 \end{pmatrix}$, $\mathbf{b} = \begin{pmatrix} 1 \\ \sqrt{2} \\ -\sqrt{2} \end{pmatrix}$

(d) $Q = \begin{pmatrix} \dfrac{1}{2} & \dfrac{1}{\sqrt{2}} & 0 & \dfrac{1}{2} \\ \dfrac{1}{2} & 0 & \dfrac{1}{\sqrt{2}} & -\dfrac{1}{2} \\ \dfrac{1}{2} & 0 & -\dfrac{1}{\sqrt{2}} & -\dfrac{1}{2} \\ \dfrac{1}{2} & -\dfrac{1}{\sqrt{2}} & 0 & \dfrac{1}{2} \end{pmatrix}$

$R = \begin{pmatrix} 1 & 1 & 0 \\ 0 & 1 & 1 \\ 0 & 0 & 1 \\ 0 & 0 & 0 \end{pmatrix}$, $\mathbf{b} = \begin{pmatrix} 2 \\ -2 \\ 0 \\ 2 \end{pmatrix}$

2. Let

$$A = \begin{pmatrix} D \\ E \end{pmatrix} = \begin{pmatrix} d_1 & & & \\ & d_2 & & \\ & & \ddots & \\ & & & d_n \\ \hline e_1 & & & \\ & e_2 & & \\ & & \ddots & \\ & & & e_n \end{pmatrix} \qquad \text{and} \qquad \mathbf{b} = \begin{pmatrix} b_1 \\ b_2 \\ \vdots \\ b_{2n} \end{pmatrix}$$

Use the normal equations to find the solution $\mathbf{x}$ to the least squares problem.

3. Let

$$A = \begin{pmatrix} 1 & 1 \\ \epsilon & 0 \\ 0 & \epsilon \end{pmatrix}$$

where ϵ is a small scalar.
(a) Determine the singular values of A exactly.
(b) Suppose that ϵ is sufficiently small that $1 + \epsilon^2$ gets rounded off to 1 on your calculator. Determine the eigenvalues of the calculated $A^T A$ and compare the square roots of these eigenvalues to your answers in part (a).

4. Show that the pseudoinverse A^+ satisfies the four Penrose conditions.

5. Let B be any matrix that satisfies Penrose conditions (1) and (3) and let $\mathbf{x} = B\mathbf{b}$. Show that $\mathbf{x}$ is a solution to the normal equations $A^T A \mathbf{x} = A^T \mathbf{b}$.

6. If $\mathbf{x} \in R^m$, we can think of $\mathbf{x}$ as an $m \times 1$ matrix. Define

$$X = \frac{1}{\|\mathbf{x}\|_2^2} \mathbf{x}^T$$

X is a $1 \times m$ matrix. Show that X and $\mathbf{x}$ satisfy the four Penrose conditions and consequently that

$$\mathbf{x}^+ = X = \frac{1}{\|\mathbf{x}\|_2^2} \mathbf{x}^T$$

7. Let A be an $m \times n$ matrix and let $\mathbf{b} \in R^m$. Show that $\mathbf{b} \in R(A)$ if and only if

$$\mathbf{b} = AA^+\mathbf{b}$$

8. Let A be an $m \times n$ matrix with singular value decomposition $U\Sigma V^T$ and suppose that A has rank r, where $r < n$. Let $\mathbf{b} \in R^m$. Show that a vector $\mathbf{x} \in R^n$ minimizes $\|\mathbf{b} - A\mathbf{x}\|_2$ if and only if

$$\mathbf{x} = A^+\mathbf{b} + c_{r+1}\mathbf{v}_{r+1} + \cdots + c_n\mathbf{v}_n$$

where $c_{r+1}, \ldots, c_n$ are scalars.

9. Let

$$A = \begin{pmatrix} 1 & 1 \\ 1 & 1 \\ 0 & 0 \end{pmatrix}$$

Determine A^+ and verify that A and A^+ satisfy the four Penrose conditions (see Example 1 of Section 7).

10. Given

$$A = \begin{pmatrix} 1 & 2 \\ -1 & -2 \end{pmatrix} \quad \text{and} \quad \mathbf{b} = \begin{pmatrix} 6 \\ -4 \end{pmatrix}$$

(a) Compute the singular value decomposition of A and use it to determine A^+.
(b) Use A^+ to find a least squares solution to the system $A\mathbf{x} = \mathbf{b}$.
(c) Find all solutions to the least squares problem $A\mathbf{x} = \mathbf{b}$.

11. Show each of the following.
(a) $(A^+)^+ = A$
(b) $(AA^+)^2 = AA^+$
(c) $(A^+A)^2 = A^+A$

12. Let $A_1 = U\Sigma_1 V^T$ and $A_2 = U\Sigma_2 V^T$, where

$$\Sigma_1 = \begin{pmatrix} \sigma_1 & & & & & & \\ & \ddots & & & & & \\ & & \sigma_{r-1} & & & & \\ & & & 0 & & & \\ & & & & \ddots & & \\ & & & & & & 0 \end{pmatrix}$$

and

$$\Sigma_2 = \begin{pmatrix} \sigma_1 & & & & & & \\ & \ddots & & & & & \\ & & \sigma_{r-1} & & & & \\ & & & \sigma_r & & & \\ & & & & 0 & & \\ & & & & & \ddots & \\ & & & & & & 0 \end{pmatrix}$$

and $\sigma_r = \epsilon > 0$. What are the values of $\|A_1 - A_2\|_F$ and $\|A_1^+ - A_2^+\|_F$? What happens to these values as we let $\epsilon \to 0$?

13. Let $A = XY^T$, where X is an $m \times r$ matrix, Y^T is an $r \times n$ matrix, and X^TX and Y^TY are both nonsingular. Show that the matrix

$$B = \left[Y(Y^TY)^{-1}(X^TX)^{-1}X^T \right]$$

satisfies the Penrose conditions and hence must equal A^+. Thus A^+ can be determined from any factorization of this form.

BIBLIOGRAPHY

A. LINEAR ALGEBRA

1. Anton, Howard. *Elementary Linear Algebra*, 4th ed. New York: John Wiley & Sons, Inc., 1984.
2. Gantmacher, F. R. *The Theory of Matrices*, 2 vols. New York: Chelsea Publishing Company, Inc., 1960.
3. Kolman, Bernard. *Elementary Linear Algebra*, 4th ed. New York: Macmillan Publishing Company, 1986.
4. Lancaster, Peter. *Theory of Matrices*, New York: Academic Press, Inc., 1969.
5. Pullman, N. J. *Matrix Theory and Its Applications*: *Selected Topics*. New York: Marcel Dekker, Inc., 1976.

B. APPLIED LINEAR ALGEBRA

6. Bellman, Richard. *Introduction to Matrix Analysis*, 2nd ed. New York: McGraw-Hill Book Company, 1970.
7. Fletcher, T. J. *Linear Algebra Through Its Applications*. New York: Van Nostrand Reinhold Company, 1972.
8. Noble, Ben, and James W. Daniel. *Applied Linear Algebra*, 2nd ed. Englewood Cliffs, N.J.: Prentice-Hall, Inc., 1977.
9. Strang, Gilbert. *Linear Algebra and Its Applications*, 2nd ed. New York: Academic Press, Inc., 1980.
10. Williams, Gareth. *Linear Algebra with Applications*. Boston: Allyn and Bacon, Inc., 1984.

C. NUMERICAL LINEAR ALGEBRA

11. Atkinson, Kendall E. *An Introduction to Numerical Analysis*. New York: John Wiley & Sons, Inc., 1978.
12. Conte, S. D., and C. deBoor. *Elementary Numerical Analysis*: *An Algorithmic Approach*, 3rd ed. New York: McGraw-Hill Book Company, 1980.
13. Dahlquist, G., and A. Bjorck. *Numerical Methods*. Englewood Cliffs, N.J.: Prentice-Hall, Inc., 1974.
14. Dongarra, J. J., C. B. Moler, J. R. Bunch, and G. N. Stewart, *Linpack Users' Guide*. Philadelphia: SIAM, 1979.
15. Forsythe, G. E., and C. B. Moler. *Computer Solution of Linear Algebraic Systems*. Englewood Cliffs, N.J.: Prentice-Hall, Inc., 1967.
16. Forsythe, G. E., M. Malcom, and C. B. Moler. *Computer Methods for Mathematical Computations*. Englewood Cliffs, N.J.: Prentice-Hall, Inc., 1977.
17. Golub, Gene H., and Charles F. Van Loan. *Matrix Computations*. Baltimore: Johns Hopkins University Press, 1983.
18. Lawson, Charles L., and Richard J. Hanson. *Solving Least Squares Problems*. Englewood Cliffs, N.J.: Prentice-Hall, Inc., 1974.
19. Parlett, B. N. *The Symmetric Eigenvalue Problem*. Englewood Cliffs, N.J.: Prentice-Hall, Inc., 1980.
20. Stewart, G. W. *Introduction to Matrix Computations*. New York: Academic Press, Inc., 1973.
21. Wilkinson, J. H. *Rounding Errors in Algebraic Processes*. Englewood Cliffs, N.J.: Prentice-Hall, Inc., 1963.
22. Wilkinson, J. H. *The Algebraic Eigenvalue Problem*. Oxford: Clarendon Press, 1965.
23. Wilkinson, J. H., and C. Reinsch. *Handbook for Automatic Computation*, Vol. II: *Linear Algebra*. New York: Springer-Verlag, 1971.

D. BOOKS OF RELATED INTEREST

24. Cheney, E. W. *Introduction to Approximation Theory*. New York: Mc-Graw-Hill Book Company, 1966.
25. Chiang, Alpha C. *Fundamental Methods of Mathematical Economics*. New York: McGraw-Hill Book Company, 1967.
26. Courant, R., and D. Hilbert. *Methods of Mathematical Physics*, Vol. I. New York: Wiley-Interscience, 1953.
27. Kreider, D. L., R. G. Kuller, and D. R. Ostberg. *Elementary Differential Equations*. Reading, Mass.: Addison-Wesley Publishing Company, Inc., 1968.
28. Rivlin, T. J. *The Chebyshev Polynomials*. New York: Wiley-Interscience, 1974.

References 1 and 3 are elementary linear algebra texts. References 2, 4, and 5 are written on a more advanced level. References 6 through 10 all have good selections of applications. References 15 and 23 contain ALGOL programs, while references 14, 15, 16, and 18 contain FORTRAN programs. Extended bibliographies are included in the following references: 2, 8, 13, 16, through 19, and 22.

ANSWERS TO SELECTED EXERCISES

SECTION 1

1. (a) $(11, 3)$;   (b) $(4, 1, 3)$;   (c) $(-2, 0, 3, 1)$;   (d) $(-2, 3, 0, 3, 1)$

2. (a) $\begin{pmatrix} 1 & -3 \\ 0 & 2 \end{pmatrix}$ (b) $\begin{pmatrix} 1 & 1 & 1 \\ 0 & 2 & 1 \\ 0 & 0 & 3 \end{pmatrix}$ (c) $\begin{pmatrix} 1 & 2 & 2 & 1 \\ 0 & 3 & 1 & -2 \\ 0 & 0 & -1 & 2 \\ 0 & 0 & 0 & 4 \end{pmatrix}$

3. (a) One solution. The two lines intersect at the point $(3, 1)$.

(b) No solution. The lines are parallel.

(c) Infinitely many solutions. Both equations represent the same line.

(d) No solution. Each pair of lines intersect in a point, however, there is no point that is on all three lines.

4. (a) $\begin{pmatrix} 1 & 1 & | & 4 \\ 1 & -1 & | & 2 \end{pmatrix}$ (b) $\begin{pmatrix} 1 & 2 & | & 4 \\ -2 & -4 & | & 4 \end{pmatrix}$

(c) $\begin{pmatrix} 2 & -1 & | & 3 \\ -4 & 2 & | & -6 \end{pmatrix}$ (d) $\begin{pmatrix} 1 & 1 & | & 1 \\ 1 & -1 & | & 1 \\ -1 & 3 & | & 3 \end{pmatrix}$

6. (a) $(1, -2)$; (b) $(3, 2)$; (c) $(\frac{1}{2}, \frac{2}{3})$; (d) $(1, 1, 2)$; (e) $(-3, 1, 2)$;
(f) $(-1, 1, 1)$; (g) $(1, 1, -1)$; (h) $(4, -3, 1, 2)$

SECTION 2

1. Row echelon form: (a), (c), (d), (g), and (h). Reduced row echelon form: (c), (d), and (g).

2. (a) Inconsistent; (b) consistent $(4, -1)$; (c) consistent, infinitely many solutions; (d) consistent $(4, 5, 2)$; (e) inconsistent; (f) consistent, $(5, 3, 2)$.

3. (a) $(-2, 5, 3)$; (b) $\varnothing$; (c) $\{(2 + 3\alpha, \alpha, -2) | \alpha \text{ real}\}$;
(d) $\{(5 - 2\alpha - \beta, \alpha, 4 - 3\beta, \beta) | \alpha, \beta \text{ real}\}$;
(e) $\{(3 - 5\alpha + 2\beta, \alpha, \beta, 6) | \alpha, \beta \text{ real}\}$; (f) $\{(\alpha, 2, -1) | \alpha \text{ real}\}$.

5. (a) $(5, 1)$; (b) inconsistent; (c) $(0, 0)$;

(d) $\left\{ \left(\dfrac{5 - \alpha}{4}, \dfrac{1 + 7\alpha}{8}, \alpha \right) \middle| \alpha \text{ real} \right\}$; (e) $\{(8 - 2\alpha, \alpha - 5, \alpha)\}$;

(f) inconsistent; (g) inconsistent; (h) inconsistent;

(i) $(0, \frac{3}{2}, 1)$; (j) $\{(-5 - \alpha, 3, 1, \alpha)\}$;

(k) $\left\{ \left(\frac{15}{4} - \frac{5}{8}\alpha - \beta, -\frac{1}{4} - \frac{1}{8}\alpha, \alpha, \beta \right) \right\}$;

(l) $\left\{ \left(1 + \frac{2}{7}\alpha, \frac{3}{7}\alpha, \alpha \right) \right\}$

6. (a) $(0, -1)$; (b) $\left\{ \left(\frac{3}{4} - \frac{5}{8}\alpha, -\frac{1}{4} - \frac{1}{8}\alpha, \alpha, 3 \right) | \alpha \text{ is real} \right\}$;
(c) $\{(0, \alpha, -\alpha)\}$; (d) $\left\{ \alpha\left(-\frac{4}{3}, 0, \frac{1}{3}, 1 \right) \right\}$

8. (a) $(5, 3, 2)$; (b) $(2, 4, 2)$; (c) $(2, 0, -2, -2, 0, 2)$

SECTION 3

1. (a) $\begin{pmatrix} 6 & 2 & 8 \\ -4 & 0 & 2 \\ 2 & 4 & 4 \end{pmatrix}$ (b) $\begin{pmatrix} 4 & 1 & 6 \\ -5 & 1 & 2 \\ 3 & -2 & 3 \end{pmatrix}$ (c) $\begin{pmatrix} 3 & 2 & 2 \\ 5 & -3 & -1 \\ -4 & 16 & 1 \end{pmatrix}$

(d) $\begin{pmatrix} 3 & 5 & -4 \\ 2 & -3 & 16 \\ 2 & -1 & 1 \end{pmatrix}$ (f) $\begin{pmatrix} 5 & 5 & 8 \\ -10 & -1 & -9 \\ 15 & 4 & 6 \end{pmatrix}$ (h) $\begin{pmatrix} 5 & -10 & 15 \\ 5 & -1 & 4 \\ 8 & -9 & 6 \end{pmatrix}$

2. (a) $\begin{pmatrix} 15 & 19 \\ 4 & 0 \end{pmatrix}$ (c) $\begin{pmatrix} 19 & 21 \\ 17 & 21 \\ 8 & 10 \end{pmatrix}$ (d) $\begin{pmatrix} 36 & 10 & 56 \\ 10 & 3 & 16 \end{pmatrix}$

(b) and (e) are not possible.

3. (a) 3×3 (b) 1×2

4. (a) $\begin{pmatrix} 3 & 2 \\ 2 & -3 \end{pmatrix} \begin{pmatrix} x_1 \\ x_2 \end{pmatrix} = \begin{pmatrix} 1 \\ 5 \end{pmatrix}$; (b) $\begin{pmatrix} 1 & 1 & 0 \\ 2 & 1 & -1 \\ 3 & -2 & 2 \end{pmatrix} \begin{pmatrix} x_1 \\ x_2 \\ x_3 \end{pmatrix} = \begin{pmatrix} 5 \\ 6 \\ 7 \end{pmatrix}$;

(c) $\begin{pmatrix} 2 & 1 & 1 \\ 1 & -1 & 2 \\ 3 & 2 & -1 \end{pmatrix} \begin{pmatrix} x_1 \\ x_2 \\ x_3 \end{pmatrix} = \begin{pmatrix} 4 \\ 2 \\ 0 \end{pmatrix}$

11. $A = A^2 = A^3 = A^n$

12. $A^{2n} = I$, $A^{2n+1} = A$

16. Monday 575, Tuesday 936, Wednesday 457.8, Thursday 110.5, Friday 457.8

17. 4500 married, 5500 single

18. $b = a_{22} - \dfrac{a_{12}a_{21}}{a_{11}}$

SECTION 4

1. (a) Type I; (b) not an elementary matrix; (c) type III; (d) type II

3. (a) $\begin{pmatrix} -2 & 0 \\ 0 & 1 \end{pmatrix}$ (b) $\begin{pmatrix} 1 & 0 & 0 \\ 0 & 0 & 1 \\ 0 & 1 & 0 \end{pmatrix}$ (c) $\begin{pmatrix} 1 & 0 & 0 \\ 0 & 1 & 0 \\ 0 & 2 & 1 \end{pmatrix}$

4. (a) $\begin{pmatrix} 0 & 0 & 1 \\ 0 & 1 & 0 \\ 1 & 0 & 0 \end{pmatrix}$ (b) $\begin{pmatrix} 1 & -3 \\ 0 & 1 \end{pmatrix}$ (c) $\begin{pmatrix} \frac{1}{2} & 0 & 0 \\ 0 & 1 & 0 \\ 0 & 0 & 1 \end{pmatrix}$

5. (b) (i) $(0, -1, 1)^T$; (ii) $(-4, -2, 5)^T$; (iii) $(0, 3, -2)^T$

6. (a) $\begin{pmatrix} 0 & 1 \\ 1 & 1 \end{pmatrix}$ (b) $\begin{pmatrix} 3 & -5 \\ -1 & 2 \end{pmatrix}$ (c) $\begin{pmatrix} -4 & 3 \\ \frac{3}{2} & -1 \end{pmatrix}$

(d) $\begin{pmatrix} \frac{1}{3} & 0 \\ -1 & \frac{1}{3} \end{pmatrix}$ (e) $\begin{pmatrix} 1 & -1 & 0 \\ 0 & 1 & -1 \\ 0 & 0 & 1 \end{pmatrix}$ (f) $\begin{pmatrix} 3 & 0 & -5 \\ 0 & \frac{1}{3} & 0 \\ -1 & 0 & 2 \end{pmatrix}$

(g) $\begin{pmatrix} 2 & -3 & 3 \\ -\frac{3}{5} & \frac{6}{5} & -1 \\ -\frac{2}{5} & -\frac{1}{5} & 0 \end{pmatrix}$ (h) $\begin{pmatrix} -\frac{1}{2} & -1 & -\frac{1}{2} \\ -2 & -1 & -1 \\ \frac{3}{2} & 1 & \frac{1}{2} \end{pmatrix}$

SECTION 5

1. (a) $(I \; A^{-1})$ (b) $\begin{pmatrix} I \\ A^{-1} \end{pmatrix}$ (c) $\begin{pmatrix} A^T A & A^T \\ A & I \end{pmatrix}$

(d) $AA^T + I$ (e) $\begin{pmatrix} I & A^{-1} \\ A & I \end{pmatrix}$

3. (a) $A\mathbf{b}_1 = \begin{pmatrix} 3 \\ 3 \end{pmatrix}$, $A\mathbf{b}_2 = \begin{pmatrix} 4 \\ -1 \end{pmatrix}$

(b) $(1 \; 1)B = (3 \quad 4), (2 \quad -1)B = (3 \quad -1)$

(c) $AB = \begin{pmatrix} 3 & 4 \\ 3 & -1 \end{pmatrix}$

4. (a) $\left(\begin{array}{cc|cc} 3 & 1 & 1 & 1 \\ 3 & 2 & 1 & 2 \\ \hline 1 & 1 & 1 & 1 \\ 1 & 2 & 1 & 1 \end{array} \right)$ (b) $\left(\begin{array}{cc|cc} 1 & 1 & 1 & 1 \\ 0 & 1 & 0 & 0 \\ \hline 3 & 1 & 1 & 1 \\ 0 & 1 & 0 & 1 \end{array} \right)$

(c) $\left(\begin{array}{cc|cc} 2 & 2 & 2 & 2 \\ 2 & 4 & 2 & 2 \\ \hline 3 & 1 & 1 & 1 \\ 3 & 2 & 1 & 2 \end{array} \right)$ (d) $\left(\begin{array}{cc|cc} 1 & 1 & 1 & 1 \\ 1 & 2 & 1 & 1 \\ \hline 3 & 2 & 1 & 2 \\ 3 & 1 & 1 & 1 \end{array} \right)$

5. (b) $\left(\begin{array}{ccc|c} 0 & 2 & 0 & -2 \\ 8 & 5 & 8 & -5 \\ 3 & 2 & 3 & -2 \\ \hline 5 & 3 & 5 & -3 \end{array} \right)$ (d) $\left(\begin{array}{cc} 3 & -3 \\ 2 & -2 \\ 1 & -1 \\ \hline 5 & -5 \\ 4 & -4 \end{array} \right)$

8. $A^2 = \begin{pmatrix} B & O \\ O & B \end{pmatrix}$, $A^4 = \begin{pmatrix} B^2 & O \\ O & B^2 \end{pmatrix}$

9. (a) $\begin{pmatrix} O & I \\ I & O \end{pmatrix}$ (b) $\begin{pmatrix} I & O \\ -B & I \end{pmatrix}$

11. $A\mathbf{x} = x_1\mathbf{a}_1 + x_2\mathbf{a}_2 + \cdots + x_n\mathbf{a}_n$

CHAPTER 2

SECTION 1

1. (a) Even; (b) odd; (c) even; (d) even; (e) odd
2. $\det(A) = 3 \cdot 2 \cdot 4 - 3 \cdot 6 \cdot 0 - 1 \cdot 0 \cdot 4 + 1 \cdot 6 \cdot 0 + 5 \cdot 0 \cdot 0 - 5 \cdot 2 \cdot 0$
$= 24$
3. (a) 1; (b) 4; (c) 0; (d) 58; (e) -39;
   (f) 0; (g) 8; (h) 20
4. (a) 2; (b) -4; (c) 0; (d) 0
5. $-x^3 + ax^2 + bx + c$
6. $\lambda = 6$ or -1

SECTION 2

1. (a) -24; (b) 30; (c) -1
2. (a) 10; (b) 20
3. (a), (e), and (f) are singular while (b), (c), and (d) are nonsingular.
6. (a) 20; (b) 108; (c) 160; (d) $\frac{5}{4}$
7. $\det(A) = u_{11}u_{22}u_{33}$
10. $(n! - 1)$ additions, $(n - 1)n!$ multiplications

SECTION 3

1. (a) $\det(A) = -7$, $\text{adj } A = \begin{pmatrix} -1 & -2 \\ -3 & 1 \end{pmatrix}$, $A^{-1} = \begin{pmatrix} \frac{1}{7} & \frac{2}{7} \\ \frac{3}{7} & -\frac{1}{7} \end{pmatrix}$;

   (c) $\det(A) = 3$, $\text{adj } A = \begin{pmatrix} -3 & 5 & 2 \\ 0 & 1 & 1 \\ 6 & -8 & -5 \end{pmatrix}$, $A^{-1} = \frac{1}{3}\text{adj } A$

2. (a) $(\frac{5}{7}, \frac{8}{7})$; (b) $(\frac{11}{5}, -\frac{4}{5})$; (c) $(4, -2, 2)$; (d) $(2, -1, 2)$;
   (e) $(-\frac{2}{3}, \frac{2}{3}, \frac{1}{3}, 0)$
3. (a) $\det(A) = 0$, so A is singular.
   (b)

$$\text{adj } A = \begin{pmatrix} -1 & 2 & -1 \\ 2 & -4 & 2 \\ -1 & 2 & -1 \end{pmatrix} \quad \text{and} \quad A \text{ adj } A = \begin{pmatrix} 0 & 0 & 0 \\ 0 & 0 & 0 \\ 0 & 0 & 0 \end{pmatrix}$$

12. DO YOUR HOMEWORK.

CHAPTER 3

SECTION 1

2. V is not a vector space. Axiom 6 does not hold.
6. If $\mathbf{x} + \mathbf{y} = \mathbf{x}$ for all $\mathbf{x}$ in the vector space, then $\mathbf{0} = \mathbf{0} + \mathbf{y} = \mathbf{y}$.
7. If $\mathbf{x} + \mathbf{y} = \mathbf{x} + \mathbf{z}$, then $-\mathbf{x} + (\mathbf{x} + \mathbf{y}) = -\mathbf{x} + (\mathbf{x} + \mathbf{z})$ and the conclusion follows using axioms 1, 2, 3, and 4.

SECTION 2
1. (a) and (c) are subspaces; (b) and (d) are not.
2. (b) and (c) are subspaces; (a) and (d) are not.
3. (a), (b), (d), and (e) are subspaces; (c) and (f) are not.
4. (a) $\{(0,0)^T\}$; (b) $\mathbb{S}\{(-2,1,0,0)^T,(3,0,1,0)^T\}$; (c) $\mathbb{S}\{(1,1,1)^T\}$;
   (d) $\mathbb{S}\{(-5,0,-3,1)^T,(1,-1,0,0)^T\}$
5. Only the set in part (c) is a subspace of P_4.
6. (a), (b), and (d) are subspaces.
7. (a), (c), and (e) are spanning sets.
8. (a) and (b) are spanning sets.
9. (b) and (c)

SECTION 3
1. (a) and (e) are linearly independent; (b), (c), and (d) are linearly dependent.
2. (a) and (e) are linearly independent; (b), (c), and (d) are not.
3. (a) and (b) are all of 3-space; (c) a plane through $(0,0,0)$;
   (d) a line through $(0,0,0)$; (e) a plane through $(0,0,0)$
4. (a) linearly independent; (b) linearly independent; (c) linearly dependent
5. (a) and (b) are linearly dependent while (c) and (d) are linearly independent.
8. When α is an odd multiple of $\pi/2$. If the graph of $y = \cos x$ is shifted to the left or right by an odd multiple of $\pi/2$, one obtains the graph of either $\sin x$ or $-\sin x$.

SECTION 4
1. Only in parts (a) and (b) do they form a basis.
2. Only in part (a) do they form a basis.
3. (c) 2
4. 1
5. (c) 2; (d) a plane through $(0,0,0)$ in 3-space
6. (b) $\{(1,1,1)^T\}$, dimension 1; (c) $\{(1,0,1)^T,(0,1,1)^T\}$, dimension 2
7. (a) $\{E_{11}, E_{22}\}$; (b) $\{E_{11}, E_{21}, E_{22}\}$; (d) $\{E_{12}, E_{21}, E_{22}\}$;
   (e) $\{E_{11}, E_{22}, E_{21} + E_{12}\}$
8. 2
9. (a) 3; (b) 3; (c) 2; (d) 2
10. (a) $\{x, x^2\}$; (b) $\{x-1, (x-1)^2\}$; (c) $\{x(x-1)\}$

SECTION 5
2. (a) 3; (b) 3; (c) 2
3. The following examples are all in reduced row echelon form:

(a) $\begin{pmatrix} 0 & 0 & 0 \\ 0 & 0 & 0 \\ 0 & 0 & 0 \end{pmatrix}$ (b) $\begin{pmatrix} 1 & 0 & 0 \\ 0 & 0 & 0 \\ 0 & 0 & 0 \end{pmatrix}$ (c) $\begin{pmatrix} 1 & 0 & 0 \\ 0 & 1 & 0 \\ 0 & 0 & 0 \end{pmatrix}$

(d) $\begin{pmatrix} 1 & 0 & 0 \\ 0 & 1 & 0 \\ 0 & 0 & 1 \end{pmatrix}$

10. (b) $n - 1$
15. If $\mathbf{x}_j$ is a solution to $A\mathbf{x} = \mathbf{e}_j$ for $j = 1, \ldots, m$ and $X = (\mathbf{x}_1, \mathbf{x}_2, \ldots, \mathbf{x}_m)$, then $AX = I_m$.

CHAPTER 4

SECTION 1

1. (a) Reflection about x_2 axis; (b) reflection about the origin;
   (c) reflection about the line $x_2 = x_1$; (d) the length of the vector is halved; (e) projection onto x_2 axis
4. All except (c) are linear transformations from R^3 into R^2.
5. (b) and (c) are linear transformations from R^2 into R^3.
6. (a), (b), and (d) are linear transformations.
7. (a) and (c) are linear transformations from P_2 into P_3.
8. $L(e^x) = e^x - 1$ and $L(x^2) = x^3/3$
9. (a) and (c) are linear transformations from $C[0, 1]$ into R^1.

SECTION 2

1. (a) $\begin{pmatrix} -1 & 0 \\ 0 & 1 \end{pmatrix}$ (b) $\begin{pmatrix} -1 & 0 \\ 0 & -1 \end{pmatrix}$ (c) $\begin{pmatrix} 0 & 1 \\ 1 & 0 \end{pmatrix}$

   (d) $\begin{pmatrix} \frac{1}{2} & 0 \\ 0 & \frac{1}{2} \end{pmatrix}$ (e) $\begin{pmatrix} 0 & 0 \\ 0 & 1 \end{pmatrix}$

2. (a) $\begin{pmatrix} 1 & 1 & 0 \\ 0 & 0 & 0 \end{pmatrix}$ (b) $\begin{pmatrix} 1 & 0 & 0 \\ 0 & 1 & 0 \end{pmatrix}$ (c) $\begin{pmatrix} -1 & 1 & 0 \\ 0 & -1 & 1 \end{pmatrix}$

3. (a) $\begin{pmatrix} 0 & 0 & 1 \\ 0 & 1 & 0 \\ 1 & 0 & 0 \end{pmatrix}$ (b) $\begin{pmatrix} 1 & 0 & 0 \\ 1 & 1 & 0 \\ 1 & 1 & 1 \end{pmatrix}$ (c) $\begin{pmatrix} 0 & 0 & 2 \\ 3 & 1 & 0 \\ 2 & 0 & -1 \end{pmatrix}$

4. (a) $(0,0,0)^T$; (b) $(2, -1, -1)^T$; (c) $(-15, 9, 6)^T$

5. (a) $\begin{pmatrix} \dfrac{1}{\sqrt{2}} & \dfrac{1}{\sqrt{2}} \\ -\dfrac{1}{\sqrt{2}} & \dfrac{1}{\sqrt{2}} \end{pmatrix}$ (b) $\begin{pmatrix} 0 & 1 \\ 1 & 0 \end{pmatrix}$ (c) $\begin{pmatrix} \sqrt{3} & -1 \\ 1 & \sqrt{3} \end{pmatrix}$

   (d) $\begin{pmatrix} 0 & 1 \\ 0 & 0 \end{pmatrix}$

6. $\begin{pmatrix} 1 & 0 \\ 0 & 1 \\ 1 & 1 \end{pmatrix}$

7. (b) $\begin{pmatrix} 0 & 0 & 1 \\ 0 & 1 & -1 \\ 1 & -1 & 0 \end{pmatrix}$

8. (a) $\begin{pmatrix} 1 & 1 & 1 \\ 2 & 0 & 1 \\ 0 & -2 & -1 \end{pmatrix}$ (b) (i) $7y_1 + 6y_2 - 8y_3$; (ii) $3y_1 + 3y_2 - 3y_3$;
   (iii) $y_1 + 5y_2 + 3y_3$

9. $\begin{pmatrix} 1 & \frac{1}{2} \\ 1 & 0 \end{pmatrix}$

10. $\begin{pmatrix} 1 & \frac{1}{2} & \frac{1}{2} \\ -2 & 0 & 0 \end{pmatrix}$ (a) $\begin{pmatrix} \frac{1}{2} \\ -2 \end{pmatrix}$ (d) $\begin{pmatrix} 5 \\ -8 \end{pmatrix}$

11. $\begin{pmatrix} 1 & 1 & 0 \\ 0 & 1 & 2 \\ 0 & 0 & 1 \end{pmatrix}$

SECTION 3

1. (a) $\begin{pmatrix} 1 & -1 \\ 1 & 1 \end{pmatrix}$ (b) $\begin{pmatrix} 1 & 2 \\ 2 & 5 \end{pmatrix}$ (c) $\begin{pmatrix} 0 & 1 \\ 1 & 0 \end{pmatrix}$

2. (a) $\begin{pmatrix} \frac{1}{2} & \frac{1}{2} \\ -\frac{1}{2} & \frac{1}{2} \end{pmatrix}$ (b) $\begin{pmatrix} 5 & -2 \\ -2 & 1 \end{pmatrix}$ (c) $\begin{pmatrix} 0 & 1 \\ 1 & 0 \end{pmatrix}$

3. (a) $S_2 = (v_1, v_2) = \begin{pmatrix} 5 & 7 \\ 2 & 3 \end{pmatrix}$

4. (a) $\begin{pmatrix} \frac{5}{2} & \frac{7}{2} \\ -\frac{1}{2} & -\frac{1}{2} \end{pmatrix}$ (b) $\begin{pmatrix} 11 & 14 \\ -4 & -5 \end{pmatrix}$ (c) $\begin{pmatrix} 2 & 3 \\ 3 & 4 \end{pmatrix}$

5. (a) $\begin{pmatrix} 2 & 0 & -1 \\ -1 & 2 & -1 \\ 0 & -1 & 1 \end{pmatrix}$ (b) (i) $(1, -4, 3)^T$, (ii) $(0, -1, 1)^T$,

   (iii) $(2, 2, -1)^T$

6. (a) $\begin{pmatrix} 1 & -1 & -2 \\ 1 & 1 & 0 \\ 1 & 0 & 1 \end{pmatrix}$ (b) $\begin{pmatrix} 7 \\ 5 \\ -2 \end{pmatrix}$

7. (a) $\begin{pmatrix} 2 & 2 \\ -1 & 1 \end{pmatrix}$ (b) $\begin{pmatrix} \frac{1}{4} & -\frac{1}{2} \\ \frac{1}{4} & \frac{1}{2} \end{pmatrix}$

8. $\begin{pmatrix} 1 & -1 & 0 \\ 0 & 1 & -1 \\ 0 & 0 & 1 \end{pmatrix}$

SECTION 4

1. For the matrix A, see the answers to Exercise 1 of Section 2.

   (a) $B = \begin{pmatrix} 0 & 1 \\ 1 & 0 \end{pmatrix}$ (b) $B = \begin{pmatrix} -1 & 0 \\ 0 & -1 \end{pmatrix}$ (c) $B = \begin{pmatrix} 1 & 0 \\ 0 & -1 \end{pmatrix}$

   (d) $B = \begin{pmatrix} \frac{1}{2} & 0 \\ 0 & \frac{1}{2} \end{pmatrix}$ (e) $B = \begin{pmatrix} \frac{1}{2} & \frac{1}{2} \\ \frac{1}{2} & \frac{1}{2} \end{pmatrix}$

2. (a) $\begin{pmatrix} 1 & 1 \\ -1 & -3 \end{pmatrix}$ (b) $\begin{pmatrix} 1 & 0 \\ -4 & -1 \end{pmatrix}$

3. (a) $\begin{pmatrix} 1 & 1 & 0 \\ 1 & 0 & 1 \\ 0 & 1 & 1 \end{pmatrix}$ (b) $\begin{pmatrix} 2 & 1 & 1 \\ 1 & 2 & 1 \\ 1 & 1 & 2 \end{pmatrix}$

4. (a) $\begin{pmatrix} 0 & 0 & 0 \\ 0 & 1 & 0 \\ 0 & 0 & 1 \end{pmatrix}$ (b) $\begin{pmatrix} 1 & 1 & 0 \\ 1 & 2 & -2 \\ 1 & 0 & 1 \end{pmatrix}$

5. (a) $\begin{pmatrix} 0 & 0 & 2 \\ 0 & 1 & 0 \\ 0 & 0 & 2 \end{pmatrix}$ (b) $\begin{pmatrix} 0 & 0 & 0 \\ 0 & 1 & 0 \\ 0 & 0 & 2 \end{pmatrix}$ (c) $\begin{pmatrix} 1 & 0 & 1 \\ 0 & 1 & 0 \\ 0 & 0 & 1 \end{pmatrix}$

   (d) $a_1 x + a_2 2^n (1 + x^2)$

6. (a) $\begin{pmatrix} 1 & 0 & 0 \\ 0 & 1 & 1 \\ 0 & 1 & -1 \end{pmatrix}$ (b) $\begin{pmatrix} 0 & 0 & 0 \\ 0 & 0 & 1 \\ 0 & 1 & 0 \end{pmatrix}$ (c) $\begin{pmatrix} 0 & 0 & 0 \\ 0 & 1 & 0 \\ 0 & 0 & -1 \end{pmatrix}$

CHAPTER 5

SECTION 1

1. (a) $0°$;   (b) $90°$

2. (a) $\sqrt{14}$ (scalar projection), $(2,1,3)^T$ (vector projection);   (b) $0, \mathbf{0}$;

   (c) $\dfrac{14\sqrt{13}}{13}, (\frac{42}{13}, \frac{28}{13})^T$;   (d) $\dfrac{8\sqrt{21}}{21}, (\frac{8}{21}, \frac{16}{21}, \frac{32}{21})^T$

3. (a) $\mathbf{p} = (3,0)^T$, $\mathbf{x} - \mathbf{p} = (0,4)^T$, $\mathbf{p}^T(\mathbf{x} - \mathbf{p}) = 3 \cdot 0 + 0 \cdot 4 = 0$

   (c) $\mathbf{p} = (3,3,3)^T$, $\mathbf{x} - \mathbf{p} = (-1,1,0)^T$, $\mathbf{p}^T(\mathbf{x} - \mathbf{p}) = -1 \cdot 3 + 1 \cdot 3 + 0 \cdot 3$
   $= 0$

4. $(1.8, 3.6)$

5. $(1.4, 3.8)$

6. 0.4

7. (a) $2x + 4y + 3z = 0$;   (c) $z - 4 = 0$

8. $\frac{5}{3}$

9. $\frac{8}{7}$

SECTION 2

1. (a) $\{(3,4)^T\}$ basis for $R(A^T)$, $\{(-4,3)^T\}$ basis for $N(A)$, $\{(1,2)^T\}$ basis
   for $R(A)$, $\{(-2,1)^T\}$ basis for $N(A^T)$

   (d) Basis for $R(A^T)$: $\{(1,0,0,0)^T, (0,1,0,0)^T, (0,0,1,1)^T\}$
   Basis for $N(A)$: $\{(0,0,-1,1)^T\}$
   Basis for $R(A)$: $\{(1,0,0,1)^T, (0,1,0,1)^T, (0,0,1,1)^T\}$
   Basis for $N(A^T)$: $\{(1,1,1,-1)^T\}$

2. (b) The orthogonal complement is spanned by $(-5,1,3)^T$

3. $\{(-1,2,0,1)^T, (2,-3,1,0)^T\}$ is one basis for $S^\perp$.

4. (a) $\mathbf{N} = (8,-2,1)^T$;   (b) $8x - 2y + z = 7$

8. $\dim N(A) = n - r$, $\dim N(A^T) = m - r$

SECTION 3

2. (a) 1;   (b) $1/\pi$;   (c) $\frac{1}{6}$

8. (a) $\|\mathbf{x}\|_1 = 7$, $\|\mathbf{x}\|_2 = 5$, $\|\mathbf{x}\|_\infty = 4$

   (b) $\|\mathbf{x}\|_1 = 4$, $\|\mathbf{x}\|_2 = \sqrt{6}$, $\|\mathbf{x}\|_\infty = 2$

   (c) $\|\mathbf{x}\|_1 = 3$, $\|\mathbf{x}\|_2 = \sqrt{3}$, $\|\mathbf{x}\|_\infty = 1$

9. $\|\mathbf{x} - \mathbf{y}\|_1 = 5$, $\|\mathbf{x} - \mathbf{y}\|_2 = 3$, $\|\mathbf{x} - \mathbf{y}\|_\infty = 2$

10. $\|\mathbf{x}\|_2 = 2$, $\|\mathbf{y}\|_2 = 6$, $\|\mathbf{x} + \mathbf{y}\|_2 = 2\sqrt{10}$

12. (a) $\sqrt{\pi}$;   (b) $\sqrt{\pi}$;   (c) $\sqrt{2\pi}$

13. (a) $\dfrac{\sqrt{10}}{2}$;   (b) $\dfrac{\sqrt{34}}{4}$

22. (a) Not a norm;   (b) norm;   (c) norm

SECTION 4

1. (a) $\|A\|_F = \sqrt{2}$, $\|A\|_\infty = 1$, $\|A\|_1 = 1$

   (b) $\|A\|_F = 5$, $\|A\|_\infty = 5$, $\|A\|_1 = 6$

   (c) $\|A\|_F = \|A\|_\infty = \|A\|_1 = 1$

   (d) $\|A\|_F = 7$, $\|A\|_\infty = 6$, $\|A\|_1 = 10$

   (e) $\|A\|_F = 9$, $\|A\|_\infty = 10$, $\|A\|_1 = 12$

2. 2
4. $\|I\|_1 = \|I\|_\infty = 1$, $\|I\|_F = \sqrt{n}$
6. (a) 11;   (b) $(-1, 1, -1)^T$
11. (a) $\|A\mathbf{x}\|_\infty \le \|A\mathbf{x}\|_2 \le \|A\|_2 \|\mathbf{x}\|_2 \le \sqrt{n} \|A\|_2 \|\mathbf{x}\|_\infty$

SECTION 5

1. (a) $(2, 1)^T$;   (c) $(1.6, 0.6, 1.2)^T$
2. (a) $\mathbf{p} = (3, 1, 0)^T$, $\mathbf{r} = (0, 0, 2)^T$ (c) $\mathbf{p} = (3.4, 0.2, 0.6, 2.8)^T$,
   $\mathbf{r} = (0.6, -0.2, 0.4, -0.8)^T$
3. (a) $\{(1 - 2\alpha, \alpha)^T | \alpha \text{ real}\}$;   (b) $\{(2 - 2\alpha, 1 - \alpha, \alpha)^T | \alpha \text{ real}\}$
4. (a) $\mathbf{p} = (1, 2, -1)^T$, $\mathbf{b} - \mathbf{p} = (2, 0, 2)^T$;
   (b) $\mathbf{p} = (3, 1, 4)^T$, $\mathbf{b} - \mathbf{p} = (-5, -1, 4)^T$
5. (a) $y = 1.8 + 2.9x$
6. $0.6 + 1.7x + 1.2x^2$

SECTION 6

1. (a) and (d)
2. (b) $\mathbf{x} = -\dfrac{\sqrt{2}}{3}\mathbf{x}_1 + \dfrac{5}{3}\mathbf{x}_2$; $\|\mathbf{x}\| = \left[\left(-\dfrac{\sqrt{2}}{3}\right)^2 + \left(\dfrac{5}{3}\right)^2\right]^{1/2} = \sqrt{3}$
3. $\mathbf{p} = (\frac{23}{18}, \frac{41}{18}, \frac{8}{9})^T$; $\mathbf{p} - \mathbf{x} = (\frac{5}{18}, \frac{5}{18}, -\frac{10}{9})^T$
4. (b) $c_1 = y_1 \cos\theta + y_2 \sin\theta$, $c_2 = -y_1 \sin\theta + y_2 \cos\theta$
5. (c) (i) 0;   (ii) $-\dfrac{\pi}{2}$;   (iii) 0;   (iv) $\dfrac{\pi}{8}$
6. (b) $Q\mathbf{x}$ can be obtained by a rotation of an angle θ followed by a reflection about the x_2 axis.
16. (b) (i) $(\sqrt{3}, \sqrt{6})^T$;   (ii) $\frac{1}{2}(4\sqrt{3}, -\sqrt{6})^T$;   (iii) $\frac{1}{6}(8\sqrt{3}, -\sqrt{6})^T$
18. (b) $\|1\| = \sqrt{2}$, $\|x\| = \dfrac{\sqrt{6}}{3}$;   (c) $l(x) = \dfrac{9}{7}x$

SECTION 7

1. (a) $\left\{\left(-\dfrac{1}{\sqrt{2}}, \dfrac{1}{\sqrt{2}}\right)^T, \left(\dfrac{1}{\sqrt{2}}, \dfrac{1}{\sqrt{2}}\right)^T\right\}$;

   (b) $\left\{\left(\dfrac{2}{\sqrt{5}}, \dfrac{1}{\sqrt{5}}\right)^T, \left(-\dfrac{1}{\sqrt{5}}, \dfrac{2}{\sqrt{5}}\right)^T\right\}$

2. (a) $\begin{pmatrix} -\dfrac{1}{\sqrt{2}} & \dfrac{1}{\sqrt{2}} \\ \dfrac{1}{\sqrt{2}} & \dfrac{1}{\sqrt{2}} \end{pmatrix} \begin{pmatrix} \sqrt{2} & \sqrt{2} \\ 0 & 4\sqrt{2} \end{pmatrix}$ (b) $\begin{pmatrix} \dfrac{2}{\sqrt{5}} & -\dfrac{1}{\sqrt{5}} \\ \dfrac{1}{\sqrt{5}} & \dfrac{2}{\sqrt{5}} \end{pmatrix} \begin{pmatrix} \sqrt{5} & 4\sqrt{5} \\ 0 & 3\sqrt{5} \end{pmatrix}$

3. $\left\{\left(\frac{1}{3}, \frac{2}{3}, -\frac{2}{3}\right)^T, \left(\frac{2}{3}, \frac{1}{3}, \frac{2}{3}\right)^T, \left(-\frac{2}{3}, \frac{2}{3}, \frac{1}{3}\right)^T\right\}$

4. $u_1(x) = \dfrac{1}{\sqrt{2}}$, $u_2(x) = \dfrac{\sqrt{6}}{2}x$, $u_3(x) = \dfrac{3\sqrt{10}}{4}\left(x^2 - \dfrac{1}{3}\right)$

5. (a) $\left\{ \dfrac{1}{3}(2,1,2)^T, \dfrac{\sqrt{2}}{6}(-1,4,-1)^T \right\}$

(b) $Q = \begin{pmatrix} \dfrac{2}{3} & \dfrac{-\sqrt{2}}{6} \\[2mm] \dfrac{1}{3} & \dfrac{2\sqrt{2}}{3} \\[2mm] \dfrac{2}{3} & \dfrac{-\sqrt{2}}{6} \end{pmatrix}$ $R = \begin{pmatrix} 3 & \dfrac{5}{3} \\[2mm] 0 & \dfrac{\sqrt{2}}{3} \end{pmatrix}$ (c) $\mathbf{x} = \begin{pmatrix} 9 \\ -3 \end{pmatrix}$

7. (b) $\begin{pmatrix} \dfrac{3}{5} & -\dfrac{4}{5\sqrt{2}} \\[2mm] \dfrac{4}{5} & \dfrac{3}{5\sqrt{2}} \\[2mm] 0 & \dfrac{1}{\sqrt{2}} \end{pmatrix} \begin{pmatrix} 5 & 1 \\ 0 & 2\sqrt{2} \end{pmatrix}$ (c) $(2.1, 5.5)^T$

8. $\left\{ \left(-\dfrac{1}{\sqrt{2}}, \dfrac{1}{\sqrt{2}}, 0, 0 \right)^T, \left(\dfrac{\sqrt{2}}{3}, \dfrac{\sqrt{2}}{3}, -\dfrac{\sqrt{2}}{2}, \dfrac{\sqrt{2}}{6} \right)^T \right\}$

9. $\left\{ \left(\dfrac{4}{5}, \dfrac{2}{5}, \dfrac{2}{5}, \dfrac{1}{5} \right)^T, \left(\dfrac{1}{5}, -\dfrac{2}{5}, -\dfrac{2}{5}, \dfrac{4}{5} \right)^T, \left(0, \dfrac{1}{\sqrt{2}}, -\dfrac{1}{\sqrt{2}}, 0 \right)^T \right\}$

SECTION 8

1. (a) $T_4 = 8x^4 - 8x^2 + 1,\ T_5 = 16x^5 - 20x^3 + 5x$

(b) $H_4 = 16x^4 - 48x^2 + 12,\ H_5 = 32x^5 - 160x^3 + 120x$

2. $p_1(x) = x,\ p_2(x) = x^2 - \dfrac{4}{\pi} + 1$

4. $p(x) = (\sinh 1)P_0(x) + \dfrac{3}{e}P_1(x) + 5\left(\sinh 1 - \dfrac{3}{e} \right)P_2(x)$

$p(x) \approx 0.9963 + 1.1036x + 0.5367x^2$

6. (a) $U_0 = 1,\ U_1 = 2x,\ U_2 = 4x^2 - 1$

11. $p(x) = (x-2)(x-3) + (x-1)(x-3) + 2(x-1)(x-2)$

13. $1 \cdot f\left(-\dfrac{1}{\sqrt{3}} \right) + 1 \cdot f\left(\dfrac{1}{\sqrt{3}} \right)$

14. (a) degree 3 or less;   (b) the formula gives the exact answer for the first integral. The approximate value for the second integral is 1.5, while the exact answer is $\dfrac{\pi}{2}$.

CHAPTER 6

SECTION 1

1. (a) $\lambda_1 = 5$, the eigenspace is spanned by $(1,1)^T$

$\lambda_2 = -1$, the eigenspace is spanned by $(1,-2)^T$

(b) $\lambda_1 = 3$, the eigenspace is spanned by $(4,3)^T$

$\lambda_2 = 2$, the eigenspace is spanned by $(1,1)^T$

(c) $\lambda_1 = \lambda_2 = 2$, the eigenspace is spanned by $(1,1)^T$

(d) $\lambda_1 = 3 + 4i$, the eigenspace is spanned by $(2i,1)^T$

$\lambda_2 = 3 - 4i$, the eigenspace is spanned by $(-2i,1)^T$

(e) $\lambda_1 = 2 + i$, the eigenspace is spanned by $(1, 1 + i)^T$
   $\lambda_2 = 2 - i$, the eigenspace is spanned by $(1, 1 - i)^T$
(f) $\lambda_1 = \lambda_2 = \lambda_3 = 0$, the eigenspace is spanned by $(1, 0, 0)^T$
(g) $\lambda_1 = 2$, the eigenspace is spanned by $(1, 1, 0)^T$
   $\lambda_2 = 1$, the eigenspace is spanned by $(1, 0, 0)^T, (0, 1, -1)^T$
(h) $\lambda_1 = 1$, the eigenspace is spanned by $(1, 0, 0)^T$
   $\lambda_2 = 4$, the eigenspace is spanned by $(1, 1, 1)^T$
   $\lambda_3 = -2$, the eigenspace is spanned by $(-1, -1, 5)^T$
(i) $\lambda_1 = 2$, the eigenspace is spanned by $(7, 3, 1)^T$
   $\lambda_2 = 1$, the eigenspace is spanned by $(3, 2, 1)^T$
   $\lambda_3 = 0$, the eigenspace is spanned by $(1, 1, 1)^T$
(j) $\lambda_1 = \lambda_2 = \lambda_3 = -1$, the eigenspace is spanned by $(1, 0, 1)^T$
(k) $\lambda_1 = \lambda_2 = 2$, the eigenspace is spanned by $\mathbf{e}_1$ and $\mathbf{e}_2$
   $\lambda_3 = 3$, the eigenspace is spanned by $\mathbf{e}_3$
   $\lambda_4 = 4$, the eigenspace is spanned by $\mathbf{e}_4$
(l) $\lambda_1 = 3$, the eigenspace is spanned by $(1, 2, 0, 0)^T$
   $\lambda_2 = 1$, the eigenspace is spanned by $(0, 1, 0, 0)^T$
   $\lambda_3 = \lambda_4 = 2$, the eigenspace is spanned by $(0, 0, 1, 0)^T$

8. β is an eigenvalue of B if and only if $\beta = \lambda - \alpha$ for some eigenvalue λ of A.

11. $\lambda_1 = 6$, $\lambda_2 = 2$

17. If $A\mathbf{z} = \lambda\mathbf{z}$, then $A\bar{\mathbf{z}} = \bar{A}\bar{\mathbf{z}} = \overline{A\mathbf{z}} = \overline{\lambda\mathbf{z}} = \bar{\lambda}\bar{\mathbf{z}}$

23. $\lambda_1 \mathbf{x}^T \mathbf{y} = (A\mathbf{x})^T \mathbf{y} = \mathbf{x}^T A^T \mathbf{y} = \lambda_2 \mathbf{x}^T \mathbf{y}$

SECTION 2

1. (a) $\mathbf{Y} = \begin{pmatrix} c_1 e^{2t} + c_2 e^{3t} \\ c_1 e^{2t} + 2c_2 e^{3t} \end{pmatrix}$ (b) $\begin{pmatrix} -c_1 e^{-2t} - 4c_2 e^{t} \\ c_1 e^{-2t} + c_2 e^{t} \end{pmatrix}$

(c) $\begin{pmatrix} 2c_1 + c_2 e^{5t} \\ c_1 - 2c_2 e^{5t} \end{pmatrix}$ (d) $\begin{pmatrix} -c_1 e^{t}\sin t + c_2 e^{t}\cos t \\ c_1 e^{t}\cos t + c_2 e^{t}\sin t \end{pmatrix}$

(e) $\begin{pmatrix} -c_1 e^{3t} \sin 2t + c_2 e^{3t} \cos 2t \\ c_1 e^{3t} \cos 2t + c_2 e^{3t} \sin 2t \end{pmatrix}$ (f) $\begin{pmatrix} -c_1 + c_2 e^{5t} + c_3 e^{t} \\ -3c_1 + 8c_2 e^{5t} \\ c_1 + 4c_2 e^{5t} \end{pmatrix}$

2. (a) $\mathbf{Y} = \begin{pmatrix} e^{-3t} + 2e^{t} \\ -e^{-3t} + 2e^{t} \end{pmatrix}$ (b) $\mathbf{Y} = \begin{pmatrix} e^{t}\cos t + 2e^{t}\sin 2t \\ e^{t}\sin 2t - 2e^{t}\cos 2t \end{pmatrix}$

(c) $\mathbf{Y} = \begin{pmatrix} -6e^{t} + 2e^{-t} + 6 \\ -3e^{t} + e^{-t} + 4 \\ -e^{t} + e^{-t} + 2 \end{pmatrix}$ (d) $\mathbf{Y} = \begin{pmatrix} -2 - 3e^{t} + 6e^{2t} \\ 1 + 3e^{t} - 3e^{2t} \\ 1 + 3e^{2t} \end{pmatrix}$

4. (a) $\begin{pmatrix} -2c_1 e^{t} - 2c_2 e^{-t} + c_3 e^{\sqrt{2}t} + c_4 e^{-\sqrt{2}t} \\ c_1 e^{t} + c_2 e^{-t} - c_3 e^{\sqrt{2}t} - c_4 e^{-\sqrt{2}t} \end{pmatrix}$ (b) $\begin{pmatrix} -c_1 - c_2 e^{t} + c_3 e^{4t} \\ -2c_1 + c_2 e^{t} + 2c_3 e^{4t} \\ c_1 + 3c_3 e^{4t} \end{pmatrix}$

5. $y_1(t) = -e^{2t} + e^{-2t} + e^{t}$, $y_2(t) = -e^{2t} - e^{-2t} + 2e^{t}$

7. $x_1(t) = \cos t + 3\sin t + \dfrac{1}{\sqrt{3}}\sin\sqrt{3}\,t$

$\qquad x_2(t) = \cos t + 3\sin t - \dfrac{1}{\sqrt{3}}\sin\sqrt{3}\,t$

9. (a) $m_1 x_1''(t) = -kx_1 + k(x_2 - x_1)$
$m_2 x_2''(t) = -k(x_2 - x_1) + k(x_3 - x_2)$
$m_3 x_3''(t) = -k(x_3 - x_2) - kx_3$

(b) $\begin{pmatrix} 0.1 \cos 2\sqrt{3}\,t + 0.9 \cos \sqrt{2}\,t \\ -0.2 \cos 2\sqrt{3}\,t + 1.2 \cos \sqrt{2}\,t \\ 0.1 \cos 2\sqrt{3}\,t + 0.9 \cos \sqrt{2}\,t \end{pmatrix}$

10. $p(\lambda) = (-1)^n(\lambda^n - a_{n-1}\lambda^{n-1} - \cdots - a_1\lambda - a_0)$

SECTION 3

5. $\det(B - \lambda I) = \det(S^{-1}AS - \lambda S^{-1}S) = \det(S^{-1}(A - \lambda I)S) = \det(A - \lambda I)$. Since B and A have the same characteristic polynomial, they must have the same eigenvalues. Thus similar matrices have the same eigenvalues.

16. (a) $A = \begin{pmatrix} 0.70 & 0.20 & 0.10 \\ 0.20 & 0.70 & 0.10 \\ 0.10 & 0.10 & 0.80 \end{pmatrix}$

(c) The membership of all three groups will approach 100,000 as n gets large.

17. (a) $\begin{pmatrix} e & e-1 \\ 0 & 1 \end{pmatrix}$ (b) $\begin{pmatrix} 2e - e^{-1} & 2e - 2e^{-1} \\ -e + e^{-1} & -e + 2e^{-1} \end{pmatrix}$ (c) $\begin{pmatrix} e & e \\ 0 & e \end{pmatrix}$

18. (a) $\begin{pmatrix} e^{-t} \\ e^{-t} \end{pmatrix}$ (b) $\begin{pmatrix} -3e^t - e^{-t} \\ e^t + e^{-t} \end{pmatrix}$ (c) $\begin{pmatrix} 3e^t - 2 \\ 2 - e^{-t} \\ e^{-t} \end{pmatrix}$

SECTION 4

1. (b) and (f) are Hermitian while (b), (c), (e), and (f) are normal.
8. (b) $\|U\mathbf{x}\|^2 = (U\mathbf{x})^H U\mathbf{x} = \mathbf{x}^H U^H U\mathbf{x} = \mathbf{x}^H\mathbf{x} = \|\mathbf{x}\|^2$
9. U is unitary, since $U^H U = (I - 2\mathbf{u}\mathbf{u}^H)^2 = I - 4\mathbf{u}\mathbf{u}^H + 4\mathbf{u}(\mathbf{u}^H\mathbf{u})\mathbf{u}^H = I$
13. $\lambda_1 = 1$, $\lambda_2 = -1$, $\mathbf{u}_1 = (1, 1)^T$, $\mathbf{u}_2 = (-1, 1)^T$

$A = 1\begin{pmatrix} 1 & 1 \\ 1 & 1 \end{pmatrix} + (-1)\begin{pmatrix} 1 & -1 \\ -1 & 1 \end{pmatrix}$

SECTION 5

1. (a) $\begin{pmatrix} 3 & -\frac{5}{2} \\ -\frac{5}{2} & 1 \end{pmatrix}$ (b) $\begin{pmatrix} 2 & \frac{1}{2} & -1 \\ \frac{1}{2} & 3 & \frac{3}{2} \\ -1 & \frac{3}{2} & 1 \end{pmatrix}$

3. (a) $Q = \frac{1}{\sqrt{2}}\begin{pmatrix} 1 & 1 \\ 1 & -1 \end{pmatrix}$, $\dfrac{(x')^2}{4} + \dfrac{(y')^2}{12} = 1$, ellipse

(d) $Q = \frac{1}{\sqrt{2}}\begin{pmatrix} 1 & 1 \\ -1 & 1 \end{pmatrix}$, $\left(y' + \dfrac{\sqrt{2}}{2}\right)^2 = -\dfrac{\sqrt{2}}{2}(x' - \sqrt{2})$ or

$(y'')^2 = -\dfrac{\sqrt{2}}{2}x''$, parabola

5. (a) Positive definite; (b) indefinite; (c) negative definite; (d) indefinite
6. (a) Minimum; (b) saddle point; (c) saddle point; (d) local maximum

SECTION 6

1. (a) $\det(A_1) = 2$, $\det(A_2) = 3$, positive definite
(b) $\det(A_1) = 3$, $\det(A_2) = -10$, not positive definite
(c) $\det(A_1) = 6$, $\det(A_2) = 14$, $\det(A_3) = -38$, not positive definite
(d) $\det(A_1) = 4$, $\det(A_2) = 8$, $\det(A_3) = 13$, positive definite

2. $a_{11} = 3$, $a_{22}^{(1)} = 2$, $a_{33}^{(2)} = \frac{4}{3}$

3. (a) $\begin{pmatrix} 1 & 0 \\ \frac{1}{2} & 1 \end{pmatrix} \begin{pmatrix} 4 & 0 \\ 0 & 9 \end{pmatrix} \begin{pmatrix} 1 & \frac{1}{2} \\ 0 & 1 \end{pmatrix}$

(b) $\begin{pmatrix} 1 & 0 \\ -\frac{1}{3} & 1 \end{pmatrix} \begin{pmatrix} 9 & 0 \\ 0 & 1 \end{pmatrix} \begin{pmatrix} 1 & -\frac{1}{3} \\ 0 & 1 \end{pmatrix}$

(c) $\begin{pmatrix} 1 & 0 & 0 \\ \frac{1}{2} & 1 & 0 \\ \frac{1}{4} & -1 & 1 \end{pmatrix} \begin{pmatrix} 16 & 0 & 0 \\ 0 & 2 & 0 \\ 0 & 0 & 4 \end{pmatrix} \begin{pmatrix} 1 & \frac{1}{2} & \frac{1}{4} \\ 0 & 1 & -1 \\ 0 & 0 & 1 \end{pmatrix}$

(d) $\begin{pmatrix} 1 & 0 & 0 \\ \frac{1}{3} & 1 & 0 \\ -\frac{2}{3} & 1 & 1 \end{pmatrix} \begin{pmatrix} 9 & 0 & 0 \\ 0 & 3 & 0 \\ 0 & 0 & 2 \end{pmatrix} \begin{pmatrix} 1 & \frac{1}{3} & -\frac{2}{3} \\ 0 & 1 & 1 \\ 0 & 0 & 1 \end{pmatrix}$

4. (a) $\begin{pmatrix} 2 & 0 \\ 1 & 3 \end{pmatrix} \begin{pmatrix} 2 & 1 \\ 0 & 3 \end{pmatrix}$ (b) $\begin{pmatrix} 3 & 0 \\ -1 & 1 \end{pmatrix} \begin{pmatrix} 3 & -1 \\ 0 & 1 \end{pmatrix}$

(c) $\begin{pmatrix} 4 & 0 & 0 \\ 2 & \sqrt{2} & 0 \\ 1 & -\sqrt{2} & 2 \end{pmatrix} \begin{pmatrix} 4 & 2 & 1 \\ 0 & \sqrt{2} & -\sqrt{2} \\ 0 & 0 & 2 \end{pmatrix}$

(d) $\begin{pmatrix} 3 & 0 & 0 \\ 1 & \sqrt{3} & 0 \\ -2 & \sqrt{3} & \sqrt{2} \end{pmatrix} \begin{pmatrix} 3 & 1 & -2 \\ 0 & \sqrt{3} & \sqrt{3} \\ 0 & 0 & \sqrt{2} \end{pmatrix}$

SECTION 7

1. (a) $\lambda_1 = 4$, $\lambda_2 = -1$, $\mathbf{x}_1 = (3, 2)^T$ (b) $\lambda_1 = 8$, $\lambda_2 = 3$, $\mathbf{x}_1 = (1, 2)^T$
(c) $\lambda_1 = 7$, $\lambda_2 = 2$, $\lambda_3 = 0$, $\mathbf{x}_1 = (1, 1, 1)^T$.
2. (a) $\lambda_1 = 3$, $\lambda_2 = -1$, $\mathbf{x}_1 = (3, 1)^T$
(b) $\lambda_1 = 2 = 2 \exp(0)$, $\lambda_2 = -2 = 2 \exp(\pi i)$, $\mathbf{x}_1 = (1, 1)^T$
(c) $\lambda_1 = 2 = 2 \exp(0)$, $\lambda_2 = -1 + \sqrt{3}\, i = 2 \exp\left(\frac{2\pi i}{3}\right)$,

$\lambda_3 = -1 - \sqrt{3}\, i = 2 \exp\left(\frac{4\pi i}{3}\right)$, $\mathbf{x}_1 = (4, 2, 1)^T$

3. $x_1 = 70{,}000$; $x_2 = 56{,}000$; $x_3 = 44{,}000$
4. $x_1 = x_2 = x_3$
5. $(I - A)^{-1} = I + A + \cdots + A^{m-1}$

6. (a) $(I - A)^{-1} = \begin{pmatrix} 1 & -1 & 3 \\ 0 & 0 & 1 \\ 0 & -1 & 2 \end{pmatrix}$

(b) $A^2 = \begin{pmatrix} 0 & -2 & 2 \\ 0 & 0 & 0 \\ 0 & 0 & 0 \end{pmatrix}$, $A^3 = \begin{pmatrix} 0 & 0 & 0 \\ 0 & 0 & 0 \\ 0 & 0 & 0 \end{pmatrix}$

7. (b) and (c) are reducible.

CHAPTER 7

SECTION 1
1. (a) 0.231×10^4;　(b) 0.326×10^2;　(c) 0.128×10^{-1};
   (d) 0.824×10^5
2. (a) $\epsilon = -2$; $\delta \approx -8.7 \times 10^{-4}$;　(b) $\epsilon = 0.04$; $\delta \approx 1.2 \times 10^{-3}$;
   (c) $\epsilon = 3.0 \times 10^{-5}$; $\delta \approx 2.3 \times 10^{-3}$;　(d) $\epsilon = -31$; $\delta \approx -3.8 \times 10^{-4}$
3. (a) 0.10101×2^5,　(b) 0.10100×2^{-1}　(c) 0.10111×2^4,
   (d) -0.11010×2^{-3}
4. (a) $10{,}420$, $\epsilon = -0.0018$, $\delta \approx -1.7 \times 10^{-7}$;　(b) 0, $\epsilon = -8$, $\delta = -1$;
   (c) 1×10^{-4}, $\epsilon = 5 \times 10^{-5}$, $\delta = 1$;
   (d) $82{,}190$, $\epsilon = 25.7504$, $\delta \approx 3.1 \times 10^{-4}$
5. (a) 0.1043×10^6;　(b) 0.1045×10^6;　(c) 0.1045×10^6
6. $\epsilon = (0.00001)_2 = \frac{1}{32}$
7. 23

SECTION 2
1. $A = \begin{pmatrix} 1 & 0 & 0 \\ 2 & 1 & 0 \\ -3 & 2 & 1 \end{pmatrix} \begin{pmatrix} 1 & 1 & 1 \\ 0 & 2 & -1 \\ 0 & 0 & 3 \end{pmatrix}$
2. (a) $(2, -1, 3)^T$;　(b) $(1, -1, 3)^T$;　(c) $(1, 5, 1)^T$
3. (a) n^2 multiplications and $n(n - 1)$ additions
   (b) n^3 multiplications and $n^2(n - 1)$ additions
7. $5n - 4$ multiplications/divisions, $3n - 3$ additions/subtractions
8. (a) $[(n - j)(n - j + 1)]/2$ multiplications; $[(n - j - 1)(n - j)]/2$ additions
   (c) It requires on the order of $\frac{2}{3}n^3$ additional multiplications/divisions to compute A^{-1} given the LU factorization.

SECTION 3
1. (a) $(1, 1, -2)$;
   (b) $\begin{pmatrix} 0 & 0 & 1 \\ 1 & 0 & 0 \\ 0 & 1 & 0 \end{pmatrix} \begin{pmatrix} 1 & 0 & 0 \\ 2 & 1 & 0 \\ 0 & 3 & 1 \end{pmatrix} \begin{pmatrix} 1 & 2 & -2 \\ 0 & 1 & 8 \\ 0 & 0 & -23 \end{pmatrix}$
2. (a) $(1, 2, 2)$;　(b) $(4, -3, 0)$;　(c) $(1, 1, 1)$
3. $P = \begin{pmatrix} 0 & 0 & 1 \\ 1 & 0 & 0 \\ 0 & 1 & 0 \end{pmatrix}$,　$L = \begin{pmatrix} 1 & 0 & 0 \\ \frac{1}{2} & 1 & 0 \\ -\frac{1}{2} & -\frac{1}{3} & 1 \end{pmatrix}$

   $U = \begin{pmatrix} 2 & 4 & -6 \\ 0 & 6 & 9 \\ 0 & 0 & 5 \end{pmatrix}$　and　$\mathbf{x} = \begin{pmatrix} 6 \\ -\frac{1}{2} \\ 1 \end{pmatrix}$
4. $P = Q = \begin{pmatrix} 0 & 1 \\ 1 & 0 \end{pmatrix}$, $PAQ = LU = \begin{pmatrix} 1 & 0 \\ \frac{1}{2} & 1 \end{pmatrix} \begin{pmatrix} 4 & 2 \\ 0 & 2 \end{pmatrix}$
5. $\hat{\mathbf{c}} = P\mathbf{c} = (-4, 6)^T$, $\mathbf{y} = L^{-1}\hat{\mathbf{c}} = (-4, 8)^T$,
   $\mathbf{z} = U^{-1}\mathbf{y} = (-3, 4)^T$, $\mathbf{x} = Q^T\mathbf{z} = (4, -3)^T$.
6. Error $\dfrac{-2000\epsilon}{0.6} \approx -3333\epsilon$. If $\epsilon = 0.001$, then $\delta = -\frac{2}{3}$
7. $(1.667, 1.001)$

8. $(5.002, 1.000)$

9. $(5.001, 1.001)$

SECTION 4

1. $\text{cond}_\infty A = 400$

2. (a) $(-0.48, 0.8)$;   (b) $(-2.902, 2.0)$

3. (a) $A_n^{-1} = \begin{pmatrix} 1-n & n \\ n & -n \end{pmatrix}$,   (b) $\text{cond}_\infty A_n = 4n$,

    (c) $\lim_{n\to\infty} \text{cond}_\infty A_n = \infty$

5. $\text{cond}_\infty(A) = 28$

6. (a) $\mathbf{r} = (-0.1, -0.9, -0.6)^T$, $\|\mathbf{r}\|_\infty = 0.9$ (b) 3.6;

    (c) $\mathbf{x} = (1, -2, 3)^T$, $\delta = 0.1$

7. $\text{cond}_1(A) = 6$

8. 0.3

SECTION 5

1. (a) $\begin{pmatrix} \dfrac{1}{\sqrt{2}} & \dfrac{1}{\sqrt{2}} \\ -\dfrac{1}{\sqrt{2}} & \dfrac{1}{\sqrt{2}} \end{pmatrix}$ (b) $\begin{pmatrix} \dfrac{\sqrt{3}}{2} & -\dfrac{1}{2} \\ \dfrac{1}{2} & \dfrac{\sqrt{3}}{2} \end{pmatrix}$ (c) $\begin{pmatrix} -\dfrac{4}{5} & \dfrac{3}{5} \\ -\dfrac{3}{5} & -\dfrac{4}{5} \end{pmatrix}$

2. (a) $\begin{pmatrix} \dfrac{3}{5} & 0 & \dfrac{4}{5} \\ 0 & 1 & 0 \\ \dfrac{4}{5} & 0 & -\dfrac{3}{5} \end{pmatrix}$ (b) $\begin{pmatrix} \dfrac{1}{\sqrt{2}} & -\dfrac{1}{\sqrt{2}} & 0 \\ -\dfrac{1}{\sqrt{2}} & -\dfrac{1}{\sqrt{2}} & 0 \\ 0 & 0 & 1 \end{pmatrix}$

    (c) $\begin{pmatrix} 1 & 0 & 0 \\ 0 & \dfrac{1}{2} & \dfrac{\sqrt{3}}{2} \\ 0 & \dfrac{\sqrt{3}}{2} & -\dfrac{1}{2} \end{pmatrix}$ (d) $\begin{pmatrix} 1 & 0 & 0 \\ 0 & -\dfrac{\sqrt{3}}{2} & \dfrac{1}{2} \\ 0 & \dfrac{1}{2} & \dfrac{\sqrt{3}}{2} \end{pmatrix}$

3. $H = I - \dfrac{1}{\beta} \mathbf{v}\mathbf{v}^T$ for the given β and $\mathbf{v}$.

    (a) $\beta = 9$, $\mathbf{v} = (-1, -1, 4)^T$,   (b) $\beta = 7$, $\mathbf{v} = (-1, 2, 3)^T$

    (c) $\beta = 18$, $\mathbf{v} = (-2, 4, -4)^T$

4. (a) $\beta = 9$, $\mathbf{v} = (0, -1, 4, 1)^T$,

    (b) $\beta = 15$, $\mathbf{v} = (0, 0, -5, -1, 2)^T$

5. (a) $H_2 H_1 A = R$, where $H_i - I - \dfrac{1}{\beta_i} \mathbf{v}_i \mathbf{v}_i^T$, $i = 1, 2$, and $\beta_1 = 6$, $\beta_2 = 5$

$$\mathbf{v}_1 = \begin{pmatrix} -2 \\ 2 \\ -2 \end{pmatrix} \qquad \mathbf{v}_2 = \begin{pmatrix} 0 \\ -1 \\ 3 \end{pmatrix}, \qquad R = \begin{pmatrix} 3 & 4 & -2 \\ 0 & 5 & 10 \\ 0 & 0 & -5 \end{pmatrix}$$

$$\mathbf{b}^{(1)} = H_2 H_1 \mathbf{b} = (11, 5, 5)^T$$

    (b) $\mathbf{x} = (-1, 3, -1)^T$

6. It takes three multiplications, two additions, and one square root to determine H. It takes four multiplications/divisions, one addition, and one square root to determine G. The calculation of GA requires $4n$ multiplications and $2n$ additions while the calculation of HA requires $3n$ multiplications/divisions and $3n$ additions.

7. (a) $n - k + 1$ multiplications/divisions, $2n - 2k + 1$ additions
   (b) $n(n - k + 1)$ multiplications/divisions, $n(2n - 2k + 1)$ additions
8. (a) $4(n - k)$ multiplications, $2(n - k)$ additions
   (b) $4n(n - k)$ multiplications, $2n(n - k)$ additions
9. (a) rotation;   (b) rotation;   (c) Givens transformation;
   (d) Givens transformation

SECTION 6

1. $\mathbf{x}^{(1)} = (1.1, 1.2)^T$, $\mathbf{x}^{(2)} = (0.99, 0.98)^T$
2. $\mathbf{x}^{(1)} = (3, 2, 1)^T$, $\mathbf{x}^{(2)} = (0, 1, 1)^T$, $\mathbf{x}^{(3)} = (1, 1, 1)^T$, $\mathbf{x}^{(4)} = (1, 1, 1)^T$
3. $\mathbf{x}^{(1)} = (1.10, 0.98)^T$, $\mathbf{x}^{(2)} = (1.0020, 0.9996)^T$
4. (a) $\mathbf{x}^{(1)} = (1.2, 1.1, 1.1)^T$;   (b) $\mathbf{x}^{(1)} = (1.20, 1.08, 0.972)^T$
5. The iteration scheme will converge for the matrices in parts (b), (c), (d), and (e).
9. $B = (D - \omega L)^{-1}[(1 - \omega)D + \omega U]$, $\mathbf{c} = \omega(D - \omega L)^{-1}\mathbf{b}$

SECTION 7

2. (a) $\sigma_1 = \sqrt{10}$, $\sigma_2 = 0$;   (b) $\sigma_1 = 3$, $\sigma_2 = 2$;   (c) $\sigma_1 = 4$, $\sigma_2 = 2$;
   (d) $\sigma_1 = 3$, $\sigma_2 = 2$, $\sigma_3 = 1$. The matrices U and V are not unique. The reader may check his or her answers by multiplying out $U\Sigma V^T$.
3. (b) Rank of $A = 2$, $\|A\|_2 = 3$, $A' = \begin{pmatrix} 1.2 & -2.4 \\ -0.6 & 1.2 \end{pmatrix}$

SECTION 8

1. (a) $\mathbf{u}_1 = \begin{pmatrix} 1 \\ 1 \end{pmatrix}$;   (b) $A_2 = \begin{pmatrix} 2 & 0 \\ 0 & 0 \end{pmatrix}$;   (c) $\lambda_1 = 2$, $\lambda_2 = 0$
   The eigenspace corresponding to λ_1 is spanned by $\mathbf{u}_1$.
2. (a) $\mathbf{v}_1 = \begin{pmatrix} 3 \\ 5 \\ 3 \end{pmatrix}$, $\mathbf{u}_1 = \begin{pmatrix} 0.6 \\ 1.0 \\ 0.6 \end{pmatrix}$, $\mathbf{v}_2 = \begin{pmatrix} 2.2 \\ 4.2 \\ 2.2 \end{pmatrix}$, $\mathbf{u}_2 = \begin{pmatrix} 0.52 \\ 1.00 \\ 0.52 \end{pmatrix}$, $\mathbf{v}_3 = \begin{pmatrix} 2.05 \\ 4.05 \\ 2.05 \end{pmatrix}$
   (b) $\lambda_1' = 4.05$;   (c) $\lambda_1 = 4$, $\delta = 0.0125$
3. (b) A has no dominant eigenvalue.
4. $A_2 = \begin{pmatrix} 3 & -1 \\ -1 & 1 \end{pmatrix}$, $A_3 = \begin{pmatrix} 3.4 & 0.2 \\ 0.2 & 0.6 \end{pmatrix}$
   $\lambda_1 = 2 + \sqrt{2} \approx 3.414$, $\lambda_2 = 2 - \sqrt{2} \approx 0.586$
5. (b) $H = I - \dfrac{1}{\beta}\mathbf{v}\mathbf{v}^T$, where $\beta = \frac{1}{3}$ and $\mathbf{v} = \left(-\frac{1}{3}, -\frac{2}{3}, \frac{1}{3}\right)^T$.
   (c) $\lambda_2 = 3$, $\lambda_3 = 1$, $HAH = \begin{pmatrix} 4 & 0 & 3 \\ 0 & 5 & -4 \\ 0 & 2 & -1 \end{pmatrix}$

SECTION 9

1. (a) $(\sqrt{2}, 0)^T$;   (b) $(1 - 3\sqrt{2}, 3\sqrt{2}, -\sqrt{2})^T$;·   (c) $(1, 0)^T$;
   (d) $(1 - \sqrt{2}, \sqrt{2}, -\sqrt{2})^T$
2. $\hat{x}_i = \dfrac{d_i b_i + e_i b_{n+i}}{d_i^2 + e_i^2}$, $i = 1, \ldots, n$
3. (a) $\sigma_1 = \sqrt{2 + \epsilon^2}$, $\sigma_2 = \epsilon$;   (b) $\lambda_1' = 2$, $\lambda_2' = 0$, $\sigma_1' = \sqrt{2}$, $\sigma_2' = 0$

9. $A^+ = \begin{pmatrix} \frac{1}{4} & \frac{1}{4} & 0 \\ \frac{1}{4} & \frac{1}{4} & 0 \end{pmatrix}$

10. (a) $A^+ = \begin{pmatrix} \frac{1}{10} & -\frac{1}{10} \\ \frac{2}{10} & -\frac{2}{10} \end{pmatrix}$ (b) $A^+ \mathbf{b} = \begin{pmatrix} 5 \\ -5 \end{pmatrix}$

(c) $\left\{ \mathbf{y} \mid \mathbf{y} = \begin{pmatrix} 5 \\ -5 \end{pmatrix} + \alpha \begin{pmatrix} -2 \\ 1 \end{pmatrix} \right\}$

12. $\|A_1 - A_2\|_F = \epsilon$, $\|A_1^+ - A_2^+\|_F = 1/\epsilon$. As $\varepsilon \to 0$, $\|A_1 - A_2\|_F \to 0$ and $\|A_1^+ - A_2^+\|_F \to \infty$.

INDEX

403